Nanotechnologies for the Life Sciences
Volume 8

Nanomaterials for Biosensors

Edited by
Challa S. S. R. Kumar

1807–2007 Knowledge for Generations

Each generation has its unique needs and aspirations. When Charles Wiley first opened his small printing shop in lower Manhattan in 1807, it was a generation of boundless potential searching for an identity. And we were there, helping to define a new American literary tradition. Over half a century later, in the midst of the Second Industrial Revolution, it was a generation focused on building the future. Once again, we were there, supplying the critical scientific, technical, and engineering knowledge that helped frame the world. Throughout the 20th Century, and into the new millennium, nations began to reach out beyond their own borders and a new international community was born. Wiley was there, expanding its operations around the world to enable a global exchange of ideas, opinions, and know-how.

For 200 years, Wiley has been an integral part of each generation's journey, enabling the flow of information and understanding necessary to meet their needs and fulfill their aspirations. Today, bold new technologies are changing the way we live and learn. Wiley will be there, providing you the must-have knowledge you need to imagine new worlds, new possibilities, and new opportunities.

Generations come and go, but you can always count on Wiley to provide you the knowledge you need, when and where you need it!

William J. Pesce
President and Chief Executive Officer

Peter Booth Wiley
Chairman of the Board

Nanotechnologies for the Life Sciences
Volume 8

Nanomaterials for Biosensors

Edited by
Challa S. S. R. Kumar

WILEY-VCH Verlag GmbH & Co. KGaA

The Editor

Dr. Challa S. S. R. Kumar
The Center for Advanced
Microstructures and Devices
(CAMD)
Louisiana State University
6980 Jefferson Highway
Baton Rouge, LA 70806
USA

Cover
Cover design by G. Schulz based on Micrograph Courtesy of Quantum Dot Corporation.

1st Edition 2007
1st Reprint 2008
2nd Reprint 2008
3rd Reprint 2010

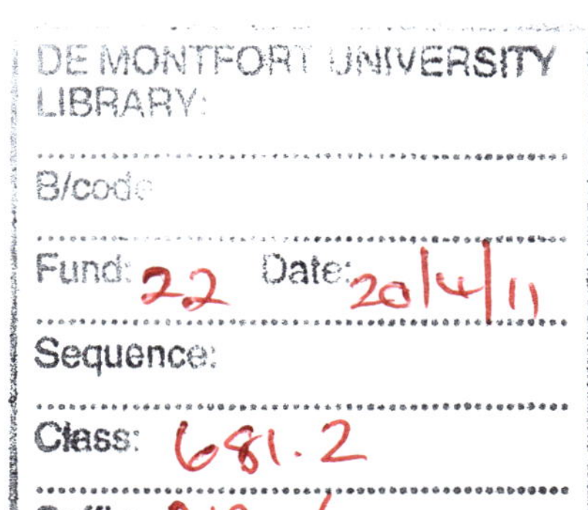

Library of Congress Card No.: applied for
British Library Cataloguing-in-Publication Data
A catalogue record for this book is available from the British Library.

Bibliographic information published by the Deutsche Nationalbibliothek
The Deutsche Nationalbibliothek lists this publication in the Deutsche Nationalbibliografie; detailed bibliographic data are available in the Internet at ⟨http://dnb.d-nb.de⟩.

Printed in the Federal Republic of Germany
Printed on acid-free paper

Typesetting Asco Typesetters, Hongkong
Printing and bookbinding buch bücher dd ag, Birkach
Cover Design Grafik-Design Schulz, Fußgönheim

ISBN 978-3-527-31388-4

Contents of the Series

Kumar, C. S. S. R. (ed.)

Nanotechnologies for the Life Sciences (NtLS)

Book Series

Vol. 1
Biofunctionalization of Nanomaterials

2005, Hardcover
ISBN-10: 3-527-31381-8
ISBN-13: 987-3-527-31381-5

Vol. 2
Biological and Pharmaceutical Nanomaterials

2005, Hardcover
ISBN-10: 3-527-31382-6
ISBN-13: 987-3-527-31382-2

Vol. 3
Nanosystem Characterization Tools in the Life Sciences

2005, Hardcover
ISBN-10: 3-527-31383-4
ISBN-13: 987-3-527-31383-9

Vol. 4
Nanodevices for the Life Sciences

2006, Hardcover
ISBN-10: 3-527-31384-2
ISBN-13: 987-3-527-31384-6

Vol. 5
Nanomaterials – Toxicity, Health and Environmental Issues

2006, Hardcover
ISBN-10: 3-527-31385-0
ISBN-13: 987-3-527-31385-3

Vol. 6
Nanomaterials for Cancer Therapy

2006, Hardcover
ISBN-10: 3-527-31386-9
ISBN-13: 987-3-527-31386-0

Vol. 7
Nanomaterials for Cancer Diagnosis

2006, Hardcover
ISBN-10: 3-527-31387-7
ISBN-13: 987-3-527-31387-7

Vol. 8
Nanomaterials for Biosensors

2006, Hardcover
ISBN-10: 3-527-31388-5
ISBN-13: 987-3-527-31388-4

Vol. 9
Tissue, Cell and Organ Engineering

2006, Hardcover
ISBN-10: 3-527-31389-3
ISBN-13: 987-3-527-31389-1

Vol. 10
Nanomaterials for Medical Diagnosis and Therapy

2006, Hardcover
ISBN-10: 3-527-31390-7
ISBN-13: 987-3-527-31390-7

Contents

Nanotechnologies for the Life Sciences Vol. 8
Nanomaterials for Biosensors. Edited by Challa S. S. R. Kumar

ISBN: 978-3-527-31388-4

Preface

As we come closer to the completion of the ten-volume series on *Nanotechnologies for the Life Sciences*, I am reminded of a statement by the great scientist Arthur C Clarke who said: "Any sufficiently advanced technology is indistinguishable from magic." This statement is particularly true in molecular biosensing based on nanomaterials where the detection limits are 'magically' becoming smaller and smaller, even reaching zeptomolar concentrations in addition to opening up possibilities for ultra-sensitive multiplexed detection. Thanks to the development of novel concepts such as bio-bar-code assays, nanomaterials-based companies are revolutionizing the commercialization of molecular diagnostics at breathtaking speeds. Therefore, on behalf of a great team of nano researchers who have been involved in the development of nanomaterials for biosensing and particularly those who have contributed to this specific volume, I am pleased to introduce you the 8th volume of the series, *Nanomaterials for Biosensors.* We have come a long way in our journey since the publication of the first volume of the series, *Biofunctionalization of Nanomaterials*, into bringing the existing knowledge base of applications of nanotechnologies in biology, biotechnology and medicine on a single platform. The eigth volume has thirteen chapters covering various aspects of biomolecular sensing using a variety of nanomaterials such as carbon nanotubes, nanowires, nanocantilevers, fullerenes, denrimers in addition to metallic and quantum dot nanoparticles. The most exciting and unique aspect of the book is that it deals with the utilization of nanomaterials not only for enhancing the capabilities in conventional biosensing platforms, but also brings out newer approaches such as biomimetic and reagent-less biosensing.

The first four chapters of the book are dedicated to various modes of biosensing using carbon-based nanomaterials. The first chapter by Padmakar D. Kichambare and Alexander Star from the University of Pittsburgh, USA, provides an overview of recent advances in biodetection using single-walled carbon nanotube field-effect transistors (NTFETs) focusing primarily on fabrication of NTFET devices and how carbon nanotubes can be effectively integrated into conventional electronics for biosensor applications, for example, antibody–antigen interactions, DNA hybridization, glucose detection and enzymatic reactions. The chapter entitled *Biosensing using Carbon Nanotube Field-effect Transistors* provides a promising outlook for novel sensing applications of carbon nanotubes in living systems as

Nanotechnologies for the Life Sciences Vol. 8
Nanomaterials for Biosensors. Edited by Challa S. S. R. Kumar

ISBN: 978-3-527-31388-4

well as new opportunities for CNT-based bioelectronics. In addition to the utility of CNTs as field effect transistors, they can also be used as material of construction as nanoelectrodes, which can be utilized as electrochemical sensing systems. The second chapter, *Carbon Nanotube-based Sensors*, contributed by Jian-Shan Ye and Fwu-Shan Sheu from the National University of Singapore, brings out the importance of various methods utilized for preparing CNT electrodes and different ways to functionalize them for biosensing applications. Particularly interesting in this chapter is the discussion on mechanistic aspects of electrocatalysis by CNTs. The chapter will be very useful for those who are interested in exploiting the electrochemistry of CNTs in molecular diagnostics. Though CNTs are the most well-studied amongst one-dimensional nanomaterials, there are a considerable number of research investigations into exploiting the potential of other 1D nanomaterials like semiconducting nanowires and cantilevers. The third chapter, therefore, is a comprehensive review on silicon nanowires, conducting polymer nanowires, metal oxide nanowires, and nanocantilevers with reference to carbon nanotubes. The chapter, *Nanotubes, Nanowires, and Nanocantilevers in Biosensor Development*, contributed by Jun Wang, Guodong Liu, and Yuehe Lin from Pacific Northwest National Laboratory in Richland, USA, provides up to date information on the development of 1D-nanomaterial-based biosensors. The fourth and final chapter on carbon nanomaterials for biosensing is *Fullerene-based Electrochemical Detection Methods for Biosensing* presented by Nikos Chaniotakis from the University of Crete, Greece. Fullerenes have not received as much attention as CNTs as suitable materials for biosensing, mainly because their physicochemical characteristics are still not very well understood. However, the chapter provides a complete picture on several possibilities for fullerenes to offer new and powerful tools as electrochemical biosensors especially in signal mediation, protein and enzyme functionalization, and light-induced switching.

Nanomaterials also offer opportunities for ultra-sensitive biomolecular sensing through their local field optical effects, which are several orders of magnitude higher than the corresponding bulk effects. Optically active metallic and quantum dot nanomaterials have opened up avenues for newer techniques such as local surface plasmon resonance (LSPR), surface-enhanced Raman scattering (SERS), surface-enhanced fluorescence (SEF), fluorescence resonance energy transfer (FRET), time-resolved fluorimetry, and others. The next two chapters in the book provide an overview of these technologies. The fifth chapter, *Optical Biosensing Based on Metal and Semiconductor Colloidal Nanocrystals* by R. Comparelli, L. Curri, P. D. Cozzoli, and M. Striccoli from the Italian National Research Council's Institute of Physicochemical Processes of CNR in Bari, focuses in general on metal and quantum dot-based optical biosensing, providing a comparative assessment of recent developments categorized into various novel and/or improved optical techniques with traditional methods. The sixth chapter is exclusively dedicated to optical biosensing by quantum dots. Authors Rumiana Bakalova, Zhivko Zhelev, Hideki Ohba, and Yoshinobu Baba from the AIST-Kyushu National Institute of Advanced Science and Technology in Saga and Nagoya University, both in Japan, present an overview on the current status and future trends of QD-based biosensor development. The chapter, *Quantum Dot-based Nanobiohybrids for Fluorescent Detection of Molecular and Cellular Biological Targets*, covers not only the basic principles of design and synthesis of highly fluorescent QDs, but also intricacies of *in vitro* and *in vivo* cellular and deep-tissue imaging.

The utility of gold nanoparticles in optical biosensing is very well known and already finding several commercial applications. However, the application of gold nanomaterials' capability as amperometric sensor is only recently being recognized as very promising and powerful tools in bio-fluid or biomaterial investigations and their associated clinical studies. The seventh chapter, *Detection of Biological Materials by Gold Nano-biosensor-based Electrochemical Methods*, provides a review on gold nanowire arrays and their utility as biosensors in bacterial detection. The authors, Juan Jiang, Manju Basu, Sara Seggerson, Albert Miller, Michael Pugia, and Subhash Basu from the university of Notre Dame in Indiana, USA, provides an in-depth analysis of the application of electrochemical impedance spectroscopy (EIS) in gold nanomaterial-based biosensing and demonstrates the technique's potential in clinical laboratories, environmental monitoring and the food industry to achieve rapid and sensitive detection. Continuing on a similar theme related to electrochemical sensing but utilizing dendrimeric nanomaterials, authors Hak-Sung Kim and Hyun C. Yoon from KAIST at Daejeon, Korea, cover various facets of bioelectrocatalytic enzyme sensors in the eighth chapter. *Dendrimer-based Electrochemical Detection Methods* is a must for readers interested in the fabrication of dendrimer-based biocomposite mono-/multilayers and their biosensing applications.

Each of the last five chapters in the book brings out several fascinating facets of nanomaterial-based biosensing very different from what we have seen so far in the first eight chapters. The author of the ninth chapter, *Coordination Biosensors: Integrated Systems for Ultrasensitive Detection of Biomarkers*, Joanne Yeh from the University of Pittsburgh Medical School, USA, presents altogether a different approach to biosensing, utilizing newer concepts to align the signal transduction centers to enhance the kinetics of reactions leading to improved sensitivity of detection. It has been observed that there is a direct electrochemical and catalytic activity of many proteins at electrodes modified with various nanomaterials such as TiO_2, ZrO_2, SiO_2 , Fe_3O_4, metal nanoparticles and carbon nanotubes. In the tenth chapter, the author Genxi Li from Nanjing University in China reviews the literature to demonstrate that nanomaterials can not only provide a friendly platform for the assembly of protein molecules but also enhance the electron-transfer process between protein molecules and the electrode. The chapter entitled *Protein-based Biosensors using Nanomaterials* brings out the advantages of combining proteins and nanomaterials to develop sensitive biosensor elements. Proteins and in general various nano-size structures in the field of life sciences provide testimony to the endless possibilities and elegant applications in our day to day world. Therefore, it is not very surprising that a new branch of science, 'Biomimetics', has roots in a variety of scientific disciplines, and the field of biosensors is not an exception. In the eleventh chapter, authors Raz Jelinek and Sofiya Kolusheva from Ben Gurion University of the Negev in Beer-Sheva, Israel, brings out the utility of concepts and methodologies from the biological world into the laboratory. The chapter, *Biomimetic Nanosensors*, provides a broader perspective on bio-inspired devices and applications related to nanomaterial-based biosensing.

As the title indicates, the twelfth chapter, *Reagentless Biosensors Based on Nanoparticles*, provides the readers with yet another novel concept in biosensing, where sensing tools are being developed based on perturbation of nanoparticle properties, without the need for reagents, in order to produce unique yet sensitive signals for biomeoclecular sensing. The author, David Benson from Wayne State University in Detroit, USA, provides a strong case for adaptation of reagentless concepts that provide sensors that can be adapted to various detection platforms. The book concludes with its thirteenth chapter, wherein the authors, Shohei Yamamura, Sathuluri Ramachandra Rao, and Eiichi Tamiya from Japan Advanced Institute of Science and Technology at Nomi, Japan, bring us closer to biosensing devices incorporating highly integrated microarray systems that can perform assays at pico- and nano-liter volume level. In this chapter, *Pico/Nanoliter Chamber Array Chips for Single-cell, DNA and Protein Analyses*, the authors discuss three very important topics – novel multiplexed PCR, cell-free protein synthesis, and high-throughput single-cell analysis systems using nanolitre microarray platforms.

Nanotechnology embodies the spirit of interdisciplinary approaches and teams. I am, therefore, very grateful to all the authors who have shared my enthusiasm and vision by contributing high-quality manuscripts keeping in tune with the theme of this volume. It is primarily due to their scholarly contributions that this book comes into existence. I am thankful to my employer, the Center for Advanced Microstructures and Devices (CAMD), for providing me with an opportunity to undertake this enormous project. No words can express the understanding of my family, friends, mentors and most importantly the readers who are now an integral part of my existence and continue to shape my life and I am indebted to them. Finally, Wiley-VCH publishers have done a remarkable job and I am grateful for their support.

September 2006
Baton Rouge

Challa S. S. R. Kumar

List of Authors

Yoshinobu Baba
Department of Applied Chemistry
Graduate School of Engineering
Nagoya University
Furo-cho, Chikusa-ku
Nagoya 464-8603
Japan

Rumiana Bakalova
On-Site Sensing and Diagnostic Research Laboratory
National Institute for Advanced Industrial Science and Technology
AIST-Kyushu
807-1 Ahuku-machi, Tosu
Saga 841-0052
Japan

Manju Basu
Department of Chemistry and Biochemistry
University of Notre Dame
Notre Dame, IN 46556
USA

Subhash Basu
Department of Chemistry and Biochemistry
University of Notre Dame
Notre Dame, IN 46556
USA

David E. Benson
Department of Chemistry
Wayne State University
Detroit, MI 48202
USA

Nikos Chaniotakis
Laboratory of Analytical Chemistry
University of Crete
71 003 Iráklion
Crete

Roberto Comparelli
CNR-IPCF Istituto per i Processi Chimico Fisici
Università di Bari c/o Dip. Chimica
Via Orabona 4
70126 Bari
Italy

Pantaleo Davide Cozzoli
CNR-IPCF Istituto per i Processi Chimico Fisici
Università di Bari c/o Dip. Chimica
Via Orabona 4
70126 Bari
Italy

Maria Lucia Curri
CNR-IPCF Istituto per i Processi Chimico Fisici
Università di Bari c/o Dip. Chimica
Via Orabona 4
70126 Bari
Italy

Juan Jiang
Department of Chemical and Biomolecular Engineering
University of Notre Dame
Notre Dame, IN 46556
USA

Raz Jelinek
Department of Chemistry and Staedler Minerva Center for Mesoscopic Macromolecular Engineering
Ben Gurion University of the Negev
Beersheva 84105
Israel

Padmakar D. Kichambare
Department of Chemistry
University of Pittsburgh
Pittsburgh, PA 15260
USA

Hak-Sung Kim
Department of Biological Sciences
Korea Advanced Institute of Science and Technology
Daejeon 305-701
Korea

Sofiya Kolusheva
Department of Chemistry and Staedler Minerva Center for Mesoscopic Macromolecular Engineering
Ben Gurion University of the Negev
Beersheva 84105
Israel

Genxi Li
Department of Biochemistry
Nanjing University
22 Hankou Rd.
Nanjing 210093
P.R. China

Yuehe Lin
Pacific Northwest National Laboratory
Richland, WA 99352
USA

Guodong Liu
Pacific Northwest National Laboratory
Richland, WA 99352
USA

Albert Miller
Department of Chemical and Biomolecular Engineering
University of Notre Dame
Notre Dame, IN 46556
USA

Hideki Ohba
On-Site Sensing and Diagnostic Research Laboratory
National Institute for Advanced Industrial Science and Technology
AIST-Kyushu
807-1 Ahuku-machi, Tosu
Saga 841-0052
Japan

Michael Pugia
Diagnostic Division
Bayer Corporation
Elkhart, IN 46515
USA

Ramachandra Rao Sathuluri
School of Materials Science
Japan Advanced Institute of Science and Technology
1-1 Asahidai, Nomi city
Ishikawa 923-1292
Japan

Sara Seggerson
Department of Chemistry and Biochemistry
University of Notre Dame
Notre Dame, IN 46556
USA

Fwu-Shan Sheu
Department of Biological Sciences
National University of Singapore
14 Science Drive 4
Singapore 117543
Singapore

Alexander Star
Department of Chemistry
University of Pittsburgh
Pittsburgh, PA 15260
USA

Marinella Striccoli
CNR-IPCF Istituto per i Processi Chimico Fisici
Università di Bari c/o Dip. Chimica
Via Orabona 4
70126 Bari
Italy

Eiichi Tamiya
School of Materials Science
Japan Advanced Institute of Science and Technology
1-1 Asahidai, Nomi city
Ishikawa 923-1292
Japan

Jun Wang
Pacific Northwest National Laboratory
Richland, WA 99352
USA

Shohei Yamamura
School of Materials Science
Japan Advanced Institute of Science and Technology
1-1 Asahidai, Nomi city
Ishikawa 923-1292
Japan

Jian-Shan Ye
Department of Biological Sciences
National University of Singapore
14 Science Drive 4
Singapore 117543
Singapore

Joanne I. Yeh
University of Pittsburgh
School of Medicine
Department of Structural Biology
3501 5th Avenue
Pittsburgh, PA 15260
USA

Hyun C. Yoon
Department of Biotechnology
Ajou University
Suwon 443-749
Korea

Zhivko Zhelev
On-Site Sensing and Diagnostic Research Laboratory
National Institute for Advanced Industrial Science and Technology
AIST-Kyushu
807-1 Ahuku-machi, Tosu
Saga 841-0052
Japan

1 Biosensing using Carbon Nanotube Field-effect Transistors

Padmakar D. Kichambare and Alexander Star

1.1 Overview

This chapter covers recent advances in biodetection using single-walled carbon nanotube field-effect transistors (NTFETs). In particular, we describe fabrication of NTFET devices and their application for electronic detection of biomolecules. A typical NTFET fabrication process consists of combination of chemical vapor deposition (CVD) and complementary metal oxide semiconductor (CMOS) processes. The NTFET devices have electronic properties comparable to traditional metal oxide semiconductor field-effect transistors (MOSFETs) and readily respond to changes in the chemical environment, enabling a direct and reliable pathway for detection of biomolecules with extreme sensitivity and selectivity. We address the challenges in effective integration of carbon nanotubes into conventional electronics for biosensor applications. We also discuss in detail recent applications of NTFETs for label-free electronic detection of antibody–antigen interactions, DNA hybridization, and enzymatic reactions.

1.2 Introduction

The interplay between nanomaterials and biological systems forms an emerging research field of broad importance. In particular, novel biosensors based on nanomaterials have received considerable attention [1–4]. Integration of one-dimensional (1D) nanomaterials, such as nanowires, into electric devices offers substantial advantages for the detection of biological species and has significant advantages over the conventional optical biodetection methods [5]. The first advantage is related to size compatibility: Electronic circuits in which the component parts are comparable in size to biological entities ensure appropriate size compatibility between the detector and the detected biological species. The second advantage to developing nanomaterial based electronic detection is that most biological processes involve electrostatic interactions and charge transfer, which are directly

Nanotechnologies for the Life Sciences Vol. 8
Nanomaterials for Biosensors. Edited by Challa S. S. R. Kumar

ISBN: 978-3-527-31388-4

detected by electronic nanocircuits. Nanowire-based electronic devices, therefore, eventually integrate the biology and electronics into a common platform suitable for electronic control and biological sensing as well as bioelectronically driven nanoassembly [6].

One promising approach for the direct electrical detection of biomolecules uses nanowires configured as field-effect transistors (FETs). FETs readily change their conductance upon binding of charged target biomolecules to their receptor linked to the device surfaces. For example, recent studies by Lieber's group have demonstrated the use of silicon nanowire FETs for detecting proteins [7], DNA hybrids [8], and viruses [9]. This biodetection approach may allow in principle selective detection at a single particle levels [10, 11]. Nanowires hold the possibility of very high sensitivity detection owing to the depletion or accumulation of charge carriers, which are caused by binding of a charged biomolecules at the surface. This surface binding can affect the entire cross-sectional conduction pathway of these nanostructures. For some nanowires, such as hollow carbon nanotubes, every atom is on the surface and exposed to the environment; even small changes in the charge environment can drastically change their electrical properties. Thus, among different nanomaterials, carbon nanotubes have a great potential for biosensing.

Among numerous applications of carbon nanotubes [12–14], carbon nanotube based sensing technology is rapidly emerging into an independent research field. As for any new research field, there is no yet consensus in the literature about the exact sensing mechanism. In this chapter, in addition to selected examples of carbon nanotube based sensors, we address the controversial carbon nanotube sensing mechanism.

To date, sensor applications of carbon nanotubes have been summarized and discussed in several excellent review articles [15–17], which primarily focus on carbon nanotube based electrochemical sensors. This chapter covers only recent advances in biodetection using carbon nanotube field-effect transistors (NTFETs). It is divided into two large sections: NTFET fabrication and their sensor applications. Section 1.3 gives a detailed description of NTFET device structure, its fabrication method and introduces device characteristics. This section also addresses technical challenges in effective integration of carbon nanotubes into CMOS electronics. Section 1.4, which focuses on sensor applications of NTFETs, is divided into several subsections. Before discussing NTFET application for biological detection we describe the effect of environmental conditions on NTFET device characteristics. We give selected examples of NTFET sensitivity for small molecules, mobile ions, and water (relative humidity). The effect of these factors should be well understood before NTFET biodetection is reviewed. We also briefly describe the operation of NTFETs in conducting media, which is particularly important for biosensor applications. Then we briefly summarize interactions of carbon nanotubes with biomolecules (e.g., polysaccharides, DNA and proteins) to set a stage for the subsequent subsections that describe in great details recent applications of NTFETs for label-free electronic detection of proteins, antibody–antigen interactions, DNA hybridization, and enzymatic reactions.

1.3 Carbon Nanotube Field-effect Transistors (NTFETs)

1.3.1 Carbon Nanotubes

Since their discovery by Iijima over a decade ago [18], interest in carbon nanotubes has grown considerably [19]. Recent advances in the synthesis and purification of carbon nanotubes have turned them into commercially available materials. Subsequently, several experiments have been undertaken to study the physical and electrical properties of carbon nanotubes on the individual and macroscopic scale [20–23]. On the macroscopic scale, spectroscopic and optical absorption measurements have been carried out to test the purity of the carbon nanotubes [24, 25]. For electronic transport measurements it is particularly interesting to perform experiments on isolated, individual carbon nanotubes. The properties of carbon nanotubes depend strongly on physical aspects such as their diameter, length, and presence of residual catalyst [12]. The properties measured from a large quantity of nanotubes could be an average of all nanotubes in the sample, so that the unique characteristics of individual carbon nanotubes could be shadowed. Experiments on individual nanotubes are very challenging due to their small size, which prohibits the application of well-established testing techniques. Moreover, their small size also makes their manipulation rather difficult. Specialized techniques are needed to mount or grow an individual carbon nanotube on the electrode with sub-micron precision.

Carbon nanotubes are hollow cylinders made of sheets of carbon atoms and can be divided into single-walled carbon nanotubes (SWNTs) and multi-walled carbon nanotubes (MWNTs). SWNTs possess a cylindrical nanostructure with a high aspect ratio, formed by rolling up a single graphite sheet into a tube (Fig. 1.1). SWNTs are, typically, a few nanometers in diameter and up to several microns long. MWNTs consist of several layers of graphene cylinders that are concentrically nested like rings of a tree trunk, with an interlayer spacing of 3.4 Å [26]. Because of their unique properties, carbon nanotubes have become a material that has generated substantial interest on nanoelectronic devices and nanosensors [27, 28]. These properties are largely dependent upon physical aspects such as diameter, length, presence of catalyst and chirality. For example, SWNT can be metallic or semiconducting, depending upon the intrinsic band gap and helicity [29]. Semiconducting SWNTs can be used to fabricate FET devices, as demonstrated by Dekker and coworkers [30]. In addition, semiconducting SWNTs exhibit significant conductance changes in response to the physisorption of different gases [24, 31, 32]. Therefore, SWNT-based nanosensors can be fabricated based on FET layout, where the solid-state gate is replaced by adsorbed molecules that modulate the nanotube conductance [33]. Since semiconducting SWNTs have a very high mobility and, because all their atoms are located at the surface, they are the perfect nanomaterial for sensors. These sensors offer several advantages for the detection of biological species. First, carbon nanotubes form the conducting channel in a transistor configura-

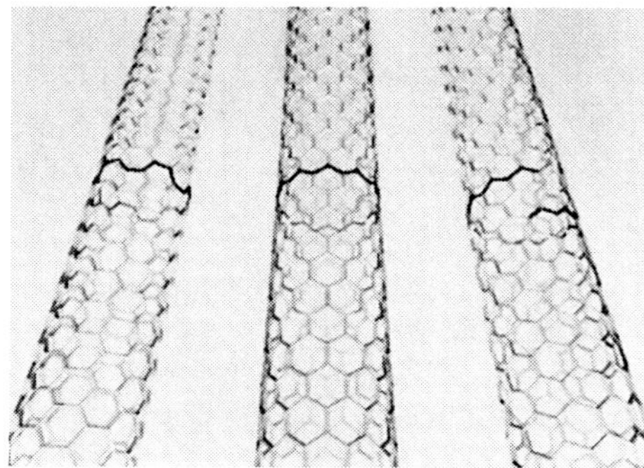

Fig. 1.1. A seamlessly rolled-up single graphite sheet forms a single-walled carbon nanotube (SWNT). SWNTs are, typically, a few nanometers in diameter and up to several micrometers long. They can be either metallic or semiconducting, depending on their helicity and diameter. Semiconducting SWNTs are used for the fabrication of nanotube field-effect transistors (NTFETs). (Adapted with permission from Ref. [4], © Wiley-VCH Verlag).

tion. Second, the nanotubes are typically located on the surface of the supporting substrate and are in direct contact with the environment. This device geometry contrasts with traditional metal oxide semiconductor field-effect transistors (MOSFETs) where the conducting channel is buried in the bulk material in which the depletion layer is formed. Lastly, all of the electrical current flows at the surface of nanotubes. All these remarkable characteristics lead to a FET device configuration that is extremely sensitive to minute variations in the surrounding environment.

1.3.2 Nanotube Synthesis

Several synthesis methods are used to produce carbon nanotubes [34]. The three most commonly used methods are the arc discharge, laser ablation, and chemical vapor deposition (CVD) techniques. While the arc and laser methods can produce large quantities of carbon nanotubes they lead to resilient contaminants, including pyrolytic and amorphous carbon [35, 36], which are difficult to remove from the sample. Such impurities result in low recovery yield for the carbon nanotube product. However, recent advances in scaling up these methods, as well as development new fabrication methods such as high pressure carbon monoxide (HiPCO), have created commercial supplies of carbon nanotubes with more than 90% purity with competitive prices. In contrast, the less scalable CVD process offers the best chance of obtaining controllable routes for the selective production of carbon nanotubes with defined properties [37]. CVD is catalytically driven, wherein a metal catalyst is used in conjunction with the thermal decomposition of hydrocarbon feedstock gases to produce carbon nanotubes. In most cases, the resultant growth of nanotubes occurs on a fixed substrate within the process. Figure 1.2 illustrates a typical CVD process for the generation of SWNTs. SWNTs are synthesized by the

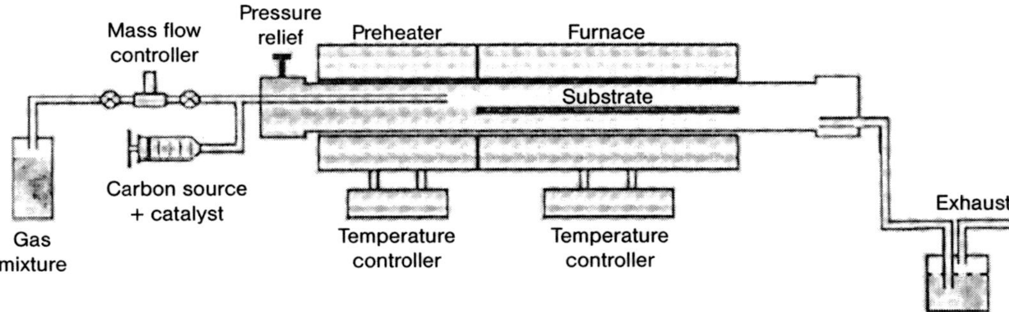

Fig. 1.2. Schematic of a chemical vapor deposition (CVD) reactor that uses a two-zone furnace. Carbon nanotubes grow on the substrate placed inside the quartz tube. (Reprinted with permission from Ref. [34], © 2001, CRC Press).

reaction of a hydrocarbon (e.g., CH_4) vapor over a dispersed Fe catalyst. The synthesis apparatus consists of a quartz tube reactor inside a combined preheater and furnace set-up. The preheat section is operated at ~200 °C. The catalysts are deposited and then hydrocarbon vapors are carried into the reaction zone of the furnace. An Ar/(10%)-H_2 carrier gas is used that controls the partial pressure inside the quartz tube reactor. Reaction temperatures are typically in the range 900–1000 °C. The SWNTs grow on the substrates (Fig. 1.3) and form thick mats that are readily harvested. This process produces highly pure SWNTs at a yield approaching 50% conversion of all hydrocarbon feedstock into carbon nanotube product. Similarly, a CVD processes sometimes utilize a feed of hydrocarbon-catalyst liquid for the production of nanotubes, and for this purpose a syringe pump is used to allow the continuous injection of this solution into a preheat section. Various gaseous feed-stocks are used to produce nanotubes, ranging from CH_4 to C_2H_2. A wide range of transition metals and rare earth promoters have been investigated for the synthesis of SWNTs by CVD. In general, a transition metal is the major component in the catalyst particles used regardless of the catalyst support. The most common metals found to be successful in the growth of SWNTs are Fe, Co, and Ni [38, 39]. However, bimetallic catalysts consisting of Fe/Ni, Co/Ni, or Co/Pt [40] are re-

Fig. 1.3. Transmission electron microscopy (TEM) image of an SWNT synthesized by chemical vapor deposition (CVD). (Reprinted with permission from Ref. [37], © 2001, The American Chemical Society).

ported to give the best yield of nanotubes, in addition to some rare earth metals that have also been studied [41].

Despite challenges, the understanding the growth mechanism of SWNTs is crucial for, ultimately, tailoring the production of SWNTs with known lengths, diameters, helical structures and placement of SWNTs at the desired location. Presently, efforts are underway to understand the mechanism of catalytic growth of SWNTs on surfaces and the role of impurities and to increase nanotube yield by varying the substrate, catalyst, and growth conditions [42]. Directional growth of SWNTs has been achieved by electric fields [43, 44], gas flow [45], lattice directions [46], and atomic steps [47].

1.3.3
Fabrication of NTFETs

To date carbon nanotubes have been used to fabricate various devices, including nanotube-based mechanical devices [48] and field emission devices [13]. This section focuses specifically on fabrication of carbon nanotube field-effect transistors (NTFETs). Figure 1.4(a) shows a schematic drawing of NTFET. A semiconducting carbon nanotube is contacted by source and drain electrodes while the gate elec-

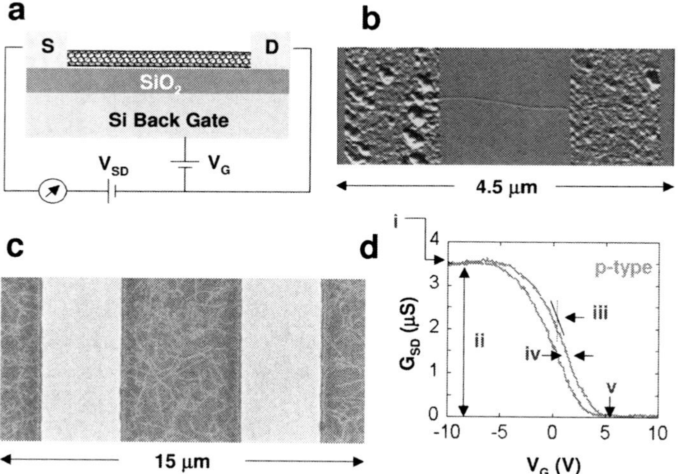

Fig. 1.4. (a) Schematic representation of a nanotube field-effect transistor (NTFET) device with a semiconducting SWNT contacted by two Ti/Au electrodes, representing the source (S) and the drain (D) with a Si back gate separated by a SiO_2 insulating layer in a transistor-configured circuit. (b) Atomic force microscopy (AFM) image of a typical NTFET device consisting of a single semiconducting SWNT. (c) Scanning electron microscopy (SEM) image of a typical NTFET device consisting of a random array of carbon nanotubes. (d) Typical NTFET transfer characteristic – dependence of the source-drain conductance (G_{SD}) on the gate voltage (V_G) – (i) maximum conductance, (ii) modulation, (iii) transconductance, (iv) hysteresis, and (v) threshold voltage.

trode, which is electrically insulated from the nanotube channel, is used to manipulate the nanotube's conductivity. Depending on the particular method of nanotube fabrication, a NTFET can be structured in different ways [49]. However, most publications on nanotube transistors report the use of a degenerately doped Si-substrate with a comparatively thick (100–500 nm) thermally grown oxide layer [30, 50–53]. Silicon substrates are readily available and can be used with both bulk-produced nanotubes and nanotubes grown directly on the Si-substrate by CVD. If doped highly, the Si substrate stays conductive even at low temperatures, making it usable as a so-called back-gate with the SiO_2 as a stable, if low-k, gate dielectric. Bulk produced nanotubes (laser ablation [54] or HiPCO [55]) are usually purified and deposited onto Si substrates by suspending them in organic solvents (e.g., chloroform, dichloroethane, etc.) and then spin-coating or drop casting on the substrates. In this approach the nanotubes create a random network over the substrate surface. Alignment of the nanotubes is possible, if an AC dielectric field is applied during deposition [56]. Generally, two different configurations of NTFETs, regarding source and drain contacts, are possible. By patterning the contacts before nanotube deposition [30] one can contact nanotubes in bulk, whereas by depositing the contacts onto the nanotubes one contacts only the ends of nanotubes [57], because depositing contact material on top of the nanotube normally destroys the nanotubes underneath the contacts. The second configuration usually guarantees a lower contact resistance than what is achievable in bulk-contacted devices [58]. Room temperature NTFET manufacturing methods, while compatible with CMOS, are limited by ability to disperse effectively carbon nanotubes in the solution and deposit them without further aggregation on device surfaces [59]. Nanotube bundle formation may decrease the semiconducting character of NTFET due to the occasional presence of nanotube bundles containing metallic nanotubes.

For CVD grown carbon nanotubes metal contacts are deposited onto the nanotubes [60, 61], because typical contact materials cannot withstand CVD temperatures, thus making it impossible to grow carbon nanotubes on CMOS structures. Often, the metal contacts are annealed to lower contact resistance [62]. Several studies have tried to optimize the material used for the contacts, including Cr/Au [49] and Pt [30], but only the Cr/Au contacts have been used widely. In this type of contact the chromium layer is a thin (1–3 nm) adhesion layer that facilitates adhesion of the gold to SiO_2. An adhesion layer of Ti [63], especially when annealed, allows deposition of smooth films of many metals onto carbon nanotubes because Ti forms titanium carbide at the interface with the nanotube. For this reason, Ti/Au-contacts are another frequently used combination of contact materials. Many publications investigating Schottky barriers between a nanotube and its contacts [61, 64] have employed such contacts. Palladium (Pd) is another material investigated that wets nanotubes well and has been used recently to produce NTFETs with ohmic contacts, i.e., contacts without Schottky barriers [52, 63].

Depending on the number of carbon nanotubes connecting the source and drain electrodes, there are two different device architectures. In the first device architecture, a single nanotube connects the source and drain (Fig. 1.4b). These devices have been used for biosensing with excellent sensitivity. However, there is substantial variation between the different devices that are fabricated and this variation is reflected in the electronic characteristics of individual nanotubes. In addition, the interface between the nanotube and the metallic contact may vary from device to device. Specialized techniques are needed either to mount or grow an individual carbon nanotube at a predetermined location. Placement is difficult and impractical for mass fabrication of NTFETs. For example, although the process of attaching a carbon nanotube strand via arc-discharge or contact method to sharp metal probe is fast, simple and economical it suffers from low yield. Therefore, it is difficult to determine the quality of carbon nanotube strand attached to metal tip unless examined under SEM. When checked under SEM a large percentage of the metal probes have multiple nanotubes attached or clusters of amorphous carbon accompanying the carbon nanotubes [65]. Hence random networks of SWNTs have been explored as an alternative [66].

Nanotube networks take up more space than individual SWNTs, but they are much easier to fabricate and show great promise towards simple mass fabrication of NTFETs. In this second configuration, the devices contain a random array of nanotubes between source and drain electrodes (Fig. 1.4c). In this configuration, current flows along several conducting channels that determine the overall device resistance. The device characteristics depend on the number of nanotubes and density of the nanotube network. It is reported that the conductance drops are associated with junctions formed by crossed semiconducting and metallic nanotubes. Local conductance is more dependent on the number of connections to the specific area; clusters of nanotubes with many paths to the electrode have significantly higher conductance than those parts of the network connected through fewer paths. Areas with low conductance typically only have two to three connections to the network, thus it is likely that these connections are dominated by the presence of highly resistive metallic/semiconducting junctions. When a sufficient back gate voltage is applied to the sample, current flow through the semiconducting tubes is suppressed. Using this technique, differences between metallic and semiconducting SWNT can be distinguished. This type of device configuration, containing a network of conducting nanotube channels, is less sensitive than devices made of single nanotubes. In both types of device configurations, the parameter used for detection is the transfer characteristic – the dependence of either the source-drain current (I_{SD}) or conductance (G_{SD}) (for a fixed source-drain voltage V_{SD}) on the gate voltage (V_G) (Fig. 1.4d).

NTFETs can operate as p-type or n-type transistors. The mode of operation can be changed from the pristine p-type to n-type by either adding electron donor molecules (n-doping) or removing adsorbed oxygen by annealing the contacts under vacuum [67]. Polymer-gated NTFETs can also tune their modes of operation: a change in the chemical group of the polymer changes the NTFET from p-type to n-type [68, 69]. Oxygen doping was attributed to the fact that the oxygen interacts with the nanotube–metal junction and causes the p-type characteristic for NTFETs in air by pinning the metal's Fermi level near the nanotube's valance band maximum [33]. However, there is no apparent consensus in the literature about the exact mechanism of chemical sensitivity of NTFETs.

1.4
Sensor Applications of NTFETs

Before discussing NTFET applications for biological detection we first describe the effect of small molecules, relative humidity, and conductive liquid media on NTFET devices characteristics. Effect of these factors should be well understood before NTFET biodetection is reviewed.

1.4.1
Sensitivity of NTFETs to Chemical Environment

Generally, the molecular species in the ambient environment have a significant impact on the electrical properties of NTFETs. The conductance of semiconducting SWNTs can be substantially increased or decreased by exposure to NO_2 or NH_3 [24]. Exposure to NH_3 effectively shifts the valance band of the nanotube away from the Fermi level, resulting in hole depletion and reduced conductance. In contrast, on exposure to NO_2 molecules the conductance of nanotubes increases by three orders of magnitude [70]. Here, exposure of the initially depleted sample to NO_2 resulted in the nanotube Fermi level shifting closer to the valence band. This caused enriched hole carriers in the nanotube and enhanced sample conductance. These results show that molecular gating effects can shift the Fermi level of semiconducting SWNTs and modulate the resistance of the sample by several orders of magnitude.

The electronic properties of SWNTs are also extremely sensitive to air or oxygen exposure [33]. Isolated semiconducting nanotubes can be converted into apparent metals through room temperature exposure to oxygen. As the surrounding medium was cycled between vacuum and air, a rapid and reversible change in the SWNT resistance occurred in step with the changing environment. Initially, in a pure atmospheric pressure oxygen environment, the thermoelectric power (TEP) S was positive with a magnitude of nearly $+20$ mV K^{-1}. This relatively large positive TEP is consistent with that reported for pristine SWNTs near room temperature [71]. As oxygen was gradually removed from the chamber, the TEP changed continuously from positive to negative, with a final equilibrium value of approximately -10 μV K^{-1}. When oxygen was reintroduced into the chamber, the TEP reversed sign and once again became positive. These dramatic 10–15% variations in R and change in sign of the TEP demonstrate that SWNTs are exceptionally sensitive to oxygen.

In the carbon nanotubes sensors mentioned above, chemical sensing experiments have been conducted with devices in which both nanotubes and nanotube–metal contacts were directly exposed to the environment. The sensing could be dominated by the interaction of molecules with the metal contacts or the contact interfaces. Adsorbed molecules would modify the metal work functions and, thereby, the Schottky barrier [72, 73]. Heinze et al. [64] have assigned the effect of oxygen to the Schottky barrier. Recently, a new device architecture has been studied in which the interface between the metallic contacts and nanotubes is covered by a

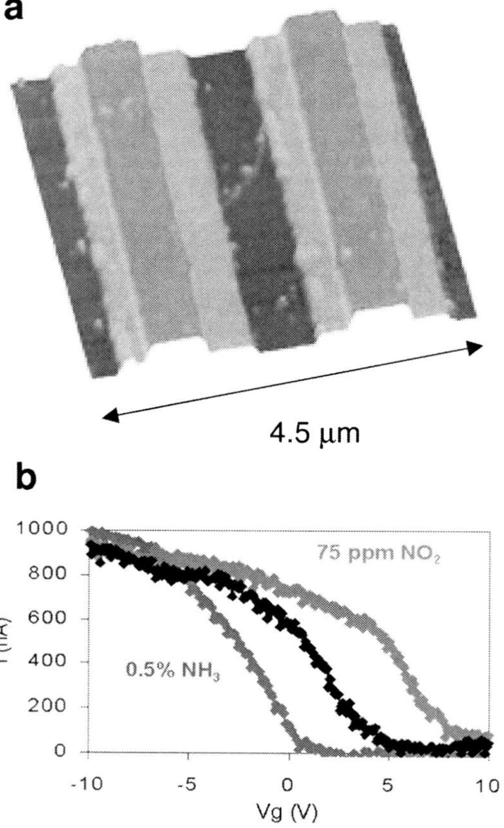

Fig. 1.5. (a) AFM image of a contact passivated NTFET device covered with poly(ethylene imine). (b) I_{SD}–V_G dependence for the device in vacuum (center curve), as well as in NH_3 and NO_2 gases. (Adapted with permission from Ref. [74], © 2003 American Institute of Physics).

passivation layer, referred to as contact-passivated [74]. In this configuration, with the junction isolated and only the central length of the nanotube channels exposed, the contacts should be isolated from the effect of chemicals. At the same time, the section of the device that is open to the environment can be doped via charge transfer. NTFETs with such configuration have been investigated by measuring sensitivity to NH_3, NO_2, and poly(ethylene imine) (Fig. 1.5).

The NTFET devices were fabricated using SWNTs grown by CVD on 200 nm of silicon dioxide on doped silicon from iron nanoparticles as described in Section 1.3.1. These particles were exposed to flowing hydrocarbon to grow carbon nanotubes, and after growth optical lithography was used to pattern electrical leads (35 nm titanium capped with 5 nm gold) on top of the nanotubes. Contact passivation was achieved by 70 nm silicon monoxide layer. Source and drain electrodes were separated by nearly a micrometer. The dependence of the source-drain current (I_{SD}) as function of the gate voltage (V_G) was measured from +10 to −10 V using a semiconductor parameter analyzer in air/water/gas mixtures. The low concentrations of gas mixtures could be introduced to the devices by mixing different proportions of air and gases. The contact-passivated devices demonstrated NH_3 and NO_2 sensitivity similar to regular NTFETs. Poly(ethylene imine) also produced negative threshold shifts of tens of volts, despite being in contact with only the center region of devices. Thus, the NTFET sensor character was preserved despite isolating Schottky barriers.

Several groups have reported that NTFET fabricated on SiO_2/Si substrates exhibits hysteresis in current versus gate-voltage characteristics and attributed the hysteresis to charge traps in bulk SiO_2, oxygen-related defect trap sites near nanotubes, or the traps at the SiO_2/Si interface. It is mentioned that thermally grown SiO_2 surface consists of Si-OH silanol groups and is hydrated by a network of water molecules that are hydrogen bonded to the silanols. The CVD nanotube growth condition (900 °C) may dehydrate the surface and condense to form Si–O–Si siloxanes. When such a surface is exposed to and stored in ambient air, the surface siloxanes on the substrate react with water and gradually revert to Si–OH, after which the substrate becomes rehydrated. Heating under dry conditions significantly removes water and reduces hysteresis in the transistors.

Kim et al. have reported that the hysteresis in electrical characteristics of NTFETs is due to charge trapping by water molecules around the nanotubes, including SiO_2 surface-bound water proximal to the nanotubes [75]. They have demonstrated that coating nanotube devices with PMMA can afford nearly hysteresis-free NTFETs [75]. This passivation is attributed to two factors. First, the ester groups of poly(methyl methacrylate) (PMMA) can hydrogen bond with silanol group on SiO_2. Baking at 150 °C combined with the polymer–SiO_2 interaction can significantly remove the silanol-bound water. Second, PMMA is hydrophobic and can keep water in the environment from permeating the PMMA and adsorbing on the nanotube in a significant manner.

Bradley et al. have attributed hysteresis in NTFET devices to cation diffusion [76], based on the following experiments. First, NTFET devices that exhibit very small hysteresis were fabricated. Subsequently, these devices were modified by the addition of an electrolyte coating that created mobile ions on the surface of the device and resulted in the large hysteresis. Experiments were also conducted to explore possible mechanisms for cation-induced hysteresis by varying the humidity that changes the hydration layer around the nanotubes, thus leading to the increase of the ionic mobility. The hysteresis has been found to be sensitive to humidity on sub-second time scales, showing promise as a humidity sensor [77].

Sensitivity of NTFETs to charges as well as NTFET operation in conducting liquid media is important for biosensor design where the sensor should operate in physiological buffers with complex mixtures of biomolecules. Figure 1.6 shows a typical transfer characteristic of NTFET measured in air and water using the silicon and water as the gate electrode, respectively. The change in device characteris-

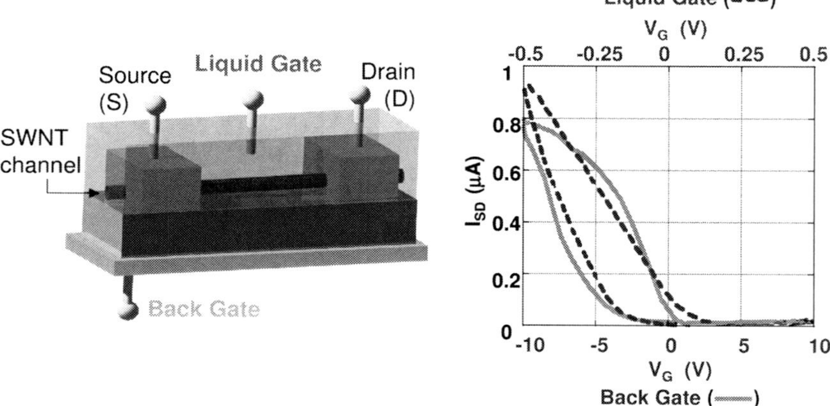

Fig. 1.6. (a) Detection in liquid with NTFET devices by using either the back gate or liquid gate configuration. (b) NTFET transfer characteristics in air (solid line), using the back gate, and in water (dashed line), using the liquid gate. Note the different *x*-scales for the back and liquid gates. (Adapted with permission from Ref. [93], © 2003, The American Physical Society.)

tics upon exposure to a water/gas mixture is reflected in the transfer characteristics. Saline or electrolytes can also gate NTFETs and give high transconductance [62, 78].

1.4.2
Bioconjugates of Carbon Nanotubes

Numerous reports demonstrate the ability to chemically functionalize nanotubes for biological applications [79, 80]. Such chemistry is readily transferable to many applications, ranging from sensors [81, 82] to electronic devices [83]. SWNTs are chemically stable and highly hydrophobic. Therefore, they require surface modification to establish effective SWNT–biomolecule interaction.

So far, two methods of exohedral functionalization of SWNTs have been developed – namely covalent and noncovalent. While covalent modifications [84] are often effective at introducing functionality, they impair the desirable mechanical and electronic properties of SWNTs. Noncovalent modifications [85], however, not only improve the solubility of SWNTs in water, but they also constitute nondestructive processes, which preserve the primary structures of the SWNTs, along with their unique mechanical and electronic properties.

Previously, it has been shown that polysaccharides such as starch [86, 82, 83], gum Arabic [84], and the β-1,3-glucans, curdlan and schizophyllan [85], will solubilize SWNTs in water. It has been proposed that at least some of these polymers achieve their goal by wrapping themselves in helical fashion around SWNTs (Fig.

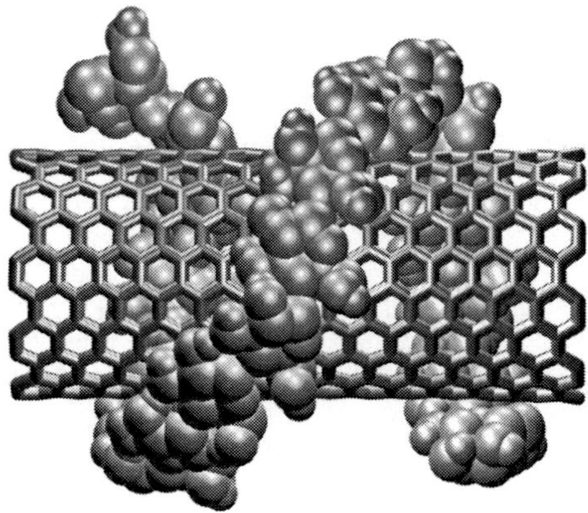

Fig. 1.7. Molecular model of SWNT wrapped in an amylose coil. (Reprinted from Ref. [79], © 2002, The American Chemical Society.)

1.7). Solubilization of the SWNTs with cyclodextrins (CD), which are macrocyclic polysaccharides, has been also investigated [86]. The observed aqueous solubility of SWNTs with γ-CD is unlikely due to encapsulation because the inner cavity dimensions of this CD are far too small to allow it to thread onto even the smallest diameter SWNTs. More recently, however, it has been shown [87] that η-CD, which has 12 α1,4-linked D-glucopyranose residues and therefore is large enough, does thread onto SWNTs in water, not only solubilizing the NTs but also permitting some partial separations according to their diameters.

Nucleic acids, such as single-stranded DNA, short double-stranded DNA, and some total RNA can also disperse SWNTs in water [88, 89]. Molecular modeling has shown [20] that the non-specific DNA–SWNT interactions in water are from the nucleic acid–base stacking on the nanotube surface, resulting in the hydrophilic sugar–phosphate backbone pointing to the exterior to achieve the solubility in water. The mode of interaction could be helical wrapping or simple surface adsorption. The charge differences among the DNA–SWNT conjugates, which are associated with the negatively charged phosphate groups of DNA and the different electronic properties of SWNTs, have allowed post-production preparation of samples enriched in metallic and semiconducting SWNTs.

Various proteins can also strongly bind to the nanotube exterior surface via non-specific adsorption. Proteins such as streptavidin and HupR crystallize in helical fashion, resulting in ordered arrays of proteins on the nanotube surface [90]. Mechanistically, the non-specific adsorption of proteins onto the nanotube surface may be more complicated than the widely attributed hydrophobic interactions. Quite possibly, the observed substantial protein adsorption is, at least in part, associated with the amino affinity of carbon nanotubes, as was demonstrated recently by monitoring the conductance change in the carbon nanotube [91]. Also, intermolecular interactions involving aromatic amino acids, i.e., histidine and tryptophan, in the polypeptide chains of the proteins can contribute to the observed affinity of the peptides to carbon nanotubes [92].

1.4.3 Protein Detection

Carbon nanotube interactions with proteins have been explored by NTFET devices [91]. In NTFET devices, the ability to measure the electronic properties of the nanotube allowed to query the electronic state of the immobilization substrate. In that work two types of measurements of the device transfer characteristics were performed. In the first measurement, referred to as a substrate-gate transfer characteristic, the current through the drain contact (at fixed source-drain bias) was monitored while a variable gate voltage was applied through a metallic gate buried underneath the SiO_2 substrate. In the second measurement, referred to as liquid-gate transfer characteristics, the device was immersed in a buffer solution and a variable gate voltage was applied through a platinum electrode. The current was passed through the drain contact and a silver reference electrode in the solution. During these measurements, the assembly was shaken gently, using a lab rotator at 3 Hz. The effect of protein adsorption was studied with both measurements. Devices were incubated with streptavidin (40 nM) in 15 mM phosphate buffer at 25 °C. Liquid-gate transfer characteristics were measured continually during the incubations. After 10 h, the devices were rinsed with distilled water and blown dry, and the substrate-gated transfer characteristics of the dried devices were measured.

These results were discussed in terms of a simple model in which adsorbed streptavidin coats the single-walled nanotube (Fig. 1.8). The gradual shift in the threshold voltage is assumed to result from the slow accumulation of a full monolayer of adsorbed protein. This coverage-dependent threshold shift is analogous

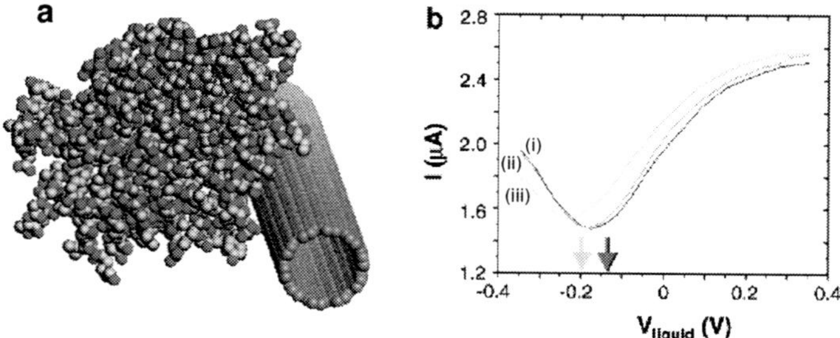

Fig. 1.8. (a) Size comparison between a carbon nanotube and a streptavidin molecule. (b) Current versus gate voltage for a nanotube device; $V_{SD} = 10$ mV. (ii) In phosphate buffer before streptavidin addition. (i) same conditions, to measure the uncertainty in the threshold voltage. (iii) After 10 h of incubation with streptavidin. Arrows indicate the threshold voltages for the three curves [the arrow for (i) is behind that for (ii)]. (Adapted with permission from Ref. [91], © 2003, The American Chemical Society.)

to the concentration-dependent shift observed when such devices are exposed to aqueous ammonia [93]. The protein adsorbate equilibrates over several hours so that only the full monolayer can be conclusively determined. Such protein monolayers form under various conditions at interfaces that permit protein crystallization, including sidewalls of MWNTs [90, 94]. The results support the proposal that conductance changes are due to charge injection or field effects caused by proteins adsorbed solely along the lengths of the nanotubes.

Protein adsorption on NTFET leads to appreciable changes in the electrical conductance of the devices that can be exploited for label-free detection of biomolecules with a high potential for miniaturization. For example, Dai and coworkers [95] have used a sensor design consisting of an array of four NTFET sensors on SiO_2/Si chips. Each NTFET consists of multiple SWNTs connected roughly in parallel across two closely spaced bridging metal electrodes. Three types of devices with different surface functional groups were prepared for the investigation of the biosensing: (1) unmodified as-made devices, (2) devices fabricated with mPEG-SH SAMs formed on, and only on, the metal contact electrodes and, lastly, (3) devices with mPEG-SH SAMs on the metal contacts and a Tween 20 coating on the carbon nanotubes. Electrical conductance of these devices upon the addition of various protein molecules was monitored. While device type 1 showed a significant conductance change with protein adsorption, device type 2 with an mPEG-SH SAM on the metal electrodes did not give any conductance change, except in the case of the protein avidin. It was reported that the metal–nanotube interface or contact region is highly susceptible to modulation by adsorbed species [64]. Modulation of metal work function can alter the Schottky barrier of the metal–nanotube interface, thus leading to a significant change in the nature of contacts and, consequently, a change in the conductance of the devices.

In situ detection of a small number of proteins by directly measuring the electron transport properties of a single SWNT has been reported by Nagahara and coworkers [96]. Cytochrome *c* (cytc) adsorption onto individual NTFET has been detected via the changes in the electron transport properties of the transistors. The adsorption of cytc induces a decrease in the conductance of the NTFET devices, corresponding to a few tens of molecules. This experiment was carried out by measuring the conductance versus electrochemical potential of the SWNT with respect to a reference electrode inserted in the solution, and observed a negative shift in the conductance versus potential plot upon protein adsorption. The number of adsorbed proteins has been estimated from this shift.

1.4.4 Detection of Antibody–Antigen Interactions

Specific sensitivity can be achieved by employing recognition layers that induce chemical reactions and modify the transfer characteristics. In this two-layer architecture carbon nanotubes function as extremely sensitive transducers while the recognition layer provides chemical selectivity and prevents non-specific binding that is common for complex biological samples.

Following this design, nanotubes have been functionalized to be biocompatible and to be capable of recognizing proteins. This functionalization has involved noncovalent binding between a bifunctional molecule and a nanotube to anchor a bioreceptor molecule with a high degree of control and specificity. Star and coworkers have fabricated [97] NTFET devices sensitive to streptavidin using a biotin-functionalized carbon nanotube bridging two microelectrodes (source and drain, Fig. 1.9a). The SWNT in the NTFET device was coated with a mixture of two polymers, poly(ethyleneimine) and poly(ethylene glycol). The former provided amino groups for the coupling of biotin–*N*-hydroxysuccinimidyl ester (Fig. 1.9b) and the latter prevented the nonspecific adsorption of proteins on the functionalized carbon nanotube. Figure 1.9(c) shows an AFM image of the device after its exposure to streptavidin labeled with gold nanoparticles (10 nm). Lighter dots represent gold nanoparticles and indicate the presence of streptavidin bound to the

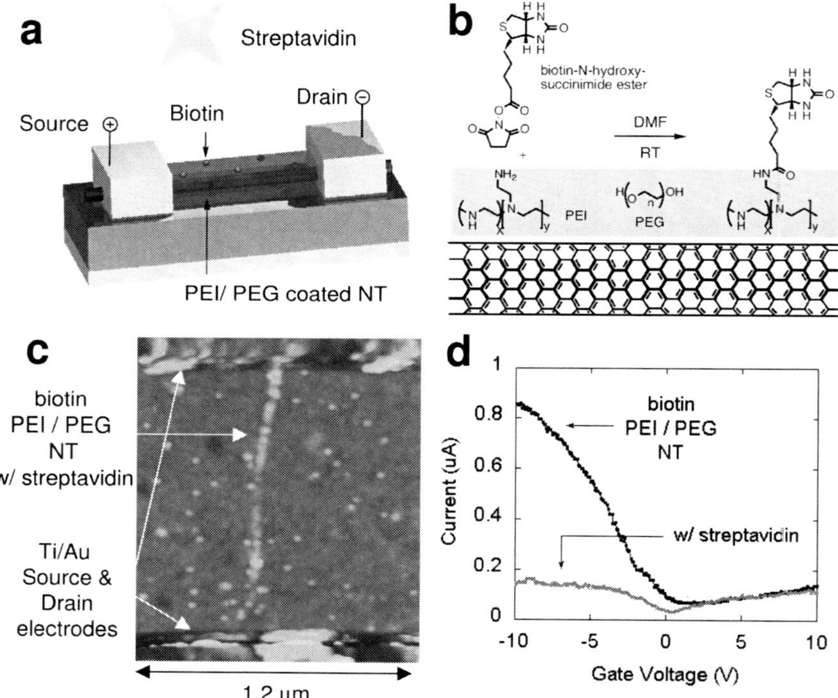

Fig. 1.9. (a) Schematic of NTFET coated with a biotinylated polymer layer for specific streptavidin binding. (b) Biotinylation reaction of the polymer layer (PEI/PEG) on the side-wall of the SWNT. (c) AFM image of the polymer-coated and biotinylated NTFET device after exposure to streptavidin labeled with gold nanoparticles (10 nm in diameter). (d) Source-drain current dependence on gate voltage of the NTFET device based on SWNTs functioned with biotin in both the absence and presence of streptavidin. (Adapted with permission from Ref. [97], © 2003, The American Chemical Society.)

biotinylated carbon nanotube. The source-drain current dependence on the gate voltage of the NTFET shows a significant change upon the streptavidin binding to the biotin-functionalized carbon nanotube (Fig. 1.9d). The experiments reveal the specific binding of the streptavidin, which occurs only at the biotinylated interface.

The mechanism of the biodetection was explained in terms of the effect of the electron doping of the carbon nanotube channel upon the binding of the charged streptavidin molecules. Dai and coworkers [98] have also analyzed specific antigen–antibody interactions using NTFET devices. In particular, they have studied the affinity binding of 10E3 mAbs antibody (a prototype target of the autoimmune response in patients with systematic lupus erythematosus and mixed connective tissue disease) to human auto antigen U1A.

1.4.5 DNA Detection

DNA biosensors based on nucleic acid recognition processes are quickly being developed towards the goal of rapid, simple and inexpensive testing of genetic and infectious diseases. To date, there are several reports on the electrochemical detection of DNA hybridization using multi-walled carbon nanotube (MWNT) electrodes [99]. Whereas electrochemical methods rely on the electrochemical behavior of the labels, measurements of the direct electron transfer between SWNTs and DNA molecules paves the way for label-free DNA detection (Fig. 1.10). To illustrate the practical utility of this new nanoelectronic detection method, an allele-specific assay to detect the presence of SNPs using NTFETs has been recently developed [100]. This DNA assay targeted the H63D polymorphism in the human HFE gene, which is associated with hereditary hemochromatosis, a common and easily treated disease of iron metabolism [101, 102].

DNA sensing mechanism using NTFETs has been recently explored by selective attachment of DNA molecules at different device segments. Tang et al. [103] have found that DNA hybridization on gold electrodes rather than on SWNT sidewalls is mainly responsible for NTFET detection due to Schottky barrier modulation. In another approach, DNA hybridization occurs on the surface at the gate of NTFET [104]. As a result, the conductance in SWNTs was changed through the gate insulators. In the work, the 5′ end-amino modified peptide nucleic acid (PNA) oligonucleotides were covalently immobilized onto the Au surfaces of the back gate of NTFETs. PNA is a synthetic analog of DNA, in which both the phosphate and the deoxyribose of the DNA backbone are replaced by a polypeptide. PNA mimics the behavior of DNA and hybridizes with complementary DNA or RNA sequences, thus enabling PNA chips to be used in biosensors. The micro-flow chip was fabricated by using poly(dimethylsiloxane) (PDMS) prepolymer. The NTFET nanosensor array was placed onto the PDMS chip in such a way that the PNA probe-modified Au side was positioned to face the open chamber for the introduction of solutions and the electrical measurements. A PNA probe with the base sequence 5′-NH_2-ACC ACC ACT TC-3′, which was fully complementary to the tumor necro-

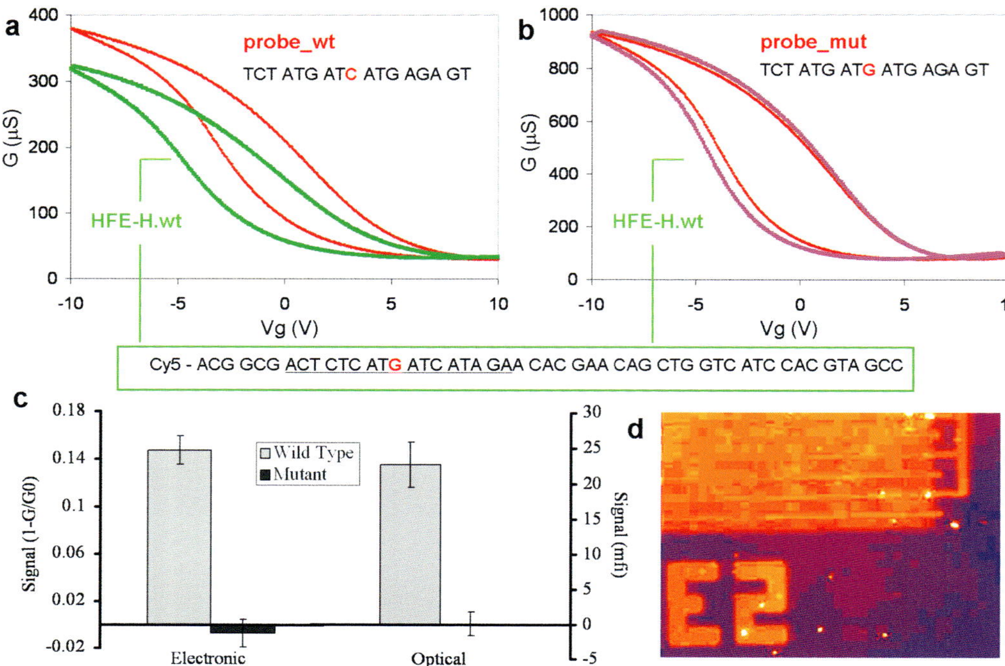

Fig. 1.10. Label-free detection of DNA hybridization using NTFET devices. (a) G–V_g curves after incubation with allele-specific wild-type capture probe and after challenging the device with wild-type synthetic HFE target (50 nM). (b) G–V_g curves in the experiment with mutant capture probe. (c) Graph with electronic $(1 - G/G_0)$ and fluorescent responses in SNP detection assays. (d) Fluorescence microscopy image of the NTFET network device, with the electrodes 10 mm apart, after incubation with Cy5-labeled DNA molecules. (Adapted with permission from Ref. [100], © 2006, The National Academy of Sciences of the USA.)

sis factor-α (TNF-α) gene sequence, was used as a model system. The base sequence for full complementary target DNA was 5′-GGT TTC GAA GTG GTG GTC TTG-3′ while the non-complementary DNA oligonucleotide sequence was 5′-CCC TAA GCC CCC AA-3′.

The electrical properties of the NTFET devices were measured at room temperature in air. First, the blank PBS solution was introduced into the PDMS-based micro flow chip, revealing that no substantial change in the source-drain current of NTFET was obtained. The current increased dramatically while monitoring in real time for about 3 h. The increase in conductance for the p-type NTFET device was consistent with an increase in negative surface charge density associated with binding of negatively charged oligonucleotides at the surface. DNA hybridization can be detected by measuring the electrical characteristics of NTFETs, and SWNT based FET can be employed for label-free, direct real time electrical detection of biomolecule binding.

1.4.6
Enzymatic Reactions

SWNTs can be made water soluble by wrapping in amylose (linear component of starch) [86]. These SWNT solutions are stable for weeks, provided nobody spits on them. Indeed, the addition of saliva, which contains α-amylase, precipitates the nanotubes as the enzyme breaks amylose down into smaller carbohydrate fragments, finally resulting in the formation of glucose. The enzymatic degradation of starch has been recently monitored electronically using NTFETs [105]. Figure 1.11(a) shows the experimental setup used for this study. NTFET devices display transconductance and source-drain current–voltage characteristics typical of the p-type device behavior. The device characteristics, i.e., the source-drain current I_{SD} as a function of the gate voltage V_G, were measured to evaluate the effect of starch deposition and the subsequent enzymatic degradation of the starch layer on the carbon nanotubes.

Starch was deposited onto the FET by soaking the silicon wafer in a 5% aqueous starch solution and the device characteristics were found to be shifted by approxi-

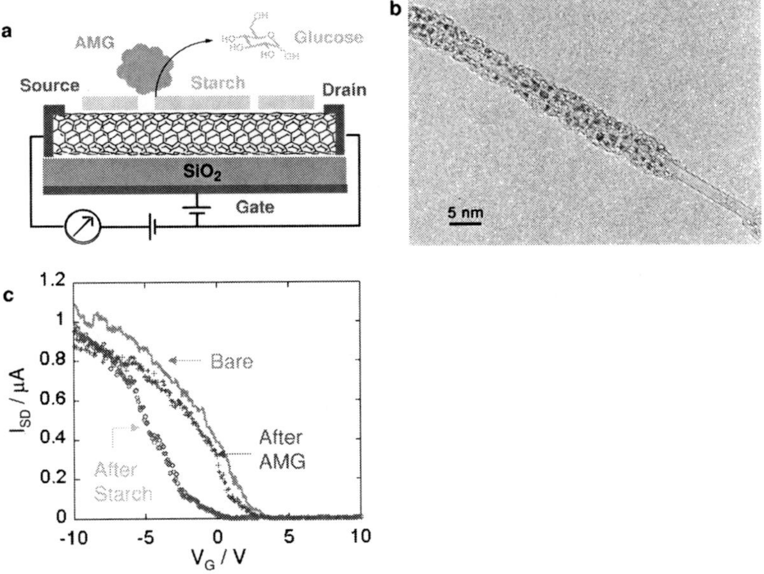

Fig. 1.11. (a) NTFET device for electronic monitoring of the enzymatic degradation of starch with amyloglucosidase (AMG) to glucose. (b) High-resolution transmission electron microscopy (HRTEM) image of a SWNT (2.0 nm diameter) after treatment with a drop of a 1% of an aqueous solution of starch. The starch had been stained with RuO_4 vapor. (c) NTFET device characteristics in the form of I_{SD}–V_G curves measured from +10 to −10 V gate voltage with a +0.6 V bias voltage before (bare) and after starch deposition, as well as after hydrolysis with AMG. (Adapted with permission from Ref. [104], © 2004, The American Chemical Society.)

mately 2 V toward more negative gate voltages. The direction of the shift equates with electron doping of the nanotube channel by the polysaccharide. Quantitatively similar doping effects have been observed when carbon nanotube FET devices were exposed to NH_3 gas, amines [106], poly(ethylene imine) (PEI) [107], and proteins [91]. After the enzyme-catalyzed reaction had been performed on the starch-functionalized devices and washed with buffer, the I_{SD} vs. V_G characteristics recovered almost completely to the trace recorded before starch deposition (Fig. 1.11). This indicates that, during the enzyme-catalyzed reaction, nearly all the starch deposited on the surface of the nanotube device is hydrolyzed to glucose which is washed off by the buffer prior to the electronic measurements.

1.4.7
Glucose Detection

The diagnosis and management of diabetes mellitus requires a tight monitoring of blood glucose levels. Dekker and coworkers have demonstrated the use of individual semiconducting SWNT as a versatile biosensor [108]. The redox enzyme glucose oxidase (GO_x) that catalyses the oxidation of β-D-glucose ($C_6H_{12}O_6$) to D-glucono-1,5-lactone ($C_6H_{10}O_6$) has been studied. The redox enzymes go through a catalytic reaction cycle where groups in the enzyme temporarily change their charge state and conformational changes occur in the enzyme that can be detected using NTFET devices.

In addition to pH sensitivity, GO_x-coated semiconducting SWNTs appeared to be sensitive to glucose, the substrate of GO_x. Figure 1.12 exhibits real-time measure-

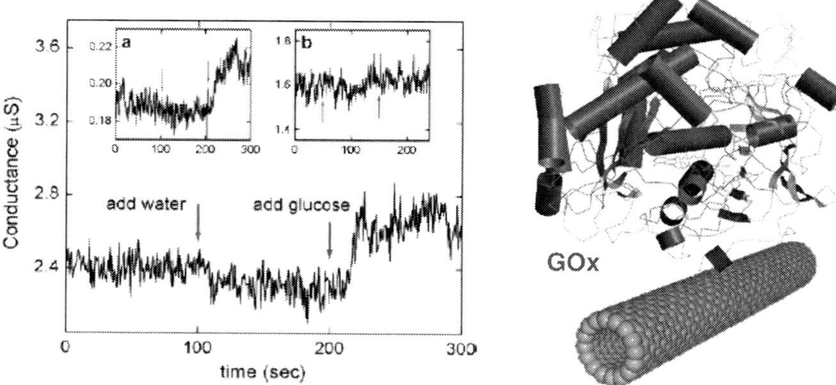

Fig. 1.12. Real time electronic response of the NTFET sensor to glucose, the substrate of glucose oxidase (GO_x). The conductance of a semiconducting SWNT with immobilized GO_x is measured as a function of time in 5 μL milli-Q water. The conductance of the GO_x-coated SWNT increases upon addition of glucose to the liquid. Inset: (a) the same measurement on a second device where the conductance was a factor of 10 lower; (b) the same measurement on a semiconducting SWNT without GO_x; no conductance increase is observed in this case. (Reprinted with permission from Ref. [107], © 2003, The American Chemical Society.) (B) Schematic of GO_x immobilized on SWNT for electronic glucose detection.

ments where the conductance of a GOx-coated semiconducting SWNT in milli-Q water has been recorded in the liquid (left-hand arrow in each graph in Fig. 1.12). No significant change in conductance was observed as a result of water addition. When 0.1 M glucose in milli-Q water was added to the liquid (right-hand arrow in each graph), however, the conductance of the tube increased by about 10%. A similar 10% conductance change was observed for another device (Fig. 1.12a inset), which had a factor 10 lower conductance. Glucose did not change the conductance of the bare SWNT but did increase the device conductance after GOx was immobilized. Inset (b) of Fig. 1.12 shows such measurement on a bare semiconducting SWNT. These measurements clearly indicate that the GOx activity is responsible for the observed increase in conductance upon glucose addition, thus rendering such nanodevices as feasible enzymatic-activity sensors.

1.5 Conclusion and Outlook

Recent advances in the rapidly developing area of biomolecule detection using carbon nanotube systems have been summarized here. SWNTs appear as structurally defined components for various electronic devices. The semiconductive properties of SWNTs are of special interest as these SWNTs have been applied to fabricate FETs for sensing applications. This area requires further development, particularly related to the fabrication of FETs based on individual SWNTs. The use of carbon nanotubes as nanocircuitry elements is particularly interesting. Biomaterials linked to nanotubes may be used as binding elements for the specific linkage of the nanotube to surface in the form of addressable structures.

Important chemical means to functionalize SWNTs with other electronic materials such as conductive polymers or nanoparticles is anticipated to generate materials of new properties and functions. The localized nanoscale contacts of SWNTs with bio-surfaces will be a major advance in understanding and exploring the new applications. The use of nanodevices to monitor various biologically significant reactions is envisioned. In future, it should be possible to connect the living cells directly to these nanoelectronic devices to measure the electronic responses of living systems. The combination of the unique electronic properties of SWNTs and catalytic features of biological system could provide new opportunities for carbon nanotubes based bioelectronics.

References

1 Park, S., Taton, T. A., Mirkin, C. A. Array-based electrical detection of DNA with nanoparticle probes., *Science* **2002**, 295, 1503–1506.

2 He, L., Musick, M. D., Nicewarner, S. R., Salinas, F. G., Benkovic, S. J., Natan, M. J., Keating, C. D. Colloidal Au-enhanced surface plasmon resonance for ultrasensitive detection of DNA hybridization., *J. Am. Chem. Soc.* **2000**, 122, 9071–9077.

3 McFarland, A. D., Van Duyne, R. P. Single silver nanoparticles as real-time optical sensors with zeptomole sensitivity., *Nano Lett.* **2003**, 3, 1057–1062.
4 Graham, A. P., Duesberg, G. S., Seidel, R. V., Liebau, M., Unger, E., Pamler, W., Kreupl, F., Hoenlein, W. Carbon nanotubes for microelectronics?, *Small* **2005**, 1, 382–390.
5 Grüner, G. Carbon nanotube transistors for biosensing applications., *Anal. & Bioanal. Chem.*, **2006**, 384, 322–335.
6 Willner, I. Biomaterials for sensors, fuel cells, and circuitry., *Science* **2002**, 298, 2407–2408.
7 Cui, Y., Wei, Q., Park, H., Lieber, C. M. Nanowire nanosensors for highly sensitive and selective detection of biological and chemical species., *Science* **2001**, 293, 1289–1292.
8 Hahm, J., Lieber, C. M. Direct ultrasensitive electrical detection of DNA and DNA sequence variations using nanowire nanosensors., *Nano Lett.* **2004**, 4, 51–54.
9 Patolsky, F., Zheng, G., Hayden, O., Lakadamyali, M., Zhung, X., Lieber, C. M. Electrical detection of single viruses, *Proc. Natl. Acad. Sci. U.S.A.* **2004**, 101, 14017–14022.
10 Weiss, S. Measuring conformational dynamics of biomolecules by single molecule fluorescence spectroscopy., *Nat. Struct. Biol.* **2000**, 7, 724–729.
11 Zhuang, X. W., Rief, M. Single-molecule folding., *Curr. Opin. Struct. Biol.* **2003**, 13, 88–97.
12 Ajayan, P. M. Nanotubes from carbon., *Chem. Rev.* **1999**, 99, 1787–1799.
13 Ajayan, P. M., Zhou, O. Z. Applications of carbon nanotubes., *Top. Appl. Phys.* **2001**, 80, 391–425.
14 Special issue on carbon nanotubes, *Acc. Chem. Res.* **2002**, vol. 35.
15 Wang, J. Carbon-nanotube based electrochemical biosensors: A review., *Electroanalysis* **2005**, 17, 7–14.
16 Lin, Y., Taylor, S., Li, H., Shiral Fernando, K. A., Qu, L., Wang, W., Gu, L., Zhou, B., Sun, Y. P. Advances toward bioapplications of carbon nanotubes., *J. Mater. Chem.* **2004**, 14, 527–541.
17 Katz, E., Willner, I. Biomolecule-functionalized carbon nanotubes: applications in nanobioelectronics., *ChemPhysChem* **2004**, 5, 1084–1104.
18 Iijima, S. Helical microtubules of graphitic carbon., *Nature (London)* **1991**, 354, 56–58.
19 Dresselhaus, M. S., Dresselhaus, G., Eklund, P. C. *Science of Fullerenes and Carbon Nanotubes*, Academic Press, San Diego, **1996**.
20 Saito, R., Dresselhaus, G., Dresselhaus, M. S. *Physical Properties of Carbon Nanotubes*, Imperial College Press, London, **1999**.
21 de Jonge, N., Lamy, Y., Schoots, K., Oosterkamp, T. H. High brightness electron beam from a multi-walled carbon nanotube, *Nature (London)* **2002**, 420, 393–395.
22 Fransen, M. J., Van Rooy, Th. L., Kruit, P. Field emission energy distributions from individual multiwalled carbon nanotubes., *Appl. Surf. Sci.* **1999**, 146, 312–327.
23 Wang, Z. L., Poncharal, P., de Heer, W. A. Measuring physical and mechanical properties of individual carbon nanotubes by in situ TEM., *J. Phys. Chem. Solids* **2000**, 61, 1025–1030.
24 Kong, J., Franklin, N. R., Zhou, C. W., Chapline, M. G., Peng, S., Cho, K., Dai, H. Nanotube molecular wires as chemical sensors., *Science* **2000**, 287, 622–625.
25 Sugie, H., Tanemura, M., Filip, V., Iwata, K., Takahashi, K., Okuyama, F. Carbon nanotubes as electron source in an x-ray tube., *Appl. Phys. Lett.* **2001**, 78, 2578–2580.
26 Baughman, R. H., Zakhidov, A., de Heer, W. A. Carbon nanotubes – the route toward applications., *Science* **2002**, 297, 787–792.
27 Dekker, C. Carbon nanotubes as molecular quantum wires., *Phys. Today* **1999**, 52, 22–28.
28 McEuen, P. L., Fuhrer, M. S., Park, H. Single-walled carbon nanotube electronics., *IEEE Trans. Nanotechnol.* **2002**, 1, 78–85.

29 Dresselhaus, M. S. Burn and interrogate., *Science* **2001**, 292, 650–651.

30 Tans, S. J., Verschueren, A. R. M., Dekker, C. Room-temperature transistor based on a single carbon nanotube., *Nature* **1998**, 393, 49–52.

31 Valentini, L., Armentano, I., Kenny, J. M., Cantalini, C., Lozzi, L., Santucci, S. Sensor for sub-ppm NO2 gas detection based on carbon nanotube thin films., *Appl. Phys. Lett.* **2003**, 82, 961–963.

32 Li, J., Lu, Y. J., Ye, Q., Cinke, M., Han, J., Meyyappan, M. Carbon nanotube sensors for gas and organic vapor detection, *Nano Lett.* **2003**, 3, 929–933.

33 Collins, P. G., Bradley, K., Ishigami, M., Zettl, A. Extreme oxygen sensitivity of electronic properties of carbon nanotubes., *Science* **2000**, 287, 1801–1804.

34 Sinnott, S. B., Andrews, R. Carbon nanotubes: Synthesis, properties, and applications., *Crit. Rev. Solid State Mater. Sci.* **2001**, 26, 145–249.

35 Kokai, F., Takahashi, K., Yudasaka, M., Yamada, R., Ichihashi, T. and Iijima, S. Laser ablation of graphite-Co/Ni and growth of single-wall carbon nanotubes in vortexes formed in an Ar atmosphere, *J. Phys. Chem. B* **2000**, 104, 6777–6784.

36 Takizawa, M., Bandow, S., Yudasaka, M., Ando, Y., Shimoyama, H., Iijima, S. Change of tube diameter distribution of single-wall carbon nanotubes induced by changing the bimetallic ratio of Ni and Y catalysts., *Chem. Phys. Lett.* **2000**, 326, 351–357.

37 Li, Y., Kim, W., Zhang, Y., Rolandi, M., Wang, W., Dai, H. Growth of single-walled carbon nanotubes from discrete catalytic nanoparticles of various sizes., *J. Phys. Chem. B* **2001**, 105, 11424–11431.

38 Iijima, S., Ichihasi, T. Single-shell carbon nanotubes of 1-nm diameter., *Nature (London)* **1993**, 363, 603–605.

39 Bethune, D. S., Kiang, C. H., de Vires, M. S., Gorman, G., Savoy, R., Vazquez, J., Beyers, R. Cobalt-catalysed growth of carbon nanotubes with single-atomic-layer walls, *Nature (London)* **1993**, 363, 605–607.

40 Seraphin, S., Zhou, D. Single-walled carbon nanotubes produced at high yield by mixed catalyst, *Appl. Phys. Lett.* **1994**, 64, 2087–2089.

41 Willems, I., Kónya, Z., Colomer, J.-F., Van Tendeloo, G., Nagaraju, N., Fonseca, A. and Nagy, J. B. Control of the outer diameter of thin carbon nanotubes synthesized by catalytic decomposition of hydrocarbon, *Chem. Phys. Lett.* **2000**, 317, 71–76.

42 Hata, K., Futaba, D. N., Mizuno, K., Namai, T., Yumura, M., Iijima, S. Water-assisted highly efficient synthesis of impurity-free single-walled carbon nanotubes., *Science* **2004**, 306, 1362–1364.

43 Joselevich, E., Lieber, C. M. Vectorial growth of metallic and semiconducting single-wall carbon nanotubes, *Nano Lett.* **2002**, 2, 1137–1141.

44 Ural, A., Li, Y., Dai, H. Electric-field-aligned growth of single-walled carbon nanotubes on surfaces., *Appl. Phys. Lett.* **2002**, 81, 3464–3466.

45 Gao, J., Yu, A., Itkis, M. E., Bekyarova, E., Zhao, B., Niyogi, S., Haddon, R. C. Large-scale fabrication of aligned single-walled carbon nanotube array and hierarchical single-walled carbon nanotube assembly., *J. Am. Chem. Soc.* **2004**, 126, 16 698–16 699.

46 Su, M., Li, Y., Maynor, B., Bwldum, A., Lu, J. P., Liu, J. Lattice-oriented growth of single-walled carbon nanotubes, *J. Phys. Chem. B* **2000**, 104, 6505–6508.

47 Ismaoh, A., Sergev, L., Wachtel, E., Joselevich, E. Atomic-step-templated formation of single wall carbon nanotube patterns, *Angew. Chem, Int. Ed.* **2004**, 43, 6140–6143.

48 Qian, D., Wagner, G. J., Liu, W. K., Yu, M. F., Ruoff, R. S. Mechanics of carbon nanotubes., *Appl. Mechanics Rev.* **2002**, 55, 495–533.

49 Durköp, T., Electronic Properties of carbon nanotubes studied in field-effect transistor geometries, PhD thesis 2004.

50 Martel, R., Schmidt, T., Shea, H. R., Hertel, T., Avouris, Ph. Single- and multi-wall carbon nanotube field-effect transistors., *Appl. Phys. Lett.* **1998**, 73, 2447–2449.

51 Fuhrer, M. S., Kim, B. M., Chen, Y. F., Durköp, T., Brintlinger, T. High-mobility nanotube transistor memory., *Nano Lett.* **2002**, 2, 755–759.

52 Javey, A., Guo, J., Wang, Q., Lundstrom, M., Dai, H. Ballistic carbon nanotube field-effect transistors, *Nature* **2003**, 424, 654–657.

53 Durköp, T., Getty, S. A., Cobas, E., Fuhrer, M. S. Extraordinary mobility in semiconducting carbon nanotubes., *Nano Lett.* **2004**, 4, 35–39.

54 Thess, A., Lee, R., Nikolaev, P., Dai, H., Petit, P., Robert, J., Xu, C., Lee, Y. H., Kim, S. G., Rinzler, A. G., Colbert, D. T., Scuseria, G. E., Tomanek, D., Fisher, J. E., Smalley, R. E. Crystalline ropes of metallic carbon nanotubes, *Science* **1996**, 273, 483–487.

55 Bronikowski, M. J., Willis, P. A., Colbert, D. T., Smith, K. A., Smalley, R. E. Gas-phase production of carbon single-walled nanotubes from carbon monoxide via the HiPco process: A parametric study., *J. Vac. Sci. Technol. A* **2001**, 19, 1800–1805.

56 Diehl, M. R., Yaliraki, S. N., Beckman, R. A., Barahona, M., Heath, J. R. Self-assembled, deterministic carbon nanotube wiring networks., *Angew. Chem. Int. Ed.* **2002**, 41, 353–356.

57 Martel, R., Derycke, V., Lavoie, C., Appenzeller, J., Chan, K. K., Tersoff, J., Avouris, Ph. Ambipolar electrical transport in semiconducting single-wall carbon nanotubes., *Phys. Rev. Lett.* **2001**, 87, 256 805-1–256 805-4.

58 Bockrath, M., Cobden, D. H., Liu, J., Rinzler, A. G., Smalley, R. E., Balents, L., McEuen, P. L. Luttinger-liquid behaviour in carbon nanotubes., *Nature* **1999**, 397, 598–601.

59 Hu, L., Hecht, D. S., Grüner, G. Percolation in transparent and conducting carbon nanotube networks., *Nano Lett.* **2004**, 4, 2513–2517.

60 Radosavljevic, M., Freitag, M., Thadani, K. V., Johnson, A. T. Nonvolatile molecular memory elements based on ambipolar nanotube field effect transistors., *Nano Lett.* **2002**, 2, 761–764.

61 Appenzeller, J., Knoch, J., Derycke, V., Martel, R., Wind, S., Avouris, Ph. Field-modulated carrier transport in carbon nanotube transistors., *Phys. Rev. Lett.* **2002**, 89, 126 801-1–126 801-4.

62 Rosenblatt, S., Yaish, Y., Park, J., Gore, J., Sazonova, V., McEuen, P. L. High performance electrolyte gated carbon nanotube transistors., *Nano Lett.* **2002**, 2, 869–872.

63 Zhang, Y., Dai, H. Formation of metal nanowires on suspended single-walled carbon nanotubes, *Appl. Phys. Lett.* **2000**, 77, 3015–3017.

64 Heinze, S., Tersoff, J., Martel, R., Derycke, V., Appenzeller, J., Avouris, Ph. Carbon nanotubes as Schottky barrier transistors., *Phys. Rev. Lett.* **2002**, 89, 106 801-1–106 801-4.

65 de Jonge, N., van Druten, N. J. Field emission from individual multiwalled carbon nanotubes prepared in an electron microscope., *Ultramicroscopy* **2003**, 95, 85–91.

66 Stadermann, M., Papadakis, S. J., Falvo, M. R., Novak, J., Snow, E., Fu, Q., Liu, J., Fridman, Y., Boland, J. J., Superfine, R., Washburn, S. Nanoscale study of conduction through carbon nanotube networks., *Phys. Rev. B* **2004**, 69, 201 402-1–201 402-3.

67 Derycke, V., Martel, R., Appenzeller, J., Avouris, Ph. Controlling doping and carrier injection in carbon nanotube transistors., *Appl. Phys. Lett.* **2002**, 80, 2773–2775.

68 Lu, C., Fu, Q., Huang, S., Liu, J. Polymer electrolyte-gated carbon nanotube field-effect transistor., *Nano Lett.* **2004**, 4, 623–627.

69 Siddons, G. P., Merchin, D., Back, J. H., Jeong, J. K., Shim, M. Highly efficient gating and doping of carbon nanotubes with polymer electrolytes., *Nano Lett.* **2004**, 4, 927–931.

70 Martel, R., Schmidt, T., Shea, H. R., Hertel, T., Avouris, Ph. Single- and

multi-wall carbon nanotube field-effect transistors., *Appl. Phys. Lett.* **1998**, 73, 2447–2449.

71 Hone, J., Ellwood, I., Muno, M., Mizel, A., Cohen, M. L., Zettl, A., Rinzler, A. G., Smalley, R. E. Thermoelectric power of single-walled carbon nanotubes., *Phys. Rev. Lett.* **1998**, 80, 1042–1045.

72 Nakanishi, T., Bachtold, A., Dekker, C. Transport through the interface between a semiconducting carbon nanotube and a metal electrode., *Phys. Rev. B* **2002**, 66, 073 307-1–073 307-4.

73 Freitag, M., Johnson, A. T., Kalinin, S., Bonnell, D. Role of single defects in electronic transport through carbon nanotube field-effect transistors., *Phys. Rev. Lett.* **2002**, 89, 216 801-1–216 801-4.

74 Bradley, K., Gabriel, J.-C. P., Star, A., Grűner, G. Short-channel effects in contact-passivated nanotube chemical sensors., *Appl. Phys. Lett.* **2003**, 83, 3821–3823.

75 Kim, W., Javery, A., Vermesh, O., Wang, Q., Li, Y., Dai, H. Hysteresis caused by water molecules in carbon nanotube field-effect transistors., *Nano Lett.* **2003**, 3, 193–198.

76 Bradley, K., Cumings, J., Star, A., Gabriel, J.-C. P., Grűner, G. Influence of mobile ions on nanotube based FET devices., *Nano Lett.* **2003**, 3, 639–641.

77 Star, A., Han, T.-R., Joshi, V., Stetter, J. R. Sensing with nafion coated carbon nanotube field-effect transistors., *Electroanalysis* **2004**, 16, 108–112.

78 Kruger, M., Buitelaar, M. R., Nussbaumer, T., Schonenberger, C., Forro, L. Electrochemical carbon nanotube field-effect transistor., *Appl. Phys. Lett.* **2001**, 78, 1291–1293.

79 Dagani, R. DNA matches., *Chem. Eng. News* **2002**, 80, 8.

80 Kim, O.-K., Je, J., Baldwin, J. W., Kooi, S., Pehrsson, P. E., Buckley, L. Solubilization of single-wall carbon nanotubes by supramolecular encapsulation of helical amylose., *J. Am. Chem. Soc.* **2003**, 125, 4426–4427.

81 Bandyopadhyaya, R., Nativ-Roth, E., Regev, O., Yerushalmi-Rozen, R. Stabilization of individual carbon nanotubes in aqueous solutions., *Nano Lett.* **2002**, 2, 25–28.

82 Numata, M., Asai, M., Kaneko, K., Hasegawa, T., Fujita, N., Kitada, Y., Sakurai, K., Shinkai, S. Curdlan and schizophyllan (beta-1,3-glucans) can entrap single-wall carbon nanotubes in their helical superstructure, *Chem. Lett.* **2004**, 33, 232–233.

83 Chen, J., Dyer, M. J., Yu, M.-F. Cyclodextrin-mediated soft cutting of single-walled carbon nanotubes., *J. Am. Chem. Soc.* **2001**, 123, 6201–6202.

84 Chen, J., Hamon, M. A., Hiu, H., Chen, Y., Rao, A. M., Eklund, P. C., Haddon, R. C. Solution properties of single-walled carbon nanotubes., *Science* **1998**, 282, 95–98.

85 Star, A., Stoddart, J. F., Steuerman, D., Diehl, M., Boukai, A., Wong, E. W., Yang, X., Chung, S. W., Heath, J. R. Preparation and properties of polymer-wrapped single-walled carbon nanotubes., *Angew Chem. Int. Ed.* **2001**, 40, 1721–1725.

86 Star, A., Steuerman, D. W., Heath, J. R., Stoddart, J. F. Starched carbon nanotubes., *Angew. Chem. Int. Ed.* **2002**, 41, 2508–2512.

87 Dodziuk, H., Ejchart, A., Anczewski, W., Ueda, H., Krinichnaya, E., Dolgonos, G., Kutner, W. Water solubilization, determination of the number of different types of single-wall carbon nanotubes and their partial separation with respect to diameters by complexation with η-cyclodextrin., *Chem. Commun.* **2003**, 986–987.

88 Zheng, M., Jagota, A., Semke, E. D., Diner, B. A., McLean, R. S., Lustig, S. R., Richardson, R. E., Tassi, N. G. DNA-assisted dispersion and separation of carbon nanotubes., *Nat. Mater.* **2003**, 2, 338–342.

89 Zheng, M., Jagota, A., Strano, M. S., Santos, A. P., Barone, P., Chou, S. G., Diner, B. A., Dresselhaus, M. S., McLean, R. S., Onoa, G. B., Samsonidze, G. G., Semke, E. D., Usrey, M., Walls, D. J. Structure-

based carbon nanotube sorting by sequence-dependent DNA assembly., *Science* **2003**, 302, 1545–1548.

90 Balavoine, F., Schultz, P., Richard, C., Mallouh, V., Ebbesen, T. W., Mioskowski, C. Helical crystallization of proteins on carbon nanotubes: A first step towards the development of new biosensors., *Angew. Chem. Int. Ed.* **1999**, 38, 1912–1915.

91 Bradley, K., Briman, M., Star, A., Grüner, G. Charge transfer from adsorbed proteins., *Nano Lett.* **2004**, 4, 253–256.

92 Wang, S., Humphreys, E. S., Chung, S.-Y., Delduco, D. F., Lustig, S. R., Wang, H., Parker, K. N., Rizzo, N. W., Subramoney, S., Chiang, Y.-M., Jagota, A. Peptides with selective affinity for carbon nanotubes, *Nat. Mater.* **2003**, 2, 196–200.

93 Bradley, K., Gabriel, J.-C. P., Briman, M., Star, A., Grüner, G. Charge transfer from ammonia physisorbed on nanotubes, *Phys. Rev. Lett.* **2003**, 91, 218 301-1–218 301-4.

94 Wilson-Kubalek, E. M., Brown, R. E., Celia, H., Milligan, R. A. Lipid nanotubes as substrates for helical crystallization of macromolecules, *Proc. Natl. Acad. Sci. U.S.A.* **1998**, 95, 8040–8045.

95 Chen, R. J., Choi, H. C., Bangsaruntip, S., Yenilmez, E., Tang, X., Wang, Q., Chang, Y. L., Dai, H. An investigation of the mechanisms of electronic sensing of protein adsorption on carbon nanotube devices., *J. Am. Chem. Soc.* **2004**, 126, 1563–1568.

96 Boussaad, S., Tao, N. J., Zhang, R., Hopson, T., Nagahara, L. A. *In situ* detection of cytochrome c adsorption with single walled carbon nanotube device., *Chem. Comm.* **2003**, 1502–1503.

97 Star, A., Gabriel, J.-C. P., Bradley, K., Grüner, G. Electronic detection of specific protein binding using nanotube FET devices., *Nano Lett.* **2003**, 3, 459–463.

98 Chen, R. J., Bangsaruntip, S., Drouvalakis, K. A., Wong Shi Kam, N., Shim, M., Li, Y., Kim, W., Utz, P. J., Dai, H. Noncovalent functionalization of carbon nanotubes for highly specific electronic biosensors., *Proc. Natl. Acad. Sci. U.S.A.* **2003**, 100, 4984–4989.

99 Li, H., Ng, H. T., Cassell, A., Fan, W., Chen, H., Ye, Q., Koehne, J., Han, J., Meyyappan, M. Carbon nanotube nanoelectrode array for ultrasensitive DNA detection, *Nano Lett.* **2003**, 3, 597–602.

100 Star, A., Tu, E., Niemann, J., Gabriel, J.-C. P., Joiner, C. S., Valcke, C. Label-free detection of DNA hybridization using carbon nanotube field-effect transistors, *Proc. Natl. Acad. Sci. U.S.A.* **2006**, 103, 921–926.

101 Limdi, J. K., Crampton, J. R. Hereditary haemochromatosis., *Q. J. Med.* **2004**, 97, 315–324.

102 Franchini, M., Veneri, D. Recent advances in hereditary hemochromatosis., *Ann. Hematol.* **2005**, 84, 347–352.

103 Tang, X., Bansaruntip, S., Nakayama, N., Yenilmez, E., Chang, Y.-I., Wang, Q. Carbon nanotube DNA sensor and sensing mechanism., *Nano Lett.* **2006**, 6, 1632–1636.

104 Maehashi, K., Matsumoto, K., Kerman, K., Takamura, Y., Tamiya, E. Ultrasensitive detection of DNA hybridization using carbon nanotube field-effect transistors., *Jpn. J. Appl. Phys.* **2004**, 43, L 1558–L 1560.

105 Star, A., Joshi, V., Han, T. R., Altoe, V. P., Grüner, G., Stoddart, J. F. Electronic detection of the enzymatic degradation of starch., *Org. Lett.* **2004**, 6, 2089–2092.

106 Kong, J., Dai, H. Full and modulated chemical gating of individual carbon nanotubes by organic amine compounds., *J. Phys. Chem B* **2001**, 105, 2890–2893.

107 Shim, M., Javey, A., Kam, N. W. S., Dai, H. Polymer functionalization for air-stable n-type carbon nanotube field-effect transistors., *J. Am. Chem. Soc.* **2001**, 123, 11 512–11 513.

108 Besteman, K., Lee, J. O., Wiertz, F. G. M., Heering, H. A., Dekker, C. Enzyme-coated carbon nanotubes as single-molecule biosensors., *Nano Lett.* **2003**, 3, 727–730.

2
Carbon Nanotube-based Sensor

Jian-Shan Ye and Fwu-Shan Sheu

2.1
Overview

Carbon nanotubes (CNTs) possess high electrical conductivity, high chemical stability, and extremely high mechanical strength and modulus. These special properties of both single-walled carbon nanotubes (SWCNTs) and multiwalled carbon nanotubes (MWCNTs) have attracted much attention in electrochemistry. To develop nanostructured macroscopic electrodes, randomly dispersed nanotubes, well-aligned CNTs, CNT paste, screen-printing CNTs, self-assembly of CNT, and CNT-packaged microelectrodes have been used. Furthermore, single and long MWCNTs, after formation, can be used as a nanoelectrode. The resultant nanotube-based electrodes have been used successfully for electroanalytical purposes such as the development of sensors and biosensors. We outline here the unique electrochemical and electrocatalytical properties of CNTs-based sensors/biosensors and discuss novel applications as well as future challenge of CNTs in electrochemical sensors.

Functionalization of CNTs is one of the most active fields in nanotube research, which provides an effective tool to broaden the electrochemical application spectrum of CNTs. In this chapter, we summarize various approaches to functionalize CNTs for the development of novel electrochemical sensors. Particular emphasis is directed to the use of lipid-functionalized CNTs for sensors and biosensors and for the synthesis of photoswitched-functional devices. Functionalization of nanotubes generates a novel, interesting class of nanomaterials, which combines the properties of the nanotubes and the functional moiety, thus offering new opportunities in the development of electrochemical sensors.

2.2
Introduction of Carbon Nanotubes

Carbon nanotubes (CNTs) have captured the imagination of researchers worldwide since they were first observed by Iijima in 1991 [1]. With 100× the tensile strength of steel, a thermal conductivity better than all but the purest diamond, and electrical conductivity similar to copper, but with the ability to carry much higher cur-

Nanotechnologies for the Life Sciences Vol. 8
Nanomaterials for Biosensors. Edited by Challa S. S. R. Kumar

ISBN: 978-3-527-31388-4

rents, CNTs are very interesting and promising nanomaterials. Their small dimensions, unique structures, strength, and remarkable physical properties make them a unique material with a wide range of potential applications.

CNTs, consisting of only sp^2 hybridized carbon atoms, are cylindrical nanostructures with a diameter ranging from 1 nm to several nanometers, and a length of tens of micrometers. They are made of graphene sheets wrapped into a hollow cylinder and capped by fullerene-like structures. There are two typical types of nanotubes, SWCNTs and MWCNTs. The latter consist of several to tens of concentric cylinders of these graphitic shells with a layer spacing of 0.3–0.4 nm. MWCNTs tend to have diameters in the range 2–500 nm, depending on the method of synthesis. An MWCNT can be considered as a mesoscale graphite system, whereas a SWCNT is truly a large single molecule. High-resolution transmission electron microscope (HRTEM) results showed that the first observed CNTs by Iijima in 1991 [1] were fullerene-like tubes consisting of coaxial multiple shells. These tubes were MWCNTs. The interlay spacing is 0.34 nm, which is slightly greater than that of graphite (0.335 nm) due to a combination of tubule curvature and van der Waals force interactions between successive graphene layers.

After two years, it was discovered that the use of transition metals as catalysts afforded CNTs with a single shell or wall only [2]. These nanotubes were SWCNTs. An ideal SWCNT can be viewed as an "extended" fullerene, and consists of a single graphite layer wrapped into one seamless hollow cylinder. Closure of the cylinder is the result of pentagon inclusion in the hexagonal carbon network of the nanotube walls during the growth process. SWCNTs normally have a narrow diameter distribution (with diameter of the order of 1 nm) but tend to assemble in nanotube bundles during the growth process [3]. This corresponds to the theoretically predicted lower limit for stable SWCNT formation based on considerations of the stress energy built into the cylindrical structure of the SWCNT.

There are three types of SWCNTs: "arm-chair" (n, m) tubes where $n = m$, "zig-zag" $(n, 0)$ tubes and chiral (all other tubes with independent n and m) tubes. Theoretical calculations indicate that the electronic properties for a single-walled nanotube will vary as a function of its diameter and helicity [4–6]. A SWCNT may behave as a semiconductor or metal, and a slight change in the chirality can transform a nanotube from a metal to a semiconductor. In general, about one-third of SWCNTs are metallic, characterized with wrapping vectors of $n - m = 3l$ $(l = 0, 1, 2 \ldots)$. All other tubes are semiconductors [7, 8].

Carbon–carbon covalent bonds are one of the strongest in nature, and a structure based on a perfect arrangement of these bonds oriented along the axis of nanotubes would produce an exceedingly strong material. Early theoretical work and recent experiments on individual nanotubes have confirmed that nanotubes are one of the stiffest structures ever made [9–11].

CNTs possess high electrical conductivity, high chemical stability, and extremely high mechanical strength and modulus. These special properties of both SWCNTs and MWCNTs have attracted much attention in electrocatalysis [12–17] and as chemical sensors/biosensors [18–24].

The present chapter only covers particular aspects related to CNTs-based electrochemical sensing systems. Different methods of preparing CNT electrodes and the possible ways to functionalize CNTs are described, with the major part devoted to the use of lipid-functionalized CNTs for sensors and biosensors and for the fabrication of photoswitched-functional devices. Furthermore, the possible mechanism of electrocatalysis by CNTs is discussed.

2.3 Growth of Carbon Nanotubes

CNTs are generally produced by three main techniques, arc discharge, laser ablation, and chemical vapor deposition (CVD). The arc discharge method, initially used for producing C_{60} fullerenes, is the most common and perhaps easiest way to produce CNTs, as it is rather simple to undertake [25]. During arc discharge, a vapor is created by an arc discharge between two carbon electrodes with or without catalyst. Nanotubes self-assemble from the resulting carbon vapor. This technique produces a mixture of components and requires the separation of nanotubes from the soot and catalytic metals present in the crude product. In the laser ablation technique [26], a high-power laser beam impinges on a volume of carbon-containing feedstock gas (methane or carbon monoxide). Presently, laser ablation produces a small amount of clean nanotubes, whereas arc discharge methods generally produce large quantities of impure material. CVD is achieved by putting a carbon source in the gas phase and using an energy source, such as plasma or a resistively heated coil, to transfer energy to a gaseous carbon molecule. CVD nanotube synthesis is essentially a two-step process, consisting of a catalyst preparation step followed by the actual synthesis of the nanotube. The catalyst is generally prepared by sputtering a transition metal onto a substrate and then using either chemical etching or thermal annealing to induce catalyst particle nucleation. Thermal annealing results in cluster formation on the substrate, from which the nanotubes will grow. Excellent alignment, as well as positional control on the nanometer scale, can be achieved by using CVD [27, 28].

Even though much progress on CNT growth has been made in the past decade, it is still challenging to produce CNTs with desired properties for specific applications. In particular, new methods are desired that can be directly integrated into device fabrication. With continued effort in the development of growth techniques, it is expected that CNTs with desired properties, quality, and quantity could be obtained for various applications.

2.4 Methods to Prepare CNTs-based Sensors and Biosensors

2.4.1 Individual MWCNTs as Nanoelectrodes

MWCNTs are tiny electrodes made of carbon, metals or semiconducting materials having typical dimensions of 1–100 nm. MWCNTs can be used as a new material for the construction of nanoelectrodes. Campbell et al. have reported the fabrication and characterization of electrodes constructed from single MWCNTs [29]. To construct a nanoelectrode, single nanotubes were attached to a mechanically cut Pt tip, using Ag epoxy, and subsequently cleaned by immersion in a fresh piranha solution (3:1 concentrated H_2SO_4:30% H_2O_2). The sigmoidal voltammetric response of these nanotubular electrodes is characteristic of steady-state radial diffusion. The limiting current of uninsulated electrodes scales linearly with the depth of immersion into electrolyte solutions. However, the walls of nanotubular electrodes can be selectively insulated with a thin layer of polyphenol so that electrochemical activity is limited to the tip region. In this case the limiting current is essentially independent of immersion depth. These nanotubular electrodes are robust, can be fabricated in high yield, and are of uniform diameter. Most importantly, their great strength and high length-to-diameter aspect ratio will be particularly valuable for applications such as scanning electrochemical microscopy (SECM) and electrochemical analysis of biological materials.

2.4.2 Randomly Distributed CNT Electrodes

To prepare CNT electrodes, randomly distributed CNTs can be castled at the surface of the electrode. Before casting, the concentrated solutions of CNTs in sulfuric acid [20], dimethylformamide [30], concentrated nitric acid, or a Nafion/water mixture [31] are prepared. Coated electrodes were are at high temperature (e.g., 200 °C for 3 h) and then readied for use after careful washing. The randomly deposited CNTs are less mechanically stable and have worse electrical contact to the underneath electrodes than the vertically aligned CNTs directly grown on conductive substrates.

2.4.3 Well-aligned Carbon Nanotube Electrodes

To construct a CNT electrode, SWCNTs are shaped into an electrode by filtering suspension of nanotubes on a membrane filter [32] to form a nanotube sheet. Another method is casting the SWCNT suspension on the surface of solid electrodes such as Pt, Au, or glassy carbon [33–36]. In contrast, to construct a MWCNT electrode, the MWCNTs are usually mixed with bromoform [12], mineral oil [15], or packed into the cavity at the tip of a microelectrode to form a CNT powder microelectrode [36] as discussed above. MWCNT electrodes prepared by these methods may suffer from mechanical instability during detection, thus limiting their practical application. Fortunately, high-density well-aligned CNTs, which are multiwalled and vertically aligned on a large area of substrates (Fig. 2.1), can be readily synthesized [37, 38]. These CNTs aligned on the substrate are very stable and can be used as photoswitched functional devices [39], electrochemical sensors/biosensors [40–44] and supercapacitors [45]. In particular, this well-aligned MWCNT electrode has been successfully used as a novel candidate for non-enzymatic glucose sensors with the resistance to toxicity by chloride ions [46].

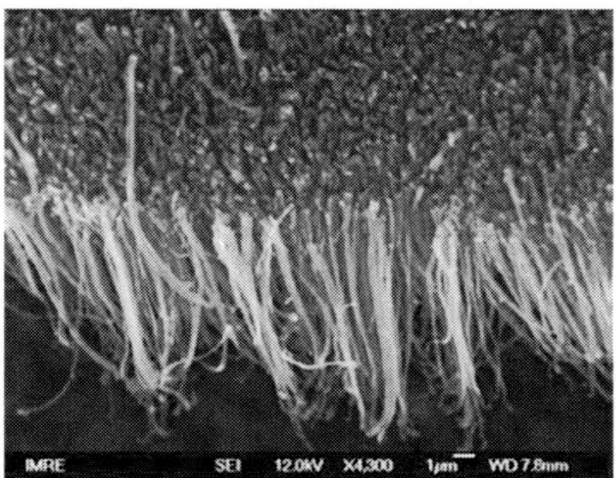

Fig. 2.1. A vertical well-aligned MWCNT electrode. (Adapted from Ref. [44] with permission.)

2.4.4 Carbon Nanotube Paste Electrodes

Carbon paste electrodes (CPEs) belong to a special group of heterogeneous carbon electrodes. To prepare CPEs, carbon paste, i.e., a mixture prepared from carbon (graphite) powder and a suitable liquid binder is packed into a suitably designed electrode body. CNT-based CPEs were first reported by Britto et al. [12]. To study the electrochemical response, nanotubes were initially dispersed in bromoform as a binder material and packed into a glass tube. The resultant electrode had randomly distributed tubes with no control over the alignment of the nanotubes. Subsequently, numerous approaches have enabled the random distribution of nanotubes on electrodes, either by dispersing the tubes with a binder, such as dihexadecyl hydrogen phosphate [47], Nafion [21], or forming the nanotube equivalent of a carbon paste [48, 49] that can be screen printed [50], forming a nanotube/Teflon composite [22].

Due to numerous advantageous properties and characteristics, CPEs are widely used in amperometry, coulometry, and potentiometry. CNTs-based CPEs are easily obtainable at minimal costs and are especially suitable for preparing an electrode materials with desired composition and, hence, with pre-determined properties. Electrodes made in this way are usually intended to be used as highly selective sensors and/or biosensors for both inorganic and organic electrochemistry. The rapid development of new nanomaterials and nanotechnologies has provided many new opportunities for electrochemical application. In particular, the immobilization and adsorption of biomolecules on electrode surfaces is of great importance and interest for biosensor and bioelectronic applications. Recent studies have demonstrated that CNTs can enhance the electrochemical reactivity of biomolecules and promote the electron-transfer reactions of proteins. These properties make CNTs an efficient material for use in a wide range of electrochemical biosensors, ranging from amperometric enzyme electrodes to DNA hybridization biosensors. Rubianes and Rivas have recently described the performance of CNT paste electrodes prepared by dispersion of MWCNTs within mineral oil [48]. The resulting paste electrode shows excellent electrocatalytic activity toward ascorbic acid, uric acid, dopamine,

3,4-dihydroxyphenylacetic acid (dopac) and hydrogen peroxide. These properties permit an important decrease in the overvoltage for the oxidation of ascorbic acid (230 mV), uric acid (160 mV) and hydrogen peroxide (300 mV) as well as a dramatic improvement in the reversibility of the redox behavior of dopamine and dopac, in comparison with classic carbon (graphite) paste electrodes (CPE). The substantial decrease in the overvoltage of the hydrogen peroxide reduction (400 mV) associated with a successful incorporation of glucose oxidase (GOD) into the composite material allows the development of a highly selective and sensitive glucose biosensor, without using any metal, redox mediator, or anti-interference membrane. Such excellent performance of CNTPEs toward hydrogen peroxide, represents a very good alternative for developing other enzymatic biosensors. To determinate homocysteine, Lawrence et al. [51] have provided an effective means by using a CNT paste electrodes. A decrease of ca. 120 mV in the overpotential for the oxidation of homocysteine, compared with a traditional carbon paste electrode, is reported along with greatly enhanced signal-to-noise characteristics. The analytical parameters have been assessed with a linear range from 5 to 200 μM, and a detection limit of 4.6 μM. Furthermore, the generic nature of this increased reactivity of the CNTP surface towards thiol moieties has been demonstrated with cysteine, glutathione, and n-acetylcysteine, providing a greatly enhanced electrochemical response compared with the carbon paste electrode. Pedano et al. have reported that CNTs paste electrodes are suitable for adsorptive stripping potentiometric measurements of trace levels of nucleic acids [52]. Compared with that obtained at carbon (graphite) paste electrode (CPE), the guanine oxidation signal is greatly enhanced due to the electroactivity inherent to CNTs. Trace ($\mu g\ L^{-1}$) levels of the oligonucleotides and polynucleotides can be readily detected following short accumulation periods, with detection limits of 2.0 $\mu g\ L^{-1}$ for a 21-base oligonucleotide and 170 $\mu g\ L^{-1}$ for calf thymus dsDNA.

SWCNTs have also been used for the development of CPEs. Ricci et al. have used Prussian blue modified-SWCNTs for successive assembling of paste electrodes [53]. The electrochemical feature of such electrodes has been fully evaluated with cyclic voltammetry (CV) and amperometric experiments. The result showed that Prussian blue-modified CNT paste electrodes have a high sensitivity towards hydrogen peroxide with a detection limit of 7.4×10^{-6} M. The Prussian blue-modified CNT paste electrode also possessed strong stability even at basic pHs (i.e., pH 9 and 10), demonstrating no significant loss of signal after three days continuous work. In addition, the loading in the paste mixture of GOD has brought a sensitive tool for the detection of glucose in a range between 0.1 and 50 mM. More recently, Antiochia et al. have studied the electrocatalytic oxidation of NADH at SWCNT CPEs for use in a redox mediator in solution and dissolved in the paste [54].

2.4.5 Screen-printing Carbon Nanotubes

Screen-printing technology is particularly attractive for the production of disposable sensors [50]. Recently, Trojanowicz et al. have successfully prepared CNT-modified screen-printed electrodes for chemical sensors and biosensors [55]. They found that MWCNTs can be used to modify working graphite ink electrodes of the three-electrode screen-printed sensing stripe. Modification has been made by evaporating on the graphite surface a solution of MWCNT in dimethylformamide. The effect of such treatment on reversibility of the electrode process of the system hexacyanoferrate(II)/(III) has been shown, along with an improvement in sensitivity of detection of the pesticide paraoxon with biosensors containing organophosphorus hydrolase immobilized by adsorption on the nanotube-modified graphite ink electrode. The catalytic sensing of methanol has also been demonstrated with the use of a screen-printed sensor modified with MWCNT and Co(II) salt present in the measuring solution. More recently, Guan et al. have prepared a disposable electrochemical biosensor for glucose monitoring [56]. The sensor was based on MWCNTs immobilized with GOD upon a screen printed carbon electrode. The effect of MWCNTs on the response of amperometric GOD electrodes for glucose was examined. Results obtained, of interest for basic and applied biochemistry, represent a first step in construction of a MWCNT-enzyme electrode biosensor with potential application in the biosensor area.

2.4.6 Self-assembly of Carbon Nanotubes

Self-assembly is the fundamental principle that generates structural organization on all scales from molecules to galaxies. Self-assembly is also a manufacturing method used to construct things at the nanometer scale. The conjugation of CNT with biomolecules and nanoparticles is an emerging field of research that has important potential applications in bionanotechnology. For example, the metallic/semiconductive properties of CNTs have been exploited to produce functional, technologically relevant devices such as sensors [57]. Electrostatic matching can be used to coat CNTs with layers of oppositely charged polyelectrolytes [58]. By using a similar strategy, Mann and his colleagues [59] have described a layer-by-layer procedure based on programmed biomolecular assembly to produce a multi-component, multi-layered CNT-based conjugate. The conjugate consists of a multi-walled CNT core coated with four functionalized layers that successively comprise protein-encapsulated iron oxide nanoparticles [biotinylated ferritin (bFn)], the tetravalent biotin-binding protein, streptavidin (SA), 24-base three-stranded biotin-terminated oligonucleotide duplexes, and oligonucleotide coupled Au nanoparticles. They demonstrated that the core/multi-shell architecture can be constructed stepwise by specific recognition processes involving biotin–SA binding or DNA duplexation, and that these interactions can be exploited for reversible assembly/disassembly of the Au nanoparticle layer. A recent study showed that CNTs can be self-assembled at the electrode surface for the development of a sensor and biosensor.

Shimoda et al. have reported the formation of macroscopically ordered CNT membranes on a substrate by self-assembly [60]. To form the ordered SWCNTs, SWCNTs produced by the laser ablation method and purified by reflux and filtration were chemically etched to short bundles by ultrasonic-assisted oxidation. After removing the acid by filtration, the processed SWCNTs were dispersed in deionized water. A thin film appears on the surface of a soaked glass substrate in the SWCNTs/water dispersion with natural vaporization of water. Transmission electron microscopy measurements show that the SWCNT bundles are uniaxially aligned. The self-assembly of SWCNTs at gold electrodes can also be achieved by thiolation [61]. Recently, effective CNT coating was obtained by self-assembling short SWCNTs at an electrode surface [23, 24]. Such vertically aligned SWCNTs act as molecular wires to allow electrical communication between the underlying electrode and a redox enzyme. Direct electron transfer between the prosthetic group of the enzyme and an electron surface obviates the need for redox mediators and is thus extremely attractive for developing reagentless sensing devices. More recently, Wang and Iqbal have reported that thin films of vertically aligned individual SWCNTs can be deposited on silicon using a CVD process [62]. Oriented SWCNT growth was achieved by employing two methods of catalyst precursor self-assembly followed by ethanol CVD. Using the silicon substrate as the working electrode in an electrochemical cell and the enzyme beta-NAD (nicotinamide adenine dinucleotide) synthetase dissolved in a buffered electrolyte solution, the enzyme was attached at the nanotube ends. This was shown using scanning electron microscopy and cyclic voltammetry. Enzyme immobilization on the 1 to 2 nm diameter tube ends of the individual SWCNTs allows for dense packing of the enzyme and utilization of the electrode as an enzymatic sensor in a biofuel cell configuration.

2.4.7 Carbon Nanotube-packaged Microelectrodes

Packaged microelectrodes are easy to prepare and have been successfully used to prepare electrodes. To fabricate CNT powder microelectrodes, a 76-μm diameter Pt microelectrode was first chemically etched to form a cavity of 10 μm deep, and the etched tip was then grounded on a flat plate (such as glass slide) with CNTs until the micro cavity was filled with CNTs [63, 64]. In this way, the direct electrochemistry of redox enzymes such as GOD [64], horseradish peroxidase [65], as well as the electrocatalytic detection of nitrite [66], cysteine [67], and hydrazine [63] have been reported.

2.5 Application of CNTs-based Electrochemical Sensors and Biosensors

2.5.1 Electrochemical and Electrocatalytical Properties of Carbon Nanotubes

CNTs combine, uniquely, high electrical conductivity, high chemical stability, and extremely high mechanical strength. These special properties of both SWCNTs and MWCNTs have attracted the interest of many researchers in the field of electrochemical sensors. Many advances in producing, modifying, characterizing and integrating CNTs into electrochemical sensing systems have been achieved [68]. CNTs, new materials for electrochemical sensing, can be either used as single probes after formation *in situ*, or even individually, when attached onto a proper transducing surface after synthesis. Both SWCNTs and MWCNTs can be used to modify several electrode surfaces in either vertically oriented "nanotube forests" or even a non-oriented way. They can be also used in sensors after mixing them with a polymer matrix to form CNT composites [69].

To find new electrocatalytic surfaces, a suitable electrode substrate, such as glassy carbon or gold, is modified with a film or layer of CNTs. Several methods have achieved the electroanalysis of different analytes by using CNT modified electrodes. Benefits of low detection limits, increased sensitivity, decreased overpotentials and resistance to surface fouling are found by these CNTs-based electrodes [69].

Electrochemistry implies the transfer of charge from one electrode to another. Due to the curvature of the carbon graphene sheet in nanotubes, the electron clouds change from a uniform distribution around the C–C backbone in graphite to an asymmetric distribution inside and outside the cylindrical sheets of the nanotube [70]. Because the electron clouds are distorted, a rich π-electron conjugation forms outside the tube, therefore making CNTs electrochemically active (Fig. 2.2).

There are numerous reports of CNT-modified electrodes for the detection of different analytes with low detection limits, decreased overpotentials, and resistance to surface fouling. However, the exact reason for the unique catalytic properties of CNTs remains unknown. In this regard, Compton and his colleagues [71, 72] addressed the question as to why CNTs are catalytic and provided definitive evidence for their electrocatalytic properties. They explored the reduction of the one-electron aqueous redox probe ferricyanide at a CNT film-modified basal plane pyrolytic graphite (BPPG) electrode and compared it with a bare BPPG electrode. Peak-to-

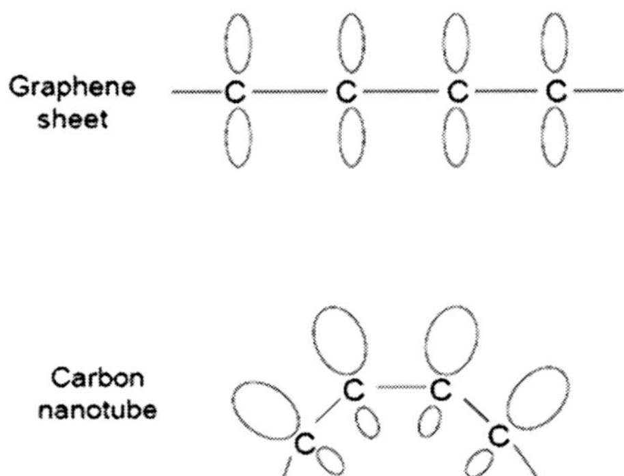

Fig. 2.2. Schematic of the electron distribution along a graphene sheet and around a carbon nanotube (CNT).

peak separations of 58 and 350 mV, respectively, were observed; the peak-to-peak separation is an indication of the heterogeneous charge transfer kinetics, showing that, in this case, CNTs exhibit fast electron transfer [71]. They further compared the electrocatalytic properties of CNTs with an edge plane pyrolytic graphite electrode (EPPG). EPPG electrodes are fabricated by taking a piece of high quality, highly ordered pyrolytic graphite (HOPG) and cutting the desired electrode geometry such that the layers of graphite lie perpendicular to the surface. Conversely, BPPG electrodes are produced by cutting the electrode geometry such that the graphite layers lie parallel to the surface. For voltammetry of ferricyanide at the EPPG electrode, a peak-to-peak separation of 78 mV was observed, suggesting that the electrochemical reaction occurs with a not dissimilar rate constant as at the CNT-modified surface. The slight difference in peak-to-peak separation of the CNTs compared with EPPGs probably reflects the slight impurities at the basal plane in the EPPG electrode. A catalytic response was also seen for the electrochemical oxidation of epinephrine where identical responses were obtained at both the CNT modified electrode and the EPPG; again, slow electrode kinetics was observed at a bare BPPG electrode. Indeed, it has been well documented that the electrode kinetics at EPPG electrodes are at least three times faster than at BPPG electrodes [73], but, seemingly, this was the first comparative approach to provide evidence for electrode kinetic enhancements at CNTs. Edge plane sites/defects are consequently introduced on the electrode surface, resulting in faster electron transfer. This occurs to the point that, after roughening for about a minute, a nearly identical response to that of the EPPG electrode can be observed. This clearly shows that edge plane sites are the dominant sites at which fast electron transfer occurs [74, 75]. The above experiments suggest that edge plane-like sites, which in CNTs occur at the ends and along the tube axis (where graphite sheets terminate at the surface of the tube), are likely to be the reason why nanotubes exhibit fast heterogeneous charge transfer, and explains why they have been widely reported as "electrocatalytic" in the electrochemical literature (Fig. 2.3) [72].

To understand the nature of the electrocatalytic properties and the electrochemical reactivity of CNT, Wang's group [76] studied the effect of electrochemical pretreatment on CNTs prepared by different processes. They found that anodic pretreatment results in a dramatic improvement in the electrochemical reactivity of ARC-produced CNT, whereas CNTs produced by CVD appear to be resistant to anodic activation. Such a dramatic difference in the pretreatment effect upon ARG and CVD-produced CNTs is illustrated using NADH, ascorbic acid, hydrazine and hydrogen peroxide model redox systems. Differences in the effect of the electrochemical pretreatments are attributed to the anodic preanodization, effectively "breaking" the basal-plane end caps of ARC-CNT, thereby exposing edge plane defects, similar to those already present in the open-end caps of CVD-CNTs. In contrast, Gooding and his colleagues [77] demonstrated the importance of oxygenated species at the ends of CNTs for their favorable electrochemical properties. Definitive evidence is presented for the favorable electrochemical properties of CNT-modified electrodes arising from the ends of SWCNTs due to oxygenated carbon species in general, and carboxylic acid moieties in particular, produced during

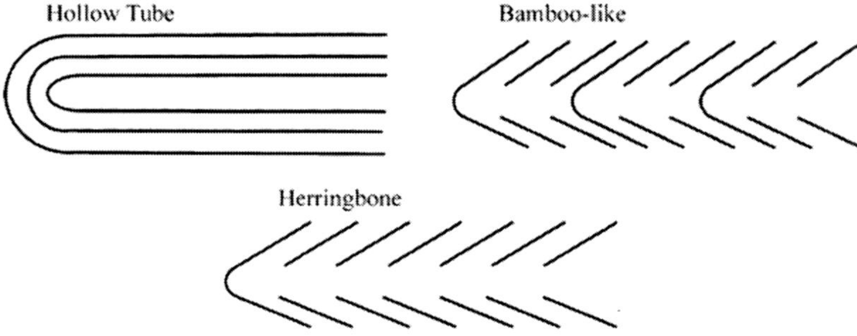

Fig. 2.3. Schematic representation of the possible variants of CNTs. In both the bamboo and herringbone variations, graphite planes are formed at an angle to the axis of the tube and, therefore, these CNTs have a higher proportion of edge plane sites/defects. (Adapted from Ref. [72] with permission.)

acid purification. Interestingly, large amounts of well-aligned CNTs with open tips have been produced recently by pyrolysis of iron(II) phthalocyanine [78]. The aligned CNTs have an average length of about 10 μm and diameters ranging from 92 to 229 nm. Some of the produced CNTs showed Y-junction structures due to the self-joint growth of two neighboring CNTs. The well-aligned CNTs indicated a bamboo-shaped multiwalled structure and fairly good crystallinity. Availability of the CNTs with open tips will make it possible to further study the electrocatalytic nature of CNTs.

2.5.2 CNTs-based Electrochemical Biosensors

For use as amperometric enzyme electrodes or DNA hybridization biosensors, CNTs need to be coupled with enzymes or ssDNA probes, respectively. Gooding and his colleagues have successfully self-assembled short SWCNTs to an electrode [24]. The vertically aligned SWCNTs act as molecular wires to allow electrical communication between the underlying electrode and a redox enzyme [23, 24] (Fig. 2.4). Such direct electron transfer between the prosthetic group of the enzyme and an electron surface makes it possible to develop reagentless sensing devices. The direct electron transfer of oxidases and dehydrogenases at CNT-modified electrodes and the dramatic decrease of the overpotential of hydrogen peroxide and NADH indicate great promise for the biosensing of glucose, lactate, cholesterol, amino acids, urate, pyruvate, glutamate, alcohol, hydroxybutyrate, etc. For example, CNT/Nafion/GOD-coated electrodes, coupling the selective reactivity of GOD to glucose with the electrocatalytic detection of hydrogen peroxide and the permselectivity of Nafion, offered a highly selective low-potential (0.05 V vs. Ag|AgCl) biosensing of glucose [21]. More interestingly, Lin and coworkers [79] reported glucose biosensors based on GOD covalently attached at the free end of vertically-aligned CNTs, which have been grown on a metal-coated silicon substrate. They

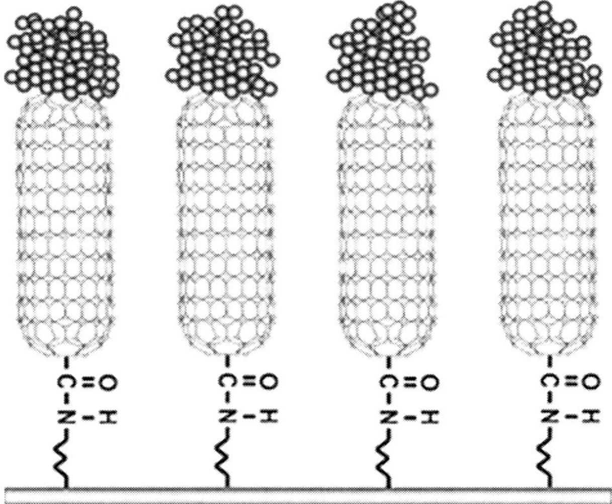

Fig. 2.4. Nanoforest of vertically aligned CNT-trees acting as molecular wires. (Adapted from Ref. [24] with permission.)

detected the reduction current of hydrogen peroxide, instead of the oxidation current, at a very low potential (−0.2 V vs. saturated calomel electrode). The low-potential reductive detection of the hydrogen peroxide led to a highly selective amperometric monitoring of the glucose substrate, along with linearity up to 30 mM, and a detection limit of 0.08 mM. Different from GOD, dehydrogenase-based amperometric devices based on the co-immobilization of dehydrogenases and their NAD^+ cofactor to electrodes for the biosensing of important substrates such as lactate, alcohol or glucose have also been described [69]. Oxidation of the product (i.e., NADH) serves as the anodic signal and regenerates the NAD^+ cofactor. Therefore, CNT-modified electrodes, offering an accelerated electron transfer of NADH along with minimization of surface fouling associated with the accumulation of NADH oxidation product [20], provide great promise as dehydrogenase-based enzyme electrodes. For example, Wang and Musameh have reported a low-potential stable detection of ethanol based on the co-immobilization of alcohol dehydrogenase and its NAD^+ cofactor within a CNT/Teflon matrix [80]. Similar advantages are expected for the biosensing of lactate or glucose in connection with lactate or glucose dehydrogenase, respectively [69]. Deo et al. [81] have reported an amperometric biosensor for organophosphorus (OP) pesticides based on a CNT-modified transducer and an organophosphorus hydrolase (OPH) biocatalyst. To prepare the CNT/OPH biosensor, a bilayer approach with the OPH layer atop of the CNT film was used. The CNT layer leads to a greatly improved anodic detection of the enzymatically generated p-nitrophenol product, including higher sensitivity and stability. The biosensor was used to measure as low as 0.15 μM paraoxon and 0.8 μM methyl parathion, with sensitivities of 25 and 6 nA μM^{-1}, respectively. Li et al. have studied a CNT-modified biosensor for monitoring total cholesterol in blood [82]. The sensor consists of a carbon working electrode and a reference electrode screen-printed on a polycarbonate substrate. Cholesterol esterase, cholesterol oxidase, peroxidase, and potassium ferrocyanide were immobilized on the screen-printed carbon electrodes. MWCNTs were added to prompt electron transfer. Experimental results show that the CNT-modified biosensor offers a reliable calibration profile and stable electrochemical properties.

The remarkable electrocatalytic activity of CNTs, together with the ability to modify CNTs for accumulating important biomolecules, make them extremely attractive for a wide range of electrochemical biosensors, ranging from enzyme-based electrochemical biosensors to DNA hybridization biosensors [69]. CNT-based electrochemical biosensors combine the specificity of enzymes with the electrocatalytic ability of nanotubes, and are expected to be extremely useful for clinical diagnostics and environmental monitoring.

2.6 Functionalization of CNTs

2.6.1 Biological Functionalization of CNTs

Biological functionalization of CNTs has come to be of significant interest in recent years due to the possibility of developing sensitive and ultrafast detection systems in biomedical sciences and biotechnological application. CNTs functionalized with biological assays could be the key to novel nano-biosensing techniques. Functionalization of CNT surfaces using proteins and antibodies could enable specific interactions and selective binding to target biomolecules with a very low sample size, often approaching a single protein. Alteration of the surface chemistry of their sidewalls can lead to covalent functionalization and, hence, to their potential application in drug delivery and chemical and biochemical sensing applications [83]. The functionalization of CNTs with proteins like streptavidin and biotin with the help of a polymer coating with poly(ethylene glycol) (PEG), and covalent interactions with its amine-terminated variant, has been demonstrated [84]. Since all the atoms in CNTs are surface atoms, binding proteins or antibodies to surfaces can greatly affect their surface states and, thus, their electrical and optical properties. This effect can be exploited as a basis for detecting biological surface reactions in a single protein or antibody attached to CNT surfaces.

2.6.2 Self-assembly of Surfactant and Lipid Molecules at CNTs

Surfactant adsorption at interfaces has been widely studied because of its importance in detergents, lubrication, and colloid stabilization [85]. CNTs are insoluble in organic solvents and in water, which at present considerably restricts their areas of use. To enhance the solubility of CNTs, detergents are mixed and shaken with nanotubes to form stable suspensions [86]. The chemical adsorption of SDS molecules on the surface of the nanotubes creates a distribution of negative charges that prevents their aggregation and induces stable suspensions in water. The SDS molecules can be oriented perpendicularly to the surface of the nanotube, forming a monolayer. The molecular organization of surfactant at the solid–liquid interface can be (a) the hydrophobic part of SDS is adsorbed on the graphite by van der Waals interactions, likely following the carbon network, and (b) the hydrophilic part of the surfactant is oriented toward the aqueous phase, forming half-cylinders on the surface of the graphite plane [87]. Because CNTs are rolled-up graphene sheets, the SDS molecules may form similar half-cylinders on the surface of the tubes, either oriented parallel or perpendicularly to the tube axis.

The self-assembly of lipids at metallic–aqueous interfaces has been well studied. To determine whether lipid molecules could adsorb and self-organize on CNTs, creating stable assemblies, Richard et al. have designed and synthesized new reagents that form lipidic "rings" made up of supramolecular half-cylinders [88]. TEM results indicated the formation of supramolecular assemblies of these molecules on the surface of the nanotubes. To explore the possibility of functionalizing the surface of CNTs in a noncovalent but permanent way with different reagents, the self-assembly of a series of molecules made of a double lipidic chain was further tested. In contrast to the single-chain lipids, no organization was detected by TEM when an aqueous solution (1 mg mL^{-1}) of the second series of molecules was directly sonicated with 1 mg of MWCNTs. To investigate whether micelles are necessary to form these supramolecular assemblies, the MWCNTs were sonicated in the presence of mixed micelles. TEM observations showed perfectly organized striations on the CNTs. The size of the striations, determined by TEM, varied from 55 to 75 Å, in perfect agreement with the length of the different lipidic chains. These results are also coherent with a half-cylinder arrangement of the double-chain lipids on the surface of the nanotubes. Hence, the formation of micelles appears to be a key step for the formation of supramolecular assemblies on the CNT surface. This process constitutes a simple, versatile protocol for the noncovalent functionalization of nanotubes.

One of the most exciting applications of CNTs is in the exploration of proteins and cells in aqueous solution. Few of these applications have yet been realized, because of the incompatibility of the CNT surface, which is hydrophobic and prone to nonspecific bioadsorption with biological components such as cells and proteins. In addition, the aqueous environment required for biological materials is not suitable for unfunctionalized CNTs [89]. In nature, cells are faced with a similar challenge of resisting nonspecific biomolecule interactions while engaging in specific molecular recognition. These functions can be simultaneously fulfilled by mucin glycoproteins, defined by their dense clusters of O-linked glycans. Zettl and associates [90] have described a biomimetic surface modification of CNTs using glycosylated polymers designed to mimic natural cell-surface mucins. A C_{18} lipid at one end of a mucin-mimic polymer is introduced to enable surface modification of

CNTs. Lipids self-assemble on the surface of CNTs through hydrophobic interactions in the presence of water [88] and lipid functionalized glycopolymers form ordered arrays on graphite surfaces [91]. The lipid-functionalized mucin-mimic is self-assembled on CNTs in a similar manner as the organization of native mucins in the cell membrane, with the glycosylated polymers projecting into the aqueous medium. CNTs modified with mucin-mimics were soluble in water, resisted non-specific protein binding, and bound specifically to biomolecules through receptor–ligand interactions. This strategy for biomimetic surface engineering provides a means to bridge CNTs and biological systems.

Several successful strategies using covalent or noncovalent chemistry have been applied to functionalize the sidewall of CNTs. Among these, noncovalent methods are attractive as they may preserve the inherent properties of the nanotubes [88, 92]. Supported bilayer lipid membranes (s-BLMs) have attracted increasing interest due to their potential application as electrochemical biosensors, molecular devices, and for investigating the photoinduced electron transfer in biomembranes [93, 94].

Recently, we [39] successfully self-assembled s-BLM on the surface of MWCNTs using Tien's method [94, 95]. The CV responses of BLMs-coated MWCNTs in PBS were studied. At bare MWCNT electrodes, a large background current i (in the range of 10^{-5} A) was observed. When the MWCNT electrodes were coated with BLMs, the background current was dramatically reduced to the range of 10^{-9} A, indicating a strong insulation effect of the lipids assembled on the surface of the MWCNTs. Since the membrane capacitance (C_m) can be obtained from the background current of the CV ($C_m = i/v$), the thickness of lipid membrane T_m on MWCNTs can be calculated from $T_m = 2.2\varepsilon_0 A/C_m$ (ε_0: vacuum dielectric permittivity, A: surface area). T_m was thus calculated to be about 4.38 nm, which is approximately double the molecular size of phosphatidylcholine – consistent with a bilayer structure of the lipid membrane [94]. The static water contact angles on the surface of MWCNTs were measured to be 6–7°, indicating a hydrophilic property of the MWCNT surface. The bilayer lipid membrane can thus be formed with the hydrophilic moiety of one lipid layer absorbed on the surface of MWCNTs, and that of another lipid layer faced to the testing PBS solution. This result agrees with that reported by Kanyó et al. [96] but differs from that found by Richard et al. [88]. Kanyó et al. reported that the surface of MWCNTs, without heating treatment, were hydrophilic [96]. In contrast, Richard et al. reported that a monolayer of synthetic lipid was self-assembled at the surface of CNTs, in which the van der Waals interaction between the hydrophobic part of lipid and the carbon network was suggested [88]. The different hydrophilicity reported may come from the different fabrication methods of the CNTs. When MWCNTs were synthesized from the chemical source of ethylenediamine, chemical elemental analysis indicated that 1.8–2.8% nitrogen existed in the MWCNTs used in the present study [37]. Functional groups containing nitrogen on the MWCNT surface may contribute partly to the hydrophilic property on the MWCNTs synthesized and used here. In addition, it is known that the carbon shell (i.e., each carbon shell of the MWCNT) is closed by various functional groups, most frequently by COOH, –OH, and CO groups. For MWCNTs, the outer shell may often contain discontinuous spots of imperfection. These local vacancies could also be closed by the functional groups mentioned above [96]. Possibilities exist for these functional groups to confer the hydrophilic surface of the MWCNTs.

2.6.3 Electrochemical Functionalization of CNTs

The electrochemical functionalization of CNTs has opened up new opportunities to fabricate novel nanostructures by improving both their solubility and processibility. Small-diameter (ca. 0.7 nm) SWCNTs are predicted to display enhanced reactivity relative to larger-diameter nanotubes due to increased curvature strain. Bahr et al. have described the derivatization of these small-diameter nanotubes via electrochemical reduction of various aryl diazonium salts [97]. The estimated degree of functionalization is as high as one out of every 20 carbons in the nanotubes bearing a functionalized moiety. The functionalizing moieties can be removed by heating in an argon atmosphere. Nanotubes derivatized with a 4-*tert*-butylbenzene moiety possessed significantly improved solubility in organic solvents. Functionalization of the nanotubes with a molecular system holds strong promise as useful building blocks for the construction of novel hybrids for nano-sensor applications.

Electrochemical functionalization of CNTs is a selective, clean, and nondestructive chemical method. But it meets difficulties in homogeneous electrografting of SWCNTs in large quantities because the reaction is often localized on a very thin film (ca. 2 μm). To solve this problem, Zhang et al. have utilized a room-temperature ionic liquid (RTIL)-supported three-dimensional network SWCNT electrode [98]. In their work, large quantities of SWCNTs were considerably untangled in RTILs so as to greatly increase the effective area of the electrode. *N*-Succinimidyl acrylate (NSA), as a model monomer, was dissolved in the supporting RTILs and was electrografted onto SWCNTs (SWCNTs-poly-NSA). As an application example, GOD was covalently anchored on the SWCNT-poly-NSA assembly, and the electrocatalytic oxidation of glucose in this assembly was investigated. RTILs have opened a new path in electrochemical functionalization of CNTs. Recently, we studied the effect of electrochemical oxidation in 0.2 M HNO_3 for MWCNTs [99]. Scanning electron microscope (SEM) and transmission electron microscope (TEM) images reveal that electrochemical oxidation increases the specific area of MWCNTs by cutting off the nanotube tips. Cyclic voltammetry and constant current charging/discharging was used to characterize the behavior of electrochemical double layer capacitors (EDLCs) of the oxidized MWCNTs in 1.0 M H_2SO_4. The specific capacitance of the oxidized-MWCNTs was remarkably improved. Electrochemical oxidation is, hence, an effective way to improve the performance of MWCNT electrodes in EDLC application. More recently, electrochemical nitration of self-assembled SWCNT sheets with $-NO_2$ groups was achieved [100]. A SWCNT sheet, used as the working electrode in 6 M aqueous solution of potassium nitrite, was anodically oxidized to form $-NO_2$ groups on the SWCNTs. Attenuated-total-reflection Fourier-transform infrared and micro-Raman spectroscopy showed the presence of chemisorbed $-NO_2$ groups, consistent with transmission electron microscope images of the nanotube bundles after functionalization.

2.6.4 Electrochemical Application of Functionalized CNTs

2.6.4.1 Application of Lipid–CNT Nanomaterials in Electrochemical Sensors

Sensors represent a most plausible and exciting application area for nanobiotechnology, and nanosensors based on CNTs are expected to emerge in the marketplace in significant volumes over the next ten years. Despite tremendous excitement recently generated by experimental breakthroughs that have led to realistic possibilities of using CNTs in electrochemical sensors, further experimental and theoretical research is necessary. The formation of stable lipid–CNT assemblies offers a simple, efficient method for the development of sensors/biosensors. The supramolecular structure of lipid–CNTs may lead to several applications in the field of nanobiotechnology. For example, it could be used for the development of molecular sensors (biosensors) for detecting the body's molecules. To actualize and optimize the full commercial potential of CNT-based electrochemical sensors, efforts must continue to be devoted to integrate the nanotube-arrays with power, miniaturized and easy-to-use electrochemical instruments for bimolecular sensing, genetic analysis, and drug discovery or screening.

The photoelectric effects of bilayer lipid membranes (BLMs) and electron mediator modified BLMs have been extensively studied, on account of their possible applications in understanding the mechanism of natural photosynthesis, in developing artificial photoelectric devices [101], and in mimicking functionalities of natural photosynthetic systems, which are represented by photoactive groups, electron donors and acceptors [102]. Various attempts have been made to realize an artificial photosynthesis and solar-energy conversion system under laboratory conditions. For example, synthetic dyes have been used to dope BLMs and the corresponding photoresponses have been investigated [103].

Recently, we have described the self-assembly of BLM and C_{60}-containing BLM at well-aligned MWCNTs for the development of a novel photoswitchable electrochemical device [39]. The lipid membrane at MWCNTs is estimated to be 4.38 nm thick, which is approximately double the molecular size of phosphatidylcholine, indicating that the BLM at the surface of nanotubes has a bilayer structure. Lipid membrane self-assembled at MWCNTs acts as an insulating layer while the incorporated C_{60} can mediate the transport of electrons as well as photocurrent across BLM (Fig. 2.5). The membrane resistance of C_{60}-BLM/MWCNTs is 369.3 Ω, which is much smaller than that of BLM/MWCNTs (3.238×10^6 Ω). Furthermore, C_{60}-BLM/MWCNTs possess photoelectric properties due to the electron mediation of C_{60} in the lipid membrane. The photoelectric conversion properties of MWCNTs, BLM/MWCNTs, and C_{60}-BLM/MWCNTs were thus studied using an amperometric technique. At the lipid interface, C_{60} transports about 30–40% of electrons, compared with that of pure MWCNTs, from MWCNTs to the redox species in solu-

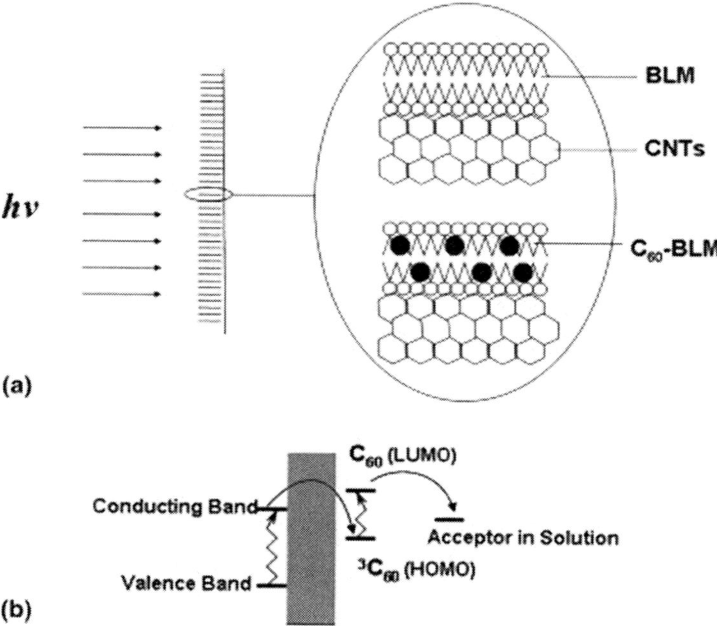

Fig. 2.5. (a) Schematic of BLM/MWCNTs and C_{60}-BLM/MWCNTs electrode for photocurrent measurement and electrochemical experiments (BLM = bilayer lipid membrane). (b) Possible mechanism of photocurrent generation at the C_{60}-BLM/MWCNTs cathode. HOMO, the highest occupied molecular orbital; LUMO, the lowest unoccupied molecular orbital. (Adapted from Ref. [39] with permission.)

tion. The successful self-assembly of BLM and incorporation of C_{60} into the BLM at MWCNTs may provide an easy way for construction of new biosensors and bioelectronic materials using BLM/MWCNTs and/or C_{60}-BLM/MWCNTs nanocomposites.

2.6.4.2 Achieving direct Electron Transfer to Redox Proteins by Functional CNTs

CNT electrodes have been successfully used to study protein electrochemistry. The novel electrochemical properties and the nanoscale size of CNTs have opened research opportunities in studying hitherto unapproachable phenomena at interfaces and bifaces for redox proteins [23, 24, 104, 105]. The direct electron transfer of proteins such as peroxidases [23, 24, 65], cytochrome *c* [15, 30, 106], myoglobin [107], catalase [108], azurin [15], and GOD [64] have been achieved by the use of CNT electrodes.

Recently, Mao and associates [109] have described the preparation and bioelectrochemical properties of functional nanohybrids through co-assembling of heme-proteins (i.e., horseradish peroxidase, hemoglobin, myoglobin and cytochrome *c*) and surfactants onto CNTs. The prepared protein–surfactant–CNT nanohybrids (Fig. 2.6) possess facilitated interfacial electron transfer of the proteins with en-

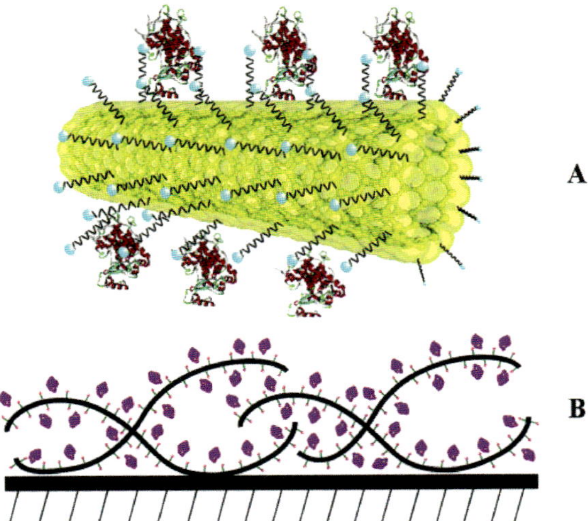

Fig. 2.6. Cartoons of (A) bioelectrochemically functional unit through co-assembling of protein and surfactant onto a single MWCNT and (B) nanohybrids consisting of single biofunctional nanotubes deposited onto the glassy carbon electrode. (Adapted from Ref. [109] with permission.)

hanced faradic responses. The enhancements are ascribed to the ability of surfactants to facilitate protein electrochemistry and to the improved portion of electroactive proteins assembled, of which the latter assignment is closely associated with the electrochemical and structural properties of the nanotubes, and the three-dimensional (3D) architecture of the CNT film confined on the glassy carbon electrode. It is proposed that the single and/or small bundles of the nanotubes in the CNT film electrode can be rationally functionalized with surfactants to be functional nanoelectrodes capable of facilitating electron transfer of proteins. The 3D confinement of these functional nanowires onto the GC electrode essentially increases the portion of electroactive proteins assembled in the nanohybrids. These properties of the protein–surfactant–CNT nanohybrids, combined with the bioelectrochemical catalytic activity, could make them useful for development of bioelectronic devices and for the investigation of protein electrochemistry at functional interfaces.

2.6.4.3 Biomolecule-functionalized CNTs for Electrochemical Sensors and Biosensors

Biomolecule-functionalized carbon nanomaterials have been of interest in electrochemical areas in both fundamental research and application. The immobilization of molecules, biomolecules or even nanoparticles on SWCNTs has been exploited, motivated by the prospects of using nanotubes as new types of sensor/biosensor materials [110–113]. In contrast, there are few studies on the modification of

MWCNTs [36, 114–117]. The conducting nature of the MWCNTs, together with their outstanding electronic and mechanical properties, make them attractive for potential applications. Hemin (iron protoporphyrin IX) is the active center of the heme-protein family, such as *b*-type cytochromes, peroxidase, myoglobin and hemoglobin. It contains a porphyrin ring, which can be immobilized at the surface of CNTs through noncovalent functionalization by π–π interaction [92]. We have successfully constructed a hemin-modified MWCNT electrode [43] and characterized its electrochemical behavior by CV. The electron-transfer coefficient (α) was found to be 0.38, with a heterogeneous electron transfer rate (k) of 2.9 s^{-1} for the adsorbed hemin. The hemin-modified MWCNT electrodes show ideal reversibility in 5 mM $K_3[Fe(CN)_6]$ in the range 0.02–1.00 V s^{-1}, indicating fast electron-transfer kinetics. CV of the hemin-modified MWCNT electrode in pH 7.4 phosphate buffer solution (PBS) clearly shows the dioxygen reduction peaks to be close to 0 V (vs. Ag|AgCl). These results are useful in the development of a novel oxygen sensor for working at a relatively low potential. To improve the sensitivity of CNT-based electrochemical sensors toward hydrogen peroxide, we functionalized MWCNTs with iron-phthalocyanines (FePc) [40]. Highly sensitive and selective glucose sensors can be constructed on FePc-MWCNTs electrodes based on the immobilization of GOD on a poly-*o*-aminophenol (POAP)-electropolymerized electrode surface. SEM images indicate that GOD enzymes trapped in POAP film tend to deposit primarily on the curved tips and evenly disperse along the sidewalls. The GOD@POAP/FePc-MWCNTs biosensor we proposed exhibits excellent performance for glucose with a rapid response (less than 8 s), a wide linear range (up to 4.0×10^{-3} M), low detection limits (2.0×10^{-7} M with signal-to-noise of 3), a highly reproducible response (RSD of 2.6%), and long-term stability (120 days). Such characteristics may be attributed to the catalytic activity of FePc-functionalized CNT, permselectivity of POAP film, as well as the large surface area of CNT materials. Aligned CNT electrode arrays have been used to achieve direct electron transfer to enzymes with redox centers close to the surface of the protein [23, 24]. Most of the electrochemistry was dominated by proteins immobilized on the electroactive ends of the nanotubes. Both horseradish peroxidase and myoglobin functionalized CNTs [23] could be used analytically to detect hydrogen peroxide.

CNTs have generated considerable recent interest in bioelectronics and bioelectrochemistry owing to their unique mechanical, electrical, and chemical properties [69, 118]. The electrocatalytic properties of these materials have been exploited as a means of promoting electron-transfer reactions of a wide range of important biomolecules [107, 119]. For example, the greatly enhanced electrochemical reactivity of hydrogen peroxide and NADH at CNT-modified electrodes makes these nanomaterials extremely attractive for numerous oxidase- and dehydrogenase-based amperometric biosensors [20, 21]. The use of CNT molecular wires offers great promise for achieving efficient electron transfer from electrode surfaces to the redox sites of enzymes [105]. Recently, Georgakilas and colleagues have described the magnetic modification of CNTs [120]. It is expected that the magnetic and catalytic properties of CNTs can be exploited for the magnetoswitchable control of electron transfer reactions with functionalized magnetic particles. Pyrene can be noncovalently attached on the CNT surface. A carboxylic derivative of pyrene is used as an interlinker for the binding of capped magnetic nanoparticles on the CNTs. The increased organophilic character of the capped nanoparticles induces high solubility in organic media for the modified CNTs.

The unique structure and the outstanding electronic properties of CNTs coupled with the specific recognition properties of DNA would indeed make CNT–DNA bioconjugates widely useful in biosensors [121]. To prepare the bioconjugates, CNTs are treated with concentrated oxidizing acids so as to cover their ends and surfaces with oxygen-containing groups such as carboxyl groups and ether groups. The pretreated CNTs are then dispersed in water or organic solvents with the aid of ultrasonic oscillation. Biomolecules can be linked to CNT-modified electrodes with the aid of 1-ethyl-3-(3-dimethylaminopropyl)carbodiimide (EDC) and/or *N*-hydroxysuccinimide (NHS) through the formation of an amide bond between the carboxylic groups on CNTs and the amino groups in biomolecules. Guo et al. have studied the electrochemical characteristics of the immobilization of DNA on the surface of MWCNTs [122, 123]. Both single strand and double strand calf thymus DNA molecules were attached to MWCNT-modified gold electrodes with the aid of EDC and NHS. The results of CV, electrochemical impedance spectroscopy (EIS), and piezoelectric quartz crystal impedance (PQCI) indicate that calf thymus DNA can be immobilized on MWCNTs via a cationic polyelectrolyte. No matter how DNA molecules were immobilized on MWCNTs, they still remained bioactive and could interact with small molecules such as chlorpromazine hydrochloride and ethidium bromide (EB). Recently, SWCNTs functionalized with amino groups have been prepared via chemical modification of carboxyl groups introduced on the CNT surface [124]. Two different approaches (amide and amine-moieties) were used to produce the amino-functionalized nanotubes. The amino-termination allows further chemistry of the functionalized SWCNTs and makes possible covalent bonding to polymers and biological systems such as DNA and carbohydrates. Wang et al. have attached CdS nanoparticles capped with octadecanethiol (ODT) to SWCNTs, which were dispersed in toluene after pretreatment with acetone, through hydrophobic interactions [125]. Then, streptavidin was anchored to CdS-SWCNTs, followed by combination with biotinylated probe DNA. The resultant SWCNTs-DNA bioconjugates were attached to a microplate through "sandwich" hybridization. Target DNA can be detected via determination of cadmium in CdS nanoparticles dissolved by nitric acid at a mercury film GCE by stripping voltammetry. The detection limit is around 40 pg mL^{-1}. With selective functionalization of oligonucleotide probes at the open ends of MWCNTs, Meyyappan's group [121] reported that a nanoelectrode array based on vertically aligned MWCNTs embedded in SiO_2 can be used for ultrasensitive DNA detection. Characteristic electrochemical behaviors are observed for measuring bulk and surface-immobilized redox species. Sensitivity is dramatically improved by lowering the nanotube density. More recently, they reported a similar approach for the fabrication of nanoelectrode arrays using vertically aligned MWCNTs embedded within a SiO_2 matrix [126]. CV and pulse voltammetry were employed to characterize the electrochemical properties of the MWCNT array. The unique graphitic structure of the

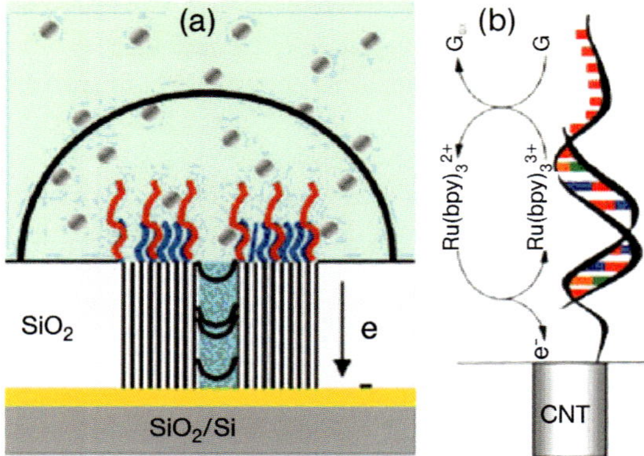

Fig. 2.7. (a) and (b) Schematic of the mechanism for DNA/RNA sensing on an inlaid MWCNT nanoelectrode combined with $Ru(bpy)_3{}^{2+}$ mediator amplified guanine oxidation. Short blue lines in (a) represent oligonucleotide probe molecules functionalized at the open end of MWCNTs, and the longer red lines represent target molecules hybridized with the probe molecules. The hemisphere represents the diffusion layer of $Ru(bpy)_3{}^{2+}$ mediators. (Adapted from Ref. [126] with permission.)

novel MWCNT nanoelectrodes were compared with model systems such as highly oriented pyrolytic graphite and glassy carbon electrodes. Low-density MWCNT nanoelectrode arrays display independent nanoelectrode behavior, showing diffusion-limited steady-state currents in cyclic voltammetry over a wide range of scan rates. Electroactive species can be detected at concentrations as low as a few nm. In addition, ultrasensitive DNA/RNA sensors have been demonstrated using the low-density MWCNT arrays with selectively functionalized oligonucleotide probes (Fig. 2.7). This platform can be widely used in analytical applications as well as fundamental electrochemical studies. In summary, biomolecule-functionalized CNT electrodes can in the future be used as a new type of miniature DNA affinity biosensors [127].

2.7
Conclusions and Future Prospects

We have addressed recent advances in electrochemistry and electrochemical application of CNTs, particularly of the functionalized CNTs with specific properties. The attractive properties of functionalized CNTs have paved the way for the construction of a wide range of electrochemical sensors/biosensors exhibiting attractive analytical behavior. In particular, functionalization of the CNT surface can result in highly soluble materials, which can be further modified with active molecules, making them compatible with biological systems. Therefore, many applications can be envisaged by using functionalized CNTs. With the creation of bifaces and interfaces at lipid–CNTs nanocomposites, progress in CNT technology may well lead to better insights into biological and physical chemistry processes. This will make it possible to find compounds more compatible with CNT technology and to facilitate more effective use in electrochemical applications. Highly selective and sensitive molecular sensors based on CNTs are set to become commonplace in the near future.

Acknowledgments

Part of the work conducted in our laboratory and reviewed in this chapter was supported by the Academic Research Grant of the National University of Singapore R-398-000-024-112, and by the Defense Science & Technology Agency of Singapore R-154-000-243-422, to F.-S. S. We are also grateful to many colleagues who have collaborated with our group over the years, contributing heavily to the development of the CNT electrochemical sensors.

References

1 S. Iijima, Helical microtubules of graphitic carbon, *Nature* 354 (**1991**) 56–58.
2 S. Iijima, T. Ichihashi, Single-shell carbon nanotubes of 1-nm diameter, *Nature* 363 (**1993**) 603–605.
3 A. Thess, R. Lee, P. Nikdaev, H. Dai, P. Petit, J. Robert, C. Xu, Y. H. Lee, S. G. Kim, A. G. Rinzler, D. T. Colbert, G. E. Scuseria, D. Tomanek, J. E. Fischer, R. E. Smalley, Crystalline ropes of metallic carbon nanotubes, *Science* 273 (**1996**) 483–487.
4 R. Saito, M. Fujita, G. Dresselhaus, M. S. Dresselhaus, Electronic structure of grapheme tubules based on C_{60}, *Phys. Rev. B* 46 (**1992**) 1804–1811.
5 R. Saito, M. Fujita, G. Dresselhaus, M. S. Dresselhaus, Electronic structure of chiral graphite tubles, *Appl. Phys. Lett.* 60 (**1992**) 2204–2206.
6 J. H. Weaver, D. M. Poirier, Solid state properties of fullerenes and fullerene-based materials, *Solid State Phys. – Adv. Res. Applicat.* 48 (**1994**) 1–108.
7 J. W. Mintmire, B. I. Dunlap, C. T. White, Are fullerene tubules metallic, *Phys. Rev. Lett.* 68 (**1992**) 631–634.
8 N. Hamada, S. Sawade, A. Oshiyama, New one-dimensional conductors: Graphitic microtubules, *Phys. Rev. Lett.* 68 (**1992**) 1579–1581.
9 B. I. Yakobson, C. J. Brabec, J. Bernholc, Nanomechanics of carbon tubes: Instabilities beyond linear response, *Phys. Rev. Lett.* 76 (**1996**) 2511–2514.
10 M. M. J. Treacy, T. W. Ebbesen, J. M. Gibson, Exceptionally high Young's modulus observed for individual carbon nanotubes, *Nature* 381 (**1996**) 678–680.
11 E. W. Wong, P. E. Sheehan, C. M. Lieber, Nanobeam mechanics: Elasticity, strength, and toughness of nanorods and nanotubes, *Science* 277 (**1997**) 1971–1975.
12 P. J. Britto, K. S. V. Santhanam, P. M. Ajayan, Carbon nanotube electrode for oxidation of dopamine, *Bioelectrochem. Bioenerg.* 41 (**1996**) 121–125.
13 E. Pennisi, Simple recipe yields fullerene tubules, *Sci. News* 142 (**1992**) 36.
14 Z. H. Wang, J. Liu, Q. L. Liang, Y. M. Wang, G. Luo, Carbon nanotube-modified electrodes for the simultaneous determination of

dopamine and ascorbic acid, *Analyst* 127 (**2002**) 653–658.

15 J. J. Davis, R. J. Coles, H. A. O. Hill, Protein electrochemistry at carbon nanotube electrodes, *J. Electroanal. Chem.* 440 (**1997**) 279–282.

16 H. X. Luo, Z. J. Shi, N. Q. Li, Z. N. Gu, Q. K. Zhuang, Investigation of the electrochemical and electrocatalytic behavior of single-wall carbon nanotube film on a glassy carbon electrode, *Anal. Chem.* 73 (**2001**) 915–920.

17 P. J. Britto, K. S. V. Santhanam, A. Rubio, J. A. Alonso, P. M. Ajayan, Improved charge transfer at carbon nanotube electrodes, *Adv. Mater.* 11 (**1999**) 154–157.

18 J. Kong, N. R. Franklin, C. Zhou, M. G. Chapline, S. Peng, K. Cho, H. Dai, Nanotube molecular wires as chemical sensors, *Science* 287 (**2000**) 622–625.

19 P. G. Collins, K. Bradley, M. Ishigami, A. Zettl, Extreme oxygen sensitivity of electronic properties of carbon nanotubes, *Science* 287 (**2000**) 1801–1804.

20 M. Musameh, J. Wang, A. Merkoci, Y. Lin, Low-potential stable NADH detection at carbon-nanotube-modified glassy carbon electrodes, *Electrochem. Commun.* 4 (**2002**) 743–746.

21 J. Wang, M. Musameh, Y. H. Lin, Solubilization of carbon nanotubes by Nafion toward the preparation of amperometric biosensors, *J. Am. Chem. Soc.* 125 (**2003**) 2408–2409.

22 J. Wang, M. Musameh, Carbon nanotube/teflon composite electrochemical sensors and biosensors, *Anal. Chem.* 75 (**2003**) 2075–2079.

23 X. Yu, D. Chattopadhyay, I. Galeska, F. Papadimitrakopoulos, J. F. Rusling, Peroxidase activity of enzymes bound to the ends of single-wall carbon nanotube forest electrodes, *Electrochem. Commun.* 5 (**2003**) 408–411.

24 J. J. Gooding, R. Wibowo, J. Q. Liu, W. R. Yang, D. Losic, S. Orbons, F. J. Mearns, J. G. Shapter, D. B. Hibbert, Protein electrochemistry using aligned carbon nanotube arrays, *J. Am. Chem. Soc.* 125 (**2003**) 9006–9007.

25 S. J. Lee, H. K. Baik, J. E. Yoo, J. H. Han, Large-scale synthesis of carbon nanotubes by plasma rotating arc discharge technique, *Diamond Relat. Mater.* 11 (**2002**) 914–917.

26 T. Guo, P. Nikolaev, A. Thess, D. T. Colbert, R. E. Smalley, Catalytic growth of single-walled nanotubes by laser vaporization, *Chem. Phys. Lett.* 243 (**1995**) 49–54.

27 Z. F. Ren, Z. P. Huang, J. W. Xu, J. H. Wang, P. Bush, M. P. Siegel, P. N. Provencio, Synthesis of large arrays of well-aligned carbon nanotubes on glass, *Science* 282 (**1998**) 1105–1107.

28 Z. F. Ren, Z. P. Huang, D. Z. Wang, J. G. Wen, J. W. Xu, J. H. Wang, L. E. Calvet, J. Chen, J. F. Klemic, M. A. Reed, Growth of a single freestanding multi-wall carbon nanotube on each nano-nickel dot, *Appl. Phys. Lett.* 75 (**1999**), 1086–1088.

29 J. K. Campbell, L. Sun, R. M. Crooks, Electrochemistry using single carbon nanotubes, *J. Am. Chem. Soc.* 121 (**1999**) 3779–3780.

30 J. X. Wang, M. X. Li, Z. J. Shi, N. Q. Li, Z. N. Gu, Direct electrochemistry of cytochrome *c* at a glassy carbon electrode modified with single-wall carbon nanotubes, *Anal. Chem.* 74 (**2002**) 1993–1997.

31 N. S. Lawrence, R. P. Deo, J. Wang, Comparison of the electrochemical reactivity of electrodes modified with carbon nanotubes from different sources, *Electroanalysis* 17 (**2005**) 65–72.

32 J. N. Barisci, G. G. Wallace, R. H. Baughman, Electrochemical studies of single-wall carbon nanotubes in aqueous solutions, *J. Electroanal. Chem.* 488 (**2000**) 92–98.

33 C. Y. Liu, A. J. Bard, F. Wudl, I. Weitz, J. R. Heath, Electrochemical characterization of films of single-walled carbon nanotubes and their possible application in supercapacitors, *Electrochem. Solid State Lett.* 2 (**1999**) 577–578.

34 C. Niu, E. K. Sichel, R. Hoch, D. Moy, H. Tennet, High power electrochemical capacitors based on carbon nanotube electrodes, *Appl. Phys. Lett.* 70 (**1997**) 1480–1482.

35 H. Luo, Z. Shi, N. Li, Z. Gu, Q. Zhuang, Investigation on the electrochemical and electrocatalytic behavior of chemically modified electrode of single wall carbon nanotube functionalized with carboxylic acid group, *Chem. J. Chin. Univ.* 21 (**2000**) 1372–1374.

36 P. F. Liu, J. H. Hu, Carbon nanotube powder microelectrodes for nitrite detection, *Sens. Actuators B* 84 (**2002**) 194–199.

37 W. D. Zhang, Y. Wen, S. M. Lin, W. C. Tjiu, G. Q. Xu, L. M. Gan, Synthesis of vertically aligned carbon nanotubes on metal deposited quartz plates, *Carbon* 40 (**2002**) 1981–1989.

38 W. D. Zhang, Y. Wen, W. C. Tjiu, G. Q. Xu, L. M. Gan, Growth of vertically aligned carbon-nanotube array on large area of quartz plates by chemical vapor deposition, *Appl. Phys. A* 74 (**2002**) 419–422.

39 J. S. Ye, H. F. Cui, W. D. Zhang, A. L. Ottova, H. T. Tien, F. S. Sheu, Self-assembly of bilayer lipid membrane at multiwalled carbon nanotubes towards the development of photo-switched functional device, *Electrochem. Commun.* 7 (**2005**) 81–86.

40 J. S. Ye, Y. Wen, W. D. Zhang, H. F. Cui, G. Q. Xu, F. S. Sheu, Electrochemical biosensing platforms using phthalocyanine-functionalized carbon nanotube electrode, *Electroanalysis* 17 (**2005**) 89–96.

41 H. F. Cui, J. S. Ye, W. D. Zhang, J. Wang, F. S. Sheu, Electrocatalytic reduction of oxygen by platinum nanoparticle/carbon nanotube composite electrode, *J. Electroanal. Chem.* 577 (**2005**) 295–302.

42 W. C. Poh, K. P. Loh, W. D. Zhang, S. Triparthy, J. S. Ye, F. S. Sheu, Biosensing properties of diamond and carbon nanotubes, *Langmuir* 20 (**2004**) 5484–5492.

43 J. S. Ye, Y. Wen, W. D. Zhang, L. M. Gan, G. Q. Xu, F. S. Sheu, Application of multi-walled carbon nanotubes functionalized with hemin for oxygen detection in neutral solution, *J. Electroanal. Chem.* 562 (**2004**) 241–246.

44 J. S. Ye, Y. Wen, W. D. Zhang, L. M. Gan, G. Q. Xu, F. S. Sheu, Selective voltammetric detection of uric acid in the presence of ascorbic acid at well-aligned carbon nanotube electrode, *Electroanalysis* 15 (**2003**) 1693–1698.

45 J. S. Ye, H. F. Cui, X. Liu, T. M. Lim, W. D. Zhang, F. S. Sheu, Preparation and characterization of well-aligned carbon nanotubes-ruthenium oxide nanocomposites for supercapacitors, *Small* 1 (**2005**) 560–565.

46 J. S. Ye, Y. Wen, W. D. Zhang, L. M. Gan, G. Q. Xu, F. S. Sheu, Non-enzymatic glucose detection using multi-walled carbon nanotube electrode, *Electrochem. Commun.* 6 (**2004**) 66–70.

47 K. B. Wu, S. S. Hu, J. J. Fei, W. Bai, Mercury-free simultaneous determination of cadmium and lead at a glassy carbon electrode modified with multi-wall carbon nanotubes, *Anal. Chim. Acta* 489 (**2003**) 215–221.

48 M. D. Rubianes, G. A. Rivas, Carbon nanotubes paste electrode, *Electrochem. Commun.* 5 (**2003**) 689–694.

49 F. Valentini, A. Amine, S. Orlanducci, M. L. Terranova, G. Palleschi, Carbon nanotube purification: Preparation and characterization of carbon nanotube paste electrodes, *Anal. Chem.* 75 (**2003**) 5413–5421.

50 J. Wang, M. Musameh, Screen-printed carbon nanotube electrodes, *Analyst* 129 (**2004**) 1–2.

51 N. S. Lawrence, R. P. Deo, J. Wang, Detection of homocysteine at carbon nanotube paste electrodes, *Talanta* 63 (**2004**) 443–449.

52 M. L. Pedano, G. A. Rivas, Adsorption and electrooxidation of nucleic acids at carbon nanotubes paste electrodes, *Electrochem. Commun.* 6 (**2004**) 10–16.

53 F. Ricci, A. Amine, D. Moscone, G. Palleschi, Prussian blue modified

carbon nanotube paste electrodes: A comparative study and a biochemical application, *Anal. Lett.* 36 (**2003**) 1921–1938.

54 R. Antiochia, I. Lavagnini, F. Magno, Electrocatalytic oxidation of NADH at single-wall carbon-nanotube-paste electrodes: Kinetic considerations for use of a redox mediator in solution and dissolved in the paste, *Anal. Bioanal. Chem.* 381 (**2005**) 1355–1361.

55 M. Trojanowicz, A. Mulchandani, M. Mascini, Carbon nanotubes-modified screen-printed electrodes for chemical sensors and biosensors, *Anal. Lett.*, 37 (**2004**) 3185–3204.

56 W. J. Guan, Y. Li, Y. Q. Chen, X. B. Zhang, G. Q. Hu, Glucose biosensor based on multi-wall carbon nanotubes and screen printed carbon electrodes, *Biosens. Bioelectron.* 21 (**2005**) 508–512.

57 E. Katz, I. Willner, Biomolecule-functionalized carbon nanotubes: Applications in nanobioelectronics, *ChemPhysChem* 5 (**2004**) 1085–1104.

58 A. B. Artyukhin, O. Bakajin, P. Stroeve, A. Noy, Layer-by-layer electrostatic self-assembly of polyelectrolyte nanoshells on individual carbon nanotube templates, *Langmuir* 20 (**2004**) 1442–1448.

59 M. Li, E. Dujardin, S. Mann, Programmed assembly of multi-layered protein/nanoparticle-carbon nanotube conjugates, *Chem. Commun.* 39 (**2005**) 4952–4954.

60 H. Shimoda, L. Fleming, K. Horton, O. Zhou, Formation of macro-scopically ordered carbon nanotube membranes by self-assembly, *Phys. B Condens. Matter* 323 (**2002**) 135–136.

61 N. O. V. Plank, R. Cheung, R. J. Andrews, Thiolation of single-wall carbon nanotubes and their self-assembly, *Appl. Phys. Lett.* 85 (**2004**) 3229–3231.

62 Y. B. Wang, Z. Iqbal, Vertically oriented single-wall carbon nanotube/enzyme on silicon as biosensor electrode, *JOM* 57 (**2005**) 27–29.

63 Y. D. Zhao, W. D. Zhang, H. Chen, Q. M. Luo, Anodic oxidation of hydrazine at carbon nanotube powder microelectrode and its detection, *Talanta* 58 (**2002**) 529–534.

64 Y. D. Zhao, W. D. Zhang, H. Chen, Q. M. Luo, Direct electron transfer of glucose oxidase molecules adsorbed onto carbon nanotube powder microelectrode, *Anal. Sci.* 18 (**2002**) 939–941.

65 Y. D. Zhao, W. D. Zhang, H. Chen, Q. M. Luo, S. F. Y. Li, Direct electrochemistry of horseradish peroxidase at carbon nanotube powder microelectrode, *Sens. Actuators, B* 87 (**2002**) 168–172.

66 Y. D. Zhao, W. D. Zhang, Q. M. Luo, S. F. Y. Li, The oxidation and reduction behavior of nitrite at carbon nanotube powder microelectrodes, *Microchem. J.* 75 (**2003**) 189–198.

67 Y. D. Zhao, W. D. Zhang, H. Chen, Q. M. Luo, Electrocatalytic oxidation of cysteine at carbon nanotube powder microelectrode and its detection, *Sens. Actuators, B* 92 (**2003**) 279–285.

68 A. Merkoci, M. Pumera, X. Llopis, B. Perez, M. del Valle, S. Alegret, New materials for electrochemical sensing VI: Carbon nanotubes, *Trends Anal. Chem.* 24 (**2005**) 826–838.

69 J. Wang, Carbon-nanotube based electrochemical biosensors: A review, *Electroanalysis* 17 (**2005**) 7–14.

70 J. Li, in *Carbon Nanotube Applications: Chemical and Physical Sensors, in Carbon Nanotubes Science and Application*, Ed. M. Meyyappan, CRC Press, New York, 2005, 213–233.

71 C. E. Banks, R. R. Moore, T. J. Davies, R. G. Compton, Investigation of modified basal plane pyrolytic graphite electrodes: Definitive evidence for the electrocatalytic properties of the ends of carbon nanotubes, *Chem. Commun.* 16 (**2004**) 1804–1805.

72 C. E. Banks, R. G. Compton, New electrodes for old: From carbon nanotubes to edge plane pyrolytic graphite, *Analyst* 131 (**2006**) 15–21.

73 C. E. Banks, T. J. Davies, G. G. Wildgoose, R. G. Compton, Electrocatalysis at graphite and carbon nanotube modified electrodes: Edge-

plane sites and tube ends are the reactive sites, *Chem. Commun.* 7 (**2005**) 829–841.

74 C. E. Banks, R. G. Compton, Exploring the electrocatalytic sites of carbon nanotubes for NADH detection: An edge plane pyrolytic graphite electrode study, *Analyst* 130 (**2005**) 1232–1239.

75 T. J. Davies, R. R. Moore, C. E. Banks, R. G. Compton, The cyclic voltammetric response of electrochemically heterogeneous surfaces, *J. Electroanal. Chem.* 574 (**2004**) 123–152.

76 M. Musameh, N. S. Lawrence, J. Wang, Electrochemical activation of carbon nanotubes, *Electrochem. Commun.* 7 (**2005**) 14–18.

77 A. Chou A. T. Bocking, N. K. Singh, J. J. Gooding, Demonstration of the importance of oxygenated species at the ends of carbon nanotubes for their favourable electrochemical properties, *Chem. Commun.* 7 (**2005**) 842–844.

78 T. J. Lee, S. I. Jung, C. Y. Park, Y. H. Choa, C. J. Lee, Synthesis of well-aligned carbon nanotubes with open tips, *Carbon* 43 (**2005**) 1341–1346.

79 Y. H. Lin, F. Lu, Y. Tu, Z. F. Ren, Glucose biosensors based on carbon nanotube nanoelectrode ensembles, *Nano Lett.* 4 (**2004**) 191–195.

80 J. Wang, M. Musameh, A reagentless amperometric alcohol biosensor based on carbon-nanotube/Teflon composite electrodes, *Anal. Lett.* 36 (**2003**) 2041–2048.

81 R. P. Deo, J. Wang, I. Block, A. Mulchandani, K. A. Joshi, M. Trojanowicz, F. Scholz, W. Chen, Y. H. Lin, Determination of organophosphate pesticides at a carbon nanotube/organophosphorus hydrolase electrochemical biosensor, *Anal. Chim. Acta* 530 (**2005**) 185–189.

82 G. Li, J. M. Liao, G. Q. Hu, N. Z. Ma, P. J. Wu, Study of carbon nanotube modified biosensor for monitoring total cholesterol in blood, *Biosen. Bioelectron.* 20 (**2005**) 2140–2144.

83 J. J. Davis, K. S. Coleman, B. R. Azamian, C. B. Bagshaw, M. L. H. Green, Chemical and biochemical sensing with modified single walled carbon nanotubes, *Chem. Eur. J.* 9 (**2003**) 3732–3739.

84 M. Shim, N. W. S. Kam, R. J. Chen, Y. M. Li, H. J. Dai, Functionalization of carbon nanotubes for biocompatibility and biomolecular recognition, *Nano Lett.* 2 (**2002**) 285–288.

85 A. C. Zettlemoyer, *J. Colloid Interface Sci.*, 28 (**1968**) 343.

86 J. M. Bonard, T. Stora, J. P. Salvetat, F. Maier, T. Stockli, C. Duschl, L. Forro, W. A. deHeer, A. Chatelain, Purification and size-selection of carbon nanotubes, *Adv. Mater.* 9 (**1997**) 827–831.

87 S. Manne, J. P. Cleveland, H. E. Gaub, G. D. Stucky, P. K. Hansma, Direct visualization of surfactant hemimicelles by force microscopy of the electrical double layer, *Langmuir* 10 (**1994**) 4409–4413.

88 C. Richard, F. Balavoibe, P. Schultz, T. W. Ebbesen, C. Mioskowski, Supramolecular self-assembly of lipid derivatives on carbon nanotubes, *Science* 300 (**2003**) 775–778.

89 Y. Lin, S. Taylor, H. P. Li, K. A. S. Fernando, L. W. Qu, W. Wang, L. R. Gu, B. Zhou, Y. P. Sun, Advances toward bioapplications of carbon nanotubes, *J. Mater. Chem.*, 14 (**2004**) 527–541.

90 X. Chen, G. S. Lee, A. Zettl, C. R. Bertozzi, Biomimetic engineering of carbon nanotubes by using cell surface mucin mimics, *Angew. Chem. Int. Ed.* 43 (**2004**) 6111–6116.

91 N. B. Holland, Y. X. Qiu, M. Ruegsegger, R. E. Marchant, Biomimetic engineering of non-adhesive glycocalyx-like surfaces using oligosaccharide surfactant polymers, *Nature* 392 (**1998**) 799–801.

92 X. B. Wang, Y. Q. Liu, W. F. Qiu, D. Zhu, Immobilization of tetra-tert-butylphthalocyanines on carbon nanotubes: A first step towards the development of new nanomaterials, *J. Mater. Chem.* 12 (**2002**) 1636–1639.

93 C. M. Drain, Self-organization of self-assembled photonic materials into

functional devices: Photo-switched conductors, *Proc. Natl. Acad. Sci. U.S.A.* 99 (**2002**) 5178–5182.

94 H. T. Tien, L. G. Wang, X. Wang, A. L. Ottova, Electronic processes in supported bilayer lipid membranes (s-BLMs) containing a geodesic form of carbon (fullerene C_{60}), *Bioelectrochem. Bioenerg.* 42 (**1997**) 161–167.

95 H. T. Tien, Z. Salamon, A. L. Ottova, Lipid bilayer-based sensors and biomolecular electronics, *Crit. Rev. Biomed. Eng.* 18 (**1991**) 323–340.

96 T. Kanyó, Z. Kónya, A. Kukovecz, F. Berger, I. Dékány, I. Kiricsi, Quantitative characterization of hydrophilic-hydrophobic properties of MWNTs surfaces, *Langmuir* 20 (**2004**) 1656–1661.

97 J. L. Bahr, J. Yang, D. V. Kosynkin, M. J. Bronikowski, R. E. Smalley, J. M. Tour, Functionalization of carbon nanotubes by electrochemical reduction of aryl diazonium salts: A bucky paper electrode, *J. Am. Chem. Soc.* 123 (**2001**) 6536–6542.

98 Y. J. Zhang, Y. F. Shen, J. H. Li, L. Niu, S. J. Dong, A. Ivaska, Electrochemical functionalization of single-walled carbon nanotubes in large quantities at a room-temperature ionic liquid supported three-dimensional network electrode, *Langmuir* 21 (**2005**) 4797–4800.

99 J. S. Ye, X. Liu, H. F. Cui, W. D. Zhang, F. S. Sheu, T. M. Lim, Electrochemical oxidation of multi-walled carbon nanotubes and its application to electrochemical double layer capacitors, *Electrochem. Commun.* 7 (**2005**) 249–255.

100 Y. B. Wang, S. V. Malhotra, F. J. Owens, Z. Iqbal, Electrochemical nitration of single-wall carbon nanotubes, *Chem. Phys. Lett.* 407 (**2005**) 68–72.

101 W. S. Xia, C. H. Huang, L. B. Gan, H. Li, X. S. Zhao, Photoelectric response of a bilayer lipid membrane doped with an azo pyridinium compound containing a rare-earth-metal complex, *J. Chem. Soc., Faraday Trans.* 92 (**1996**) 769–772.

102 K. Uosaki, T. Kondo, X. Q. Zhang, M. Yanagida, Very efficient visible light-induced uphill electron transfer at a self-assembled monolayer with a porphyrin-ferrocene-thiol linked molecule, *J. Am. Chem. Soc.* 119 (**1997**) 8367–8368.

103 H. Fujiwara, Y. Yonezawa, Photoelectric response of a black lipid-membrane containing an amphiphilic azobenzene derivative, *Nature* 351 (**1991**) 724–726.

104 J. Q. Liu, A. Chou, W. Rahmat, M. N. Paddon-Row, J. J. Gooding, Achieving direct electrical connection to glucose oxidase using aligned single walled carbon nanotube arrays, *Electroanalysis* 17 (**2005**) 38–46.

105 F. Patolsky, Y. Weizmann, I. Willner, Long-range electrical contacting of redox enzymes by SWCNT connectors, *Angew. Chem.* 43 (**2004**) 2113–2117.

106 G. Wang, J. J. Xu, H. Y. Chen, Interfacing cytochrome *c* to electrodes with a DNA – carbon nanotube composite film, *Electrochem. Commun.* 4 (**2002**) 506–509.

107 G. C. Zhao, L. Zhang, X. W. Wei, Z. S. Yang, Myoglobin on multi-walled carbon nanotubes modified electrode: Direct electrochemistry and electrocatalysis, *Electrochem. Commun.* 5 (**2003**) 825–829.

108 L. Wang, J. X. Wang, F. M. Zhou, Direct electrochemistry of catalase at a gold electrode modified with single-wall carbon nanotubes, *Electroanalysis* 16 (**2004**) 627–632.

109 Y. M. Yan, W. Zheng, M. N. Zhang, L. Wang, L. Su, L. Q. Mao, Bioelectrochemically functional nanohybrids through co-assembling of proteins and surfactants onto carbon nanotubes: Facilitated electron transfer of assembled proteins with enhanced faradic response, *Langmuir* 21 (**2005**) 6560–6566.

110 R. J. Chen, Y. G. Zhang, D. W. Wang, H. J. Dai, Noncovalent sidewall functionalization of single-walled carbon nanotubes for protein immobilization, *J. Am. Chem. Soc.* 123 (**2001**) 3838–3839.

111 S. E. Kooi, U. Schlecht, U. Burghard, K. Kern, Electrochemical modification of single carbon nanotubes, *Angew. Chem. Int. Ed.* 41 (**2002**) 1353–1355.

112 H. C. Choi, M. Shim, S. Bangsaruntip, H. J. Dai, Spontaneous reduction of metal ions on the sidewalls of carbon nanotubes, *J. Am. Chem. Soc.* 124 (**2002**) 9058–9059.

113 J. Chen, H. Liu, W. A. Weimer, M. D. Halls, D. H. Waldeck, G. C. Walker, Noncovalent engineering of carbon nanotube surfaces by rigid, functional conjugated polymers, *J. Am. Chem. Soc.* 124 (**2002**) 9034–9035.

114 S. Hermans, J. Sloan, D. S. Shephard, B. F. G. Johnson, M. L. H. Green, Bimetallic nanoparticles aligned at the tips of carbon nanotubes, *Chem. Commun.* 3 (**2002**) 276–277.

115 C. Downs, J. Nugent, P. M. Ajayan, D. J. Duquette, K. S. V. Santhanam, Efficient polymerization of aniline at carbon nanotube electrodes, *Adv. Mater.* 11 (**1999**) 1028–1031.

116 J. H. Chen, Z. P. Huang, D. Z. Wang, S. X. Yang, W. Z. Li, J. G. Wen, Z. F. Ren, Electrochemical synthesis of polypyrrole films over each of well-aligned carbon nanotubes, *Synth. Met.* 125 (**2002**) 289–294.

117 L. Dunsch, P. Janda, K. Mukhopadhyay, H. Shinohara, Electrochemical metal deposition on carbon nanotubes, *New Diam. Front. Carbon Technol.* 11 (**2001**) 427–435.

118 C. N. R. Rao, B. C. Satishkumar, A. Govindaraj, M. Nath, Nanotubes, *ChemPhysChem* 2 (**2001**) 78–105.

119 F. H. Wu, G. C. Zhao, X. W. Wei, Electrocatalytic oxidation of nitric oxide at multi-walled carbon nanotubes modified electrode, *Electrochem. Commun.* 4 (**2002**) 690–694.

120 V. Georgakilas, V. Tzitzios, D. Gournis, D. Petridis, Attachment of magnetic nanoparticles on carbon nanotubes and their soluble derivatives, *Chem. Mater.* 17 (**2005**) 1613–1617.

121 J. Li, H. T. Ng, A. Cassell, W. Fan, H. Chen, Q. Ye, J. Koehne, J. Han, M. Meyyappan, Carbon nanotube nanoelectrode array for ultrasensitive DNA detection, *Nano Lett.* 3 (**2003**) 597–602.

122 M. L. Guo, J. H. Chen, L. H. Nie, S. Z. Yao, Electrostatic assembly of calf thymus DNA on multi-walled carbon nanotube modified gold electrode and its interaction with chlorpromazine hydrochloride, *Electrochim. Acta* 49 (**2004**) 2637–2643.

123 M. L. Guo, J. H. Chen, D. Y. Liu, L. H. Nie, S. Z. Yao, Electrochemical characteristics of the immobilization of calf thymus DNA molecules on multi-walled carbon nanotubes, *Bioelectrochemistry* 62 (**2004**) 29–35.

124 T. Ramanathan, F. T. Fisher, R. S. Ruoff, L. C. Brinson, Amino-functionalized carbon nanotubes for binding to polymers and biological systems, *Chem. Mater.*, 17 (**2005**) 1290–1295.

125 J. Wang, G. Liu, M. R. Jan, Q. Zhu, Electrochemical detection of DNA hybridization based on carbon-nanotubes loaded with CdS tags, *Electrochem. Commun.* 5 (**2003**) 1000–1004.

126 J. Koehne, J. Li, A. M. Cassell, H. Chen, Q. Ye, H. T. Ng, J. Han, M. Meyyappan, The fabrication and electrochemical characterization of carbon nanotube nanoelectrode arrays, *J. Mater. Chem.* 14 (**2004**) 676–684.

127 J. Li, J. E. Koehne, A. M. Cassell, H. Chen, H. T. Ng, Q. Ye, W. Fan, J. Han, M. Meyyappan, Inlaid multi-walled carbon nanotube nanoelectrode arrays for electroanalysis, *Electroanalysis* 17 (**2005**) 15–27.

3
Nanotubes, Nanowires, and Nanocantilevers in Biosensor Development

Jun Wang, Guodong Liu, and Yuehe Lin

3.1
Introduction

Recent developments in designing and synthesizing conducting one-dimensional (1D) nanostructured materials, including carbon nanotubes (CNTs) and nanowires, etc., have attracted much attention across scientific and engineering disciplines because of their great potential for replacing conventional bulk materials in micro- and nanoelectronic devices [1, 2] and in chemical [3, 4] and biological sensors [4–9]. Various methodologies and technologies have been developed to fabricate 1D conducting nanomaterials [10, 11]. For example, chemical methods, including catalytic vapor deposition (CVD) have been widely used to synthesize CNTs [12, 13] and silicon nanowires [14, 15] on catalytically patterned substrates at desired sites with controlled orientations. Microfabrication [10, 11], soft lithography techniques [10, 11], and electric fields [16], etc. have been employed to fabricate highly oriented 1D nanomaterials. Various electrochemical techniques have been developed to fabricate 1D conducting polymer nanowires [17, 18]. New techniques are being explored to synthesize new 1D nanomaterials.

One-dimensional nanostructured materials as building blocks for biosensors are promising because of their unique electronic, optical, chemical, and mechanical properties, which are intrinsically associated with their low dimensionality and the quantum confinement effect. Therefore, 1D nanomaterials have broad applications in developing various types of biosensors, e.g., electrical [19, 20], optical [21], and mechanical [22] (nanocantilever biosensors). For example, CNTs exhibit excellent electronic properties. They can be used as molecular wires for facilitating electron transfer on the surface [23]. One-dimensional nanomaterials have a high aspect ratio, which makes them exhibit extreme sensitivity and superior response. Owning to their high sensitivity, 1D nanomaterials can be used to construct label-free biosensors. This will be very attractive and avoid exhausting and complicated labeling. With the ever-decreasing sizes of these 1D nanostructures, the "bottom-up" chemical approach is playing an increasingly important role because of its capability to make much smaller features compared to the "top-down" approach. So, combined with bottom-up techniques, 1D nanomaterials are ideal building blocks for

Nanotechnologies for the Life Sciences Vol. 8
Nanomaterials for Biosensors. Edited by Challa S. S. R. Kumar

ISBN: 978-3-527-31388-4

constructing miniature biochips. The detection mechanism with these tiny nanosensors is based on chemical interactions between the surface atoms of 1D nanostructured materials and adsorbed molecules. This interaction will provide a direct electronic readout within a few seconds of electron-donating or -withdrawing molecules adsorbing onto the nanomaterials. A miniature biochip based on 1D nanomaterials will be implanted in the body and will detect multiple biological molecules *in vivo*! In general, these 1D nanomaterials are expected to play an important role in the development of various emerging technologies that will improve the way we live.

Biosensors based on 1D nanomaterials have shown great advantages. However, real applications in biological diagnosis are a long way off. A major challenge remains to fully exploit the 1D nanostructure with one lateral dimension between 1 and 100 nm. For example, 1D nanostructured material-based biosensors need to bind biological recognition molecules onto the device. It is a great challenge to individually address high-density biomolecule nanoarrays. There is also a need for deconvolution of noise from the signals. To analyze proteomic signatures, a major challenge will be to identify signatures from low-concentration molecular species in the presence of an extremely high concentration of non-specific proteins.

As a branch of nanotechnology, 1D nanomaterial-based biosensor development has made great progress. Each year, thousands of articles on nano-related biosensors are published, and many reviews have appeared in different journals [24, 25]. The present chapter introduces reviews on biosensor development based on 1D nanomaterials, CNTs, semiconducting nanowires, and some cantilevers. The chapter is comprehensive – previous reviews on nanomaterials-based biosensor development have focused on one of 1D nanomaterials, e.g., either carbon nanotubes or nanowires. The emphasis here, however, is on CNTs and electrochemical/electronic biosensor developments. Section 3.2 gives a detailed description of carbon nanotubes-based biosensor development, from fabrication of carbon nanotubes, the strategies for construction of carbon nanotube based biosensors to their bioapplications. In the section on the applications of CNTs based biosensors, various detection principles, e.g., electrochemical, electronic, and optical method, and their applications are reviewed in detail. Section 3.3 introduces the method for synthesis of semiconducting nanowires, e.g., silicon nanowires, conducting polymer nanowires and metal oxide nanowires and their applications in DNA and proteins sensing. Section 3.4 simply describes the development for nanocantilever-based biosensors and their application in DNA and protein diagnosis. Each section starts with a brief introduction and then goes into details. Finally, Section 3.5 summarizes the development of 1D nanomaterials based biosensors.

3.2 Carbon Nanotubes in Biosensor Development

CNTs, rolled graphene sheets, were discovered in 1991 by Iijima [26] following the historical finding of the new fascinating member of the carbon family –

"Buckyball" fullerene (carbon nanocage) and other nanocarbon particles [27]. CNTs have basic sp^2 carbon units that comprise a seamless structure with hexagonal honeycomb lattices, being several nanometers in diameter and up to hundreds of microns long. CNTs can be divided into two major groups, i.e., single-wall CNTs (SWCNTs) and multiwall CNTs (MWCNTs). SWCNTs represent a single graphite sheet rolled flawlessly, demonstrating a tube diameter of 1 to 2 nm, whereas MWCNTs show concentric and closed graphite tubules with diameters ranging from 2 to 50 nm and an interlayer distance approximately 0.34 nm [28]. Typical SWCNTs have an open-ended nanostraw or a capped nanohorn tubular structure. Because of the highly oriented architectures, these novel nanostructures exhibit physicochemical properties different from those of bulk graphite and diamond and thus provide their unique electronic, chemical, thermal, and mechanical properties [29, 30]. Since the early 1990s, CNT science has been one of the fastest growing areas of research in chemistry, physics, materials, and life technologies. The important properties and possible potential applications of CNTs have been reviewed recently [31–35].

The first type of CNT-based sensor was prepared by using MWCNTs mixed with bromoform as a binder packed into a glass capillary. This modified electrode exhibited remarkable improvement with regard to the electrochemical oxidation of dopamine [36]. Since then, work has concentrated on their electrocatalytic performance towards the redox behaviors of biomolecules [37, 38], especially towards the fabrication of effective, prototype deoxyribonucleic acid (DNA) and glucose biosensors [39, 40]. Recent summaries of the preparation conditions, interferences, interfacing, comparison, or analytical promise of the CNT-based sensors can be found in more specific reviews [41–43].

3.2.1 Preparation and Purification of CNTs

CNTs synthesis has mainly involved three major methods: the carbon arc-discharge method, or electric arc discharge (EAD) [26, 44], the laser vaporization of a graphite electrode or laser ablation (LA) [45], and a chemical method, CVD [11, 12]. EAD uses a direct current arc between carbon electrodes within a noble gas, like argon or helium [26]. In CVD, the CNTs are formed by the decomposition of the gaseous hydrocarbon at 700–900 °C and atmospheric pressure [46]. Materials produced by the EAD and LA protocols are in the forms of porous membranes and powders that require further processing. CNTs can be grown directly on substrates by the CVD process. Among these three techniques, the CVD is the most promising synthesis route for economically producing large quantities of CNTs. This is because the catalyst-involved CVD can use a lower temperature to form CNTs than the other two techniques. In addition, the catalyst can be deposited on a substrate, which allows for ordered synthesis and the formation of novel structures. CNT-based research for sensor applications only gained momentum after these highly oriented, large-scale productions emerged.

MWCNTs were first made by Yacaman et al. [47] followed by others [12, 13, 45]. Their experimental set-up usually consists of a high-temperature oven in which the catalysts are placed onto a highly resistant ceramic or metal plate. The nature and yield of the deposit obtained in the reaction are controlled by varying different parameters, such as the nature of the metals and the supports, the hydrocarbon sources, the gas flows, the reaction temperature, and the reaction time. By selecting the proper conditions, both the physical (e.g., length, shape, diameter) and chemical (e.g., number of defects, graphitization) properties of MWCNTs can be designed in advance. The choice of catalysts is vital in growing a good quality of CNTs and has been a subject for several research groups. Supported Co, Ni, and Fe catalysts were found to be the most active in the CVD growth thus far, although the metals and supports demand different temperature ranges [48, 49].

The first SWCNTs were reported by Iijima and Ichihashi when employing the EAD set-up with a low product yield [50]. This synthesis was significantly improved in 1996 when Dai et al. demonstrated that LA can be an effective way to grow highly uniform tubes that have a greater tendency to form aligned bundles than those prepared using EAD [45]. These ordered CNTs were fabricated by laser vaporization of a carbon target in a furnace at 1100 to 1200 °C, which was a much lower temperature than that was previously thought necessary for nanotube formation. A Co-Ni catalyst assists the growth of the nanotubes, presumably because it prevents the ends form being "capped" during synthesis. In a later work, the same group showed that high quality SWCNTs could be produced by CVD decomposition of methane on supported transition metal oxide catalysts [51]. The experimental set-up was similar to apparatus generally producing multiwalled nanotubes; however, some factors, including the catalyst composition, the support, and the hydrocarbon, were different. Fe_2O_3 was found to be significantly more efficient in SWCNT production than CoO or NiO. Methane was used instead of the generally applied acetylene or ethylene because of its kinetic stability at high temperatures. Since there is no pyrolytic decomposition, the carbon atoms needed for the growth of nanotubes are produced by a catalytic reaction from the methane on the metal surfaces. The reaction time was dramatically reduced from the usually applied hour(s) to 10 min, and thus it prevents the outer surface of the nanotubes from being coated with amorphous carbon. In the following year, Colomer et al. proved, after viewing the nanotube bundles by transmission electron microscopy, that the best yield of SWCNTs is obtained by using a Fe-Co binary mixture supported by alumina [52]. Other groups have explored the possibility of extending their successful approach for MWCNTs into SWCNTs growth, mainly focusing on CVD [12, 13, 46–49].

High-purity CNTs can be, theoretically, achieved by optimizing the synthetic routes, and this should be viewed as part of the overall performance of the proposed preparation protocol. Nevertheless, the as-synthesized CNT materials unavoidably contain significant amounts of impurities, including amorphous carbon, graphite particles, and metal catalysts. The purification schemes that have been developed usually take advantage of differences in the aspect ratio [53] and oxidation rate [54] between the nanotubes and the impurities. Zhou et al. proposed a method that can reach final product purity over 95% [55]. The protocol combines hydrogen peroxide reflux with filtration that can effectively remove most of the impurities. Recently, Chiang et al. modified a gas-phase purification technique previously reported by Smalley and others [53] that uses a combination of high-temperature oxidations and repeated extractions with hydrochloric acid. This improved procedure significantly reduces the amount of impurities (catalyst and non-nanotube forms of carbon) within the nanotubes, increasing their stability remarkably. The onset of decomposition of the purified nanotubes (determined by thermal gravimetric analysis in air) is more than 300 °C higher than that of the crude nanotubes. Transmission electron microscopy analysis of the purified nanotubes reveals near complete removal of iron catalyst particles. The iron content of the nanotubes was reduced from 22.7 wt.% in the crude nanotubes to less than 0.02 wt.% in the final product. Nanotubes purified by this method can be readily dispersed in common organic solvents, in particular *N,N*-dimethylformamide, using prolonged ultrasonic treatment. These dispersions can then be used to incorporate single-wall CNTs into polymer films.

3.2.2
Construction of CNT-based Biosensors

Interest in exploring CNTs in biosensor fabrication has grown exponentially since the inception of CNTs in 1991 [26] and the first CNT-based sensor report [36]. Following the preparation and purification of a large quantity of CNTs as discussed above, the immobilization of these sensing layers onto the transducer support is introduced below. Basically, these CNT immobilizations have been based on dispersion, solubilization, adsorption, functionalization, composite entrapment, and other surface anchoring protocols.

3.2.2.1 Dispersion and Stabilization by Oxidative Acids

The well-ordered, all-carbon hollow CNTs possess unique walls and ends and thus resemble the sensing properties of basal planes of pyrolytic graphite (through their walls) and of edge planes of pyrolytic graphite (through their open ends). This sensing mechanism, in addition to their high surface-to-volume ratio, high chemical and thermal stability, high tensile strength, and elastic nanostructures, has made them excellent candidates for sensor fabrication. However, the spontaneous coagulation and the lack of solubility of CNTs in aqueous media is a major challenge for their application. To prevent coagulation of the as-synthesized CNTs, oxidative acid treatments are usually explored, including refluxing and sonication in a concentrated mixture containing sulfuric acid and nitric acid, as reported by Smalley's group [56]. This procedure, while occasionally generating surface defects and tube shortening, can produce abundant carboxylated sites on the CNT walls and caps. A dark stable aqueous suspension of CNTs can be achieved after removing access acids. More recently, Mallouk's group have reported that a stable SWCNT aqueous dispersion at concentrations above 0.3% can be obtained as a result of the hydrogel formation. This follows the treatments by a mixture containing $H_2SO_4 + (NH_4)_2S_2O_8 + P_2O_5$ and a subsequent mixture with $H_2SO_4 + KMnO_4$ [57]. Similar approaches have also been reported for a stable aqueous MWCNT hydrogel formation [58].

3.2.2.2 Dispersion by Surfactant Interaction

To preserve the intact CNT nanostructures after dispersion, a noncovalent stabilization/immobilization might be attractive. Simple physical stabilization, such as centrifugation, filtration, distillation, and sonication, in coupling with simple immobilization, including evaporation, casting, or spin coating, can be directly applied for respective sensor fabrications [35, 36]. Nevertheless, the hydrophobicity of the CNT walls, which accounts for vast majority of the tubes, is a major barrier when it comes to dispersing and manipulating the sensor surface and anchoring in a controlled manner.

Considering the relatively hydrophilic caps of CNTs, the noncovalent surfactant- and polymer-assisted aqueous dispersion may offer an alternative to overcome the drawbacks of the simple physical stabilization [59, 60]. The systematic study of the SWCNT dispersion in various surfactants has been reported and tabulated by Sun et al. [41], with the CNT solubilities ranging 10–50 mg mL^{-1}.

3.2.2.3 Polymer-assisted Solubilization

Because of the pseudo-amphiphilic feature of CNTs, due to their hydrophilic carboxylated ends and hydrophobic sidewalls, various ionic and nonionic polymers, such as poly(*p*-phenylenevinylene) [59], poly{(m-phenylenevinylene)-*co*-[2,5-dioctyloxy-(*p*-phenylene)-vinylene]}, [60] or poly(ethylene oxide)-poly(propylene oxide)-poly(ethylene oxide) triblock polymers, have been reported to "wrap" CNT in polymeric chains to facilitate dispersing and to stabilize the nanotubes, without impairing their physical properties [59, 60]. Wang et al. developed a method that directly applied Nafion polymer-assisted CNT dispersion in sensor fabrication [61]. Similar to other polymers used to wrap and solubilize CNTs, Nafion bears a polar side chain and can produce a CNT suspension in phosphate buffer or alcohol solution. Increasing the Nafion content from 0.1 to 5 weight percent (wt.%) results in dramatic enhancement of the solubility of both single-wall and multiwall CNTs (Fig. 3.1). A homogeneous solution of the Nafion/CNT complex is observed in Nafion solution, but no such solubilization is observed in ethanol or phosphate-buffer solutions containing no Nafion. The CNT/Nafion association does not impair the electrocatalytic properties of CNTs with respect to the redox reaction of hydrogen peroxide. The Nafion-induced solubilization of CNT thus permits various applications, including the modification of electrode surfaces for preparing amperometric biosensors, and field-effect transistors [62].

3.2.2.4 CNT Adsorption on the Transducer Substrate

Although the dispersion and distribution of CNTs in aqueous media have proved to be challenging, some non-polar organic solvents such as *N,N*-dimethylformamide (DMF) cause less coagulation of the tubes and thus permit a greater extent of dis-

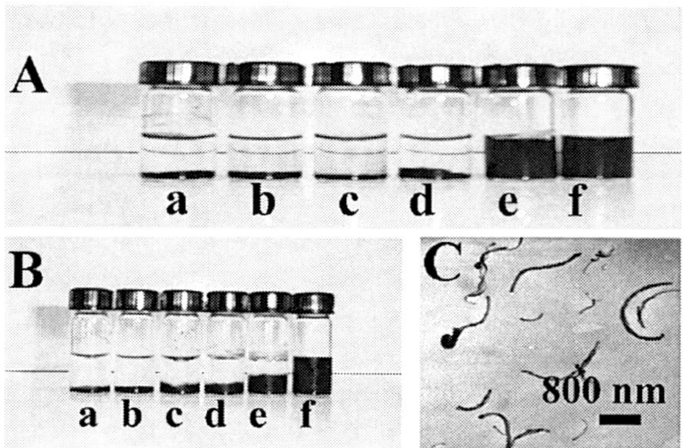

Fig. 3.1. Photographs of vials containing 0.5 mg mL^{-1} SWCNT (A) and MWCNT (B) in different solutions: phosphate buffer (0.05 M, pH 7.4) (a), 98% ethanol (b), 10% ethanol in phosphate buffer (c), 0.1% Nafion in phosphate buffer (d), 0.5% Nafion in phosphate buffer (e), and 5% Nafion in ethanol (f). (C) TEM image of a 0.5% Nafion solution containing MWCNT (0.3 mg mL^{-1}). (Reprinted with permission from Ref. [61]. © 2003 American Chemical Society.)

persion. This organic solubility offers the possibility for directly coating or spin casting of the CNT organic solution onto the sensor substrate and subsequent solvent evaporation. Re-immersion of these resultant CNT-based sensors into aqueous media showed no loss of operational performance and thus provides proof of strong adsorption on the surface. The first CNT-based sensor by Britto and co-authors was based on nanotubes solubilized in bromoform as a binder material following packing into a glass tube to complete the sensor construction [36]. Presently, many sensors still use this approach because of the ease of fabrication. The most often used substrates are glassy carbon, gold, platinum, carbon fiber, and glass. Lately, a protocol derived to co-incorporate some recognition reagent has been reported [63, 64]. Although these protocols are almost the simplest and most convenient ways to fabricate CNT-based sensors, the non-specific CNT adsorption needs to be addressed to gain greater control over their random distribution if a highly reproducible and stable sensor is demanded.

3.2.2.5 **Surface Functionalization of CNTs**

Biosensors need specific surface recognition towards targets, and have thus promoted the modification of CNTs. These modifications usually start with the CNTs' sidewalls, ends, and defects, which are rich in nanotube-bound carboxylic groups. The latter are the nondestructive outcome of oxidative acid pretreatment on CNTs. The external-added functional molecules can be as small as simple amino acids or as large as protein macromolecules. Linkages between the nanotubes and the functional components, with or without coupling agents, are based on carboxylate chemistry via amidation and esterification, as well as ionic interaction schemes.

Accordingly, these functionalizations can be covalent or noncovalent bonding in nature.

Liu et al. have used dicyclohexylcarbodiimide (DCC), a coupling reagent which converts the carboxylate ends of the SWCNTs into carbodiimide leaving groups, to react with amines on cysteamine ($NH_2CH_2CH_2SH$) [65]. The single-walled nanotubes with free thiol terminal groups then covalently attached onto substrate gold surface through a self-assembly process. Although atomic force microscopy (AFM) and transmission electron microscopy (TEM) have revealed different surface morphology of the resultant modified surfaces [66], this functionalization protocol offers control over the spatial distribution, lengths, and other surface patterns of the nanotubes aligned on the substrate by adjusting the assembled amount and time. Gooding and coauthors employed a similar carbodiimide-activated conjugating method in immobilization of microperoxidase (MP-11) onto the perpendicularly aligned nanotubes that were pre-anchored on the cysteamine-modified gold electrode [66]. Other direct bonding or electrostatic complexing approaches include using such as bridging metal ions to connect a polyelectrolyte-modified surface and the carboxylic acid terminated tubes [67] and complexing with an oppositely charged polyelectrolyte [66]. While these approaches provide various patterned nanotubes, a major concern lies in that the aligned tubes have little support and, therefore, the electrodes may lack robustness [67].

To elucidate the covalent, electrostatic, and nonspecific contributions to protein–SWCNT interaction, Davis et al. carried out the experiments of amidation both in the presence and absence of the coupling reagent DCC [69, 70]. They discovered that glucose oxidase is adsorbed along the length of CNTs randomly distributed on a glassy carbon electrode. Though coupling can be controlled, to a degree, through variation of tube oxidative pre-activation chemistry, careful control experiments and observations made by AFM suggest that immobilization is strong, physical, and does not require covalent bonding. Figures 3.2 and 3.3(a) exhibit their proposed protein–nanotube conjugates, which were readily characterized at the molecular level by AFM. Ferrocene monocarboxylic acid behaves as a mediator to promote charge transfer communication between electrode surface and the enzyme molecules. Under such conditions, the glucose signal was $10\times$ greater than if only glucose oxidase was adsorbed onto the glassy carbon electrode without CNTs (Fig. 3.3b). This approach demonstrated the possible device application; protein attachment appears to occur with retention of the native biological structure. The role of nanotubes in this proposed glucose sensor was to provide (1) the high-aspect ratio electrode to which high capacity glucose oxidase loading was achieved, and thus greater signals were generated from more active enzyme interfacing and (2) provide possible direct electrical communication between a redox-active biomolecule and the delocalized π system of its CNT support [71].

3.2.2.6 Composite Entrapment and CNTs Bulky Electrode Material

The first nanotube composites were prepared by Ajayan and coauthors by mechanically mixing MWCNTs and epoxy resin [72]. Because CNTs themselves could be

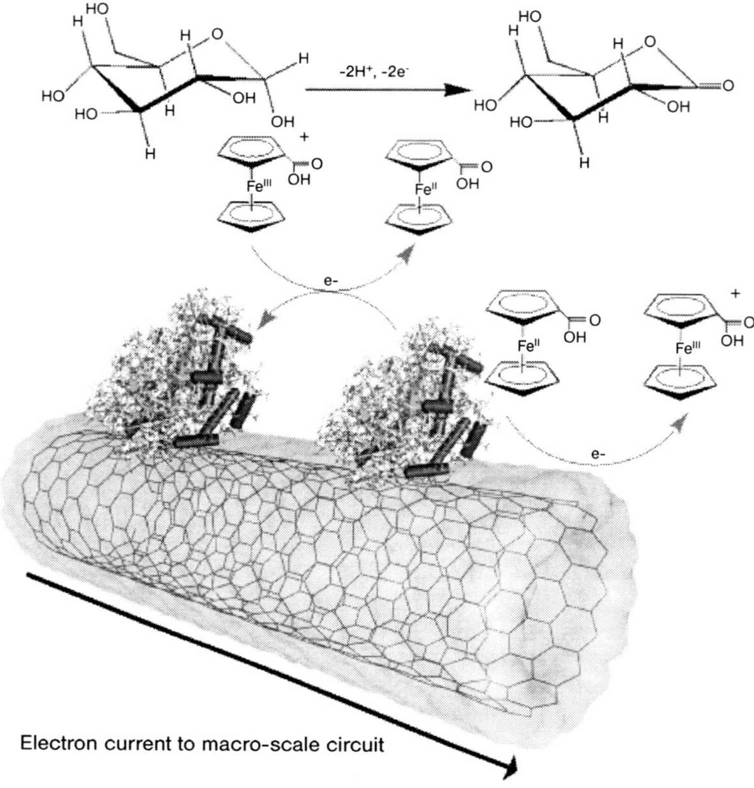

Fig. 3.2. Schematic representation of the "SWCNT Glucose Biosensor." Solution-phase D-glucopyranose is turned over by oxidase enzymes immobilized on the nanotubes. This redox process at the enzyme flavin moieties is "communicated" to the nanotube π system through the diffusive mediator ferrocene monocarboxylic acid. The redox action of the ferrocenes at the nanotube surface ultimately generates a quantifiable catalytic current that is characteristic of substrate detection and turnover. (Reprinted with permission from Ref. [70]. © 2003 Wiley-VCH.)

viewed as an extreme form of conducting polymer, the combination of nanotubes with conventional π-conjugated conducting polymers offers new electronic properties and high surface area capacity. This well suits the integration of substrate and transducer when making biosensors.

Wallace et al. combined MWCNTs and glucose oxidase to be embedded into polypyrrole (Ppy) with 0.1 M $NaClO_4$ as supporting electrolyte [73]. Such an enzyme electrode retains its stability at 70% after 3 days storage in the dry state at 4 °C. The use of these 3D electrodes offers advantages in that large accessible enzyme loadings can be obtained within an ultrathin layer. The iron-loaded nanotube tips (generated from CNT preparation) also contribute partial catalytic capacity toward H_2O_2 oxidation. A biosensor based on Ppy/DNA composite covered CNT under-

a)
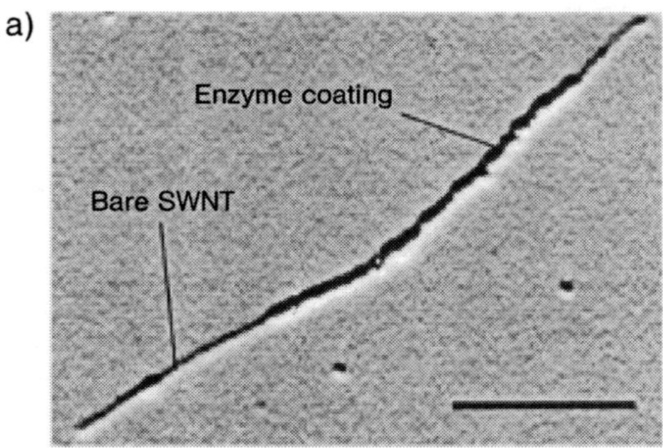

b)
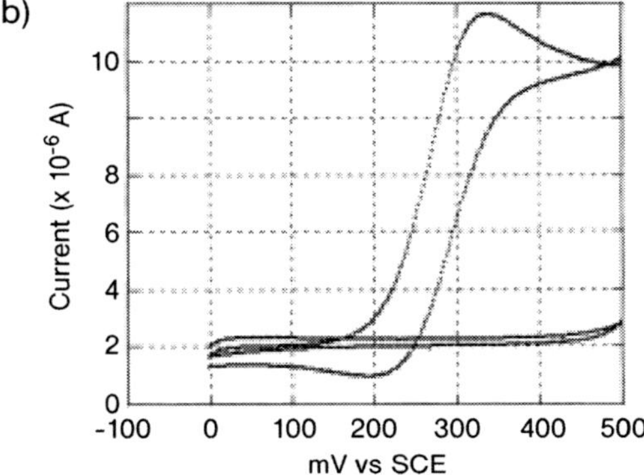

Fig. 3.3. (a) TMAFM amplitude micrograph of a GOX-modified SWCNT in which a high degree of enzyme loading is apparent. The scale bar is 200 nm. (b) Voltammetric response of such nanotubes in the absence (lower curves) and presence (upper curves) of the substrate, β-D-glucose. (TMAFM = tapping-mode atomic force microscopy). (Reprinted with permission from Ref. [70]. © 2003 Wiley-VCH.)

layers was recently provided for DNA sensing [74]. By applying an impedance technique to this two-layer-based sensor, the complementary DNA target can be detected down to 5×10^{-11} M.

Sol–gel chemistry involves the hydrolysis and condensation of suitable alkoxysilane precursors and has been widely employed for the preparation of inorganic materials (monolithic, hybrid, composites, and chromatographic stationary phase) suitable for various applications. Recently, Bachas and coworkers have applied a CNT sol–gel composite as an enzyme-friendly platform to develop biosensors [75]. Using L-amino acid oxidase as a model enzyme, the biosensors were made in aqueous sol–gel processes involving methyltrimethoxysilane, ethyltrimethoxysilane and propyltrimethoxysilane as precursors. Aliquot amounts of MWCNTs and enzyme were added into the sol when the hydrolysis took place. The resultant sensor proved to be stable and retained more than 50% of its response after 1 month of testing. In such an immobilization protocol, the porous –Si–O– sol–gel network encapsulates biomolecules while the use of CNTs, as the conductive part of the composite, facilitated fast electron transfer rates.

Because of the unique sp^2 hybrid surface structures, CNTs themselves could be viewed as an extreme form of conducting polymer. The CNT bulky material may be applied directly towards sensor construction without binders or other auxiliary components. Wang and co-authors have developed a simple approach for preparing effective CNT-based biosensors from CNT/Teflon composite material by hand-mixing a certain amount of CNTs in the dry-state with granular Teflon to obtain a desired composition of CNT/Teflon [76]. Carbon composites, based on the dispersion of graphite powder within an insulator, offer convenient bulk modification for the preparation of reagentless and renewable biosensors. Wang et al.'s approach relies on CNTs as the sole conductive component rather than as the modifier cast on other electrode surfaces. The bulk of CNT/Teflon composites hence serve as a reservoir for the enzymes. By comparing the sensors' performance against their respective composite compositions, the CNT content of 40–60 wt.% has been suggested. CNT/Teflon composites combine advantages of CNTs and bulk composite electrodes that permit a wide range of applications without the need for a graphite surface. Certain amounts of enzymes (e.g., glucose oxidase and alcohol dehydrogenase) and cofactor (e.g., NAD^+) can be mixed with the CNT/Teflon composite and used as electrode materials, depending upon specific needs. These biosensor interfacings displayed a marked electrocatalytic action toward hydrogen peroxide and β-nicotinamide adenine dinucleotide (NADH) and, hence, this is promising for the development of biosensors for glucose (in connection with oxidase enzymes) and ethanol (in connection with dehydrogenase enzyme), respectively [76]. Similar approaches have been developed for CNT-based sensor fabrication using nanotube–mineral oil paste, reagent-embedded CNT paste, or powders to determine DNA, glucose, cysteine, and other biomolecules.

Composites consisting of CNTs and other nanotubes or nanoparticles were reported recently for enhancing the catalytic capacity of the sensing devices. Luong's group [77] and Yao's group [78] used platinum nanoparticles combining with CNTs to construct a biosensing platform for glucose oxidase. Wang et al. employed semiconductor CdS nanoparticle-tagged CNTs for DNA hybridization detection [79].

3.2.2.7 More Sophisticated Surface Tailoring Based on Combination of Co-adsorption, Integration, Prohibition, Spacing, Linkage, Sandwich, Tagging, and other Anchoring Approaches

Biosensors involve biomolecules and biorecognition reactions. Their optimal operational performance depends on the maximization of the desired signals while minimizing the side reaction. DNAs, enzymes, antigens, and other biomolecules usually bear charges, depending on the medium pH. CNT biomodifications based on non-specific interaction with DNAs or proteins can be achieved through the sidewall electrostatic interaction, hydrogen bonding, and other mechanisms, as well as the insertion of smaller biomolecules into the tubular channel. The nature of these noncovalent bondings are complicated and were proposed by Dai et al. as mainly the results of hydrophobic interaction and the π-stacking of the conjugated pyrenyl group of 1-pyrenebutanoic, succinimidyl ester, or coating with some surfactants, such as Triton [80, 81]. More specific binding to functionalize the CNTs can take advantage of covalent bonding, DNA hybridization, coupling agents, and antigen–antibody interactions. Biosensing by these approaches was reported for the CNTs attachment of decorated glucose oxidase [82], thiolated DNA [83], amine-terminated DNA [84], and peptide nucleic acid (PNA) – a DNA mimic [85].

The non-specific adsorption of proteins on nanotubes is not always desirable, especially when tested in real biofluid samples that contain many co-existing proteins. More sophisticated sensors, therefore, need to address issues like target-recognition enhancement, blockage of undesired interference (the co-existing proteins' non-specific adsorption on the nanotube surfaces), long-term storage, etc. Accordingly, CNT surface engineering might be a combination of different tailoring techniques. Dai et al. have presented a typical example [81] (Fig. 3.4). In their approach U1A, a protein involved in the splicing of message ribonucleic acid (mRNA), was covalently linked to Tween-20, a surfactant. The complex was then noncovalently cast onto the single-walled CNT surface. The latter was as-grown on a quartz wafer that was *in situ* monitored by its conductance and frequency response during the sensing measurement.

This CNT modification with the adsorption of biotinylated Tween-20 allowed streptavidin recognition by the specific biotin–streptavidin interaction, but provided resistance toward other protein adsorptions. Under such design, the sensor could detect the binding of 10E3, a specific antibody for recognition of U1A, at concentrations as low as 1 nM (Fig. 3.4), while showing no response toward other existing proteins such as streptavidin, avidin, bovine serum albumin (BSA), a-glucosidase, staphylococcal protein A (SpA), and immunoglobulin G (IgG) (Fig. 3.4C). The blocking mechanism toward coexisting proteins was proposed as the formation of a nearly uniform layer of surfactant Tween-20 through the favorable hydrophobic interaction on the nanotube surface, with the poly(ethylene glycol) (PEG) segments extending into the aqueous media to provide the observed protein resistance. Noticeably, this protein-resistant assembly may be covalently conjugated to specific antigens to allow sensitive detection of antibodies, or *vice versa* [81]. This proposed real-time immunosensor can thus compare favorably with a standard fluorescence-based assay with immobilized antigens on planar arrays.

Wang et al. have recently developed a CNT-based amplified bioelectronic protocol that uses DNA for linking particles to CNTs [63]. As can been seen in Fig. 3.5(a), the preparation is based on the sandwich hybridization (a) or antigen–antibody (b) binding along with magnetic separation of the analyte-linked magnetic-bead/CNT assembly (A), followed by enzymatic amplification (B), and chronopotentiometric stripping detection of the product at the CNT-modified electrode (C). TEM observations (Fig. 3.5D) indicate that the hybridization event leads to crosslinking of the alkaline phosphatase (ALP)-loaded CNTs and the magnetic beads (with the DNA duplex acting as "glue"). In this new bioaffinity assay, CNTs play a dual amplification role in both the recognition and transduction events, namely as carriers for

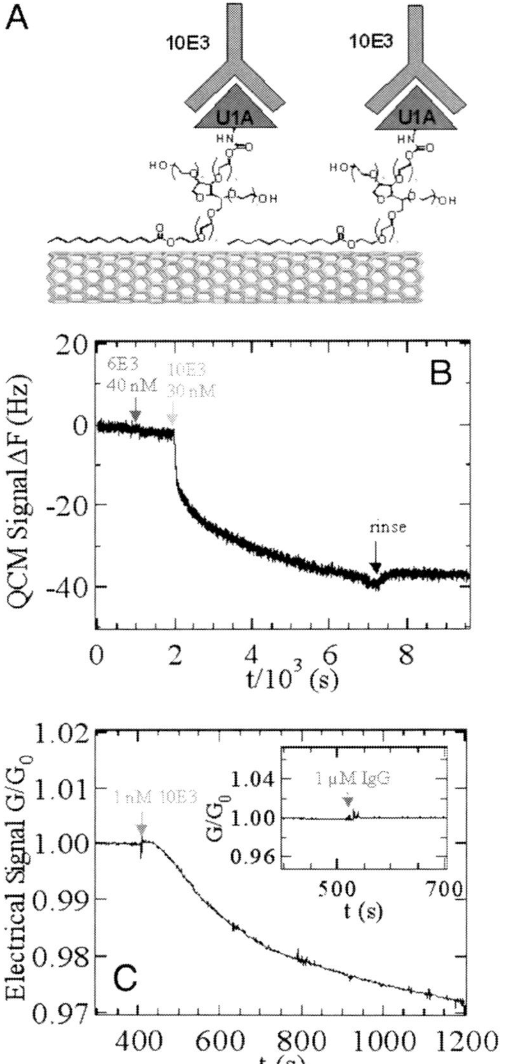

Fig. 3.4. Specific detection of mAbs binding to a recombinant human autoantigen. (A) Scheme for specific recognition of 10E3 mAb with a nanotube device coated with a U1A antigen–Tween conjugate. (B) QCM frequency shift vs. time curve showing selective detection of 10E3 while also showing rejection of the antibody 6E3, which recognizes the highly structurally related autoantigen TIAR. (C) Conductance vs. time curve of a device, revealing a specific response to ≤1 nM 10E3 while rejecting polyclonal IgG at a much greater concentration of 1 μM (inset). (Reprinted with permission from Ref. [81]. © 2003 of National Academy of Sciences, U.S.A.)

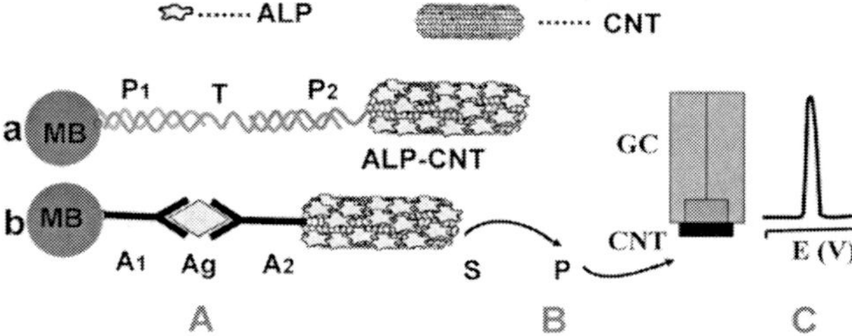

Fig. 3.5. Schematic representation of the analytical protocol: (A) Capture of the ALP-loaded CNT tags to the streptavidin-modified magnetic beads by a sandwich DNA hybridization (a) or Ab-Ag-Ab interaction (b). (B) Enzymatic reaction. (C) Electrochemical detection of the product of the enzymatic reaction at the CNT-modified glassy carbon electrode. (D) TEM image of the magnetic beads-DNA-CNT assembly produced following a 20-min hybridization with the 10 pg mL^{-1} target sample. Micrographs were taken with a Hitachi H7000 instrument operated at 75 kV after washing the DNA-linked CNT/particle assembly with autoclaved water, placing a 5-μL drop of the aggregate sample onto a carbon-coated copper grid (3 mm diameter, 200-mesh), and allowing it to dry. MB, Magnetic beads; P_1, DNA probe 1; T, DNA target; P_2, DNA probe 2; Ab_1, first antibody; Ag, antigen; Ab_2, secondary antibody; S and P, substrate and product, respectively, of the enzymatic reaction; GC, glassy carbon electrode; CNT, carbon nanotube layer. (Reprinted with permission from Ref. [63]. © 2003 American Chemical Society.)

numerous enzyme tags and for accumulating the product of the enzymatic reactions. With such an assembly, the extraordinarily low detection limits were reported for DNA and IgG of 1 and 500 fg mL^{-1}, respectively.

3.2.3
CNT-based Electrochemical Biosensors

Soluble CNTs have been electrochemically and quantum-chemically characterized for their bulk properties [86]. Results showed that the electronic states are not strongly affected when the nanotubes are functionalized. The electronic properties of CNTs range from metallic to semiconductive, depending on the nanotube's own diameter and chirality. These subtle electronic properties offer various electrochemical features for CNT-based sensors after functionalization of the nanotubes.

3.2.3.1 Direct Electrochemistry of Biomolecules on Carbon Nanotubes

Recently, direct electrochemical communications between redox-active macrobiomolecules and conventional electrode substrates mediated by CNTs have received attention because of their potential to lead a mechanistic study of the structure–function relationship of these biomolecules and their guidance toward biosensor design. Wang et al. reported the direct electrochemistry of cytochrome *c* at electrochemically activated SWCNT-modified electrodes [23]. Gooding et al. and Rusling et al. have used aligned nanotubes to promote the direct electrochemistry of redox-active proteins [66, 67]. These studies demonstrate that direct electrochemistry of redox-active biomacromolecules can be improved through the use of CNTs.

In addition to the aforementioned charge transfer promotion, there have been reports on the electrocatalytic behavior of CNTs toward some small biomolecules such as cysteine and hemocysteine [87], ascorbic acid [88], uric acid [89], dopamine [90]. The mechanisms of this surface mediation are not currently well understood, and the terminology of electrocatalysis should be exercised cautiously for different electrode material and media. The promoted electrochemical reactions for hydrogen peroxide, NADH, and quinones enable possible sensing schemes for more than 800 enzymes that involve these substrates, products, coenzymes, and cofactors.

Several hundred enzymatic reactions of NAD^+/NADH-dependent dehydrogenases have NADH as a cofactor. The electrochemical oxidation of NADH has thus been the subject of numerous studies related to the development of amperometric biosensors. Problems inherent to such anodic detection are the large overpotential encountered for NADH oxidation at ordinary electrodes and surface fouling associated with the accumulation of reaction products. CNTs have thus been examined in recent work [91] as the new electrode material to alleviate these problems.

Lin et al. have employed CNTs that had been pretreated with dispersion in concentrated sulfuric acid to cast a glassy carbon electrode [91]. Figure 3.6 shows a typical hydrodynamic voltammogram of 1×10^{-4} M NADH in a physiological medium (0.05 M phosphate solution, pH 7.4). This voltammogram demonstrates an electrocatalytic behavior of the CNT coating towards NADH with varying potentials, as evidenced by the MWCNT-coated electrode (B) responding to NADH over

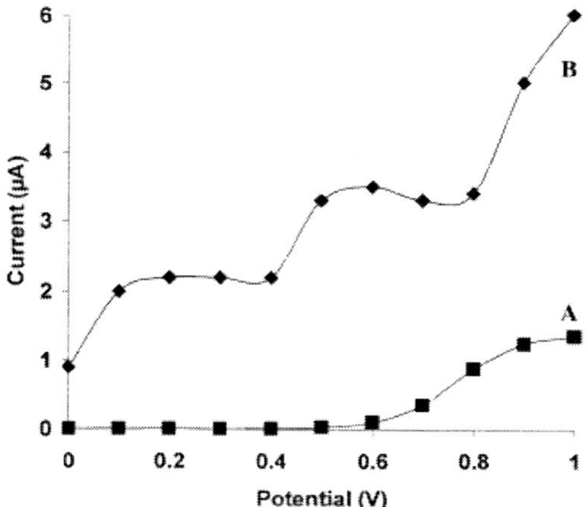

Fig. 3.6. Hydrodynamic voltammograms for 1×10^{-4} M NADH at unmodified (A) and MWCNT-modified (B) GC electrodes. Operating conditions: stirring rate, 500 rpm; electrolyte, phosphate buffer (0.05 M, pH 7.4). (Reprinted with permission from Ref. [91]. © 2002 Elsevier.)

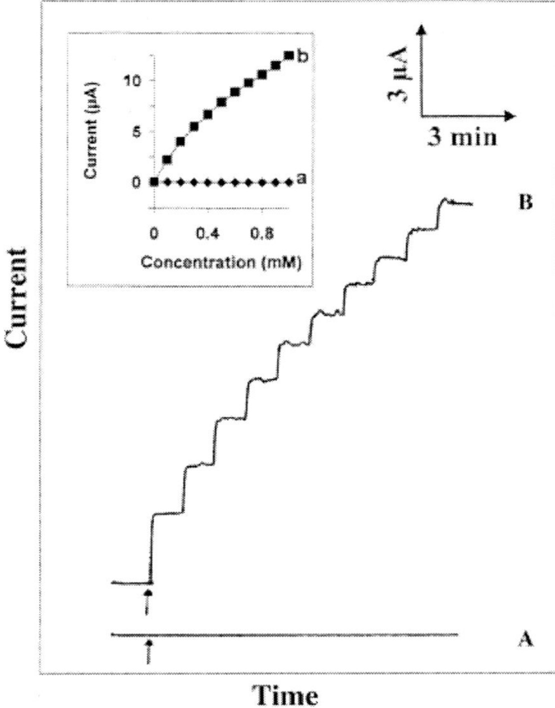

Fig. 3.7. Current–time recordings obtained after increasing the NADH concentration (by 1×10^{-4} M at each step) at unmodified (A) and MWCNT-modified (B) GC electrodes. Inset: the corresponding calibration curve. Operating conditions: potential, +0.3 V. (Reprinted with permission from Ref. [91]. © 2002 Elsevier.)

the entire 0.0–1.0 V range, whereas the bare electrode (A) responds only at potentials higher than +0.6 V. The modified electrode yields an approximately three-fold larger NADH peak than does the unmodified electrode. Figure 3.7 shows that successive additions of 1×10^{-4} M NADH result in increasing response detected at the CNT-modified electrode (B) but no response at the unmodified electrode (A) when the detection potential was kept low (i.e., 0.3 V). Evidently, the electrocatalytic action of CNT enables the fast response (i.e., 10 s to reach the steady state) to the change of NADH concentrations at the low-detection potential. The amperometric response of 5×10^{-3} M NADH appears to be very stable; the decay of the signal is less than 10% and 25% after a 60-min period at the MWCNT-modified and SWCNT-modified electrodes, compared with 75% and 53% at the graphite-coated and acid-treated electrodes, respectively. This shows the capability of CNTs to resist the fouling effects and prevent the diminishing of signals in successive cyclic voltammetric detections. The oxygen-rich groups on the CNT surface, introduced during the acid dispersion, are perhaps responsible for such electrocatalytic behavior for the oxidation of NADH. The resistance to fouling of CNT-based electrodes has yet to be understood.

The CNT-coating offers remarkably decreased overvoltage for the NADH oxidation as well as reducing the surface fouling effects of the electrodes. These characteristics indicate the great promise of CNTs for developing highly sensitive, low potential, and stable amperometric biosensors based on dehydrogenase enzymes.

3.2.3.2 Enzyme/CNTs Biosensors

Most reported CNT-based biosensors for glucose thus far involve the enzymatic reaction by glucose oxidase (GOx) [61, 69, 70, 76, 78]. Wang and Musameh have employed CNT/Teflon-based electrodes, which are immobilized with GO_x enzyme [76]. Figure 3.8 compares the amperometric response to successive additions of 2 mM

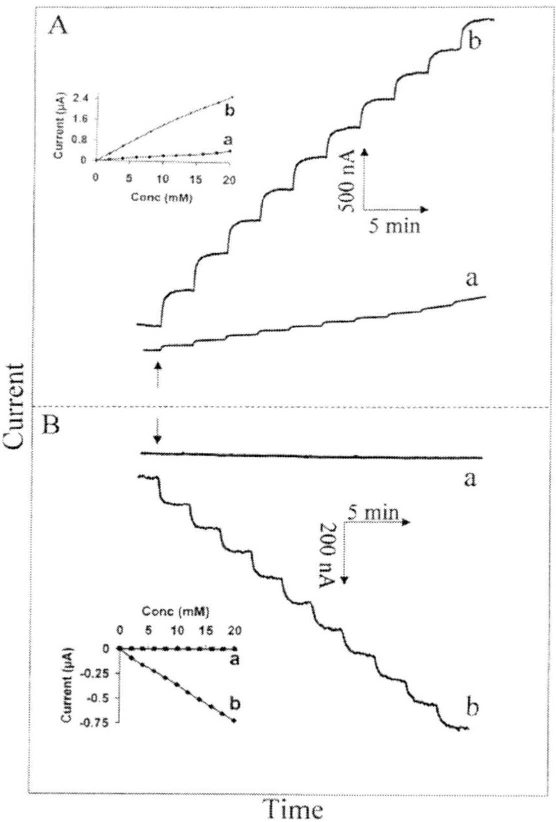

Fig. 3.8. Current–time recordings for successive 2-mM additions of glucose at graphite/Teflon/GO_x (a) and the MWCNT/Teflon/GO_x (b) electrodes measured at +0.6 (A) and +0.1 V (B). Electrode composition, 30:69:1 wt.% carbon/Teflon/GO_x. (Reprinted with permission from Ref. [76]. © 2003 American Chemical Society.)

glucose at the graphite/Teflon/GO_x (a) and the MWCNT/Teflon/GO_x (b) electrodes using operating potentials of +0.6 V (A) and +0.1 V (B). The CNT-based bioelectrode offers substantially larger signals, especially at low potential, reflecting the electrocatalytic activity of CNTs. Such low-potential operation of the CNT-based biosensor results in a highly linear response (over the entire 2–20 mM range) and a slower response time (~1 min vs. 25 s at +0.6 V). The glucose biocomposite based on single-wall CNTs results in a more sensitive but slower response than that based on multiwall CNTs. The low-potential detection also leads to high selectivity (i.e., effective discrimination against coexisting electroactive species). Despite the absence of external (permselective) coating, the glucose response at +0.1 V was not affected by adding the common acetaminophen and uric acid interferences at 0.2 mM [76]. A similar addition of ascorbic acid resulted in a large interference, reflecting the accelerated oxidation of this compound at the CNT surface [88].

Recently, Sheu et al. have reported the non-enzymatic detection of glucose [92]. This approach uses MWCNTs but needs a strong basic media that may not be compatible with the bioenvironment.

3.2.3.3 DNA and Protein Biosensors

Interest in the detection of DNA has grown aggressively because of its importance in the diagnosis and treatment of genetic disease, drug discovery, and anti-bioterrorism efforts. The completion of the Human Genome Project offers an abundance of gene mapping and screening. A hybridization recognition scheme, based on the Watson–Crick base pair principle, is the central point when constructing a DNA sensor. The unique electric, thermal, chemical, mechanical, and 3D spatial properties of CNTs make them a natural choice as transducers for hybridization-based DNA sensors. Different attachment protocols for DNA probes (single strand DNA molecules) onto CNTs were introduced, either by unmodified or surface-confined nanotubes [40, 63, 93].

Wang and coworkers have developed an ultrasensitive DNA biosensor based on a "dual amplification route" by using CNTs both as recognition sites and transducers, namely as carriers for numerous enzyme tags and for accumulating the product of the enzymatic reaction [63]. Figure 3.9 displays typical chronopotentiograms for extremely low target DNA concentrations (0.01 to 100 pg mL^{-1}; a–e). Well-defined α-naphthol signals are observed for these low DNA concentrations in connection with 20-min hybridization. The resulting plot of response vs. log[Target] (shown as inset) is linear and suitable for quantitative work. The favorable response of the 5 fg mL^{-1} DNA target (B) indicates a remarkably low detection limit of around 1 fg mL^{-1} (54 aM), i.e., 820 copies or 1.3 zmol in the 25-μL sample. Such a low detection limit compares favorably with the lowest values of 5 zmol (3000 copies) and 25 amol reported for electrical DNA detection. The smaller signal observed in a control experiment for a huge (~10^6) excess of a non-complementary oligonucleotide (Fig. 3.9C vs. B) reflects the high selectivity associated with the effective magnetic separation. The amplified electrical signal is coupled to a good reproducibility. Two series of six repetitive measurements of

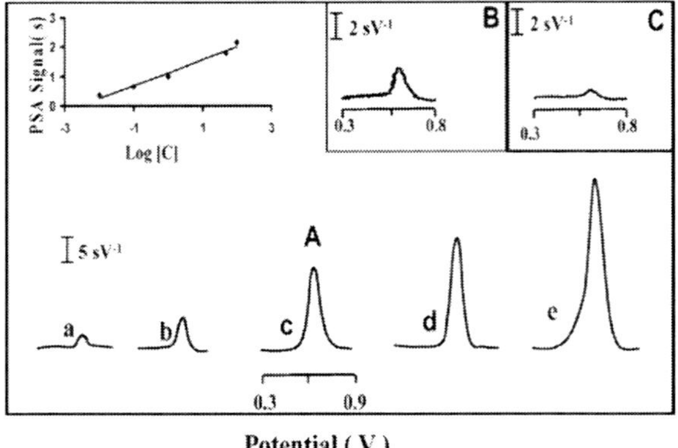

Fig. 3.9. Chronopotentiometric signals for increasing levels of the DNA target: (a) 0.01, (b) 0.1, (c) 1, (d) 50, (e) 100 pg mL^{-1}. Also shown (inset) is the resulting calibration plot (A), and the response for 5 fg mL^{-1} target DNA (B) and 10 ng mL^{-1} non-complementary (NC) oligonucleotide (C). Sample volume, 25 (B) and 50 μL (C). (Reprinted with permission from Ref. [63]. © 2003 American Chemical Society.)

1 pg mL^{-1} DNA target or 0.8 ng mL^{-1} IgG yielded reproducible signals with relative standard deviations of 5.6% and 8.9%, respectively [63].

Noticeably, Wang's "dual amplification" mode for the DNA sensor as above has also been applied to an immunosensor for IgG because of the similarity between DNA hybridization and antigen–antibody interaction. An extremely low detection for IgG was reported, as 500 fg mL^{-1} [63]. In contrast, DNA, RNA, proteins, and enzymes all bear multiple charges, and their adsorption onto the nanotubes is expected to change their electronic properties. These changes, upon adsorption, can be readily transduced into measurable sensing signals [38, 94]. DNA has already been used as a template to localize CNTs to make new building blocks or alignments in electronics and bioelectronics, such as field-effect transistors (FETs) [95]. These functionalized FETs, coupled with advanced sensor array techniques, can therefore serve as a new direction in CNT-based bioassays [77, 80].

3.2.3.4 **Immunosensors**

Electrochemical immunosensors based on CNTs have been designed [20]. Vertically aligned arrays of single CNTs called SWCNT forests have been developed for amperometric enzyme-linked immunoassays of proteins by Rusling and coworkers [20]. A prototype amperometric immunosensor was evaluated based on the adsorption of antibodies onto perpendicularly oriented assemblies of SWCNT forests. The forests were self-assembled from oxidatively shortened SWCNTs onto Nafion/iron oxide coated pyrolytic graphite electrodes. Anti-biotin antibodies strongly adsorbed to the SWCNT forests. They found that the detection limit for horseradish peroxidase (HRP) labeled biotin was 2.5 pmol mL^{-1} (2.5 nM) in the presence of a soluble mediator. Unlabeled biotin was detected in a competitive approach with a detection limit of 16 nmol mL^{-1} and a relative standard deviation of 12%. The immunosensor showed low non-specific adsorption of biotin-HRP (approx. 0.1%) when blocked with bovine serum albumin. The biosensor platform is also being developed to accommodate flow-through sensor design with direct electron transfer detection of the enzyme label on a CNT matrix. Traditional electrochemical immunosensors were based on mediated electron transfer. However, efficient direct electron transfer will offer some advantages, such as a simple possibility for a reagentless immunoassay. CNTs exhibit excellent electrical properties, and they are a suitable candidate for developing such a reagentless immunoassay.

3.2.4 Flow-injection Analysis

Flow detectors have been widely used in process chemistry and online monitoring. It is the core part for real time, *in situ* analysis and for the integration of separation and detection – "lab-on-a-chip." Microfabricated fluidic devices, particularly used as sensors for capillary electrophoresis (CE) and liquid chromatography (LC), have gained steadily growing attention in recent years [64]. Wang et al. have presented a CNT-based detector for a conventional as well as a miniaturized CE system [64]. CE combines the advantages of high performance, design flexibility, reagent economy, high throughput, miniaturization, and automation. It demands a detector with high sensitivity, inherent miniaturization (of both the detector and control instrumentation), and compatibility with advanced micromachining technologies, and low cost and power requirements [64]. Conventional detectors are based on gold, platinum, and various forms of carbon. A CNT-based electrode offers an alternative for low potential detection, it imparts enhanced sensitivity, and it leads to long-term stability. In this approach, Wang et al. used a Nafion/CNT-coated screen-printed electrode for end-column amperometric detection. After anodic pretreatment (3 min at +1.5 V in 1 M sulfuric acid), this sensor showed substantial electrocatalytic behavior and resistance to surface fouling toward hydrazine, phenol, tyrosine, purine, and amino acids, when compared with the bare surface. The CNT microchip detector also displayed well-defined concentration dependence.

Typical flow detection for glucose, based on the CNT/Nafion/GO_x modified gas chromatography (GC) electrode, can be seen in Fig. 3.10 (from the work by Wang's group) [61]. Figure 3.10 compares the amperometric responses for relevant physiological levels of glucose, ascorbic acid, acetaminophen, and uric acid at the CNT/Nafion/GO_x modified GC electrode (B) and Nafion/GO_x modified GC electrode (A). In Figure 3.10, the accelerated electron-transfer reaction of hydrogen peroxide at the CNT/Nafion/GO_x modified GC electrode allows for glucose measurements at very low potentials (i.e., −0.05 V) where interfering reactions are minimized. As a result, a well-defined glucose signal (d) is observed, while the signals of acetaminophen (a), uric acid (b), and ascorbic acid (c) are negligible. No such discrimination is obtained at the Nafion/GO_x biosensor (without the CNT) (A) held at +0.80 V, where large oxidation peaks are observed for all interferences, indicating that the permselective (charge exclusion) properties of Nafion are not adequate to

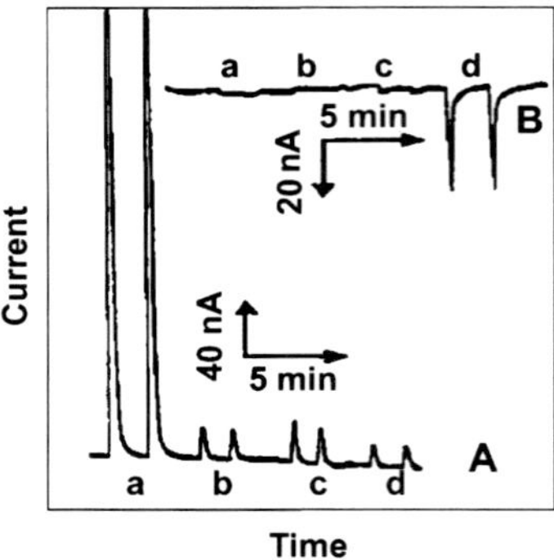

Fig. 3.10. Flow-injection signals for 2×10^{-4} M acetaminophen (a), 2×10^{-4} M ascorbic acid (b), 2×10^{-4} M uric acid (c), and 1×10^{-2} M glucose (d) at a Nafion/GO_x-modified GC electrode (A) at +0.8 V, and at a MWCNT/Nafion/GO_x-modified GC electrode (B) at −0.05 V, and flow rate of 1.25 mL min^{-1}. (Reprinted with permission from Ref. [61]. © 2003 American Chemical Society.)

fully eliminate anionic interferences. In short, the coupling of the permselective properties of Nafion with the electrocatalytic action of CNT allows for glucose detection with effective discrimination against both neutral and anionic redox constituents. Similarly, the CNT/Nafion-coated electrodes have also been demonstrated to dramatically improve the signal of dopamine in the presence of the common ascorbic acid interference [61].

Jin et al. used CNT-based sensors for LC detection [97, 98]. To overcome the aggregation of nanotubes in aqueous and common organic solution, nitric acid was used to treat CNTs to introduce carboxyl groups in the open ends of the MWCNTs. Such functionalized CNTs were cast onto the glassy carbon as the flow detector. Various biofluids containing neurotransmitters were tested using this protocol. The results from real samples such as plasma agreed with those from other methods.

In terms of the mechanism of the CNT-based flow sensor, Ghosh et al. recently found that the flow of a liquid on single-walled CNT bundles induces a voltage in the sample along the direction of the flow [98]. The magnitude of the voltage depended sensitively on the ionic conductivity and on the polar nature of the liquid. The nonlinear response of flow velocity and the voltage was attributed to a direct forcing of the free charge carriers in the nanotubes by the fluctuating Coulombic field of the liquid flowing past the nanotubes. Their work highlighted the potential of a CNT-based device as sensitive flow sensors and for energy conversion [98].

3.2.5
Carbon Nanotube Array-based Biosensors

Control over the orientation, distribution, and effective sensing sites of the nanotubes have been the subject of CNT-based biosensor research [56, 66]. Several approaches have been reported to achieve these controlled-density aligned CNTs and CNT arrays. The versatile approach is that of self-assembling aligned nanotube arrays by using oxidative acid treated single walled nanotubes [99]. These shortened tubes then react with DCC to introduce a carbodiimide leaving a group that allows reaction with thiols. Finally, these thiolated nanotubes are self-assembled on the gold surface to form aligned CNT arrays. Alternatively, aligned CNT arrays can be grown off a surface by using pyrolyzing CVD of relevant catalysts and carbon materials, followed by transferring the tubes onto a substrate support [73]. The third approach is to grow directly aligned CNTs onto an electrode surface by using CVDs or plasma, offering a controllable size and a given location for the catalyst spots that allows the growth of a given numbers of nanotubes [93].

Although vertically aligned CNTs have good material properties (e.g., good electrical conductivity, the capability to promote electron transfer reactions) and are of the right size (20 to 200 nm) for nanoelectrode arrays (NEAs), they lack the right spacing, having little support, and, therefore, the electrodes may lack robustness. To make each nanotube work as an individual nanoelectrode, the spacing needs to be sufficiently larger than the diameter of the nanotubes to prevent diffusion layer overlap from the neighboring electrodes [100]. Ren and co-authors recently developed a nonlithography method that allows the fabrication of low site density aligned CNT arrays with an interspacing of more than several micrometers [101]. Figure 3.11 demonstrates the manufacturing process of such aligned CNT arrays: Ni nanoparticles were randomly deposited on a 1-cm^2 Cr-coated silicon substrate

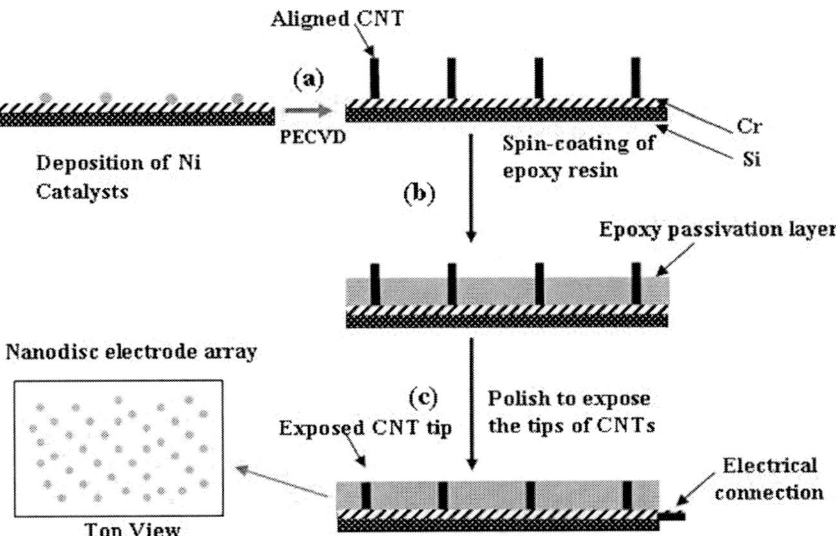

Fig. 3.11. Fabrication scheme of a low-site-density aligned CNT nanoelectrode array. (Reprinted with permission from Ref. [101]. © 2003 American Chemical Society.)

(Fig. 3.11a) by applying a pulse current to the substrate in a $NiSO_4$ electrolyte solution. The size and the site density of the Ni nanoparticles were controlled by the amplitude and the duration of the pulse current. On these Ni particles, the CNTs were grown (Fig. 3.11b) in the plasma-enhanced chemical vapor deposition (PECVD) system at 650 °C for 8 min with 160 sccm NH_3 and 40 sccm C_2H_2 gases with a total pressure of 15 Torr and a plasma intensity of 170 W. The aligned CNT arrays had a site density of 1×10^6–3×10^6 cm^{-2}, a length of 10 to 12 μm, and a diameter of 50–80 nm. A thin layer of Epon epoxy resin 828 (Miller-Stephenson Chemical Co., Inc., Sylmar, CA) was coated on the surface by magnetron sputtering to insulate the Cr layer. This was followed by applying m-phenylenediamine (MPDA) as a hardener. After these steps, the CNTs were half-embedded in the polymer resin, and the protruding part of the CNTs beyond the polymer resin was mechanically removed by polishing with a lens, followed by ultrasonication in water. Then the electronic connection was made on the CNT-Si substrate to make the CNT nanoelectrode arrays (Fig. 3.11c). Finally, the electrode arrays were pretreated by electrochemical etching in 1.0 M NaOH at 1.5 V for 90 s before electrochemical characterizations [101]. Results showed that, within these low site density CNTs, the NEAs consist of millions of nanoelectrodes, with each electrode being less than 100 nm in diameter. There is no degradation of these sensors for several weeks because of the excellent stability of the epoxy layer. Since the total current of the loosely packed electrode arrays is proportional to the total number of individual electrodes, having the number of the electrodes up to millions is highly desirable. The size reduction of each individual electrode and the increased total number of the electrodes result in improved signal-to-noise ratio (S/N) and detection limits.

Ng et al. have developed a soft lithography-mediated selective CVD template approach in preparing the multiwalled CNT membrane [102]. This membrane can be integrated with a flexible elastomeric polydimethylsiloxane framework to fabricate microsensing devices. The presented sensor design can be developed into a generic platform for electrochemical detection and gas sensing, as well as other general purpose sensory systems.

The unmodified and surface-confined aligned CNTs and CNT arrays by different preparation protocols discussed above have all been applied to the study of protein interaction [101, 103], DNA hybridization [104], and enzyme catalysis [66, 102]. Lin et al. have reported typical results for glucose biosensing using a CNT array (Fig. 3.12) [103]. The GO_x molecules were attached to the broken tips of the CNTs via carbodiimide chemistry by forming amide linkages between their amine residues and carboxylic acid groups on the CNT tips. Fig. 3.12(a) and (b) compare amperometric responses for 5 mM glucose (G), 0.5 mM ascorbic acid (AA), 0.5 mM acetaminophen (AC), and 0.5 mM uric acid (UA) at the GO_x-modified NEA and the potentials of +0.4 V (a) and −0.2 V (b). Well-defined cathodic and anodic glucose responses are obtained at this aligned CNT/GO_x-based biosensor at both potentials. However, the glucose detection at a lower operating potential (−0.2 V) is significantly less influenced by the interferences, indicating high selectivity towards the glucose substrate. Such a highly selective response to glucose is obtained

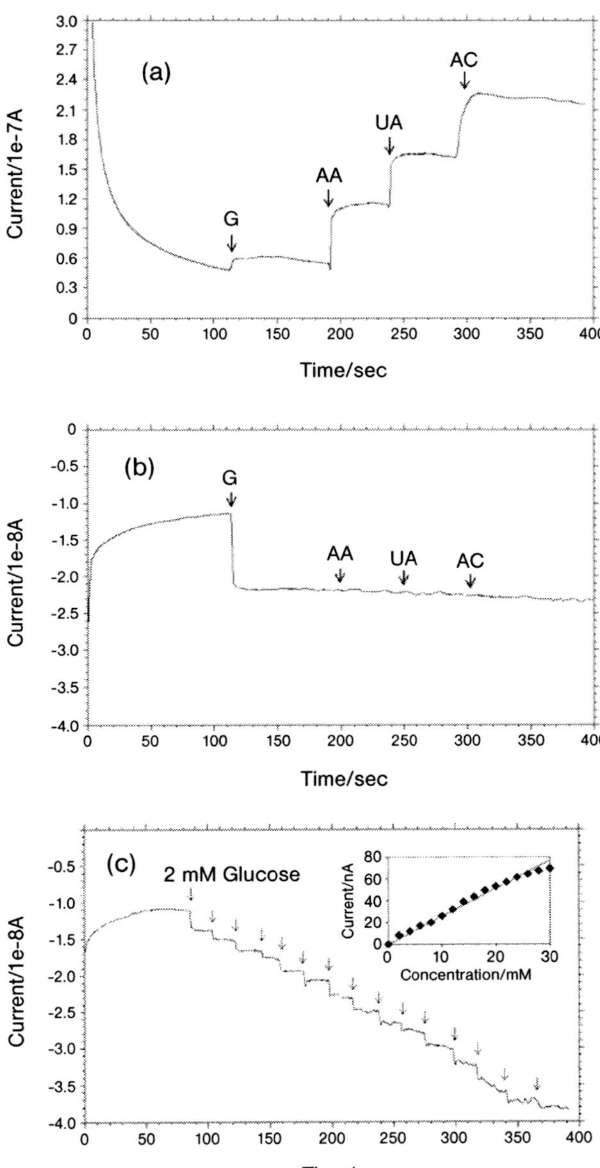

Fig. 3.12. (a, b) Amperometric responses for 5 mM glucose (G), 0.5 mM ascorbic acid (AA), 0.5 mM acetaminophen (AC), and 0.5 mM uric acid (UA) at a GO_x-modified, CNT-nanoelectrode array and potentials of +0.4 (a) and −0.2 V (b). Electrolyte: 0.1 M phosphate buffer/0.1 M NaCl (pH 7.4). (c) Amperometric response at the GO_x-modified, CNT-nanoelectrode array for each successive addition of 2 mM glucose. Inset: the corresponding calibration curve. Potential: −0.2 V, other conditions are as in (a, b). (Reprinted with permission from Ref. [103]. © 2003 American Chemical Society.)

at this aligned CNT/GO_x-based biosensor without the use of mediators and permselective membranes. The amperometric response at this sensor for each successive addition of 2×10^{-3} M glucose is presented in Fig. 3.12(c) with the corresponding calibration curve in the inset. The linear response to glucose is up to 30 mM, and steady state is reached within 20 to 30 s [103]. Because of this low potential detection, CNTs eliminate perspective interferences through the preferential detection of hydrogen peroxide at the CNT-based electrodes. Such development of interference-free transducers will significantly simplify the design and fabrication of biosensors. Biosensors based on low-site-density aligned CNTs are also suitable for the highly selective detection of glucose in various biological fluids (e.g., saliva, sweat, urine, and serum) [103].

Rusling et al. reported the first example of enzymes covalently attached onto the ends of vertically oriented SWCNT forest arrays [67]. These arrays were made from their unique methodology of assembling dense orthogonally oriented arrays of shortened SWCNTs. Quasi-reversible Fe^{III}/Fe^{II} voltammetry was obtained for the iron heme enzymes myoglobin and horseradish peroxidase coupled to the carboxylated ends of the nanotube forests by amide linkages. Their observation suggested that the "trees" in the nanotube forest behaved, electrically, similarly to a metal, conducting electrons from the external circuit to the enzymes. Accordingly, the electrochemically manifested peroxidase activity of myoglobin and horseradish peroxidase attached to the CNT array was demonstrated, showing analytical promise for hydrogen peroxide with a detection limit down to ~100 nM in buffer solutions. The covalently attached enzymes kept their activity for weeks in these prototype SWCNT-forest array biosensors [67]. Gooding and co-authors also observed the direct charge transfer between redox-active enzymes and the aligned CNTs' surface [66]. Their mechanistic study revealed that the rate of electron transfer remains the same regardless of the lengths of the tubes. These findings enable electroactive molecules to be located several hundred nanometers from a macroscopic substrate electrode with no loss in performance. This might guide future modification of the CNT array for sensor applications.

While most of the array studies have focused on their electrochemical mode, because of the unique electronic properties of individual nanotubes and nanotube arrays, optical sensing based on CNT arrays has also been carried and has been reviewed by Xu [21].

3.2.6 Chemiluminescence

Miscellaneous CNT sensing applications have been reported recently by using different transducer techniques with various nanotube platforms. As a traditional technique, fluorescence spectroscopy has been widely employed in sensor and bioassay application because of its high sensitivity. However, there have been few reports regarding its study on CNTs, partly because of the limited aqueous solubility or dispersion of CNTs [105, 106]. By using amine-terminated oligonucleotide-functionalized CNTs, Hazani and co-authors were able to enhance nanotube solu-

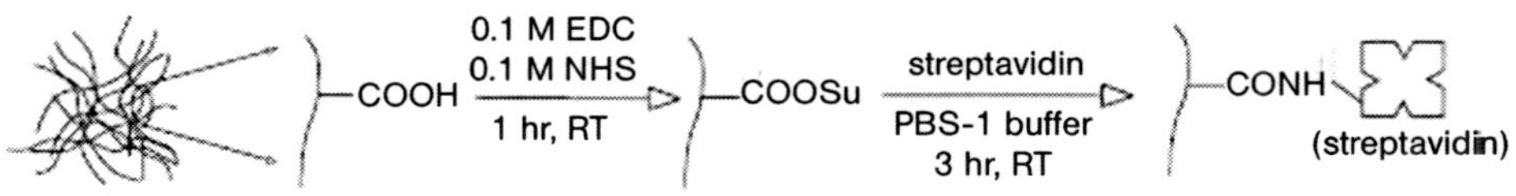

Fig. 3.13. Reaction scheme of immobilization of streptavidin on nanotubes by covalent coupling. (Reprinted with permission from Ref. [105]. Copyright 2003 Wiley-VCH.)

bility to facilitate the confocal fluorescence imaging of the DNA hybridization from a fluorescence dye-tagged complementary sequence [107]. Baker et al. used thiol-terminated oligonucleotide-modified CNTs for the similar assay [108].

A more direct application of CNT based biosensors has been reported by Wohlstadter et al. in their electrogenerated chemiluminescence (ECL) study of immunoassays for α-fetoprotein (AFP) [105]. Nanotubes possess several characteristics that make them attractive for ECL-based assays: First, they are conductive or semiconductive and hence can act as electrodes to generate ECL in aqueous solutions; second, nanotubes can be surface-functionalized as we discussed before; finally, their high surface area-to-volume ratio and sp^2 network make them feasible for immobilizing biomolecules as well as their associated electrochemistry. Wohlstadter et al. first mixed nanotubes with poly(vinyl acetate) (EVA) to form a nanotube–EVA composite. This composite was then etched with strong acid to produce a densely packed nanotube sheet with available surface carboxylic acid groups. As can be seen in Fig. 3.13, streptavidin was then immobilized on the nanotube sheet by carbodiimide-activated coupling, followed by the attachment of a biotinylated mAb for AFP via the specific biotin–streptavidin interaction. Figure 3.14 shows the whole assay based on this sensor. The capture of AFP resulted in the binding of a $Ru(bpy)_3{}^{3+}$ labeled antibody on to the EVA-MWCNTs composite electrode. As the $Ru(bpy)_3{}^{3+}$ is chemiluminescent, binding is transduced by the release of light upon applying the electrode at potentials more positive than +1 V. This immunosensor with electrochemiluminescence detection was sensitive at AFP concentrations as low as 0.1 nM, with a linear range up to 30 nM [105].

By using chemiluminescent $Ru(bpy)_3{}^{3+}$ as a marker, Dong et al. found that the CNT/Nafion composite possesses ECL sensitivity two orders of magnitude more than that at the silica/Nafion composite and three orders of magnitude more than that at pure Nafion films, again proving that the solubility enhancement for CNTs plays a key role in developing a CNT-based fluorescence sensor [106].

3.2.7
Field-effect Transistor and Bioelectronics

Advances in electronic detection based on 1D nanomaterials provide the ability for label-free and real time, yet sensitive and selective, sensing biomolecules. The development of CNTs with unique electronic properties has been spotlighted for future solid-state nanoelectronics. Therefore, CNT-based molecular electronics have

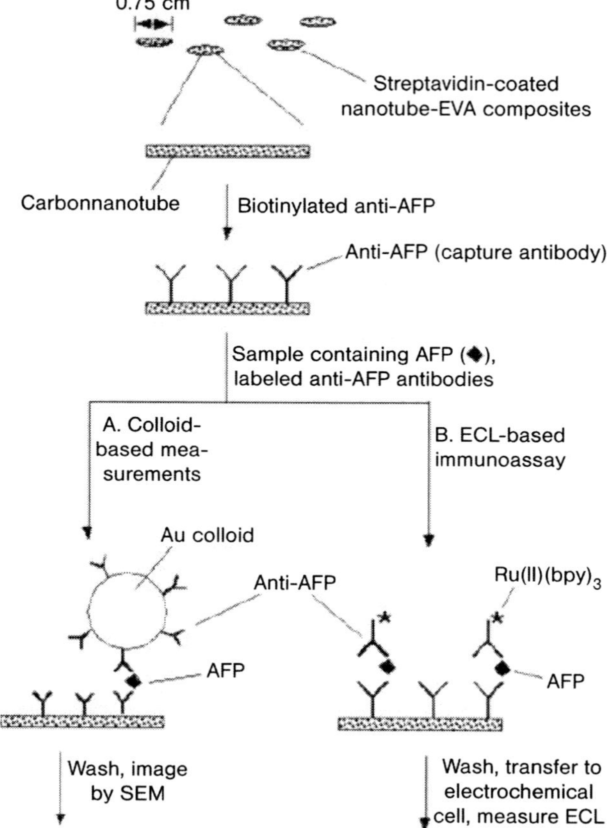

Fig. 3.14. Schematic of procedures for AFP assays based on the CNT-ECL sensor (not to scale). (Reprinted with permission from Ref. [105]. Copyright 2003 Wiley-VCH.)

received wide attention because of the semiconductor features of the nanotubes [109]. Among them, the study of CNT field-effect transistors (CNTFETs) is the core study to compare with the silicon-based transistors. The first CNTFETs were demonstrated by Tans et al. and Martel et al., respectively, in observing the CNTs' exploitable switching behavior [110, 133]. Since then, efforts have been made to improve the electrical characteristics of the CNTFETs. Presently, those explorable features of CNTFETs are the ballistic (scattering-free) and spin-conserving transport of electrons along the nanotubes, their ability to display metallic conducting as well as semiconducting behavior, and their access to the energy gap, which depends on the tube diameter and the rolling orientation of the tubes. CNTs also have extraordinarily high thermal conductivity. These unique properties make them behave similarly to conventional metal-oxide semiconductor field effect transistors

(MOSFETs) as well as differently, with a change from Schottky-barrier modulation at the contacts to bulk switching. These responding multiplicities, coupled with different CNT assembling approaches, offer various possible CNTFETs applications in biological diagnosis, e.g., proteins [5, 81] and cancer cell [111]. Chen and coworkers have demonstrated an exploration of SWCNT as a platform for investigating surface–protein and protein–protein binding and developing a highly specific electronic biosensor. They put SWCNT on a junction (drain and source) as shown in Fig. 3.15 [81]. The SWCNT was modified with polyethylene, which can reduce the non-specific interaction of proteins. A specific receptor was conjugated onto polyethylene-modified SWCNT. Therefore, the device can be highly specific in detecting proteins such as 10E3 m Ab (Fig. 3.4). The detection limit of this method was found to be about 340 ng mL^{-1} or 1.0 nM.

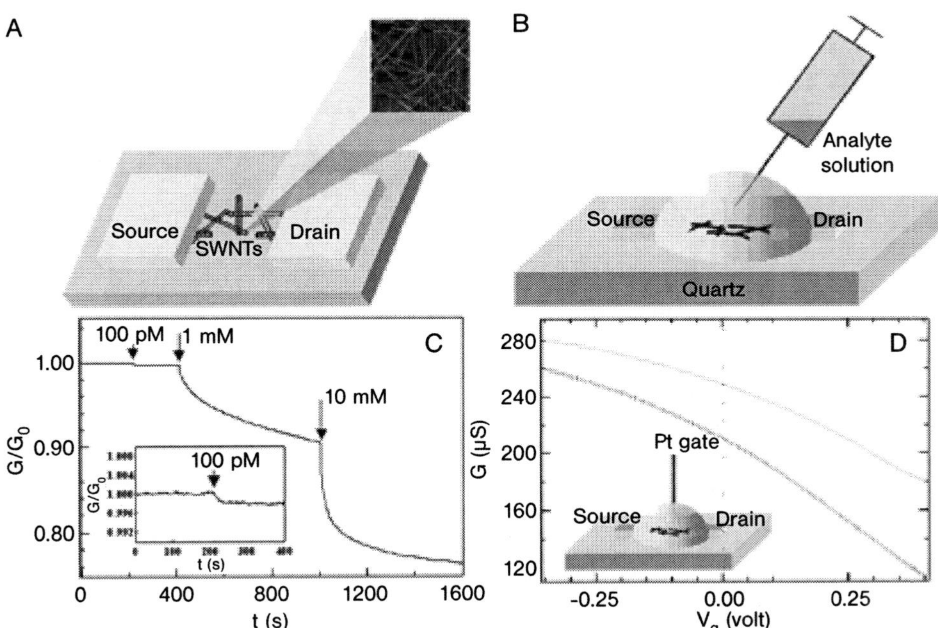

Fig. 3.15. Carbon nanotubes as electronic devices for sensing in aqueous solutions. (A) Schematic view of the electronic sensing device consisting of interconnected nanotubes bridging two metal electrode pads. An AFM image of a portion of the nanotube network (0.5 μm on a side) is shown. (B) Schematic setup for sensing in solution. (C) Conductance (G) evolution of a device for electronic monitoring of SA adsorption on nanotubes. The conductance is normalized by the initial conductance G_0. Inset: Sensitivity to a 100-pM protein solution. (D) Electrical conductance (G) vs. gate voltage (V_g) for a device in a 10-mM phosphate buffer solution. The gate voltage is applied through a Pt electrode immersed in the solution (inset). The upper (solid) and lower (broken) curves are the G–V_g characteristics for the device before and after SA binding, respectively. The shift in the two curves suggests a change in the charge environment of the nanotubes. (Reprinted with permission from Ref. [81]. © 2003 NAS).

3.3 Nanowires in Biosensor Development

Recently, nanowires have been explored as building blocks to fabricate nanoscale electronic devices through self-assembly – a typical bottom-up approach for biosensing [4, 6, 9]. The underlying mechanism for a nanowire biosensor is a field effect that is transduced using a field-effect transistor [4]. This "bottom-up" approach to bionanoelectronics has several attractive features. First, the nanowires are extremely sensitive for detection of biointeractions on their surface because of their high aspect ratio. Second, the electronically switchable properties of semiconducting nanowires provide a sensing modality, a direct and label-free electrical readout, which is exceptionally attractive. Third, the size of the nanowires can be readily tuned to sub-100 nm and smaller, which can lead to a high density of a device on a chip. Therefore, it is feasible for the miniaturized devices to detect multiple samples real time *in vivo*. Fourth, the candidate materials for the nanowires are unlimited, which gives the researcher great flexibility in selecting the right materials for the functionality of the desired device. Presently, the most studied nanowires as a building block for biosensing are semiconducting nanowires, e.g., silicon nanowires [4, 6, 9], conducting polymer nanowires [17, 18], and oxide nanowires [112]. Some metal nanowires have also been developed for sensing. We will focus on semiconducting nanowires for biosensor development here.

3.3.1 Silicon Nanowire-based Biosensors

Silicon nanowires are generally fabricated by CVD [14, 15] and template etching [113, 114]. For the purpose of biosensing, the modification of silicon nanowires is through a silanization reaction on the silicon surface, which introduces the active groups, e.g., amino, carboxyl, or biotin, on the surface of the nanowires. Then the receptor, which can specifically recognize the interested analyte, will be immobilized on the surface through those active groups, the interaction between biomolecules and their counterparts on the silicon surface will introduce the conductance change caused by the field effect, and the conductance change can be electronically transduced. Thus far, silicon nanowire-based biosensors have been developed for detection DNA [4, 8, 115, 116], proteins [4, 9], and viruses [6].

Lieber's group is pioneering the development of silicon nanowire-based biosensors [4, 8]. Biological macromolecules, such as proteins and nucleic acids, are typically charged in aqueous solutions and, as such, can be detected readily by nanowire sensors when appropriate receptors are linked to the nanowire active surface. Figure 3.16 shows the real time detection of proteins and DNA based on silicon nanowires. They modified silicon nanowires with biotin, which has a strong affinity to protein streptavidin on the oxide surface of nanowires [4]. When a solution of streptavidin is delivered to a nanowire sensor device modified with a biotin receptor, they found that the conductance of nanowires increases rapidly to a constant value, and this conductance value is maintained after the addition of pure buffer

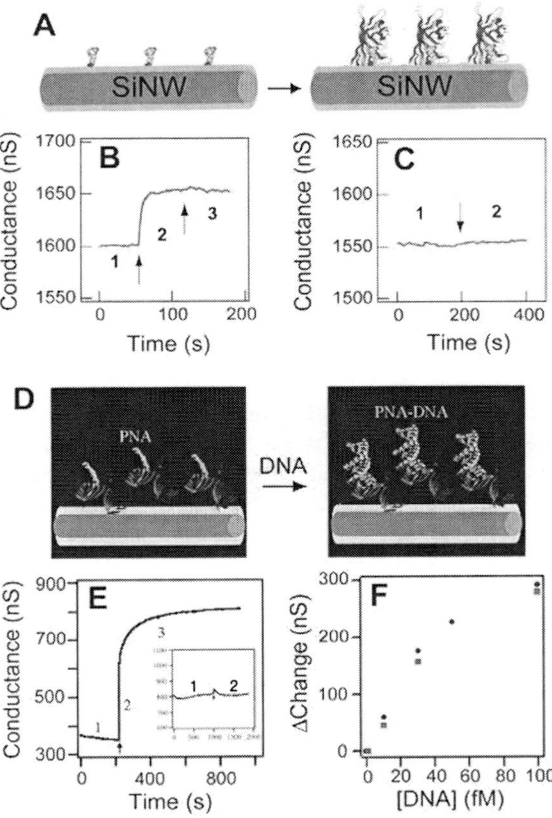

Fig. 3.16. Real-time detection of proteins and DNA. (A) Schematic of a biotin-modified Si nanowire and subsequent binding of streptavidin to the modified surface. (B) Conductance versus time for a biotin-modified Si nanowire, where Region 1 corresponds to the buffer solution, Region 2 corresponds to the addition of 250 nM streptavidin, and Region 3 corresponds to pure buffer solution. (C) Conductance versus time for an unmodified Si nanowire; Regions 1 and 2 are the same as in (B). (D) Schematic of a Si nanowire sensor surface modified with a PNA receptor before and after duplex formation with target DNA. (E) Si nanowire DNA sensing; the arrow corresponds to the addition of a 60-fM complementary DNA sample, and the inset shows the device conductance following addition of 100-fM mutant DNA. (F) Conductance versus DNA concentration; data points indicated by ■ and ● are obtained from two independent devices. (Reprinted with permission from Ref. [4], [8]. (A)–(C) 2001 © AAAS and (D)–(F) 2004 © ACS.)

(Fig. 3.16B). However, adding a streptavidin solution to unmodified silicon nanowires does not produce a change in conductance (Fig. 3.16C). The conductance change is caused by the specific interaction between biotin and streptavidin, and these results are further proved by an experiment in which blocking the streptavidin binding sites leads to an absence of response from biotin-modified silicon nanowires. They also studied the detection limit and found that the electrical detection could be carried out to 10 pM, which is below the detection level required for a number of disease marker proteins. More recently, a silicon nanowire field effect device has been investigated as a biosensor to detect the sequence of DNA [8]. First, a PNA was immobilized on the surface of the p-type Si nanowires (Fig. 3.16D). When a complementary DNA target was introduced to the device, the hybridization of DNA (Fig. 3.16D) caused the conductance of the nanowires to increase (Fig. 3.16E). PNA was used as a receptor for DNA detection because the uncharged PNA molecules have a greater affinity and stability than the corresponding DNA recognition sequence. They have used this device to detect the wild type versus the DF508 mutation site in the cystic fibrosis transmembrane receptor gene and showed that the conductance increases when adding a 60-fM wild-type DNA sample solution (Fig. 3.16E). The increase in conductance of the Si nanowire device is consistent with the increase in the negative charge density associated with the binding of negatively charged DNA at the surface, and moreover, careful control experiments show that the binding response is specific to the wild-type sequence. Further study shows that the direct electrical detection for DNA is possible down to at least the 10-fM level, and the method is reproducible (Fig. 3.16F).

The same group have also studied the use of the Si nanowire for detecting a single virus and demonstrated that the Si nanowires could be assembled to form arrays for multiplexed detection of samples [6]. Addressable arrays are fabricated by a process that uses a fluid-based assembly, such as microfluidic or Langmuir–Blodgett methods, to align and set the average spacing of nanowires over large areas for photolithography to define interconnections (Fig. 3.17A). They also fabricated a state-of-the art array containing more than 100 addressable elements (Fig. 3.17B). All the of the active nanowire sensor devices are confined to a central rectangular area on the device chip that overlaps with the microfluidic sample delivery channels, as illustrated in the figure. They further demonstrated the use of this nanowire array for detecting two types of virus at the same time. An antibody receptor that is specific either for influenza or for adenovirus was modified on p-type Si nanowires. Simultaneous conductance measurements were obtained when adenovirus, influenza, and a mixture of both viruses were delivered to the device. Other groups have also studied the use of silicon nanowires for DNA and protein analysis [115, 116].

3.3.2
Conducting Polymer Nanowire-based Biosensors

In recent years, conducting polymer-based nanostructured materials have been used extensively in resistive sensors [17]. Because of their promising properties, which include high surface areas, chemical specificities, tunable conductivities, material flexibilities, and easy processing, various methods have been developed to fabricate conducting polymer nanowires. For example, (a) polyaniline nanowires have been obtained through a facile synthesis [117] or by electrospinning methods [118]; (b) template-directed electrochemical processes have been employed to fabricate nanowire junctions that feature robust polymer electrode contacts [119]; and (c) mechanical stretching [120] and magnetic field-assisted assembly [121] pro-

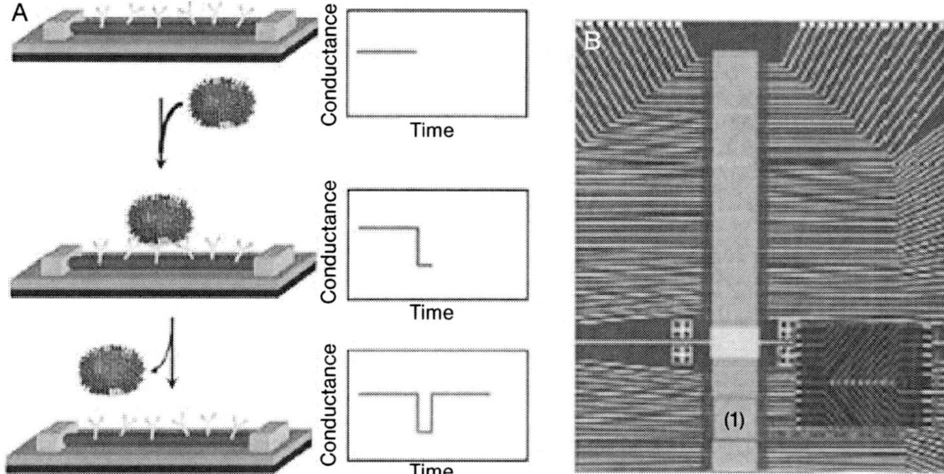

Fig. 3.17. (A) Schematic of a single virus binding and unbinding to the surface of a Si nanowire device modified with antibody receptors and the corresponding time-dependent change in conductance. (B) Optical image of the upper portion of a sensor device array, where the inset shows one row of individually addressable nanowire elements. The rectangle labeled (1) highlights the position of the microfluidic channel used to deliver samples and overlap the active elements. (Reprinted with permission from Ref. [6]. © 2004 NAS).

cesses have produced miniaturized polymer-electrode junctions. A templateless electrochemical assembly of conducting polymer nanowires has also been developed recently [17, 18]. The development of biosensors based on conducting polymer nanowires is still in its infancy. Some examples are introduced as follows.

Tao's group have demonstrated a glucose biosensor using conducting polymer/enzyme junctions and found that a unique feature can arise when shrinking a sensor to a nanometer [122]. Figure 3.18(A) shows the structure of the polymer/enzyme nanojunction sensor. The thickness of the polyaniline in the junction is 20–60 nm. The polyaniline/enzyme nanojunction was prepared by co-polymerization of monomer aniline and GOx in an aqueous solution. The signal transduction mechanism of the sensor is based on the changes in nanojunction conductance as a result of glucose oxidation induced change in the polymer redox state. Because of the small size of the nanojunction, they found that the response to glucose is fast and less than 1.0 s. However, the response time for glucose with a 10 μm gap is up to 10 min. The detection of limit for this method is at the μM level.

Myung's group has developed a facile technique for synthesizing conducting polymer nanowires by electrodeposition within channels between two electrodes on the surface of silicon wafers. They demonstrated that this technique can fabricate multiple individually addressable conducting polymer nanowires between two junctions. They also demonstrated the capability to create a scalable and high-density array by site-specific positioning of conducting polymer nanowires of the same and different composition on the same chip. Furthermore, the same group

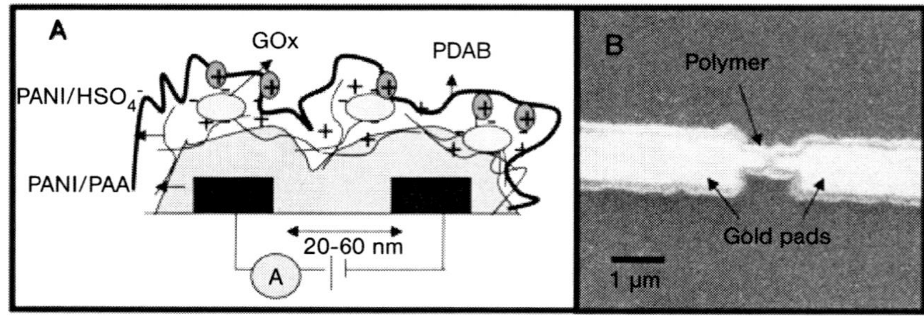

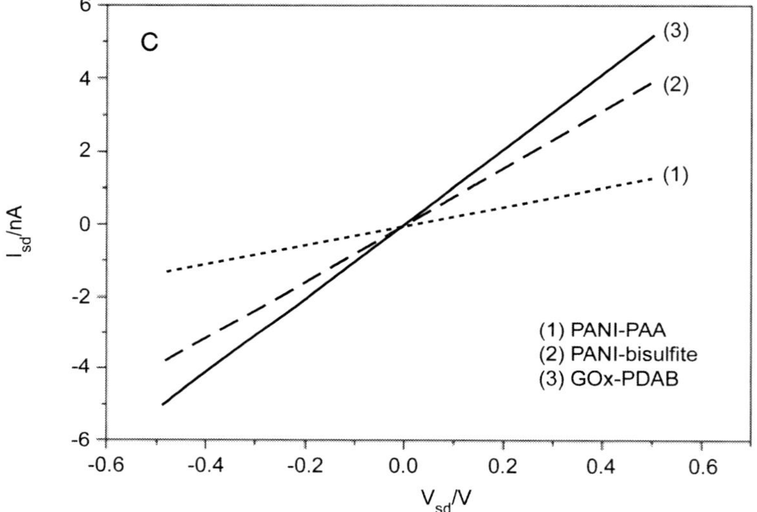

Fig. 3.18. (A) Structure of the polymer nanojunction sensor. (B) SEM image of PANI-PAA/PANI-bisulfite/GOx-PDAB films deposited on gold pads with gaps of 20–60 nm. (C) *I–V* curves obtained in air after each nanogap modification step: (1) polymerization of PANI-PAA carried out in 0.4 M aniline + 150 mg mL^{-1} PAA (MW: 2000) solution with 0.5 M Na_2SO_4 and 0.5 M H_2SO_4 by a potential sweep between −0.2 and 0.9 V vs. SCE during the first cycle and between −0.2 and 0.78 V vs. SCE during the following cycles at 0.05 V s^{-1}; (2) polymerization of PANI bisulfite in a 0.4 M aniline + 0.5 M $NaHSO_4$ solution acidified to pH 0 with H_2SO_4 by a single potential sweep from −0.2 to 0.9 V vs. SCE; (3) immobilization of GOx-PDAB by exposing the polymer nanojunction to 0.5 M Na_2SO_4 + 25 mM 1,2-diaminobenzene + 167 μM glucose oxidase in a pH 5 citric acid/Na_2HPO_4 (McIlvaine) buffer solution for 15 min, followed by electrodeposition of PDAB at +0.4 V vs. SCE for 4 min. (Reprinted with permission from Ref. [122]. © 2004 ACS).

reported bio-affinity sensing using biological functionalized conducting polymer nanowire. The device incorporated with polypyrrole nanowires made by the facile technique can be used for studying protein–protein interaction [123].

Wang et al. demonstrated template-free fabrication of polyaniline nanowire on electrode junctions by electrodeposition, and this method can be extended to synthesize other conducting polymer nanowires, e.g., polypyrrole and poly(Edot). They systematically studied the electron transport properties of these conducting polymer nanowires with an electrolyte gate [124].

3.3.3 Metal Oxide Nanowire-based Biosensors

Metal oxide nanowires (MONWs) have been used to develop biosensors [112, 125]. They can work as a valid alternative to CNTs or Si NWs. Curreli and coworkers have reported a selective functionalization of In_2O_3 nanowires for biosensor applications. They first generated a self-assembled monolayer (SAM) of 4-(1,4-dihydroxybenzene) butyl phosphonic acid (HQ-PA) on the InO_2O_3 NW surface. Oxidized HQ-PA can react with a range of functional groups, which can be easily incorporated in biomolecules. They have successfully attached DNA on the In_2O_3, and this study opens an avenue for using such metal oxide nanowires for biosensing [112].

3.4 Nanocantilevers for Biosensors

Recently, microfabricated cantilevers have successfully been used for biosensors [126–132]. The adsorption of biomolecules on the surface will induce surface stress, which can be measured with micro/nanocantilevers. Adsorption of two proteins, immunoglobulin (IgG) and albumin (BSA), on a gold surface has been studied in a buffer solution in terms of surface stress measurements. Fritz and coworkers have reported the specific transduction, via surface stress changes, of DNA hybridization and receptor–ligand binding into a direct nanomechanic response of microfabricated cantilevers [126]. Cantilevers were fabricated in arrays and functionalized with a selection of biomolecules. The differential deflections of the cantilever were responses of individual cantilevers. Figure 3.19 illustrates the DNA hybridization on nanocantilevers in solution. First, a different sequence of DNA probe was modified on the surface of cantilevers (Fig. 3.19A), and then target DNA was injected. Only the cantilever that provides the matching sequence was found to have surface stress. It has been demonstrated that a single mismatch between 12-mer oligonucleotide is detectable. This method shows important advantages in that it does not require labeling, optical excitation, or external probes. In addition, the transduction process is repeatable and enables cyclic operation. At the same time, Hansen et al. have reported a cantilever-based optical deflection assay for discrimination of DNA single-nucleotide mismatches [127].

Furthermore, Wu and coworkers have extended this research to a real biological system, disease-related proteins. They reported that microcantilevers of different geometries have been used to detect two forms of prostate-specific antigens (PSAs) over a wide range of concentration range 0.2 ng mL^{-1} to 60 μg mL^{-1} in a background of human serum albumin and human plasminogen at 1.0 mg mL^{-1}

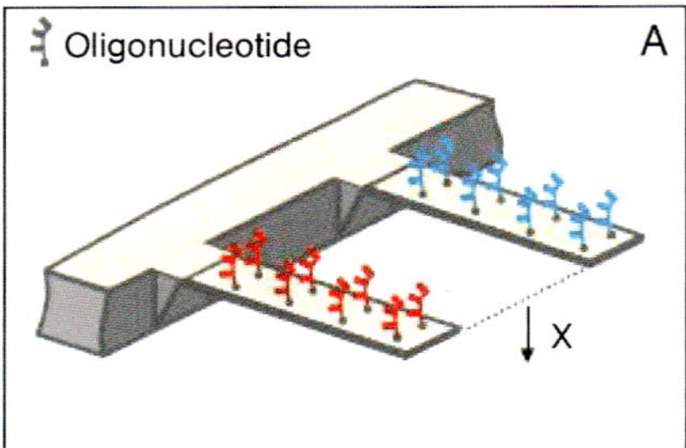

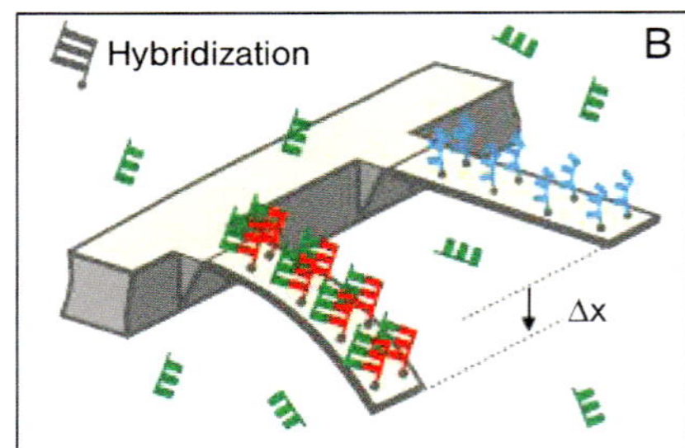

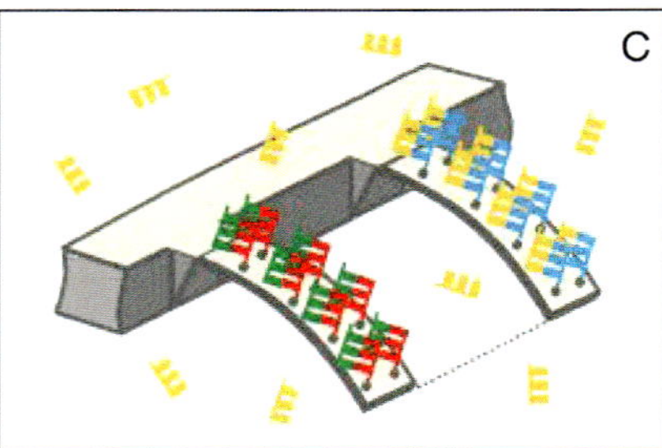

Fig. 3.19. Scheme illustrating the hybridization experiment. Each cantilever is functionalized on one side with a different oligonucleotide base sequence (red or blue). (A) The differential signal is set to zero. (B) After injection of the first complementary oligonucleotide (green), hybridization occurs on the cantilever that provides the matching sequence (red), increasing the differential signal D*x*. (C) Injection of the second complementary oligonucleotide (yellow) causes the cantilever functionalized with the second oligonucleotide (blue) to bend.

[129]. This study makes this technique a clinically relevant diagnostic technique for prostate cancer.

3.5 Summary

One-dimensional nanomaterials, such as CNTs, semiconducting nanowires, and Si-based nanocantilevers have shown promise as new detection platforms that are equal or superior to many other sensing materials. This is mainly attributed to their unique electronic, mechanic, thermal, and chemical properties. The preparation, purification, and dispersion of single-walled and multiwalled nanotubes have, especially, been reviewed. Various sensor fabrication protocols based on those 1D nanomaterials have been discussed in detail. A typical application for DNA and proteins, including enzymes and antibodies, has been described, and the respective responding mechanisms have been addressed in this comprehensive review. Various 1D nanomaterial-based biosensors have found broad application from their respective designs, including unmodified and surface-confined nanomaterials. Accordingly, biosensing based on amperometric amplification, field-effect transistors, signal-enhanced immunoassay, and non-enzymatic monitoring, and surface stress has been described for various biosensor designs. This research field is experiencing explosive growth, and new reports appear on a daily basis. We anticipate that the high orderly array design, combined with multiple biorecognition, will be the hotspot for the next stage of the 1D nanomaterial-based sensor studies.

Acknowledgments

The work performed at Pacific Northwest National Laboratory (PNNL) was supported by the Laboratory Directed Research and Development program, DOE-EMSP, SERDP (Project ID 1297), and NIH/1R01 ES010976-01A2. The authors' research described in this chapter was performed in part at the Environmental Molecular Sciences Laboratory, a national scientific user facility sponsored by the DOE Office of Biological and Environmental Research and located at PNNL. PNNL is operated for DOE by Battelle under Contract DE-AC05-76RL01830.

Glossary

Amperometry An electrochemical technique that measures electrical current at a fixed potential upon adding analyte or titrant into the measuring cell.
Biosensor Any probe designed to measure biological molecules' concentration or structures, monitor biological processes, or translate biochemical signals into quantifiable physical signals.
Carbon nanotube (CNT) A 1D fullerene with a cylindrical shape that consists of a seamless structure with hexagonal honeycomb lattices, being several nanometers in diameter and up to hundreds of microns long. CNTs can be divided into two major groups, i.e., single-wall carbon nanotubes (SWCNTs) and multiwall carbon nanotubes (MWCNTs). SWCNTs represent a single graphite sheet rolled flawlessly, demonstrating a tube diameter of 1 to 2 nm, whereas MWCNTs show concentric and closed graphite tubules with diameters ranging from 2 to 50 nm and an interlayer distance of approximately 0.34 nm.
Catalysis The acceleration of a chemical reaction by a catalyst.
Chemiluminescence A luminescence phenomenon produced by the direct transformation of chemical energy into light energy.
DNA Deoxyribonucleic acid. A naturally occurring polymer consisting of a phosphate backbone, sugar rings, and various bases. Usually it is found as single- or double-stranded deoxynucleotides.
Field effect transistor (FET) A semiconductor transistor with a region of donor material with two terminals designated as the "source" and the "drain," respectively, and an adjoining region of acceptor material in between, called the "gate."

The voltage between the gate and the substrate controls the current flow between the source and the drain by depleting the donor region of its charge carriers to a greater or lesser extent.

Enzyme electrode A type of biosensor that uses an enzyme-anchored electrode setup.

Hybridization The process of forming double stranded DNA molecules by combining two complementary single-stranded oligonucleotides.

Nafion A brand from a DuPont produced polymer that was synthesized by modifying a Teflon polymer. Like Teflon, Nafion is extremely chemically inert. However, unlike Teflon, Nafion is very ion-conductive because it contains sulfonic acid groups. These unique properties make Nafion useful in ion-exchange membranes, humidity sensors, fuel cells, etc.

Lithography A technique that creates chemical patterns on a metal or ceramic surface. It is currently used in making integrated electronic circuits, computer chips, etc.

Organophosphorus compounds Organic molecules that contain the element phosphorus. Organophosphorus (OP) compounds are very toxic and are thus widely used as pesticides and chemical-warfare agents (CWAs).

Polymer A macromolecule consisting of repeated chemical units.

Screen printing A traditional printing method that is used to print everything from T-shirts to coffee mugs and decals. It uses a squeegee to force ink through a stencil created on a mesh fabric onto some type of substrate such as a silk, metal, or stone. It is a mass production method. It has been used for fabricating sensors, such as single-use glucose strips.

Sol–gel technique A sol is a homogeneous dispersion of the solid particles (~100 to 1000 nm) in a liquid where only the Brownian motions suspend the particles. A gel is a state where both liquid and solid are dispersed in each other, which presents a solid network containing liquid components. The sol–gel technique is a low-temperature method using chemical precursors that can produce ceramics and glasses with better purity and homogeneity than high-temperature conventional processes. This technique currently has wide application in preparing electronic, optical, and electro-optic devices.

Abbreviations

1D	One-dimensional
AFM	Atomic-force microscopy
AFP	α-Fetoprotein
ALP	Alkaline phosphatase
BSA	Bovine serum albumin
CE	Capillary electrophoresis
CNT	Carbon nanotube
CNTFET	Carbon nanotube field-effect transistor
CVD	Catalytic vapor deposition

CWA	Chemical-warfare agent
DCC	Dicyclohexylcarbodiimide
DMF	*N,N*-Dimethylformamide
DNA	Deoxyribonucleic acid
DOE	U.S. Department of Energy
EAD	Electric arc discharge
ECL	Electrogenerated chemiluminescence
EVA	Poly(vinyl acetate)
FET	Field-effect transistor
GC	Gas chromatography
GOx	Glucose oxidase
HQ-PA	4-(1,4-Dihydroxybenzene) butyl phosphonic acid
HRP	Horseradish peroxidase
IgG	Immunoglobulin G
LA	Laser ablation
LC	Liquid chromatography
MONW	Metal oxide nanowire
MOSFET	Metal-oxide-semiconductor field effect transistor
MPDA	*m*-Phenylenediamine
mRNA	Message ribonucleic acid
MWCNT	Multiwall carbon nanotube
NADH	β-Nicotinamide adenine dinucleotide
NEA	Nanoelectrode array
PECVD	Plasma-enhanced chemical vapor deposition
PEG	Poly(ethylene glycol)
PNA	Peptide nucleic acid
PNNL	Pacific Northwest National Laboratory
Ppy	Polypyrrole
PSA	Prostate-specific antigen
RNA	Ribonucleic acid
SAM	Self-assembled monolayer
SpA	Staphylococcal protein A
SWCNT	Single-wall carbon nanotube
TEM	Transmission electron microscopy

References

1 Bachtold, A., Hadley, P., Nakanishi, T., Dekker, C. Logic circuits with carbon nanotube transistors, *Science* **2001**, 294, 1317–1320.

2 Rueckes, T., Kim, K., Joselevich, E., Tseng, G. Y., Cheung, C. L., Lieber, C. M. Carbon nanotube-based nonvolatile random access memory for molecular computing, *Science* **2000**, 289, 94–97.

3 Kong, J., Franklin, N. R., Zhou, C. W., Chapline, M. G., Peng, S., Cho, K. J., Dai, H. J. Nanotube molecular wires as chemical sensors, *Science* **2000**, 287, 622–625.

4 Cui, Y., Wei, Q. Q., Park, H. K., Lieber, C. M. Nanowire nanosensors for highly sensitive and selective detection of biological and chemical species, *Science* **2001**, 293, 1289–1292.
5 Star, A., Gabriel, J. C. P., Bradley, K., Gruner, G. Electronic detection of specific protein binding using nanotube FET devices, *Nano Lett.* **2003**, 3, 459–463.
6 Patolsky, F., Zheng, G. F., Hayden, O., Lakadamyali, M., Zhuang, X. W., Lieber, C. M. Electrical detection of single viruses, *Proc. Natl Acad. Sci. U.S.A.* **2004**, 101, 14 017–14 022.
7 So, H. M., Won, K., Kim, Y. H., Kim, B. K., Ryu, B. H., Na, P. S., Kim, H., Lee, J. O. Single-walled carbon nanotube biosensors using aptamers as molecular recognition elements, *J. Am Chem. Soc.* **2005**, 127, 11 906–11 907.
8 Hahm, J., Lieber, C. M. Direct ultrasensitive electrical detection of DNA and DNA sequence variations using nanowire nanosensors, *Nano Lett.* **2004**, 4, 51–54.
9 Wang, W. U., Chen, C., Lin, K. H., Fang, Y., Lieber, C. M. Label-free detection of small-molecule-protein interactions by using nanowire nanosensors, *Proc. Natl Acad. Sci. U.S.A.* **2005**, 102, 3208–3212.
10 Xia, Y. N., Yang, P. D., Sun, Y. G., Wu, Y. Y., Mayers, B., Gates, B., Yin, Y. D., Kim, F., Yan, Y. Q. One-dimensional nanostructures: Synthesis, characterization, and applications, *Adv. Mater.* **2003**, 5, 353–389.
11 Law, M., Goldberger, J., Yang, P. D. Semiconductor nanowires and nanotubes, *Ann. Rev. Mater. Res.* **2004**, 34, 83–122.
12 Nikolaev, P., Bronikovski, M. J., Bradley, K., Rohmund, F., Colbert, D. T., Smith, K. A., Smalley, R. E. Gas-phase catalytic growth of single-walled carbon nanotubes from carbon monoxide, *Chem. Phys. Lett.* **1999**, 313, 91–97.
13 Amelinckx, S., Zhang, B., Bernaerts, J., Zhang, X. F., Ivanov, V., Nagy, J. B. A formation mechanism for catalytically grown helix-shaped graphite nanotubes, *Science*, **1996**, 265, 635–639.
14 Niu, J. J., Sha, H., Wang, Y. W., Ma, X. Y., Yang, D. R. Crystallization and disappearance of defects of the annealed silicon nanowires, *Microelectron. Eng.* **2003**, 66, 65–69.
15 Wu, Y., Fan, R., Yang, P. D. Block-by-block growth of single-crystalline Si/SiGe superlattice nanowires, *Nano Lett.* **2002**, 2, 83–86.
16 Cheng, S. W., Cheung, H. F. Role of electric field on formation of silicon nanowires, *J. Appl. Phys.* **2003**, 94, 1190–1194.
17 Wang, J., Chan, S., Carlson, R. R., Luo, Y., Ge, G. L., Ries, R. S., Heath, J. R., Tseng, H. R. Electrochemically fabricated polyaniline nanoframework electrode junctions that function as resistive sensors, *Nano Lett.* **2004**, 4, 1693–1697.
18 Liang, L., Liu, J., Windisch, C. F., Exarhos, G. J., Lin, Y. H. Direct assembly of large arrays of oriented conducting polymer nanowires, *Angew. Chem. Int. Ed.* **2002**, 41, 3665–3668.
19 Wang, J., Liu, G., Jan, M. Ultrasensitive electrical biosensing of proteins and DNA: Carbon-nanotube derived amplification of the recognition and transduction events, *J. Am. Chem. Soc.* **2004**, 126, 3010.
20 O'Connor, M., Kim, S. N., Killard, A. J., Forster, R. J., Smyth, M. R. Papadimitrakopoulos, F., Rusling, J. F. Mediated amperometric immunosensing using single walled carbon nanotube forests, *Analyst* **2004**, 129, 1176–1180.
21 Xu, J. M. Highly ordered carbon nanotube arrays and IR detection, *Infra. Phys. Tech.* **2001**, 42, 485–491.
22 Fritz, J., Baller, M. K., Lang, H. P., Strunz, T., Meyer, E., Guntherodt, H. J., Delamarche, E., Gerber, C., Gimzewski, J. K. Stress at the solid-liquid interface of self-assembled monolayers on gold investigated with a nanomechanical sensor, *Langmuir* **2000**, 16, 9694–9696.
23 Wang, J. X., Li, M. X., Shi, Z. J., Li, N. Q., Gu, Z. N. Direct electro-

chemistry of cytochrome c at a glassy carbon electrode modified with single-wall carbon nanotubes, *Anal. Chem.* 2002, 74, 1993–1997.

24 Wang, J. Nanomaterial-based electrochemical biosensors, *Analyst* **2005**, 130, 421–426.

25 Kohli, P., Wirtz, M., Martin, C. R. Nanotube membrane based biosensors, *Electroanalysis* **2004**, 16, 9–18.

26 Iijima, S. N. Helical microtubules of graphitic carbon, *Nature*, **1991**, 354, 56–58.

27 Kroto, H. W., Heath, J. R., O'Brien, S. C., Curl, R. F., Smalley, R. E. C-60 – buckminsterfullerene, *Nature*, **1985**, 318, 162–163.

28 Ajayan, P. M. Carbon naonotube from carbon, *Chem. Rev.* **1999**, 99, 1787–1799.

29 Harris, P. J. F. *Carbon Nanotubes and Related Structures*, Cambridge University Press, Cambridge, UK, **1999**.

30 Thostenson, E. T., Ren, Z. F., Chou, T. W. C. Advances in the science and technology of carbon nanotubes and their composites: A review, *Compos. Sci. Technol.* **2001**, 61, 1899–1912.

31 Popov, V. N. Carbon nanotubes: Properties and application, *Mater. Sci. Eng. R: Rep.*, **2004**, 43, 61–102.

32 Kohli, P., Martin, C. R. The emerging field of nanotube biotechnology, *Nat. Rev. Drug Discov.* **2003**, 2, 29–37.

33 Sinnott, S. B. Chemical functionalization of carbon nanotubes, *J. Nanosci. Nanotechnol.* **2002**, 2, 113–123.

34 Baughman, R. H., Zakhidov, A. A., de Heer, W. A. Carbon nanotubes – the route toward applications, *Science*, **2002**, 297, 787–792.

35 Ouyang, M., Huang, J., Lieber, C. M. Fundamental electronic properties and applications of single-walled carbon nanotubes, *Acc. Chem. Res.* **2002**, 35, 1018–1025.

36 Britto, P. J., Sunthanam, K. S. V., Ayajan, P. M. Carbon nanotube electrode for oxidation of dopamine, *Bioelectrochem. Bioenerg.* **1996**, 41, 121–125.

37 Besteman, K., Lee, J., Wiertz, F. G. M., Heering, H. A., Dekker, C. Enzyme-coated carbon nanotubes as single-molecule biosensors, *Nano Lett.* **2003**, 3, 727–730.

38 Salimi, A., Banks, C. E., Compton, R. G. Abrasive immobilization of carbon nanotubes on a basal plane pyrolytic graphite electrode: Application to the detection of epinephrine, *Analyst*, **2004**, 129, 225–228.

39 Wang, J., Musameh, M. Carbon nanotube screen-printed electrochemical sensors, *Analyst*, **2004**, 129, 1–2.

40 Pedano, M. L., Rivas, G. A. Adsorption and electrooxidation of nucleic acids at carbon nanotubes paste electrodes, *Electrochem. Commun.* **2004**, 6, 10–16.

41 Lin, Y., Taylor, S., Li, H. P., Fernando, K. A. S., Qu, L. W., Wang, W., Gu, L. R., Zhou, B., Sun, Y. P. Advances toward bioapplications of carbon nanotubes, *J. Mater. Chem.* **2004**, 14, 527–541.

42 Martin, C. R., Kohli, P. The emerging field of nanotube biotechnology, *Nat. Rev. Drug Discov.*, **2003**, 2, 29–37.

43 Li, N., Wang, J., Li, M. *Rev. Anal. Chem.* **2003**, 22, 19.

44 Odom, T. W., Huang, J. L., Lieber, C. M. Single-walled carbon nanotubes – From fundamental studies to new device concepts, *Ann. N. Y. Acad. Sci.* **2002**, 960, 203–215.

45 Thess, A., Lee, R., Nikolaev, P., Dai, H., Petit, H., Robert, J., Xu, D., Lee, Y. H., Kim, S. G., Rinzler, A. G., Colbert, D. T., Scuseria, G., Tomanek, D., Fisher, J. E., Smalley, R. E. Crystalline ropes of metallic carbon nanotubes, *Science*, **1996**, 273, 483–487.

46 Xie, S., Li, W., An, Z., Chang, B., Sun, L. *Mater. Sci. Eng. A*, **2000**, 286, 11.

47 Yacaman, M. J., Yoshida, M. M., Rendon, L., Santiesteban, J. G. Catalytic growth of carbon microtubules with fullerene structure, *Appl. Phys. Lett.*, **1993**, 62, 202.

48 Ivanov, V., Nagy, J. B., Lambin, P., Lucas, A., Zhang, X. B., Zhang,

X. F., Bernaerts, D., Van Tendeloo, G., Amelinckx, S., Van Landuyt, J. The study of carbon nanotubules produced by catalytic method, *Chem. Phys. Lett.* **1994**, 223, 329–335.

49 Yudasaka, M., Kituchi, R., Matsui, T., Ohki, Y., Yoshimura, S. Specific conditions for Ni catalyzed carbon nanotube growth by chemical vapor deposition, *Appl. Phys. Lett.* **1995**, 67, 2477–2479.

50 Iijima, S., Ichihashi, T. Single-shell carbon nanotubes of 1-nm diameter, *Nature*, **1993**, 363, 603–605.

51 Kong, J., Soh, H., Cassell, A. M., Quate, C. F., Dai, H. J. Synthesis of individual single-walled carbon nanotubes on patterned silicon wafers, *Nature* **1998**, 395, 878–881.

52 Colomer, J. F., Bister, G., Willems, I., Konya, Z., Fonseca, A., Van Tendeloo, G., Nagy, J. B. Synthesis of single-wall carbon nanotubes by catalytic decomposition of hydrocarbons, *Chem. Commun.* **1998**, 1343–1344.

53 Shelimov, K. B., Esenaliev, R. O., Rinzler, A. G., Huffman, C. B., Smalley, R. E. Purification of single-wall carbon nanotubes by ultrasonically assisted filtration, *Chem. Phys. Lett.* **1998**, 282, 429–434.

54 Tohji, K., Goto, T., Takahashi, H., Shinoda, Y., Shimizu, N., Jeyadevan, B., Saito, I., Miasuoka, Y., Kasuya, A., Ohsuna, T., Hiraga, K., Nashina, Y. Purifying single-walled nanotubes, Nature **1996**, 383, 679.

55 Zhou, O., Gao, B., Bower, D., Fleming, L., Shimoda, H. Structure and electrochemical properties of carbon nanotube intercalation compounds, *Mol. Cryst. Liq. Cryst.* **2000**, 340, 541.

56 Liu, J., Rinzler, A. G., Dai, H., Hafner, J. H., Bradley, R. K., Boul, P. J., Lu, A., Iverson, T., Shelimov, K., Huffman, C. B., Rodriguez-Macias, F., Shon, Y. S., Lee, T. R., Colbert, D. T., Smalley, R. E. Fullerene pipes, *Science* **1998**, 280, 1253–1256.

57 Kovtyukhova, N. I., Mallouk, T. E., Pan, L., Dickey, E. C. Individual single-walled nanotubes and hydrogels made by oxidative exfoliation of carbon nanotube ropes, *J. Am. Chem. Soc.* **2003**, 125, 9761–9769.

58 Shaffer, M. S. P., Windle, A. H. Analogies between polymer solutions and carbon nanotube dispersions, *Macromolecules* **1999**, 32, 6864–4866.

59 Star, A., Stoddart, J. F., Steuerman, D., Diehl, M., Boukai, A., Wong, E. W., Yang, X., Chung, S., Choi, H., Heath, J. R. Preparation and properties of polymer-wrapped single-walled carbon nanotubes, *Angew. Chem. Int. Ed.* **2001**, 40, 1721–1275.

60 Riggs, J. E., Guo, Z. X., Carroll, D. L., Sun, Y. P. Strong luminescence of solubilized carbon nanotubes, *J. Am. Chem. Soc.* **2000**, 122, 5879–5880.

61 Wang, J., Musameh, M., Lin, Y. Solubilization of carbon nanotubes by Nafion toward the preparation of amperometric biosensors, *J. Am. Chem. Soc.* **2003**, 125, 2408–2409.

62 Star, A., Han, T., Joshi, V., Stetter, J. R. Sensing with Nafion coated carbon nanotube field-effect transistors, *Electroanalysis* **2004**, 16, 108–112.

63 Wang, J., Liu, G., Jan, M. R. Ultrasensitive electrical biosensing of proteins and DNA: Carbon-nanotube derived amplification of the recognition and transduction events, *J. Am. Chem. Soc.* **2004**, 126, 3010–3011.

64 Wang, J., Chen, G., Chatrathi, M. P., Musameh, M. Capillary electrophoresis microchip with a carbon nanotube-modified electrochemical detector, *Anal. Chem.* **2004**, 76, 298–302.

65 Liu, Z., Shen, Z., Zhu, T., Hou, S., Ying, L., Shi, Z., Gu, Z. Organizing single-walled carbon nanotubes on gold using a wet chemical self-assembling technique, *Langmuir* **2003**, 16, 3569–3573.

66 Gooding, J. J., Wibowo, R., Liu, J., Yang, W., Losic, D., Orbons, S., Mearns, F. J., Shapter, J. G., Hibbert, D. B. Protein electrochemistry using aligned carbon nanotube

arrays, *J. Am. Chem. Soc.* **2003**, 125, 9006–9007.

67 Yu, X., Chattopadhyay, D., Galeska, I., Papadimitrakopoulos, F., Rusling, J. F. Peroxidase activity of enzymes bound to the ends of single-wall carbon nanotube forest electrodes, *Electrochem. Commun.* **2003**, 5, 408–411.

68 Kim, B., Sigmund, W. M. Self-alignment of shortened multiwall carbon nanotubes on polyelectrolyte layers, *Langmuir* **2003**, 19, 4848–4851.

69 Azamian, B. R., Davis, J. J., Coleman, K. S., Bagshaw, C. B. B., Green, M. L. J. Bioelectrochemical single-walled carbon nanotubes, *J. Am. Chem. Soc.* **2002**, 124, 12 664–12 665.

70 Davis, J. J., Coleman, K. S., Azamian, B. R., Bagshaw, C. B., Green, M. L. H. Mical and biochemical sensing with modified single walled carbon nanotubes, *Chem. Eur. J.* **2003**, 9, 3732–3739.

71 Azamian, B. R., Coleman, K. S., Davis, J. J., Hanson, N., Green, M. L. J. Directly observed covalent coupling of quantum dots to single-wall carbon nanotubes, *Chem. Commun.* **2002**, 366.

72 Ajayan, P. M., Stephan, O., Colliex, C., Trauth, D. Aligned carbon nanotube arrays formed by cutting a polymer resin-nanotube composite, *Science* **1994**, 265, 1212–1214.

73 Gao, M., Dai, L., Wallace, G. G. Biosensors based on aligned carbon nanotubes coated with inherently conducting polymers, *Electroanalysis* **2003**, 15, 1089–1094.

74 Xu, Y., Jiang, Y., Cai, H., He, P., Fang, Y. Electrochemical impedance detection of DNA hybridization based on the formation of M-DNA on polypyrrole/carbon nanotube modified electrode, *Anal. Chim. Acta* **2004**, 516, 19–27.

75 Gavalas, V. G., Law, S. A., Christopher, B. J., Andrews, R., Bachas, Carbon, L. G. Nanotube aqueous sol-gel composites: Enzyme-friendly platforms for the development of stable biosensors, *Anal. Biochem.* **2004**, 329, 247–252.

76 Wang, J., Musameh, M. Carbon nanotube/Teflon composite electrochemical sensors and biosensors, *Anal. Chem.* **2003**, 75, 2075–2079.

77 Hrapovic, S., Liu, Y., Male, K. B., Luong, J. H. Electrochemical biosensing platforms using platinum nanoparticles and carbon nanotubes, *Anal. Chem.* **2004**, 76, 1083–1088.

78 Tang, H., Chen, J., Yao, S., Nie, L., Deng, G., Kuang, Y. Carbon nanotube aqueous sol-gel composites: Enzyme-friendly platforms for the development of stable biosensors, *Anal. Biochem.* **2004**, 331, 97–252.

79 Wang, J., Liu, G., Jan, M. R., Zhu, Q. Electrochemical detection of DNA hybridization based on carbon-nanotubes loaded with CdS tags, *Electrochem. Commun.* **2003**, 5, 1000–1004.

80 Chen, R. J., Zhang, Y., Wang, D., Dai, H. Noncovalent sidewall functionalization of single-walled carbon nanotubes for protein immobilization, *J. Am. Chem. Soc.* **2001**, 123, 3838–3839.

81 Chen, R. J., Bangsaruntip, S., Drouvalakis, K. A., Kam, N., Shim, M., Li, Y., Kim, W., Utz, P. J., Dai, H. Noncovalent functionalization of carbon nanotubes for highly specific electronic biosensors, *Proc. Natl. Acad. Sci. U.S.A.* **2003**, 100, 4984–4989.

82 Besteman, K., Lee, J.-O., Wiertz, F. G. M., Heering, H. A., Dekker, C. Enzyme-coated carbon nanotubes as single-molecule biosensors, *Nano Lett.* **2003**, 3, 727–730.

83 Baker, S. E., Cai, W., Lasseter, T. L., Weidkamp, K. P., Hamers, R. J. Covalently bonded adducts of deoxyribonucleic acid (DNA) oligonucleotides with single-wall carbon nanotubes: Synthesis and hybridization, *Nano Lett.*, **2002**, 2, 1413–1417.

84 Hazani, M., Naaman, R., Hennrich, F., Kappes, M. M. Confocal fluorescence imaging of DNA-functionalized carbon nanotubes, *Nano Lett.*, **2003**, 3, 153–155.

85 Williams, K. A., Veenhuizen, P. T. M., de la Torre, B. G., Eritja,

R., Dekker, C. Nanotechnology: Carbon nanotubes with DNA recognition, *Nature* **2002**, 420, 761.

86 Melle-Franco, M., Marcaccio, M., Paolucci, D., Paolucci, F., Georgakilas, V. D., Guldi, M., Prato, M., Zerbetto, F. Cyclic voltammetry and bulk electronic properties of soluble carbon nanotubes, *J. Am. Chem. Soc.* **2004**, 126, 1646–1647.

87 Zhao, Y., Zhang, W., Chen, H., Luo, Q. Electrocatalytic oxidation of cysteine at carbon nanotube powder microelectrode and its detection, *Sen. Actuators, B: Chem.* **2003**, 92, 279–285.

88 Wang, Z., Liu, J., Liang, Q., Wang, Y., Luo, G. Carbon nanotube-modified electrodes for the simultaneous determination of dopamine and ascorbic acid, *Analyst* **2002**, 127, 653.

89 Wang, Z., Wang, Y., Luo, G. A selective voltammetric method for uric acid detection at β-cyclodextrin modified electrode incorporating carbon nanotubes, *Analyst*, **2002**, 127, 1353.

90 Wu, K., Fei, J., Hu, S. Simultaneous determination of dopamine and serotonin on a glassy carbon electrode coated with a film of carbon nanotubes, *Anal. Biochem.* **2003**, 318, 100–106.

91 Musameh, M., Wang, J., Merkoci, A., Lin, Y. Low-potential stable NADH detection at carbon-nanotube-modified glassy carbon electrodes, *Electrochem. Commun.* **2002**, 4, 743–746.

92 Ye, J., Wen, Y., Zhang, W., Gan, L. M., Xu, G., Sheu, F. Nonenzymatic glucose detection using multi-walled carbon nanotube electrodes, *Electrochem. Commun.* **2004**, 6, 66–70.

93 Li, J., Ng, H. T., Cassell, A., Fan, W., Chen, H., Ye, Q., Koehne, J., Han, J., Meyyappan, M. Carbon nanotube nanoelectrode array for ultrasensitive DNA detection *Nano Lett.* **2003**, 3, 597–602.

94 Boussaad, S., Tao, N. J., Zhang, R., Hopson, T., Nagahara, L. A. *In situ* detection of cytochrome c adsorption with single walled carbon nanotube device, *Chem. Commun.* **2003**, 1502.

95 Xin, H., Woolley, A. T. DNA-templated nanotube localization, *J. Am. Chem. Soc.* **2003**, 125, 8710–8711.

96 Reyes, D. R., Iossifidis, D., Auroux, P. A., Manz, A. Micro total analysis systems. 1. Introduction, theory, and technology, *Anal. Chem.* **2002**, 74, 2623–2636.

97 Yamamoto, K., Shi, G., Zhou, T., Xu, F., Xu, J., Kato, T., Jin, J. Y., Jin, L. Study of carbon nanotubes–HRP modified electrode and its application for novel on-line biosensors, *Analyst*, **2003**, 128, 249.

98 Ghosh, S., Sood, A. K., Kumar, N. Carbon nanotube flow sensors, *Science*, **2003**, 299, 1042–1044.

99 Lin, Y., Lu, F., Wang, J. Disposable carbon nanotube modified screen-printed biosensor for amperometric detection of organophosphorus pesticides and nerve agents, *Electroanalysis* **2004**, 16, 145–149.

100 Morf, W. E., De Rooij, N. F. Performance of amperometric sensors based on multiple microelectrode arrays, *Sens. Actuators, B-Chem.*, **1997**, 44, 538–541.

101 Tu, Y., Lin, Y., Ren, Z. F. Nanoelectrode arrays based on low site density aligned carbon nanotubes, *Nano Lett.* **2003**, 3, 107–109.

102 Ng, H. T., Fang, A., Li, J., Li, S. F. Flexible Carbon nanotubes membrane sensory system – A generic platform, *J. Nanosci. Nanotechnol.* **2001**, 1, 375–379.

103 Lin, Y., Lu, F., Tu, Y., Ren, Z. Glucose biosensors based on carbon nanotube nanoelectrode ensembles, *Nano Lett.* **2004**, 4, 191–195.

104 He, P., Dai, L. Aligned carbon nanotube–DNA electrochemical sensors, *Chem. Commun.* **2004**, 3, 348.

105 Wohlstadter, J. N., Wilbur, J. L., Sigal, G. B., Biebuyck, H. A., Billadeau, M. A., Dong, L., Fischer, A. B., Gudibande, S. R., Jameison, S. H., Kenten, J. H., Leginus, J., Leland, J. K., Massey, R. J., Wohlstadter, S. J. Carbon nanotube-based biosensor, *Adv. Mater.* **2003**, 15, 1184–1187.

106 Guo, Z., Dong, S. Electrogenerated chemiluminescence from $Ru(Bpy)_3^{2+}$ ion-exchanged in carbon nanotube/perfluorosulfonated ionomer composite films, *Anal. Chem.* **2004**, 76, 2683–2688.

107 Hazani, M., Naaman, R., Hennrich, F., Kappes, M. M. Confocal fluorescence imaging of DNA-functionalized carbon nanotubes, *Nano Lett.* **2003**, 3, 153–155.

108 Baker, S. E., Cai, W., Lasseter, T. L., Weidkamp, K. P., Hamers, R. J. Covalently bonded adducts of deoxyribonucleic acid (DNA) oligonucleotides with single-wall carbon nanotubes: Synthesis and hybridization, *Nano Lett.* **2002**, 2, 1413–1417.

109 Dai, H. Carbon nanotubes: Synthesis, integration, and properties, *Acc. Chem. Res.* **2002**, 35, 1035–1044.

110 Tans, S. J., Verschueren, A. R. M., Dekker, C. *Nature*, **1998**, 393, 49.

111 Kam, N. W. S., O'Connell, M., Wisdom, J. A., Dai, H. J. Carbon nanotubes as multifunctional biological transporters and near-infrared agents for selective cancer cell destruction, *Proc. Natl. Acad. Soc. U.S.A.* **2005**, 102, 11 600–11 605.

112 Curreli, M., Li, C., Sun, Y. H., Lei, B., Gundersen, M. A., Thompson, M. E., Zhou, C. W. Selective functionalization of In2O3 nanowire mat devices for biosensing applications, *J. Am. Chem. Soc.* **2005**, 127, 6922–6923.

113 Melosh, N., Boukai, A., Diana, F., Gerardot, B., Badolato, A., Petroff, P., Heath, J. R. Ultrahigh density nanowire lattices and circuits, *Science*, **2003**, 300, 112.

114 Juhasz, R., Elfstrom, N., Linnros, J. Controlled fabrication of silicon nanowires by electron beam lithography and electrochemical size reduction, *Nano Lett.* **2005**, 5, 275–280.

115 Li, Z., Chen, Y., Li, X., Kamins, T. I., Nauka, K., Williams, R. S. Sequence-specific label-free DNA sensors based on silicon nanowires, *Nano Lett.* **2004**, 4, 245–247.

116 Li, Z., Rajendran, B., Kamins, T. I., Li, X., Chen, Y., Williams, R. S. Silicon nanowires for sequence-specific DNA sensing: Device fabrication and simulation, *Appl. Phys. A-Mater. Sci. Proc.* **2005**, 80, 1257–1263.

117 Huang, J. X., Virji, S., Weiller, B. H., Kaner, R. B. Polyaniline nanofibers: Facile synthesis and chemical sensors, *J. Am. Chem. Soc.* **2003**, 125, 314–315.

118 Kameoka, J., Verbridge, S. S., Liu, H. Q., Czaplewski, D. A., Craighead, H. G. Fabrication of suspended silica glass nanofibers from polymeric materials using a scanned electro-spinning source, *Nano Lett.* **2004**, 4, 2105–2108.

119 Ramanathan, K., Bangar, M. A., Yun, M. H., Chen, W. F., Mulchandani, A., Myung, N. V. Individually addressable conducting polymer nanowires array, *Nano Lett.* **2004**, 4, 1237–1239.

120 He, H. X., Li, C. Z., Tao, N. J. Conductance of polymer nanowires fabricated by a combined electrodeposition and mechanical break junction method, *Appl. Phys. Lett.* **2001**, 78, 811–813.

121 Zhang, H., Boussaad, S., Ly, N., Tao, N. J. Magnetic field-assisted assembly of metal/polymer/metal junction sensor, *Appl. Phys. Lett.*, **2004**, 84, 133–135.

122 Forzani, E. S., Zhang, H. Q., Nagahara, L. A., Amlani, I., Tsui, R., Tao, N. J. A conducting polymer nanojunction sensor for glucose detection, *Nano Lett.* **2004**, 4, 1785–1788.

123 Ramanathan, K., Bangar, M. A., Yun, M., Chen, W., Myung, N. V. Mulchandani, A. Bioaffinity sensing using biologically functionalized conducting-polymer nanowire, *J. Am. Chem. Soc.* **2005**, 127, 496–497.

124 Alam, M. M., Wang, J., Guo, Y. Y., Lee, S. P., Tseng, H. R. Electrolyte-gated transistors based on conducting polymer nanowire junction arrays, *J. Phys. Chem. B* **2005**, 109, 12 777–12 784.

125 Zhang, D. H., Liu, Z. Q., Li, C., Tang, T., Liu, X. L., Han, S., Lei, B., Zhou, C. W. Detection of NO2 down to ppb levels using individual and multiple In2O3 nanowire devices, *Nano Lett.* **2004**, 4, 1919–1924.

126 Fritz, J., Baller, M. K., Lang, H. P., Rothuizen, H., Vettiger, P., Meyer, E., Guntherodt, H. J., Gerber, C., Gimzewski, J. K. Translating biomolecular recognition into nanomechanics, *Science* **2000**, 288, 316–318.

127 Hansen, K. M., Ji, H. F., Wu, G. H., Datar, R., Cote, R., Majumdar, A., Thundat, T. Cantilever-based optical deflection assay for discrimination of DNA single-nucleotide mismatches, *Anal. Chem.* **2001**, 73, 1567–1571.

128 Stevenson, K. A., Mehta, A., Sachenko, P., Hansen, K. M., Thundat, T. Nanomechanical effect of enzymatic manipulation of DNA on microcantilever surfaces, *Langmuir* **2002**, 18, 8732–8736.

129 Wu, G. H., Datar, R. H., Hansen, K. M., Thundat, T., Cote, R. J., Majumdar, A. Bioassay of prostate-specific antigen (PSA) using microcantilevers, *Nat. Biotechnol.* **2001**, 19, 856–860.

130 Tian, F., Hansen, K. M., Ferrell, T. L., Thundat, T. Dynamic microcantilever sensors for discerning biomolecular interactions, *Anal. Chem.* **2005**, 77, 1601–1606.

131 Su, M., Li, S. U., Dravid, V. P. Microcantilever resonance-based DNA detection with nanoparticle probes, *Appl. Phys. Lett.* **2003**, 82, 3562–3564.

132 Chen, H., Han, J., Li, J., Meyyappan, M. Microelectronic DNA assay for the detection of BRCA1 gene mutations, *Biomed. Microdev.* **2004**, 6, 55–60.

133 Martel, R., Schmidt, T., Shea, H. R., Hertel, T., Avouris, Ph. Single- and multiwall carbon nanotube field-effect transistors, *Appl. Phys. Lett.* **1998**, 73, 2447–2449.

4
Fullerene-based Electrochemical Detection Methods for Biosensing

Nikos Chaniotakis

4.1
Introduction

The football-like structure of fullerenes was discovered by Harold Kroto, James Heath, Sean O'Brien, Robert Curl, and Richard Smalley in 1985 [1]. These large nanostructures are made up of only carbon atoms, and are called Fullerenes after the American architect Richard Buckminster Fuller [2]. Buckminster Fuller designed large dome-shaped structures, such as the famous geodesic dome in Montreal Canada, built in 1967 for the Expo '67. Geodesic domes have no internal supports; they are light, and rigid, made up of specific geometrical prefabricated panels that can be assembled and taken apart quickly. Similarly, fullerenes have only carbon atoms, in a football like shell, made of five- and six-membered rings that are fused together.

The diameter of C_{60} Fullerene is so small, approximately 10 Å (10^{-10} m) [3], that even the state of the art high resolution electron microscope cannot give us a clear image of it. The molecular model and the simulated HRTEM images of C_{60} fullerene [4] shown in Fig. 4.1 can help us visualize what this material actually looks like, while Fig. 4.2 can help us visualize how small it actually is. The size ratios between the earth ($d = 1.275 \times 10^{-7}$ m) and the football ($d = 0.22$ m) is approximately the same as that between the football and the C_{60} fullerene.

4.2
Aims of the Chapter

This chapter addresses the use of the nanostructured fullerenes in the area of electrochemical biosensing, and its implementation to the development of electrochemical biosensors. Therefore, the basic concepts of both the idea of biosensing and the electrochemical properties of fullerenes must be well understood. This knowledge will allow for the understanding of the usefulness of the fullerenes in biosensing, as well as the possibilities provided for future applications. Fullerenes

Nanotechnologies for the Life Sciences Vol. 8
Nanomaterials for Biosensors. Edited by Challa S. S. R. Kumar

ISBN: 978-3-527-31388-4

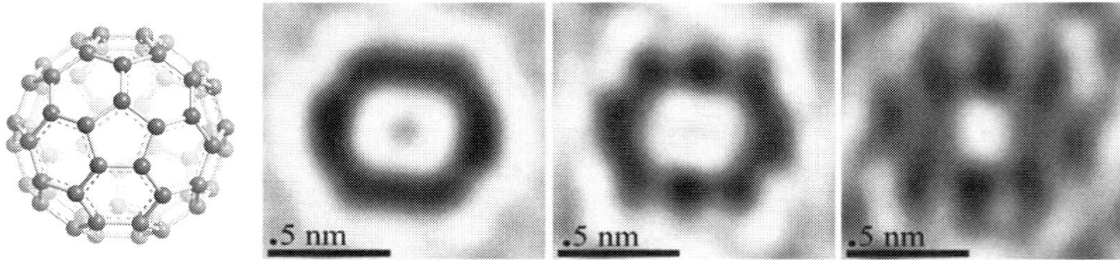

Fig. 4.1. Left: Molecular model drawing of C_{60}. Right: simulated HRTEM images focused at different planes of C_{60}. (Adopted from Ref. [4].)

are structures, or molecules, that have been at the forefront of research and public interest for more than 20 years now. Despite this, their applications are quite limited, mainly because their physicochemical characteristics are still not very well understood. The use of the fullerenes and their derivatives can provide multidimensional advances in the area of chemical sensors and biosensors in particular. Their electrochemical characteristics, combined with their unique physicochemical properties, provide the grounds for such claims, even though the current range of applications in this specific scientific area is rather limited.

This chapter provides a complete overview of fullerene-based electrochemical detection methods with specific emphasis on biosensing. Initially, it provides the background on the development of fullerenes as well as their electrochemical characteristics. This information is followed by a detailed description of the evolution and design characteristics of sensing systems based on biomolecules, the so-called biosensing. This combination provides the reader with well-rounded information on the exact role of the fullerenes in biosensing. Finally, it gives direction on the future role of these and similar nanostructures for possible applications in chemistry and biochemistry, both *in vitro*, and *in vivo*.

  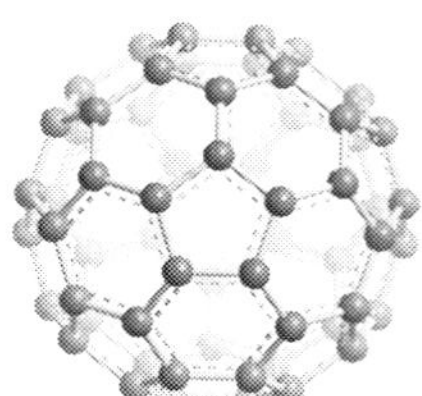

Fig. 4.2. The relative size of C_{60} to a football is approximately the same as that of a football to the earth.

4.3 Electrochemical Biosensing

The monitoring of the chemical and biochemical substances involved in the biological world is one of the important biological functions that make life possible. This is because the detection of substrates, signals and other external stimuli allow for direct interaction of a living system with the environment, and thus make life viable. Such information can also be very useful in many everyday human activities. Health issues, environmental problems, food quality and safety are some of the human activity areas that need direct and selective quantitative biochemical information. For this reason researchers for more than 50 years, and since the introduction of the oxygen electrode [5], have been using biosensing principles for the development of analytical instruments. These efforts have led to the development of a series of analytical devices called "biosensors" [6].

According to IUPAC [7], a biosensor is "A device that uses specific biochemical reactions mediated by isolated enzymes, immunosystems, tissues, organelles or whole cells to detect chemical compounds usually by electrical, thermal or optical signals." Based on this definition, electrochemical biosensing is the recognition process that is based on the use of a specific biochemical reaction or mediated by isolated enzymes, immunosystems, tissues, organelles or whole cells, and the signal of the processes involved is relayed (transduced) to the analysts using current, potential, or impedance. The transduction is based on the interaction of the activity of electrochemically active species that are in a close contact and adjacent to the working electrode, the transducer. These species are then reduced or oxidized, depending on the experimental conditions, and the resulting signal is monitored. The signal that is obtained is related to the analyte of interest, and thus the analytical information sought is obtained. While the transduction mechanism has to do with the electrode and electronics employed, the difficult task of analyte recognition is undertaken by the biosensing element. The biosensing element is usually a membrane, or a layer of material into which the enzymes, immunosystems, tissues, organelles or whole cells are immobilized.

Electrochemical biosensing methods are characterized by the fact that they can provide continuous, on-line information on the activity of an analyte. Analytes that are usually detected using biosensors are substances that can either take part in an enzyme-catalyzed reaction (catalytic biosensors), or inhibit an enzyme-catalyzed reaction (inhibition biosensors). Electrochemical biosensors can thus provide on-line and continuous quantitative biochemical information on the activity of the analyte, either within a biological system or in the environment surrounding these biological systems. This information is based on the electrochemical signals that result from a biological recognition event, and it is manifested through the electrode, the so-called transducer. The transducer can be any surface or matrix that can interact with the biological recognizing element, relaying the electrochemical recognizing process to the display and the analyst. Figure 4.3 shows a schematic diagram of all the components of a biosensor.

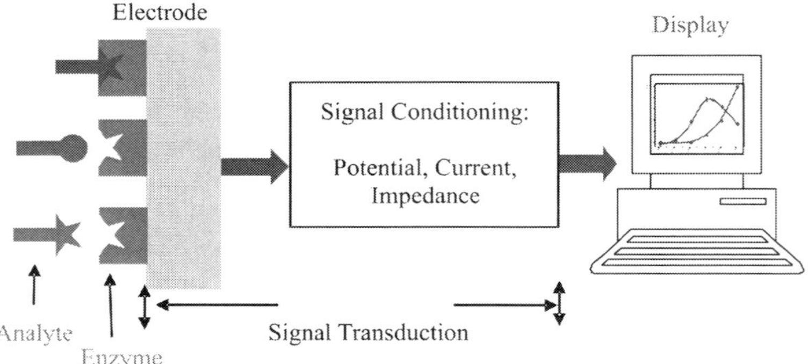

Fig. 4.3. Schematic diagram showing the main components of a biosensor. The analyte interacts with the biological recognition element (biocatalyst) and it is converted into product(s). This reaction is monitored by the signal transduction system, converting it into an electrical signal. The output from the transducer is amplified, processed, and displayed to the analyst.

The role of chemistry and biochemistry is to optimize the biological recognition element, as well as the transducer surface that is in contact with this biosensing element. The biosensing process is usually based on the what is called the lock and key concept [8] or, in a more modern term, the host–guest interaction. The host is the biological molecule or the system that is responsible for the recognition, while the guest is the analyte or the species that we are interested in recognizing and quantitatively measuring its concentration or activity in the sample.

In the electrochemical biosensors the signal transduction takes place using the electrochemical properties of the species involved in the recognition process. Current, potential and resistance are the main parameters that can be utilized for decoding the effect a stimulus has (the analyte) on a recognition element. Clearly, there are various processes involved within a biosensor, which will transform a biochemical information into analytically useful data. For this reason the proper interaction and communication between the biological part, the biochemical processes, the transducers and other materials used is very important. To this, one must add the understanding and control of the electrochemical processes involved throughout the system. Only under these conditions can a biosensor device be designed that will provide analytically useful information on the presence of a specific analyte. At the same time, the proper interaction of these disciplines and materials involved will in the long run determine the stability of the device, the selectivity over other interfering substances, the detection limit that can be achieved, as well as the size of the device and its possible applications. As shown in Fig. 4.3, the sensor element is responsible for recognition of the analyte, while the transducer translates this information into the appropriate signal for recording. These two processes should be spatially very close, especially when electrochemical detection is utilized, without any inverse effects. Electrochemical processes are those involving the three basic ohms law parameters, that is current (I), potential (V), and resistance (R). Based on these parameters, electrochemical biosensors are categorized as amperometric, potentiometric, or impedometric. The simplicity and the sensitivity of the measurement of these parameters is the basis for the wide range of application of these methods. Electrochemical biosensing has an additional advantage, namely its ability to monitor localized events taking place at the interface between the biosensing element and the system under investigation. For all these reasons, electrochemical biosensors have found a very wide range of applications, while scientists all over the world are still working on improving their characteristics, as well as on finding new applications.

4.3.1 Making a Biosensor

As an example, an outline of how a biosensor can be made in simple steps is provided here. Even though some technical terms are omitted for simplicity, the system described can be used for the measurement of an analyte such as glucose with very good analytical characteristics.

Making a biosensor has become by now a relatively simple procedure. The main components as described in Fig. 4.3 are:

1. The enzyme for the biorecognition.
2. The membrane(s) to hold the enzyme and to complete the biosensor element.
3. The transducer electrode, which includes the working and the counter electrode and, if possible, a reference electrode.
4. The electronic signal processing and display unit.

All these materials and tools are available, at a relatively low cost. A very good detailed description of the procedure can be found in the book *Biosensors. A Practical Approach* [9].

4.4 Evolution of Biosensors

Electrochemical biosensors have been in continuous development and improvement since their first appearance in 1969 [6]. The development of electrochemical biosensors are concomitant with the three "biosensor generations" that have appeared in the literature. The major factor that determines the successful development of biosensors is the communication between the biosensing element or biomolecule and the transducer. The first-generation electronic coupling between redox enzymes and electrodes achieved this communication with the native enzyme co-factor [10] (Fig. 4.4A). The second-generation replaces this native redox couple with an electrochemically active compound, the mediator [11]. The mediator can be either free in solution or immobilized with the biomolecule (Fig. 4.4B).

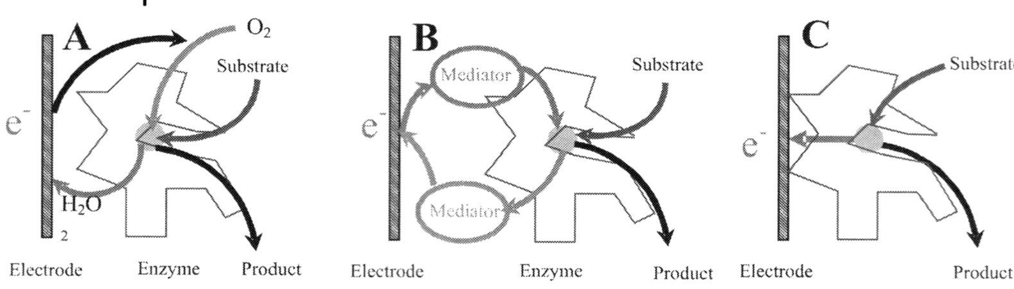

Fig. 4.4. Schematic diagram of (A) first-, (B) second-, and (C) third-generation biosensors.

Third-generation biosensors aim at direct electron transfer between the native enzyme co-factor and the electrode surface [12, 13] (Fig. 4.4C).

A mediator is a substance that acts as the "middle man" in transferring the generated signal from the active site to the transducer. For electrochemical biosensors, a mediator is a redox species (a species that can be easily reduced or oxidized) that is spatially close to both the active center and the transducer, without harmful effects to the biosensing element.

Evidently, from these schematic diagrams of the biosensors, the interaction between the analyte, the catalytic site of the biological element and transducer plays a decisive role in the performance and quality characteristics of the biosensing system. Several processes are involved. Initially, there is a diffusion process of the substrates to the active center and then another diffusion of the products from the active center to the transducer. Alternatively, the product of the reaction can provide the signal to the transducer using either a mediator, or the biomolecule itself. This is one of the main roles that fullerenes are asked to play in biosensing.

4.5 Mediation Process in Biosensors

Electrochemical detection in biosensing provides a convenient, efficient means of amplification. For such a process to occur within the closed system of a biosensor, the signal must flow to the transducer regardless of the activity of substances other than either the analyte or the product of the catalytic reaction. This task is usually undertaken by the natural coenzymes and co-factors found in the enzymes. The electrochemical reactions of the redox enzymes are known to have slow electron-transfer kinetics to and from a transducer or electrode. There are three main parameters that determine the kinetics of this electron-transfer mechanism. The first has to do with the physicochemical properties of the active site, such as charge, hydrophobicity and surface active groups. The second is related to the distance of the active site from the outer surface of the protein, while the distance of the protein from the electrode surface is the third determining parameter. For these reasons, and in order that these biocatalytic reactions can proceed at analytically useful rates, the potential used on the transducer is much larger than that required to actually oxidize or reduce the natural biomolecule mediator. Lowering the redox potential of the substrates involved in these processes will not only increase the reaction rates but, at the same time, it will improve the sensitivity and most importantly the selectivity of the biosensor [14]. Lowering the operating potential will aid in increasing the enzyme stabilization, prolonging its activity, and decreasing the rate of denaturation. To achieve this, artificial mediators or electron relays are employed that can efficiently interact with both the enzyme catalytic site and the electrode surface. Such successful mediators should have lower redox potential than that of the natural enzyme mediator so that they can act as electron acceptors during the enzymatic reaction. Also, they should be compatible with both the transducer surface as well as the biomolecule. Finally, they should be stable, while providing sites for chemical functionalization. Molecules or nanostructures that are good candidates for such electron mediation should also have fast and reversible redox states, they should be stable both in the oxidized and in the reduced form, their redox potentials should be close to those of the enzyme's active site, and they should not react with any of the reactants or products of the enzymatic reaction. The reduced form of the mediator can then transfer the electron to the electrode surface, and after they are oxidized back to its initial state are ready to proceed to the next cycle.

4.5.1
Case A: Non-mediated Biosensor

Under normal conditions the electron transfer in enzyme-catalyzed redox reactions takes place with the mediation of natural cofactors such as O_2 and NADH. Figure 4.5 shows this process for the glucose biosensor.

The electron-transfer process is carried out by the natural coenzymes found in the biosystem, and the electrons are transferred to the transducer for the signal

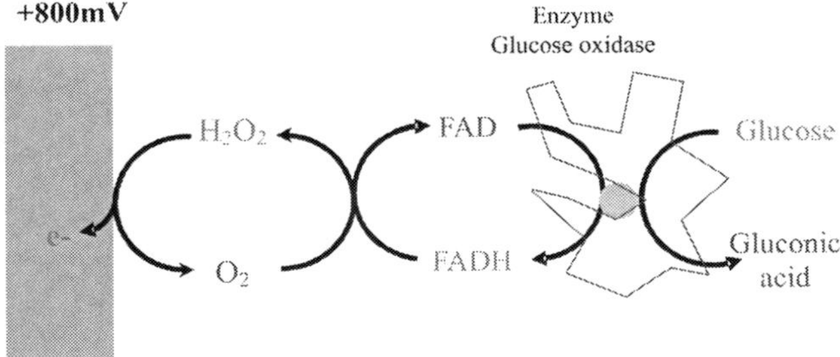

Fig. 4.5. Flowchart of the processes involved in a non-mediated electrochemical biosensor. The operating potential in the case of glucose is +800 mV.

generation. In this case, the mediation is carried out using FAD–FADH and H_2O_2–O_2 electrochemical couples. Since the system is sensitive to the activity of both peroxide and oxygen, oxygen electrodes have been extensively used as transducers in electrochemical biosensors. Evidently, close contact between the enzymes with the transducer surface is mandatory. Such heterogeneous electron-transfer reactions are energetically demanding since the catalytic site of the enzyme is deeply buried within the protein structure and protected by the hydrophobic core of the protein. This is evident by the rather large potential polarization potential of the transducer (+800 mV) required to operate the glucose biosensor (Fig. 4.5). Such large potentials have very detrimental effects on the lifetime and selectivity of the biosensor. This is why, under these potentials, the stability of the enzymes to denaturation and deactivation is drastically decreased, while at the same time all substances that can be oxidized at that potential will seriously interfere during the measurement. Based on these facts a decrease in the operating potential is a very important experimental requirement, and it can be achieved with the use of mediators.

4.5.2
Case B: Mediated Biosensor

As mentioned above, the use of mediators in the development of biosensors can provide specific solutions to operational problems of biosensors. Mediators can aid in improving the stability, the reproducibility, the selectivity and thus the range of applications of biosensors. Figure 4.6 shows the flowchart of a mediated biosensor. Numerous compounds have been proposed as mediators in biosensors. Most of them are different ferrocene derivatives or osmium complexes [15, 16]. Figure 4.7 shows two voltammograms of a glucose biosensor without (a) and with (b) the use of mediator. Note that at the operation potential of 400 mV, the signal obtained, and thus the sensitivity of the biosensor, is more than 5× higher.

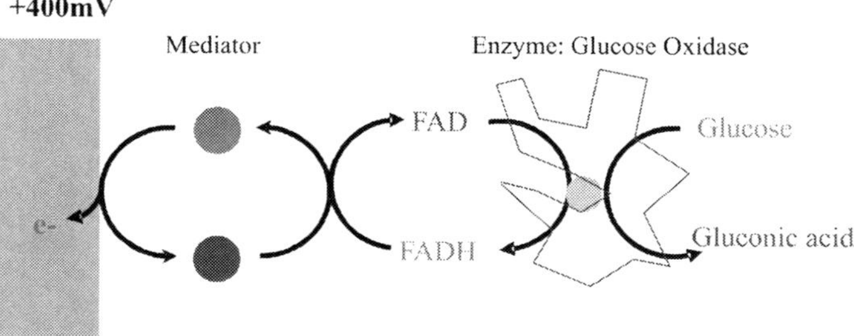

Fig. 4.6. Flowchart of the processes involved in a mediated electrochemical biosensor. Note that the operating potential has dropped to +400 mV.

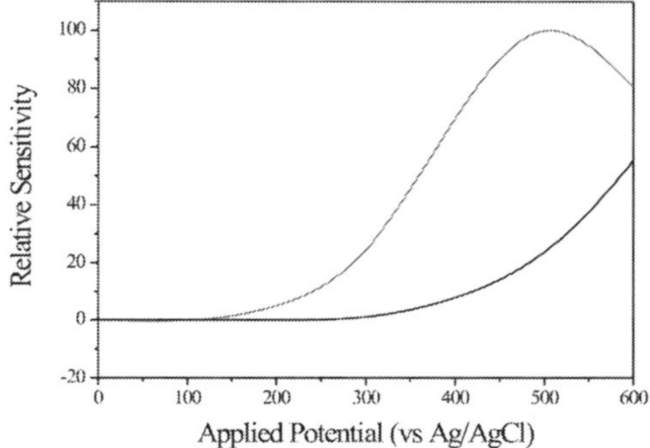

Fig. 4.7. Response of a non-mediated (a) and a mediated (b) glucose biosensor.

4.6 Fullerenes

4.6.1 Synthesis of Fullerenes

Fullerenes were discovered by Kroto and Smalley in 1985 in vaporized graphite under inert gas. Since then, several methods for the synthesis of fullerene have been reported. Many of them are based on laser ablation of carbon. Alternatively, graphite can be heated using either high power current source or, instead, an AC or DC arc discharge, to generate a "soot" of various carbon structures. Simple chromatographic separation of the soot provides high purity fullerenes [17, 18]. Based on this technology fullerenes can now be synthesized in large quantities (Fig. 4.8) [19, 20].

After the successful synthesis of C_{60} by thermal methods, fullerenes have also been synthesized using wet chemistry lab technology. A step-by-step synthesis based on polyyne is now also possible [21–23] based on the precursor shown in Fig. 4.9. The synthesis of fullerenes in a step by step procedure provides the capability to control the surface functionalization of the C_{60}. Such functionalization can add new physicochemical characteristics to these molecules, rendering them, for example, more water soluble, adding sites for covalent bonding with biomolecules, or making them more susceptible to redox reactions.

4.6.2 Biofunctionalization of Fullerenes

Fullerenes are very robust, very lipophilic and relatively unreactive nanostructures. There are a lot of instances where the physicochemical characteristics of these structures need to be altered to make them more hydrophilic, to link them to other

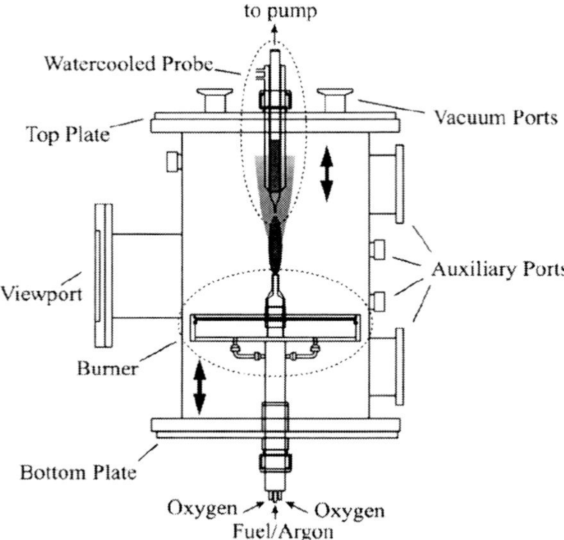

Fig. 4.8. Schematic of apparatus used to synthesize large quantities of fullerenes.

biomolecules, or immobilize them to a protein or other substrates, so that they are more suitable for a specific application. Initially, the functionalization of the fullerenes with organic and organometallic ligands has enabled them to be extensively studied in many fields. Different organic and organometallic ligands have been added by adsorption or covalent immobilization to the shell of the fullerenes, in-

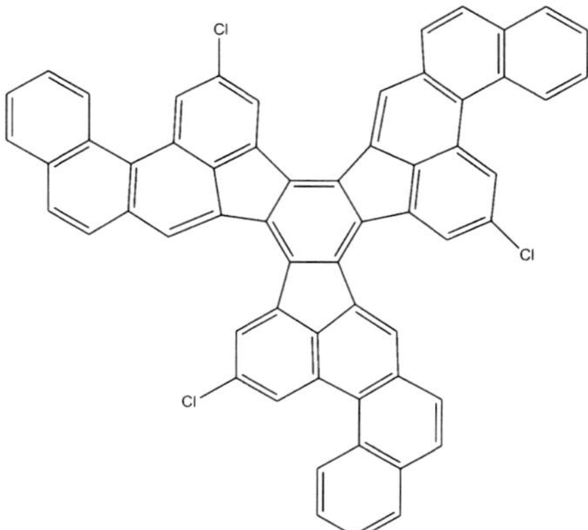

Fig. 4.9. Precursor for the chemical synthesis of C_{60}. Upon heating, chlorine is liberated, and the molecule folds up to generate C_{60}.

creasing their hydrophilicity, or altering their optical and electrochemical properties [24–31]. In addition, the ability of the fullerenes to accept and donate electrons to the species surrounding them can also be drastically influenced and elegantly controlled using these functionalization methods [32, 33]. These functionalized nanostructures have already found applications in medicine, electronics and optoelectronics [34] and, recently, biosensing applications.

One of the advantages of grafting the biomolecule onto the surface of the fullerene is the fact that these two units will be spatially very close. Such a close arrangement will allow for direct and efficient interaction of the nanostructure with the biomaterial, and thus increase the efficiency of the processes involved [35].

Various substituted C_{60} derivatives have been derived to probe DNA reactions [36] or their radical scavenging properties. Unique structures, such as that illustrated in Fig. 4.10, have proved to be very interesting both for their interaction with radicals as well as with cell walls and DNA. Similarly, very interesting water-soluble fullerene derivatives such as the one shown in Fig. 4.11 [37] have appeared in the literature. However, their application in biosensing has not been extensively evaluated. It is, for example, envisioned that such positively charged species will both protect and stabilize proteins and enzymes [38], while at the same time pro-

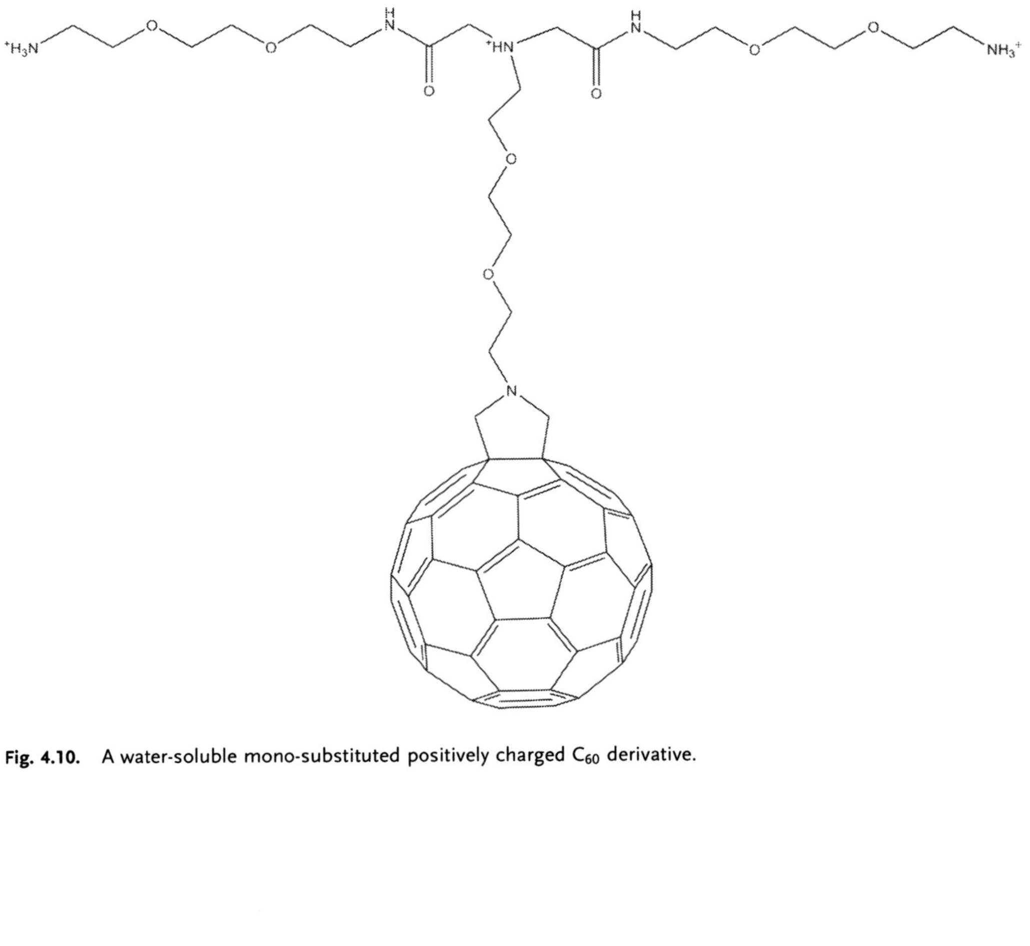

Fig. 4.10. A water-soluble mono-substituted positively charged C_{60} derivative.

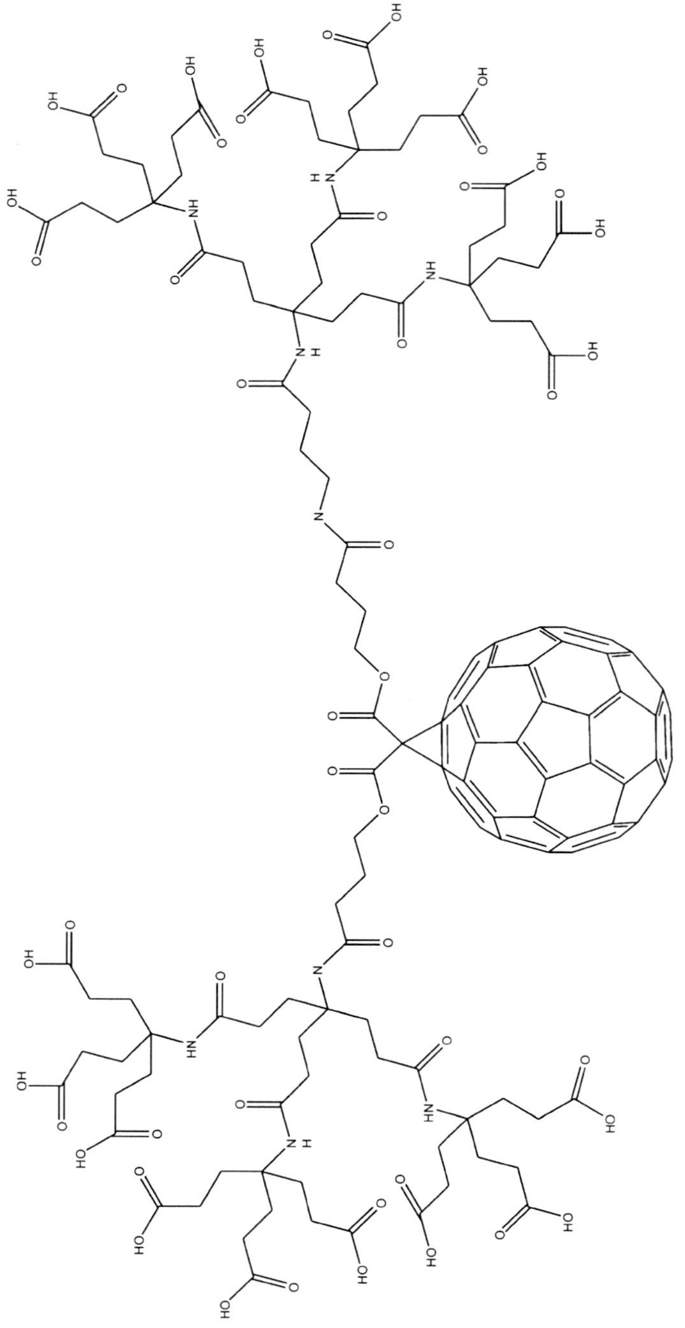

Fig. 4.11. A highly water-soluble dendro[60]fullerene.

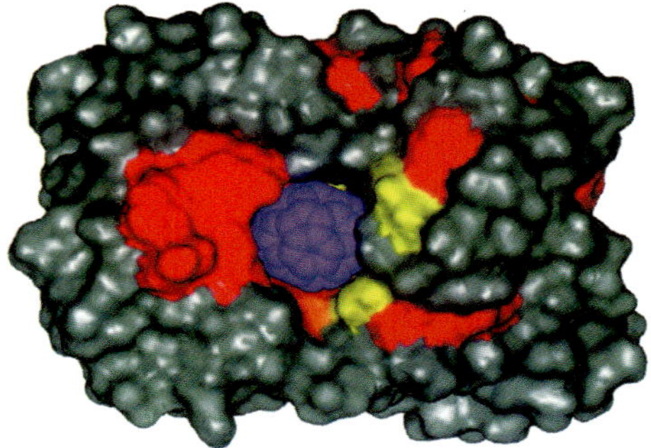

Fig. 4.12. Model of Van der Waals surface representation of C_{60} bound to the anti-fullerene antibody.

vide an efficient way for signal transduction and mediation. Finally, the discovery of C_{60} antibodies [39] (Fig. 4.12) has already paved the way for the application of such systems in piezoelectric crystal immunosensors [40].

4.6.3
Electrochemistry of Fullerenes

From the early stages of the discovery of fullerenes their electrochemical characteristics were considered to be very interesting and unique [41]. For this reason the electrochemistry of fullerenes in both solutions and films has been under intensive scientific investigation for many years [42]. Fullerene C_{60} behaves as an n-type semiconductor with a bandgap in the range of 1.6 eV, and its electron affinity is of the order of 2.7 eV [43]. Fullerenes in general have an expanded network of sp^2 hybridized bonds that allows them to undergo up to six, one-electron reversible reduction reactions between -0.61 and 1.00 V vs. SHE. Molecular orbital calculations predict very low lying triply degenerate LUMOs [44, 45]. The electron affinity of C_{60} can be explained qualitatively by considering its numerous pyracylene units, which upon receiving two electrons could go from an unstable $4n$-system to a stable aromatic $4n + 2$ system. Moreover, the formation of sp^3-like anionic centers may lower the energy of the somewhat strained fullerene surface of bent sp^2 carbons [46, 47, 48, 49].

Evidently, these early results show that the fullerene anions are stable due to the large separation between the distinct redox states and are, therefore, attractive oxidizing agents. Since then, there has been a considerable effort to clarify their electrochemical behavior in different solutions and matrices [50]. The bottom line of all these studies is that fullerenes indeed have unique electrochemical characteristics, as shown by the variety of redox couples. As carbonaceous materials, fullerenes are expected to have very unique electrochemical behavior within biological systems.

4.7
Fullerene-mediated Biosensing

As shown above, C_{60} fullerenes are nanomaterials with very unique redox characteristics. They are ideal substances for absorbing energy, taking up electrons (reduction), and releasing them (oxidation) with ease to a transducer. They are not harmful to biological materials and proteins, while they are small enough to come at least close to the active site of catalytic enzymes. At the same time they can be chemically modified, so that they can be functionalized to meet the needs of various applications.

For these reasons fullerenes were recognized very early on as materials that can play a decisive role in biocatalysis. Indeed, fullerenes have found applications in the area of biocatalysis and sensors from their early days of existence [51–54]. The idea of introducing a C_{60} chemically modified electrode to electrochemical research was first presented by Compton and coworkers in 1992 [55]. The sensor was prepared by immobilizing C_{60} films by drop coating onto surfaces of the noble metal electrodes, which were then coated with the Nafion protecting films. In this way, the amount of C_{60} required for performing electrochemical experiments was reduced and the signal enhanced compared with using C_{60} dissolved in solution. Subsequently, the electrochemical behavior of the C_{60} chemically modified electrodes (CMEs), has been widely investigated, providing the possibility of their electroanalytical applications [56]. Soon after, C_{60} fullerene film modified electrodes were investigated for the electrochemical reactions of cytochrome *c* using C_{60} fullerene deposited on glassy carbon (Fig. 4.13) [57, 58]. It was established then that the response of cytochrome *c* was quite independent of the underlying substrate (gold or glassy carbon), indicating that the fullerene film indeed acted as a promoter. Moreover, the molecular sieve type character and the possible effect of negative charges or polar groups on the surface of the films were suggested to be the reasons for the enhanced stability of the electrochemical response observed com-

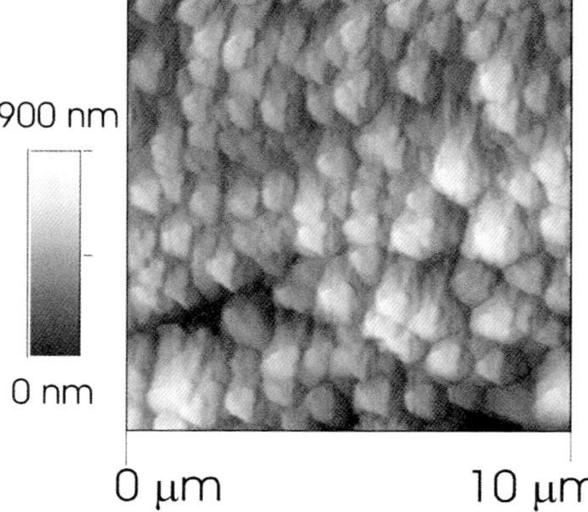

Fig. 4.13. AFM picture of a C_{60} fullerene on glassy carbon. (Adopted from Ref. [57].)

pared with bare electrodes. The rate of electron transfer to cytochrome *c* was drastically improved with the use of fullerenes as electron shuttles between the cytochrome and the carbon electrode.

These very promising initial results in biocatalysis suggested that, indeed, fullerenes can be ideal substances for electron mediation also in biosensors, since they are small, robust, biocompatible and have a wide range of oxidation/reduction potentials. In addition fullerenes are very lipophilic, while at the same time they can be chemically modified by functionalized matrix, and also with the hydrophilic proteins and the active site of the enzymes. All these characteristics make them ideal for use as mediators in electrochemical biosensing. In 1998 [59] carboxylic derivatives of C_{60} were covalently attached to a cystamine-monolayer-functionalized Au-electrode for the monitoring of biocatalysis transformations. It was then shown that the C_{60} monolayer can provide electrical communication between the electrode and a soluble glucose oxidase, GOx, with sufficiently high electron-transfer rates. The use of C_{60} as an electron mediator for electrocatalyzed biotransformations presented in this work set the stage for the direct application of C_{60} in biosensors. This was followed by reports in which fullerenes were used as mediating agents capable of charge transfer between redox enzymes and electrodes. Fullerene molecules can be used without functionalization, or after immobilization in gold electrodes with good success [60, 61].

Fullerenes has also been suggested as mediators in supported bilayer lipid membrane (s-BLM) biosensors [62–65]. s-BLMs are self-assembled systems that can be used for the design of electrochemical sensors and biosensors. S-BLMs are very thin membranes formed onto a transducer (platinum, gold, etc.) that can come in contact with the system to be analyzed. The introduction of biomolecules into these BLMs makes them suitable for the monitoring of chemical and biological species in systems such as that shown in Fig. 4.14.

Based on this idea, fullerenes have been used for the development of highly sensitive s-BLM-based chemical sensors for I^- [66]. The increase in sensitivity is because fullerenes change the electrical parameters of s-BLMs and facilitate the electron transfer of I^- at the metal surface, according to reaction scheme (1):

$$\begin{aligned} &C_{60} + I^- \rightarrow C_{60}I_3^- \\ &C_{60}I_3^- - e^- \rightarrow C_{60}I_2 + \tfrac{1}{2}I_2 \end{aligned} \qquad (1)$$

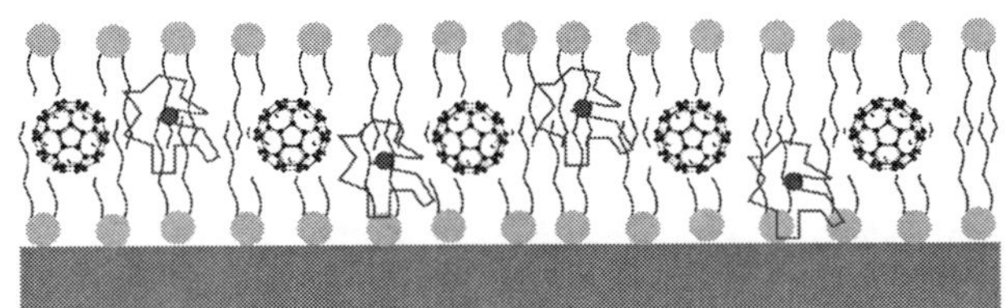

Fig. 4.14. s-BLM incorporating fullerene and enzyme for the construction of a BLM biosensor.

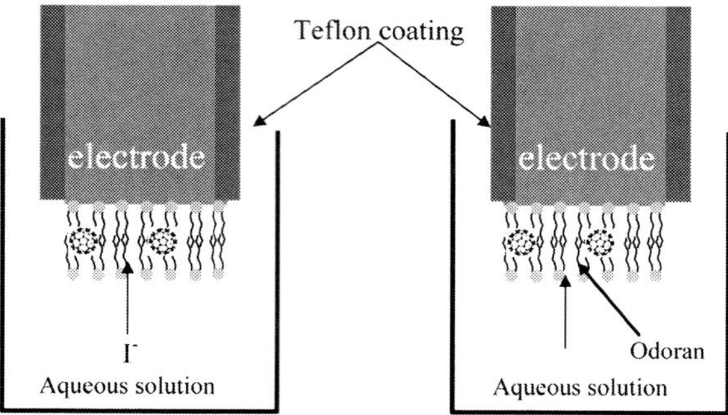

Fig. 4.15. Scheme of electrochemical sensor based on C_{60}-modified s-BLM for the detection of volatile organic compounds (odorants). (Adopted from Ref. [67].)

The same idea has been applied for the electrochemical detection of volatile organic compounds using a sensor setup such as the one shown in Fig. 4.15 [67].

In these molecular devices fullerenes can also act as light-sensitive dipoles, capable of photoinduced charge separation, which undergo redox reactions across the substrate–hydrophobic lipid bilayer–aqueous solution junctions. Based on this, light addressable devices can be developed. Indeed, recent results [68] have also indicated that the photoinduced activation or switching of electron transfer of the enzyme glucose oxidase to the transducer is possible (Fig. 4.16).

Using fullerenes as an optical nanostructure switch, the biosensor is active only under light illumination, at potentials very close to zero versus Ag/AgCl reference electrode.

The functionalized fullerenes grafted onto polyelectrolytes were shown to be significant tools in biosensor design. Polyelectrolytes are very efficient enzyme stabilization systems.

The controlled interaction with enzymes allows for the protection and stabilization of the protein from denaturization, unfolding and, thus, deactivation [69–73].

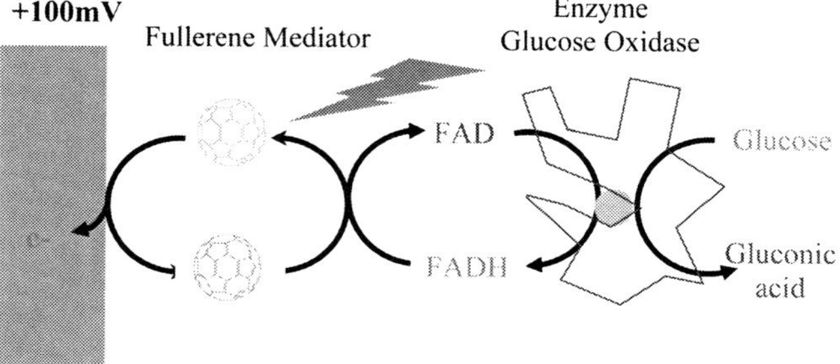

Fig. 4.16. Flowchart of the processes involved in a light-induced fullerene-mediated electrochemical biosensor. Note that the operating potential has dropped to +100 mV.

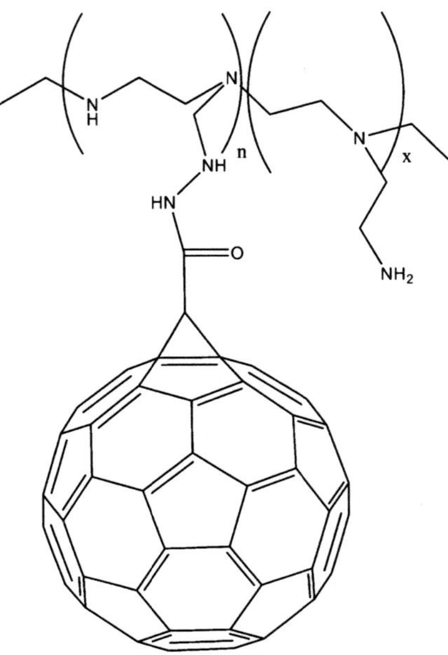

Fig. 4.17. Structure of fullerene-functionalized polyethylenimine (PEI-C_{60}). Such structures provide both stabilization and signal mediation.

Systems such as polyethylenimine functionalized with fullerenes (Fig. 4.17) via covalent bonding allow for the simultaneous stabilization of the protein, and mediation of the signal [74].

Controlled modification of the fullerene surface with biomolecules is the state of the art method to improve the catalytic redox properties of proteins (Fig. 4.18). Covalent bonding of fullerene onto proteins has significant advantages over direct

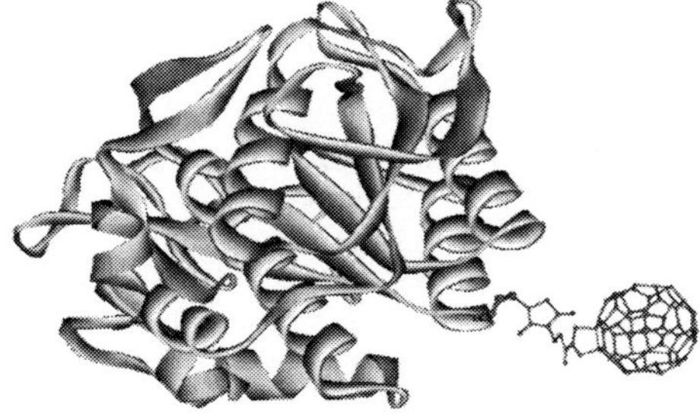

Fig. 4.18. Schematic illustration of protein conjugated to *N*-(3-maleimidopropionyl)-3,4-fullerrolidine used for the development of subtilisin-based biosensor. (Adopted from Ref. [75].)

immobilization on nonporous silica or similar matrices [75]. The small size of the C_{60} molecule makes the active site of the enzyme accessible to the substrate, and thus there is no diffusional limitation for the substrate to reach the active site imposed by the C_{60} surface. The enzyme retains its activity and behaves in a similar manner as its free solution form.

4.8 Conclusions

Even though the basic physicochemical characteristics of Fullerenes have only recently been realized, they have already contributed significantly in the area of biosensing and nanobiotechnology. In addition, the use of fullerenes and their derivatives have already provided multidimensional advances in the area of electrochemical sensors and biosensors in particular. The electrochemical characteristics of fullerenes, combined with their unique physicochemical properties lend its use in the design of novel biosensor systems, even though the current range of applications in this specific scientific area is still rather limited. Given their signal mediation, protein and enzyme functionalization and light induced switching; fullerenes can potentially provide new and powerful tools in the fabrication of electrochemical biosensors in the future.

References

1 H. Kroto, C-60 – Buckminsterfullerene. *Nature* **1985**, 318(6042), 162–163.
2 http://www.archinform.net/arch/1098.htm?ID=RQadsJnbeC0BZsIC.
3 S. E. Campbell, G. Luengo, V. I. Srdanov, F. Wudl, J. N. Israelachvili. Very low viscosity at the solid–liquid interface induced by adsorbed C60 monolayers. *Nature* **1996**, 382, 520–522.
4 A. Goel, J. B. Howard, J. B. Vander Sande. Size analysis of single fullerene molecules by electron microscopy. *Carbon* **2004**, 42, 1907–1915.
5 L. C. Clark Jr. Monitor and control of blood and tissue O_2 tensions. *Trans. Am. Soc. Artif. Intern. Organs* **1956**, 2, 41–48.
6 S. J. Updike, G. P. Hicks. The enzyme electrode. *Nature* **1967**, 214, 986–988.
7 *IUPAC Compendium of Chemical Terminology*, 2nd Edn, **1997**. http://www.iupac.org/goldbook/B00663.pdf.
8 E. Fischer. Einfluss der configuration auf die wirkung der enzyme. *Ber. Dtsch. Chem. Ges.* **1894**, 27, 2984–2993.
9 A. E. G. Cass. *Biosensors. A Practical Approach*, The Practical Approach Series, Oxford University Press, **1990**.
10 A. Karyakin, O. Gitelmacher, E. Karyakina. Prussian Blue-based first-generation biosensor. A sensitive amperometric electrode for glucose. *Anal. Chem.* **1995**, 67(14), 2419.
11 F. Scheller, F. Schubert, B. Neumann, D. Pfeiffer, R. Hintsche, I. Dransfeld, U. Wollenberger, R. Renneberg, A. Warsinke, G. Johansson, M. Skoog, X. Yang, V. Bogdanovskaya, A. Bückmann, S. Yu. Zaitsev. Second generation biosensors. *Biosensors Bioelectron.*, **1991**, 6(3), 245–253.
12 L. Gorton, A. Lindgren, T. Larsson, F. D. Munteanu, T. Ruzgas, I. Gazaryan. Direct electron transfer between heme-containing enzymes and electrodes as basis for third

generation biosensors. *Anal. Chim. Acta*, **1999**, 400(1–3), 91–108 L.

13 T. Lotzbeyer, W. Schumann, H. L. Schmidt. Minizymes. A new strategy for the development of reagentless amperometric biosensors based on direct electron-transfer processes. *Biosensors Bioelectron.*, **1997**, 12(9–10), 1–6.

14 W. Kutner, J. Wang, M. L. X. Her, R. P. Buck. Analytical aspects of chemically modified electrodes: Classification, critical evaluation and recommendations. *Pure Appl. Chem.*, **1998**, 70, 1301–1318.

15 A. Turner. Amperometric biosensors based on mediator-modified electrodes. *Methods Enzymol.*, **1988**, 137, 90–103.

16 A. Cass, G. Davis, G. Francis, A. Hill, W. Aston, J. Higgins, E. Plotkin, L. Scott, A. Turner. Ferrocene mediated enzyme electrode for amperometric determination of glucose. *Anal. Chem.*, **1984**, 56(4), 667–671.

17 E. Rohlfing. High-resolution time-of-flight mass-spectrometry of carbon and carbonaceous clusters. *J. Chem. Phys.* **1990**, 93, 7851.

18 W. Creasy, J. Zimmerman, R. Ruoff. Van de waals binding of fullerenes to a graphite plane. *J. Phys. Chem.* **1993**, 97, 973.

19 W. Krätschmer, L. D. Lamb, K. Fostiropoulos, D. R. Huffman. Solid C60: A new form of carbon. *Nature*, **1990**, 347, 354–358.

20 P. Hebgen, A. Goel, J.-B. Howard, L. C. Rainey, J. B. Vander Sande. Synthesis of fullerenes and fullerenic nanostructures in a low-pressure benzene/oxygen flame. *Proc. Combustion Institute*, **2000**, 28, 1397–1404.

21 S. W. McElvany, M. M. Ross, N. S. Gorof, F. Diederich. Cyclocarbon coalescence: Mechanisms for tailor-made fullerene formation. *Science*, **1993**, 259, 1594–1596.

22 L. Scott, M. Boorum, B. J. McMahon, S. Hagen, J. Mack, J. Blank, H. Wegner, A. de Meijere. A rational chemical synthesis of C60. *Science*, **2002**, 295, 1500–1503.

23 A. Yasuda. Chemical synthesis scheme for a C60 fullerene. *Carbon*, **2005**, 43(4), 889–892.

24 D. M. Guldi. Fullerene-porphyrin architectures; photosynthetic antenna and reaction center models. *Chem. Soc. Rev.* **2002**, 31, 22–36.

25 B. Gotschy. Magnetism in C60 charge transfer complexes. *Fullerene Sci. Technol.*, **1996**, 4, 677–698.

26 D. V. Konarev, A. Yu. Kovalevsky, A. L. Litvinov, N. V. Drichko, B. P. Tarasov, P. Coppens, R. N. Lyubovskaya. Molecular complexes of fullerenes C60 and C70 with saturated amines. *J. Solid State Chem.*, **2002**, 168, 474–485.

27 D. V. Konarev, R. N. Lyubovskaya, N. V. Drichko, E. I. Yudanova, Yu. M. Shul'ga, A. L. Litvinov, V. N. Semkin, B. P. Tarasov. Donor-acceptor complexes of fullerene C60 with organic and organometallic donors? *J. Mater. Chem.*, **2000**, 803–818.

28 M. M. Olmstead, D. A. Costa, K. Maitra, B. C. Noll, S. L. Phillips, P. M. Van Calcar, A. L. Balch, J. Interaction of curved and flat molecular surfaces. The structures of crystalline compounds composed of fullerene (C60, C60O, C70, and C120O) and metal octaethylporphyrin units. *J. Am. Chem. Soc.* **1999**, 121, 7090–7097.

29 D. V. Konarev, I. S. Neretin, Yu. L. Slovokhotov, E. I. Yudanova, N. V. Drichko, Yu. M. Shul'ga, B. P. Tarasov, L. L. Gumanov, A. S. Batsanov, J. A. K. Howard, R. N. Lyubovskaya. New molecular complexes of fullerenes C60 and C70 with tetraphenylporphyrins [M(tpp)], in which M = H2, Mn, Co, Cu, Zn, and FeCl. *Chem. Eur. J.*, **2001**, 7, 2605–2516.

30 D. V. Konarev, A. Yu. Kovalevsky, P. Coppens, R. N. Lyubovskaya. Synthesis and crystal structure of a C60 complex with a bis(ethylenedithio)-tetrathiafulvalene radical cation salt: (BEDT-TTF-I3)C60. *Chem. Commun.*, **2000**, 2357–2358.

31 G. L. Marcorin, T. D. Ros, S. Castellano, G. Stefancich,

I. Bonin, S. Miertus, M. Prato. Design and synthesis of novel [60]fullerene derivatives as potential HIV aspartic protease inhibitors. *Org. Lett.*, **2000**, 2, 3955–3958.

32 D. D. Konarev, S. S. Khasanov, A. Otsuka, G. Saito, R. N. Lyubovskaya. Crystal structure and magnetic properties of an ionic multi-component complex of fullerene (OMTTF center dot I-3)center dot C-60 – Comparison with OMTTF-I-3 salts. *Synth. Metals*, **2005**, 151(3), 231–238.

33 G. Saito, T. Teramoto, A. Otsuka, Y. Sugita, T. Ban, M. Kusunoki, K. Sakaguchi. Preparation and ionicity of C60 charge transfer complexes. *Synth. Methods*, **1994**, 64, 359–368.

34 A. Hirsch. *Fullerenes and Related Structures*, Springer, Berlin, **1998**.

35 S. Bosi, T. Da Ros, G. Spalluto, M. Prato. Fullerene derivatives: An attractive tool for biological applications, *Eur. J. Med. Chem.*, **2003**, 38(11–12), 913–923.

36 C. Cusan, T. Da Ros, G. Spalluto, S. Foley, J.-M. Janot, P. Seta, C. Larroque, M. C. Tomasini, T. Antonelli, L. Ferraro, M. Prato. Synthesis and molecular modeling studies of fullerene-trimethoxyindole-oligonucleotide conjugates as possible probes for studying photochemical reactions in DNA triple helices. *Eur. J. Org. Chem.*, **2002**, 17, 2928–2934.

37 M. Brettreich, A. Hirsch. A highly water-soluble dendro[60]fullerene. *Tetrahedron Lett.*, **1998**, 39, 2731–2734.

38 N. A. Chaniotakis. Enzyme stabilization strategies based on electrolytes and polyelectrolytes for biosensor applications. *Anal. Bional Chem.*, **2004**, 378(1), 89–95.

39 B. C. Braden, F. A. Goldbaum, B.-X. Chen, A. N. Kirschner, S. R. Wilson, B. F. Erlanger. X-ray crystal structure of an anti-Buckminster-fullerene antibody Fab fragment: Biomolecular recognition of C60. *Proc. Natl. Acad. Sci. U.S.A.*, **2000**, 97, 12 193–12 197.

40 N. Y. Pan, J. S. Shih. Piezoelectric crystal immunosensors based on immobilized fullerene C60-antibodies. *Sens. Actuators B-Chem.*, **2004**, 98(2–3), 180–187.

41 R. E. Haufler, J. Conceicao, L. P. F. Chibante, Y. Chai, N. E. Byrne, S. Flanagan, M. M. Haley, S. C. O'Brien, C. Pan, et al. Efficient production of C60 (buckminsterfullerene), C60H36, and the solvated buckide ion. *J. Phys. Chem.*, **1990**, 94(24), 8634–8636.

42 C. Jehoulet, A. J. Bard, F. Wudl. Electrochemical reduction and oxidation of C60 films. *J. Am. Chem. Soc.*, **1991**, 113(14), 5456–5457.

43 Y. Yang, F. Arias, L. Echegoyen, L. P. F. Chibante, S. Flanagan, A. Robertson, L. J. Wilson. Reversible fullerene electrochemistry correlation with the HOMO-LUMO energy difference for C60, C70, C76, C78, and C84. *J. Am. Chem. Soc.*, **1995**, 117, 7801.

44 R. C. Haddon. L. E. Brus, K. Raghavachari. Electronic structure and bonding in icosahedral C60. *Chem. Phys. Lett.*, **1986**, 125, 459.

45 R. C. Haddon. Electronic structure, conductivity and superconductivity of alkali metal doped C60. *Acc. Chem. Res.*, **1992**, 25, 127–133.

46 F. Wudl. The chemical properties of buckminsterfullerene (C60) and the birth and infancy of fulleroids. *Acc. Chem. Res.*, **1992**, 25, 157.

47 A. Hirsch. *Chemistry of the Fullerenes*, Georg Thieme, Stuttgart, **1994**.

48 B. Miller, J. M. Rosamilia, G. Dabbagh, R. Tycko, R. C. Haddon, A. J. Muller, W. Wilson, D. W. Murphy, A. F. Hebard. Photoelectrochemical behavior of C60 films. *J. Am. Chem. Soc.*, **1991**, 113(16), 6291–6293.

49 C. Jehoulet, Y. S. Obeng, Y. T. K. Kim, F. Zhou, A. J. Bard. Electrochemistry and Langmuir trough studies of fullerene C60 and C70 films. *J. Am. Chem. Soc.*, **1992**, 114(11), 4237–4247.

50 L. Echegoyen, F. Diederich, L. E. Echegoyen. *Fullerenes: Chemistry, Physics, and Technology*, eds K. M. Kadish, R. S. Ruoff, Wiley, New York, **2000**, pp. 1–52.

51 *Principles of Chemical and Biological Sensors*, ed. D. Diamond, Wiley, New York, **1998**.

52 A. Jensen, S. Wilson, D. Schuster. Biological applications of fullerenes. *Bioorg. Med. Chem.*, **1996**, 4(6), 767–779.

53 D. Gust, T. Moore, A. Moore. Fullerenes linked to photosynthetic pigments. *Res. Chem. Intermediates.* **1997**, 23(7), 621–651.

54 H. Imahori, K. Hagiwara, M. Aoki, T. Akiyama, S. Taniguchi, T. Okada, M. Shirakawa, Y. Sakata. Linkage and solvent dependence of photoinduced electron transfer in zincporphyrin-C60 dyads. *J. Am. Chem. Soc.*, **1996**, 118(47), 11 771–11 782.

55 R. G. Compton, R. A. Spackman, R. G. Wellington, M. L. H. Green, J. Turner. A fullerene(C60)-modified electrode: Electrochemical formation of tetra-butylammonium salts of C60 anions. *Electroanal. Chem.*, **1992**, 327, 337–341.

56 Á. Szûcs, M. Tölgyesi, E. Szûcs, M. Csiszár, A. Loix, L. Lamberts, J. B. Nagy, M. Novák. Electrochemistry of C60 films in aqueous solutions, in *Fullerenes Volume 5, Chapter 42: Recent Advances in the Chemistry and Physics of Fullerenes and Related Materials*, ed. K. M. Kadish, R. S. Ruoff, Electrochemical Society, New Jersey, **1997**, 68–81.

57 M. Csiszár, A. Szücs, M. Tölgyesi, A. Mechler, J. B. Nagy, M. Novak. Electrochemical reactions of cytochrome c on electrodes modified fullerene films. *J. Electroanal. Chem.*, **2001**, 497, 69–74.

58 M.-X. Li, N.-Q. Li, Z.-N. Gu, X.-H. Zhou, Y.-L. Sun, Y.-Q. Wu. Electrocatalysis by a C60-γ-cyclodextrin (1:2) and nafion chemically modified electrode of hemoglobin. *Anal. Chim. Acta*, **1997**, 356, 225.

59 F. Patolsky, G. Tao, E. Katz, I. Willner. C60-mediated bioelectro catalyzed oxidation of glucose with glucose oxidase. *J. Electroanal. Chem.*, **1998**, 454, 9.

60 V. Gavalas, N. A. Chaniotakis. Fullerene-mediated amperometric biosensors. *Anal. Chim. Acta*, **2000**, 409(1–2), 23, 131–135.

61 S. Sotiropoulou, V. Gavalas, V. Vamvakaki, N. A. Chaniotakis. Novel carbon materials in biosensor systems. *Biosensors Bioelectron.*, **2003**, 18(2–3), 211–215.

62 H. T. Tien, L. G. Wang, W. Xing, A. L. Ottova. Electronic processes in supported bilayer lipid membranes (s-BLMs) containing a geodesic form of carbon (fullerene C60). *Bioelectrochem. Bioenerget.*, **1997**, 42, 161–167.

63 H. T. Tien, A. L. Ottova. Supported planar lipid bilayers (s-BLMs) as electrochemical biosensors. *Electrochim. Acta*, **1998**, 43(23), 3587–3610.

64 K. Asaka, H. T. Tien, A. Ottova. Voltammetric study of charge transfer across supported bilayer lipid membranes (s-BLMs). *J. Biochem. Biophys. Methods*, **1999**, 40, 27–37.

65 J. S. Ye, A. Ottova, H. T. Tien, F. S. Sheu. Nitric oxide enhances the capacitance of self-assembled, supported bilayer lipid membranes. *Electrochem. Commun.*, **2001**, 3(10), 580–584.

66 H. T. Tien, R. H. Barish, L.-Q. Gu, A. L. Ottova. Supported bilayer lipid membranes as ion and molecular probes. *Anal. Sci.*, **1998**, 1(14), 3–18.

67 I. Szymanska, H. Radecka, J. Radecki, D. Kikut-Ligaj. Fullerene modified supported lipid membrane as sensitive element of sensor for odorants. *Biosensors Bioelectron.*, **2001**, 16, 911–915.

68 S. Sotiropoulou, N. A. Chaniotakis. Photoinduced electron transfer (mediation) nano switch between fullerene C60 and glucose oxidase, submitted for publication.

69 S. Sotiropoulou, V. Vamvakaki, N. A. Chaniotakis. Stabilization of enzymes in nanoporous materials for biosensor applications. *Biosensors Bioelectron.*, **2005**, 20(8), 1674–1679.

70 S. Sotiropoulou, V. Gavalas, V. Vamvakaki, N. A. Chaniotakis. Novel carbon materials in biosensor systems. *Biosensors Bioelectron.*, **2003**, 18(2–3), 211–215.

71 V. G. Gavalas, N. A. Chaniotakis. Polyelectrolyte stabilized oxidase based biosensors: Effect of diethylaminoethyl-dextran on the stabilization of glucose and lactate oxidases into porous conductive carbon. *Anal. Chim. Acta*, **2000**, 404(1), 10, 67–73.

72 V. G. Gavalas, N. A. Chaniotakis, T. D. Gibson. Improved operational stability of biosensors based on enzyme-polyelectrolyte complex adsorbed into a porous carbon electrode. *Biosensors Bioelectron.*, **1998**, 13(11), 1205–1211.

73 N. A. Cahniotakis. Enzyme stabilization strategies based on electrolytes and polyelectrolytes for biosensor applications. *Anal. Bioanal. Chem.* **2004**, 378(1), 89–95.

74 V. Gavalas. Stabilization of Amperometric Biosensors with DEAED in Porous Carbon Electrodes. Ph.D. Thesis, **2000**.

75 P. Nednoor, M. Capaccio, V. G. Gavalas, M. S. Meier, J. E. Anthony, L. G. Bachas. Hybrid nanoparticles based on organized protein immobilization on fullerenes. *Bioconjugate Chem.* **2004**, 15(1), 12–15.

5
Optical Biosensing Based on Metal and Semiconductor Colloidal Nanocrystals

Roberto Comparelli, Maria Lucia Curri, Pantaleo Davide Cozzoli, and Marinella Striccoli

5.1
Overview

Nanomaterial-based optical sensors are among the most advanced class of sensing devices produced in the recent past for the continuous, real-time monitoring of diverse analytes. Following the tremendous impact of the unusual properties of nanoscale matter on present technology, scientific publications dealing with sensing applications have concomitantly proliferated. Research on these topics embraces wide and interdisciplinary fields under intense evolution, making it difficult to provide a comprehensive scenario of the current status of their development. This chapter is specifically intended to offer a critical view on the specific potential of colloidal nanocrystals as a distinguishable class of nanosized materials for biosensing. Especially, it will emphasize how the size-dependent optical properties of nanocrystals can be flexibly tailored by the synthetic tools accessible by modern material chemistry, specifically addressing optical techniques and methods for well-defined sensing purposes.

The review of this subject will therefore be organized as follows: (a) the introduction will focus on the chemical-physical properties that justify the choice of nanocrystals as advantageous in various diagnostic approaches; (b) several optical techniques that can be used as transduction methods for biosensing will then be described in detail; (c) selected examples of specific applications and biochemical studies will be presented; and (d) in the conclusions, a few key issues regarding the commercial development of the presented techniques will be discussed.

5.2
Introduction

Current interest in both fundamental and practical scientific research is oriented towards the design and engineering of new generations of devices based on nanosized inorganic objects coupled with assembled molecules [1–3]. The tremendous

Nanotechnologies for the Life Sciences Vol. 8
Nanomaterials for Biosensors. Edited by Challa S. S. R. Kumar

ISBN: 978-3-527-31388-4

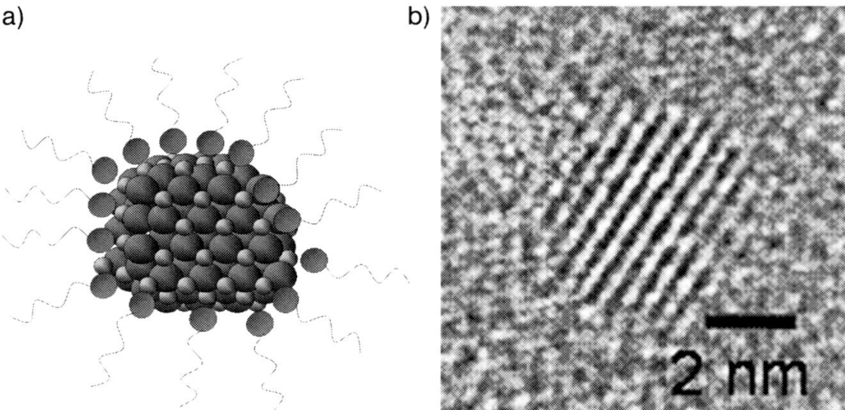

Fig. 5.1. (a) Schematic representation of single colloidal nanocrystal coated with organic molecules. (b) High-resolution transmission electron microscope picture of CdS nanocrystal.

efforts in this direction have been motivated by the recognition of size-, shape-, and composition-dependence of the optoelectronic, magnetic, and catalytic properties of matter at the nanoscale [4].

Recent advances in both physical and chemical synthetic approaches have made various nanostructured materials available for novel technological applications (Fig. 5.1).

The unique and intriguing optoelectronic properties of such nanostructured objects can find applications in the fabrication of new and original nanosensors in the wide field of biochemistry. In particular, this chapter emphasizes the impact of semiconductor and noble metal colloidal nanocrystals on the development of optically driven detection methodologies for biosensing purposes.

A "biosensor" can be broadly defined as a sensing device able to measure any property with biological significance or connected with bioactivity by means of biomolecules and/or biologically related structures. More appropriately, a biosensor is a small device that, as the result of a chemical interaction or process occurring between the analyte and the sensing element, transforms quantitative or qualitative chemical or biochemical information into an analytically useful signal. Biosensors are usually considered a subset of chemical sensors because of the peculiar transduction methods associated with them, sometimes referred to as the sensor "platforms". The fundamental understanding of the physics and chemistry of nanosized metal and semiconductors has stimulated intense efforts to the development of innovative strategies for biosensing devices. Although various sensing mechanisms, based on the changes of magnetic, electrochemical, piezoelectric, and resistivity properties of matter, have been exploited so far, optical transduction has largely been the most preferred tool in many biosensing applications because of its superior advantages and versatility. Especially, optical platforms are technologically appealing, as they offer high detection limits, little sensitivity to electro-

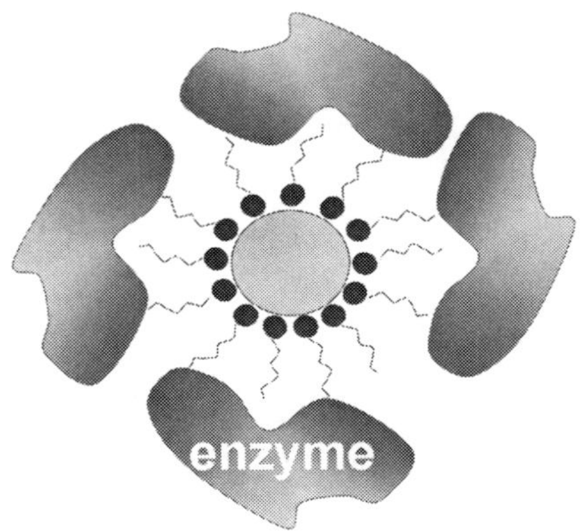

Fig. 5.2. Colloidal nanocrystals are commensurate with biomacromolecules.

magnetic noise, the possibility of remote control and information transfer through optical fibers, long lifetime, and amenability to multiplexing. Such interest is further motivated by the expectation that nanomaterial-based optical biosensors will not only replace conventional sensors, but also enable the fabrication of unprecedented sensing devices. Finally, the recent advantages in photonic technologies and the low cost availability of commercial optical fibers and lasers will certainly encourage access to new frontiers of biosensing.

Inorganic objects on the nanometer scale promise easy integration with many bio-related domains. The fact that nanoparticles share the same size regime (<50 nm) as that of many biomolecules opens access to the fabrication of novel composite nanostructures with the biological and the inorganic components suitably coupled (Fig. 5.2).

Including nanomaterials in a sensing device can be beneficial in that unprecedented transduction modes, relying on their special properties, can be, in principle, developed for any event of biological relevance.

The emerging disciplines of nanoengineering, nanoelectronics, and nanobioelectronics, require suitably sized and functionalized nanosized particles as the "building blocks" with which to construct their architectures and active elements. Especially, colloidal nanocrystals represent attractive precursors for creating various electronic and sensor components due to the ease of their fabrication and to their intrinsic robustness, allowing for further processing. Such objects are composed of a crystalline inorganic core and a protective shell of surface anchored molecules provided by a solution-phase synthesis. Nanocrystals are grown directly in a liquid solution by chemical methods that facilitate their medium-scale production in a reaction flask.

Colloidal nanoparticles differ significantly from other classes of nanostructured materials in that their surface chemistry can be conveniently manipulated to enhance compatibility with biological environments, to achieve nanocrystal immobilization onto substrates, conjugation with selected biomolecules, and/or incorporation into different matrixes (Fig. 5.3).

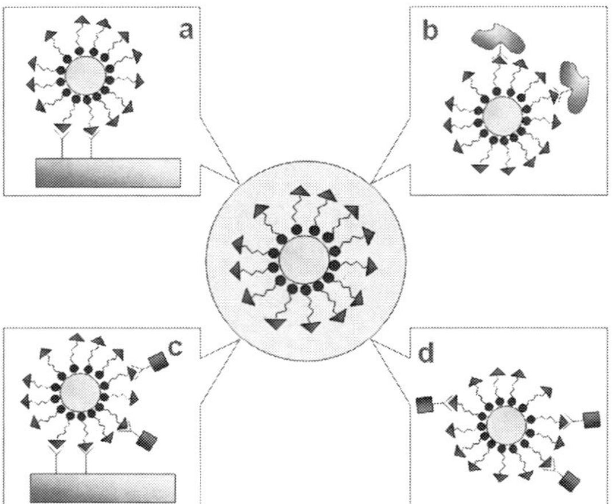

Fig. 5.3. Colloidal nanocrystals can be surface functionalized for (a) guided assembly, (b) biomolecular recognition, (c) sensing devices, and (d) solution-based receptors.

Nanoparticles have been used in biosensors for the last two decades. The fabrication and characterization methods for nanomaterials have been progressively refined to the point that deliberate modulation of their size, shape, and composition can be now obtained, thereby permitting a fine control of their properties. Colloidal techniques have long been known, but only recently has a great body of work been devoted to the synthesis of nanocrystals specifically intended for the construction of devices and complex composite nanostructures.

Today, the key goal in nanocrystal-based technology is to reproducibly synthesize nanoparticles having several properties, such as specific size, tailored shapes, crystal phase and crystallographic orientation, well-defined surface chemistry, solubility in any desired solvent, and multifunctional performances. These latter can be achieved by combining different materials in a single hybrid nanostructure, due to novel structural and electronic properties arising from interactions of the single components. A new concept of "purpose-built" nanomaterials has thus emerged, whose basic idea is to model and design materials with the proper morphology, size, shape and surface status to probe, tune and optimize the material's physical-chemical properties. The ability in tailoring all these characteristics is an essential condition to extend NCs to large-scale technological and industrial applications.

Now, the structure of nanosized noble metals and semiconductors can be systematically engineered to produce materials with specific emissive, absorptive, and light-scattering properties, which make these materials ideal for multiplexed analyte detection. The control of composition and shape can allow for the variation of additional parameters useful in the detection of target analytes.

Furthermore, techniques for surface modification and patterning have advanced to such a great extent that the binding affinity of materials for various biomolecules can be modulated by suitable chemical tools as much as desired, leading to novel generations of nanoscale arrays of biomacromolecules and small molecules together on surfaces [2a]. The overall ensemble of these capabilities has allowed the design of functional materials that can be implemented into new optical assays in biosensors based on improved modes of signal transduction.

5.3 Colloidal Nanocrystals

5.3.1 Size-dependent Optical Properties

Most nanocrystal physical properties evolve as a function of the size and/or shape, approximately following some "scaling laws" [5]. Two major reasons account for this special dependence in nanocrystals: the confinement of charge carrier motion in a small material volume, and the significant fraction of atoms residing at the surface, as compared with that in the corresponding bulk counterpart.

Now, in any material, there will be a size below which a substantial change of fundamental electrical and optical properties occurs with size: these variations will be seen when the energy spacing δ of successive quantum levels ($\delta = 4E_F/3n$, where E_F is the Fermi level of the bulk material and n is the total number of valence electrons in the nanocrystal) exceeds the thermal energy (~25 meV at room temperature). For a given temperature, this occurs at a very large size in semiconductors, as compared with metals, insulators, van der Waals, or molecular crystals.

In semiconductors, the quantum-sized effects result in a widening of the band gap and in the development of well-defined energy levels at the band edges [7]. These facts can be clearly observed in the optical absorption spectra of "quantum dots": the threshold energy for absorption is shifted to shorter wavelengths with decreasing the particle size and discrete spectral features develop in the spectrum, which correspond to the allowed optical transitions (Fig. 5.4). The positions of the lowest energy absorption peak (which is usually the most defined) as well as that of the luminescence peak are strictly correlated to the average particles size, while their widths reflect, in part, the remaining size distribution. In anisotropic nanocrystals, such as nanorods and tetrapods, the shift behavior of the exciton peak is characteristically dependent on the shortest confined dimension [8] (Fig. 5.5). Furthermore, the size-dependence of semiconductor nanocrystal band-gap provides a tool for tuning the redox potentials of photogenerated holes and electrons, allowing redox process that would be forbidden for extended solids.

In noble metals, the decrease in size below the electron mean free path gives rise to intense absorption in the visible–near-UV region [9]. The optical absorption spectra exhibit absorption peaks that do not derive from quantum confinement effects but result from the collective oscillations of the itinerant free electron gas on the particle surface, that are induced by the incident electromagnetic wave. Such resonances, referred to as surface plasmons (SPs), are seen when the wavelength

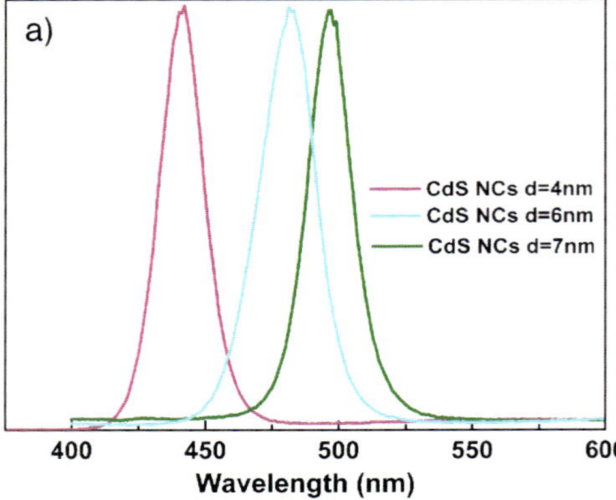

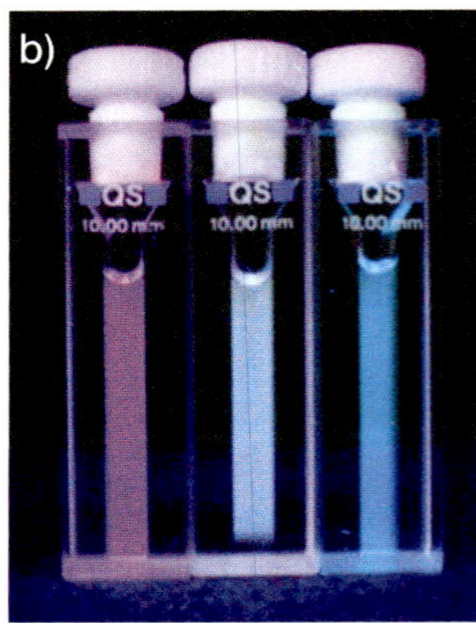

Fig. 5.4. (a) Emission spectra of CdS nanocrystals with different sizes. (b) Distinguishable emission colors of CdS nanocrystals in chloroform solution, excited with a near-UV lamp. (From Ref. [6].)

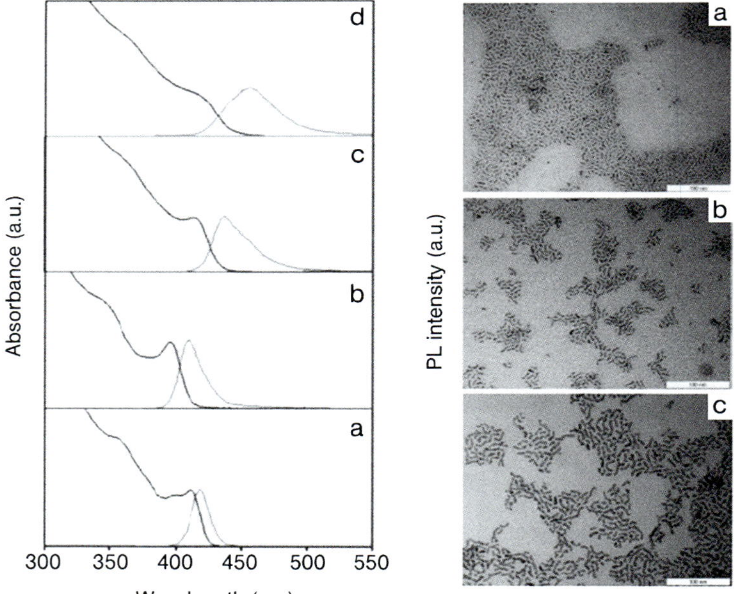

Fig. 5.5. Left-hand side: UV/Vis absorption and PL emission spectra of spherical (a) and rod-like ZnSe nanocrystals (b–d) with aspect ratios of 3, 6, and 8, respectively. Right-hand side: TEM overview of ZnSe nanorods with aspect ratios of 4, 10 and 5. (From Ref. [8].)

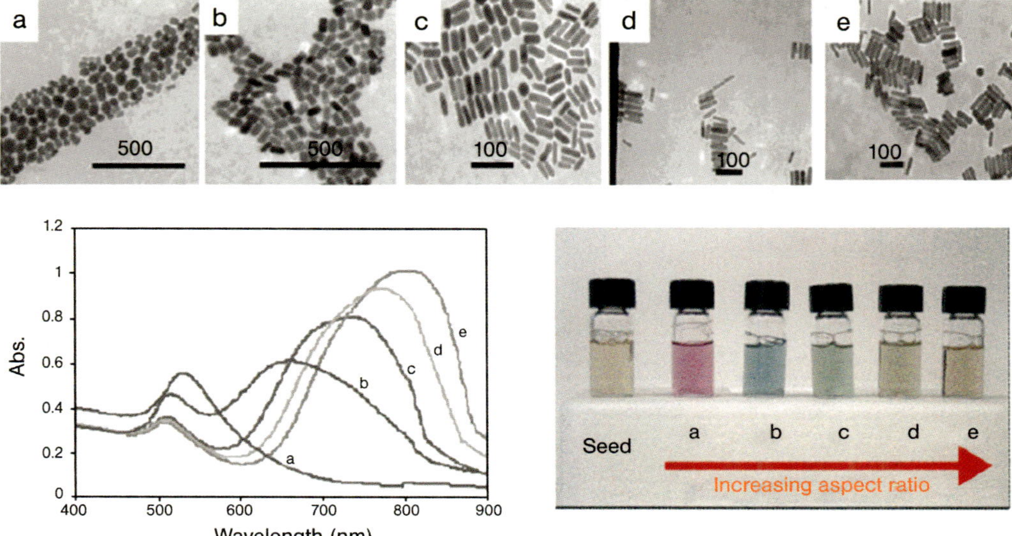

Fig. 5.6. TEM images (top), absorption spectra (left), and photographs (right) of aqueous solutions of gold nanorods of various aspect ratios. Seed sample: aspect ratio 1; sample a, aspect ratio 1.35 (0.32); sample b, aspect ratio 1.95 (0.34); sample c, aspect ratio 3.06 (0.28); sample d, aspect ratio 3.50 (0.29); sample e, aspect ratio 4.42 (0.23). Scale bars: 500 nm for a and b, 100 nm for (c–e). The value in parenthesis represents the error in the determination of the aspect ratio. (From Ref. [10].)

of incident light far exceeds the particle diameter. In the "intrinsic size" regime (<5 nm), noble metal nanoparticles do not show any plasmon absorption, while the absorption of particles larger than 50–60 nm (in the "extrinsic size" region) is broad and covers most of the visible region.

The spectrum of anisotropic particles can exhibit several bands (Fig. 5.6) corresponding to multipole SP resonances and becomes further modified by charge accumulation, molecular adsorption, solvent nature, and aggregation status of the particles (Fig. 5.7) [10, 11].

Interestingly, in the nanosized regime noble metals exhibit a considerable shift of their redox potential toward negative values, which makes them work as efficient electron-transfer mediators [12, 13] (Fig. 5.8).

In nanocrystals the number of atoms residing at the surface is a large fraction of the total and the density of bulk defects is rather low [14]. To minimize surface energy, substantial reconstructions in the atomic positions can occur, so that nanocrystal surfaces can be largely disordered, yielding spherical or ellipsoidal shapes, or can expose regular facets, as are present in extended crystals. Surface rearrangement can be more easily visible in extremely small clusters (in the 3–50 atoms range), where, as there is no clearly identifiable interior, unique bonding geometries, distinct from those of the bulk solid, are assumed.

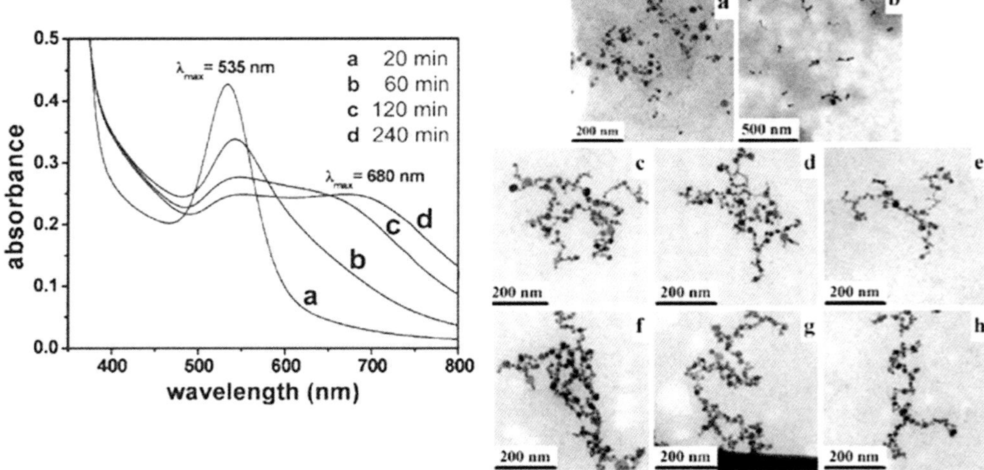

Fig. 5.7. The surface plasmon resonance of Au nanoparticles can be modified by their aggregation status. The left-hand side of the figure shows the temporal evolution of the absorption spectrum of gold upon UV-photocatalytic reduction of 10^{-3} M $HAuCl_4$ in the presence of 5×10^{-2} M TiO_2 nanorods in $CHCl_3$–EtOH. The corresponding TEM images (right-hand side) show the progress of Au nanoparticle assembly after irradiation for (a) 20, (b) 60, (c)–(e) 120, and (d)–(h) 240 min. (From Ref. [11].)

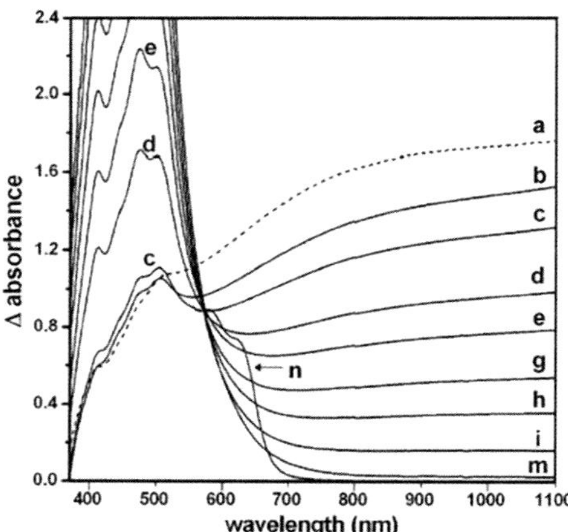

Fig. 5.8. An example of electron transfer mediated by nanosized metal nanoparticles, showing the absorption changes occurring to UV pre-irradiated TiO_2-stabilized Au NPs (a) upon successive additions (5 mM each) of Uniblue A (UBA), an organic dye, under O_2-free conditions (b–m). When all stored electrons are consumed, the absorption features of unreduced UBA can be observed (spectrum n). Conditions: $[TiO_2] = 0.1$ M, nanorods: 3×25 nm; $[Au] = 5 * 10^{-7}$ M, metal size: 15 +/− 3 nm; pre-irradiation time: 30 min. Spectra were taken using the respective photocatalyst solution as the reference. (From Ref. [12], some spectra have been removed for clarity.)

In nanocrystals having applications in opto-electronic devices, surface states invariably result in energy levels within the forbidden energy gap. They act as traps for electrons or holes, ultimately leading to the degradation of the electrical and optical properties of the material. In contrast, lower coordination surface atoms in metal and semiconductor oxides are usually beneficial toward catalysis and sensing of various chemical reactions [15]. The coordination numbers of surface atoms may vary over wide ranges because different crystallographic faces, edges, steps, point defects, and dislocations may be exposed, resulting in a substantial energetic heterogeneity. Unsaturated valences may be, therefore, completed by adsorbed target molecules, the latter affecting the overall electronic distribution of the nano-sized object. More importantly, the qualitative change in the electronic structure of nanocrystals coupled with chemisorption of suitable substrates can bestow unusual optoelectronic behavior for both the inorganic and organic component.

5.3.2 Chemical Synthesis

In colloidal approaches, a fine modulation over the NC size and shape is possible by adjusting the balance between the nucleation and the growth stage. To achieve this, in general the key ingredient is the presence of one or more organic molecules in the reactor, broadly termed as "surfactants". Surfactants are amphiphilic compounds, i.e., molecules composed by one hydrophilic part (a polar or a charged functional group) and one hydrophobic part (in the simplest case, one or several hydrocarbon chains). They can act in two main different ways, described below.

(a) *Organic Templates*: Depending on physical-chemical parameters, such as surfactant concentration, temperature, solvent polarity, additives, ionic strength, amphiphilic compounds tend to assemble into soluble aggregates [16]. These assemblies can have spherical shapes or evolve into rod-like or cylindrical aggregates, flexible bilayers, and planar bilayers. Therefore, the liquid phase can contain hydrophilic or hydrophobic compartments of nanometer size prior to the synthesis, which serve as NC templates. The addition of monomers to the growing cluster is naturally terminated once a NC fills the compartment volume, which predetermines a specified distribution of the NC population close to that of nanocompartments [17] (Fig. 5.9).

In bulk colloidal dispersions, the role of compartments is generally played by direct or reverse micelles, microemulsion droplets, or by the pores (cavities) in any bicontinuous cubic phase of a lipid. A self-assembled amphiphile layer on the interface is also capable of compartmentalizing the accepted reagents, to introduce nanocrystalline structures (so-called membrane mimetic chemistry), and to orient the nanocrystal domains in the nanoreactors [18].

(b) *Terminating agents*: The growth of NCs can be properly controlled in many cases by use of special molecules that behave as "terminating" agents. Such molecules direct the growth of nanostructures by dynamically coordinating their surface under the reaction conditions. Some examples of suitable surfactants include molecules carrying functional groups with electron-donor atoms, such as alkyl thiols,

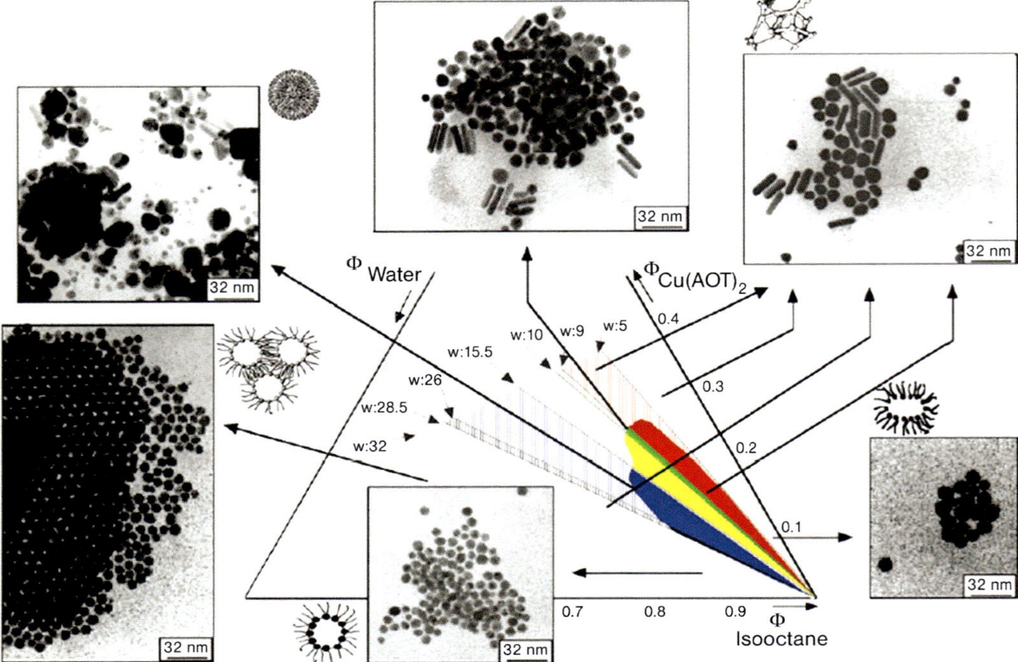

Fig. 5.9. Example of nanoparticle shape control achieved by combining the strategy of surfactant-based templating and the capping of salts or molecules. Various shapes of copper nanocrystals are obtained in the various parts of the phase diagram of $Cu(AOT)_2$–water–isooctane in the oil-rich region. (From Ref. [17].)

phosphines, phosphine oxides, phosphates, phosphonates, amides or amines, carboxylic acids, and nitrogen-containing aromatics [17–19].

The choice of surfactants varies from case to case: a molecule that binds too strongly to the surface of the quantum dot is not useful, as it would not allow the nanocrystal to grow. A weakly coordinating molecule, however, would yield large particles, or aggregates. The surfactant molecules must be chemically stable at the reaction temperature to be suitable candidates for controlling growth. At lower temperatures, or more generally, when the growth is stopped, the surfactants are more strongly bound to the surface of the NCs and provide their solubility in a wide range of solvents. More importantly, surface defects (unsaturated valences) that act as traps for electron and holes, in turn degrading their optoelectronic properties, can be effectively passivated through coordination bonds with the surface ligands.

By controlling the mixture of surfactant molecules that are present during the generation and the time growth of the nanocrystals, excellent control of their size and shape is possible for many materials (Fig. 5.10) [20]. Under appropriate conditions, the choice of suitable surfactants can indeed modulate the reactivity of the

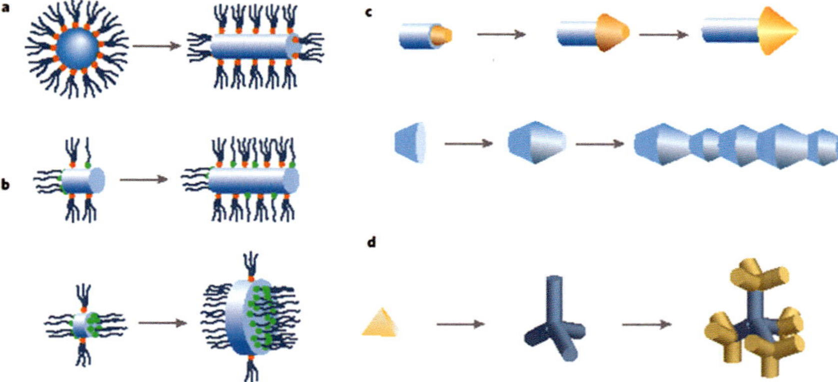

Fig. 5.10. Shape control of colloidal nanocrystals. (a) Kinetic shape control at high growth rate. High-energy facets grow more quickly than low energy facets in a kinetic regime, using one type of metal coordinating group. (b) Kinetic shape control through selective adhesion. The introduction of an organic molecule that selectively adheres to a particular crystal facet can be used to slow the growth of that side relative to others, i.e., metal coordinating groups with different affinities to nanocrystal facets are used, leading to the formation of rod- or disk-shaped nanocrystals. (c) More intricate shapes result from sequential elimination of a high-energy facet. The persistent growth of an intermediate-energy facet eventually eliminates the initial high-energy facet, forming complex structures such as an arrow- or zigzag-shaped nanocrystals. (d) Controlled branching of nanocrystals. The existence of two or more crystal structures in different domains of the same crystal, coupled with the manipulation of surface energy at the nanoscale, can be exploited to produce branched inorganic nanostructures such as tetrapods. Inorganic dendrimers can be further prepared by creating subsequent branch points at defined locations on the existing nanostructures. (From Ref. [20].)

molecular precursors and direct shape changes by adhering selectively to the various facets of the growing crystallites. Delicate control of parameters, such as strain, surface reactivity, interfacial energy, and crystal solubility at the nanoscale [8–21], opens access to various complex NC morphologies and even hybrid architectures, ranging from simple core/shell systems to individual nanostructures with linear and/or branched topology [21], with site-specific deposits of a different material [22], and hetero-groups made of magnetic, metal or fluorescent spherical NCs [21d, 23].

The surfactant coating on the NC surface provides great synthetic flexibility in that it can be exchanged with another coating of organic molecules having different functional groups or polarity.

After the synthesis, colloidal nanocrystals can be further processed (e.g., functionalized with selected biomolecules or coupled with other nanosized objects) and/or transferred to any desired substrate or object.

Many applications of the quantum-mechanical aspects of quantum dots can be found in optics. As the more general case of atoms and molecules, quantum dots can be excited either optically or electrically. Regardless of the nature of excitation, quantum dots may emit photons when they relax from the excited state to the ground state. Based on these properties, quantum dots may be used as lasing media, as single-photon sources, as optically addressable charge storage devices, or as fluorescent labels.

Colloidal quantum dots have also been used for the development of light-emitting diodes [24] and in the fabrication of photovoltaic devices [25]. In these devices, the nanocrystals are incorporated into a thin film of conducting polymer.

Chemically synthesized quantum dots fluoresce in the visible range with a wavelength tunable by the size of the colloids. The possibility of controlling the onset of absorption and the color of fluorescence by tailoring the size of the nanocrystals makes them interesting objects for the labeling of biological structures [26] as a new class of fluorescent markers. The tunability combined with extremely reduced photobleaching makes colloidal quantum dots an interesting alternative to conventional fluorescent molecules.

5.4 Nanocrystal Functionalization for Biosensing

In general, practical applications can take advantage of some unique properties of NCs, such as that of being conjugable, connectable or anyway joinable with biomaterials, thus enabling biolabeling [29], biosensing [30] and use for biomedical purposes [28c–31].

Sensing processes operated by biomaterials bound to surfaces are commonly practiced in analytical biochemistry. Actually, many recent sensing applications are the result of a natural evolution of the bioconjugation chemistry, which is based on the idea of merging biological and non-biological molecular species and/or systems, to realize novel markers for cellular biology, labeling, and biosensing.

In this perspective, the use of NC-biomaterial conjugates has been demonstrated to provide a general route for the development of optical biosensors, which exploit the newly designed nanomaterial based hybrids.

The synthesis of noble metal NCs (which possess a naturally higher bioaffinity) in the presence of biomolecules has been proposed to provide nanoparticles directly connected to biologically relevant molecules [32]. However, semiconductor nanocrystals, which are typically prepared by non-aqueous routes in organic media, require several post-synthetic treatments to enhance their compatibility with biological environments. Such processing is delicate in that it should preserve the NC optical properties over the desired spectral range.

Therefore, in the last decade several studies have provided the means to obtain water-soluble and biocompatible NCs, while keeping intact their original optical characteristics.

The plethora of nanocrystal functionalization methods reported can be mainly divided into three classes: (a) surface capping exchange (typically with bi-functional agents), (b) coating with siloxane and (c) modification of the pristine capping layer of NCs via hydrophobic interactions. Figure 5.11 gives a schematic representation describing different surface functionalization methods.

5.4.1
Surface Capping Exchange

This approach has been proposed for the functionalization of cadmium and zinc chalcogenide NCs (e.g., CdSe, CdS, ZnS, ZnSe) and of their related core–shell derivatives (CdSe@ZnS, CdSe@ZnSe, CdS@ZnS). The basic idea is the substitution of the pristine organic capping layer with bifunctional ligands. Mercapto-acids (mercapto-acetic, mercapto-succinic, mercapto-propionic, mercapto-benzoic, and so on) have been the most commonly used molecules to this purpose [34]. They interact with the NC surface by means of the thiol group while water solubility is ensured by the outermost exposed carboxylic moiety. A major limitation is that mercaptoacid-capped NCs are highly soluble in water only when the carboxylic group is deprotonated, so that the stability of the particles at acid pHs is poor. More importantly, in comparison with the original NC emission, a reduction in photoluminescence (PL) quantum yield is usually observed, the extent of which depends on the nature of the linker exchanged and on the medium. Also, progressive detachment of thiol ligands from the NC surface due to the established dynamic adsorption–desorption equilibrium can usually result in NC aggregation and precipitation over periods longer than one week [29a–35]. Despite these drawbacks, this simple procedure allows for the preparation of fresh batches in gram scale on an "as-needed" basis. Moreover, if further bioconjugation of such thiol-modified NCs is accomplished, their stability in water can be enhanced [29a].

Bifunctional agents, such as dithiothreitol (DTT) [34] and dihydrolipoic acid (DHLA) [30–36] carrying two thiol moieties per molecule, provide stronger interactions with the NC surface, thus prolonging their stability in water (up to 2 months in the case of DTT and up to 1–2 years with DHLA) [37]. This evidence suggests that polydentate thiolated ligands could be even more effective [35b]. DHLA-capped NCs are stable at $pH > 7$, while at slightly acid pHs aggregation can occur.

Mercapto-capped NCs can be bioconjugated either by replacing the capping layer with thiol-modified biomolecules (e.g., DNA [30a] or proteins [38, 39]) or by carboxylic group activation. The hydroxyl groups in DTT can be exploited to achieve bioconjugation by just making them react with nucleophilic agents, such as amines. This approach, suitable to different types of NCs [28c], does not require specific coupling agents and allows one to employ commercially available nucleotides. If NCs are stabilized by negatively charged ligands (for instance DHLA or citrate), bioconjugation can be obtained by electrostatic interactions [36a–40].

The surface capping exchange can also be performed directly with biomolecules. For instance, small biomolecules, like serotonin, can replace tri-*n*-octylphosphine oxide (TOPO) on CdSe@ZnS NC [41]. Analogously, these NCs have been functionalized with synthetic phytochelatin-related peptides, providing them with a coating of biotinylated peptides, which efficiently bind to streptavidin [42]. Lysine and biotin have been used to modify Au NPs and to bind them to various proteins [43],

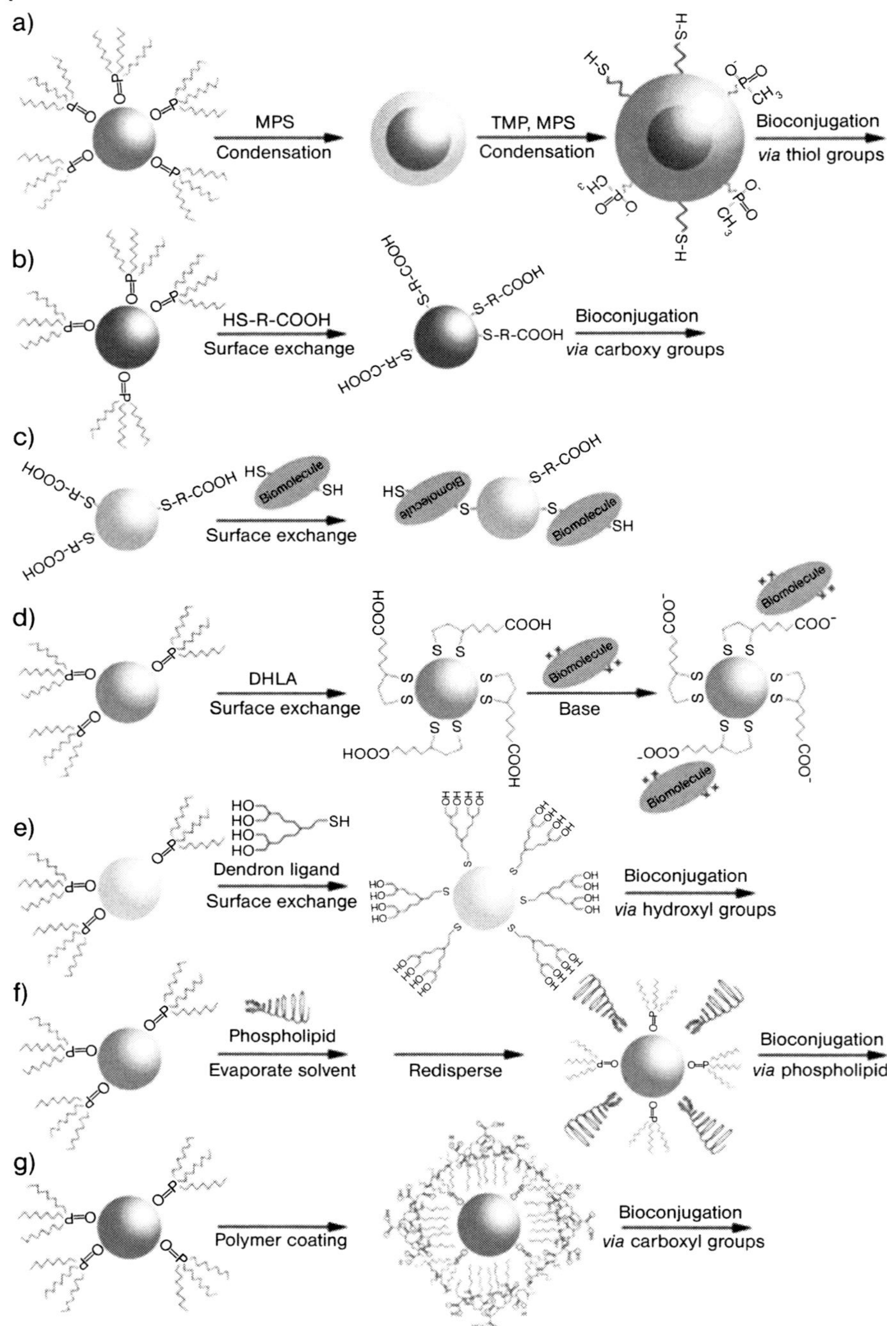

Fig. 5.11. Example strategies for nanocrystal surface modification: (a) coating with silica shell; capping exchange with (b) bifunctional linkers, (c) thiolated molecules, (d) bivalent linkers, (e) and organic dendrons. (f) and (g) Nanocrystal modification via hydrophobic interaction. (From Ref. [33].)

immunoglobulin (IgG), and BSA. This approach is particularly successful for metal particles capped by loosely bound molecules, such as citrate, as demonstrated in the case of Au [44] and Ag [45]. Conversely, in the case of CdS NCs [46] weaker interactions have been observed as well as a luminescence quenching.

Finally, organic dendrons can also replace the native ligand capping on core–shell NCs, leaving the exposed hydroxyl moieties available for bioconjugation in aqueous environment, as reported for TOPO-capped CdSe NCs and Au NPs [47].

5.4.2 Coating with a Silica Shell

Surface silanization is the procedure through which a hydrophilic silica shell is grown on NCs after displacing the original hydrophobic capping layer, ultimately providing NCs with solubility in water [26a]. In the general approach, the first step accomplishes ligand exchange with mercaptosilanes, usually mercaptopropyltrimethoxysilane (MPS). The mercapto groups (–SH) are bound to the NC surface, whereas methoxy groups ($-OCH_3$) are exposed to the solvent. Subsequently, siloxane bonds (–Si–O–Si–) can be formed by reacting the methoxysilane moieties with each other under basic conditions. The final result is a highly cross-linked silica shell possessing improved resistance against dissolution [48, 49]. To provide hydrophilicity, in a further step, molecules bearing mercaptosilane groups at one end and hydrophilic groups at the other one are attached to the shell via the formation of siloxane bonds, leading to a multilayered shell. By properly selecting the hydrophilic functional groups in the latter step, it is possible to tailor the NC surface functionality [26a–48a]. Interestingly, negatively charged particles are obtained by providing the surface with phosphonate groups (for instance, by using trihydroxysilylpropylmethylphosphonate, TMP), which prevent the NCs from agglomerating and/or precipitating by means of electrostatic repulsions. Alternatively, by replacing TMP with poly(ethylene glycol) (PEG)–silane, stable neutral NCs can be prepared that repel each other by steric hindrance [50]. Finally, trimethoxysilylpropyl trimethylammonium chloride can be suitably manipulated to confer a small positive charge to the particles [32]. Bioconjugation can occur via conversion of residual MPS thiol groups into amine or carboxyl groups and via reaction with suitable bifunctional linkers, as established in the pertinent literature [28c–31a].

Silanized NCs also do not undergo to aggregation phenomena in electrolytic solutions, while retaining their original optical properties. Unfortunately, the silica shell often suffers for being inhomogeneous [29–32]. More importantly, the chemical complexity and the delicateness of the steps involved in the silanization process render the preparation of silanized NCs extremely difficult to scale up beyond the milligram quantities per batch.

5.4.3 Surface Modification through Hydrophobic Interactions

A class of modification techniques allows one to obtain water-soluble NCs without removing the pristine capping layer. Such procedures ensure retention of the original optical properties of NCs, as the inorganic core always remains shielded from the external environment. The general strategy relies on exploiting hydrophobic interactions between the pristine NC capping layer and suitable amphiphilic molecules. These latter can intercalate and/or interdigitate with the native capping layer alkyl chains, exposing their hydrophilic head out to the solution. A successful example is represented by phospholipid micelles encapsulating CdSe@ZnS NCs [28c–31a]. The polar head of phospholipids can be used for bioconjugation by reaction with bifunctional agents, like EDC, 1-ethyl-3-(3-(dimethylamino)propyl)-carbodiimide hydrochloride. As drawbacks, aggregation can be induced to some extent by EDC [36a], and the technique is laborious and provides only milligrams quantities per batch. Despite this, the NC optical properties are fully retained after phase transfer [31b–51].

Commercially available amphiphilic polymers bearing hydrophobic alkyl side chains and hydrophilic groups, such as PEG or multiple carboxylate groups, have also been used [52]. Several polymers have been employed: octylamine-modified polyacrylic acid [52a], block copolymers [52b], and polyanhydrides [52c]. Bioconjugation is accomplished by linking biocompatible ligands to polymer-coated NCs by reaction with the hydrophilic heads. PEG-derivatized polymers can also improve biocompatibility and reduce non-specific binding. Carboxylate-functionalized polymers strongly interact with primary amines via condensation, which can be applied to most common proteins in their native form without further modification or activation. In this regard, a high reactivity of polyanhydrides toward primary ammines has been observed [52c]. Notably, NCs functionalized with high molecular weight triblock polymers [as those composed of a hydrophobic poly(butyl acrylate) segment, a hydrophobic poly(methacrylate) segment, of a hydrophilic poly(methacylic acid) segment, and of and a hydrophobic hydrocarbon chain] are so well protected that their optical properties do not change under a broad range of pH (1–14) and salt conditions (0.01 to 1 M), or even upon harsh treatments with concentrated acid [52b]. Nonetheless, both phospholipid micelles and polymer coating can not prevent the NC size from increasing to a great extent, which can limit the intracellular mobility and may preclude fluorescence resonance energy transfer studies [35b].

More recently, host–guest chemistry has been exploited to induce the phase transfer of organic capped metal and semiconductor NCs into water by means of cyclodextrins [53]. Due to their peculiar characteristics [54], cyclodextrin-modified NCs can be very appealing for biological studies. This method deals with the formation of a host–guest complex between the hydrophobic cyclodextrin cavity and the alkyl chains of the surfactants anchored on an NC surface. Among the various classes of cyclodextrins tested, α-cyclodextrins appear the most suitable for this purpose, since no aggregation or precipitation phenomena have been observed. In an alternative approach, metal NCs have also been synthesized [55] in the presence of modified cyclodextrins behaving as the capping ligands for the particles, while leaving their cavity available to perform host–guest chemistry. In this case, only a broadening of the NC size distribution has been observed during the synthesis.

5.5 Optical Techniques

Recent advances have been reported in the development of sensors based on optical detection methods, with remarkable improvements in terms of sensitivity and selectivity, This progress has been triggered by the access to new technical instrumentation and techniques, as well as by the availability of unprecedented materials. The discovery and the fundamental understanding of the peculiar optical properties of nanocrystals have led to a re-evaluation of the potential of well-assessed optical methods (colorimetric assays and fluorescence-based recognition) and, at the same time, have extended the technological opportunities.

Depending on the nature of the optical transduction signals, nanocrystal-based biosensors may be roughly divided into luminescence-based sensors and absorption-based sensors, relying on the changes of the emission and absorbance intensity, respectively [56]. Another possible classification can be made with reference to the chemical composition (metal or semiconductor) of the material used in the sensing process. Such diverse categorization can sometimes merge with each other, resulting in a lack of clarity in the definition. In this section we adopt a more schematic approach by classifying the methods according to the specific optical process that is exploited for the detection process.

5.5.1 Colorimetric Tests

The simplest method for detecting analytes is by measuring the absorption changes that occur because of some analyte–NC interactions modifying the optical properties of the nanocrystals [57]. Especially, metal nanoparticles are convenient for this purpose, as their surface plasmons are characterized by high extinction coefficients, many orders of magnitude larger than those of organic dyes [58]. The absorption spectrum, primarily reflecting the particle size, shape, composition [59], as well as the dielectric constant of the surrounding medium, can be sensitively affected by any process that alters the surface charge density, like direct adsorption of molecules, or variation in the mean interparticle distance induced by molecular recognition events (e.g., DNA hybridization, antigen–antibody interactions). The resulting shift in the plasmon maximum, and changes in band width and intensity, can be explained qualitatively in terms of particle electromagnetic interactions through simple quasi-static limit scattering models [60].

5.5.2 Fluorescence

Fluorescence is the optical phenomenon for which a molecule or a material (fluorophore), after gaining excitation by some means, can return to its minimum energy state by emitting light. Excitation and emission processes in a typical mole-

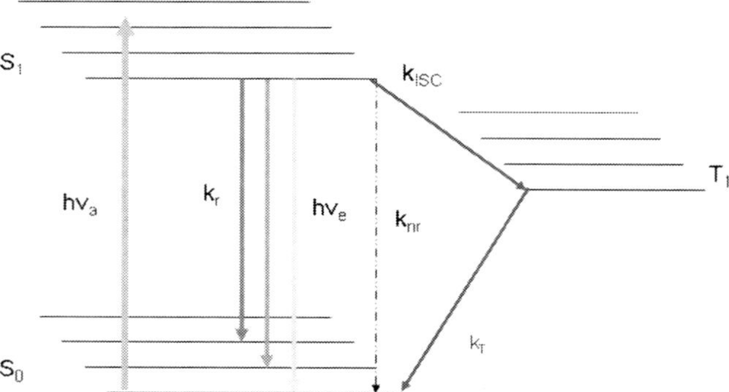

Fig. 5.12. Jablonski diagram describing the fluorescence process in a molecule. Upon absorption of a photon of energy ($h\nu_a = E_{S1} - E_{S0}$), a molecule in the ground singlet state (S0) is promoted to a vibronic sublevel of the lowest excited singlet state (S1). Nonradiative, fast relaxation brings the molecule down to the lowest S1 sublevel. Emission of a photon of energy $h\nu_e < h\nu_a$ with radiative rate k_r can take place, bringing back the molecule to one of the vibronic sublevels of the ground state. Alternatively, the molecule can undergo collisional quenching, returning to its ground state without photon emission (nonradiative rate k_{nr}). A third type of process is intersystem crossing to the first excited triplet state T1 (rate k_{ISC}). Relaxation to the ground state takes place either by photon emission (phosphorescence) or nonradiative relaxation (rate k_{nr}).

cule are represented by a Jablonski diagram, depicting the initial, final, and intermediate electronic and vibrational states of the molecule (Fig. 5.12).

In general, the molecule can be excited, by a one- or multiphoton process, from its electronic ground state S_0 to the first excited state S_1. Its return to the ground state can be accompanied by the emission of a fluorescence photon. The $S_1 \pm S_0$ transition competes with another photophysical process, called intersystem crossing, that is responsible for converting the S_1 state into the first triplet state T_1. An irreversible photobleaching can follow the latter event, ultimately preventing a fraction of molecules from contributing to the overall fluorescence signal.

In general, fast intramolecular vibrational relaxations are responsible for the lower energy of the emitted photons with respect to that associated with the incident light. The difference in energy between absorption and emission is named the Stokes shift – it accounts for the wavelength separation of the emitted fluorescence from the excitation light. This phenomenon renders fluorescence a powerful analytical tool. The delay (lifetime) between the absorption step and the detected fluorescence depends on the nature of the fluorophore and on its local environment.

The efficiency of photon absorption is proportional to the local electric field and to the dipole moment of the fluorophore. For an immobilized fluorophore, the spatial orientation of the absorbing molecule dipole can be determined by recording the emitted fluorescence as a function of the direction of the linear polarization of the excitation light. For a mobile molecule, more information is needed, because the emission dipole may change orientation significantly in a short time. Fluctuations that are faster than the fluorescence lifetime lead to a depolarized emission, while fluctuations covering timescales longer than the lifetime but shorter than the instrumental integration time lead to anticorrelation of the two orthogonal emission polarizations. Fluorescence polarization measurements have been used in analytical and clinical chemistry and as a biophysical research tool for studying the mobility of membrane lipids, domain motions in proteins, and interactions at the molecular level. Also, fluorescence polarization-based immunoassays are also extensively utilized for clinical diagnostics.

Measurement of fluorescence intensity is a widely applied technique in biological applications, as it is related to the number of photons emitted by a fluorescing sample upon excitation. Based on fluorescence wavelength and intensity level determination, various molecular or cellular targets can be easily tagged. The fluorescence intensity may be influenced by several factors, such as the excitation intensity thermal fluctuations, self-absorption, self-fluorescence, and fading due to photobleaching of the probe. However, organic fluorophores usually have emission bands with long tails extending to the red part of the visible spectrum. This issue could limit multiplexing and would require the use of interference filters in the optical readout, reducing the dynamic range of intensity levels. In contrast, the emission spectrum of semiconductor nanocrystals is narrow and Gaussian; their absorption spectrum is continuous and extended in the UV, so that a single wavelength can excite nanocrystals of different sizes, yielding several emission colors at once. Therefore, multicolor NC probes can be used to track multiple molecular targets simultaneously.

5.5.3 Fluorescence Resonance Energy Transfer

Among the most promising optical biodetection techniques, Fluorescence Resonance Energy Transfer (FRET) has been used to carry out analyte recognition and binding, while simultaneously producing useful output signals through an integrated signal transduction system [61]. FRET has been used in carefully designed sensing systems for proteins, peptides, nucleic acids and small molecules.

FRET is one of the few tools available for measuring nanometer scale distances and/or changes in distances, both *in vitro* and *in vivo*. This technique indeed relies on the distance-dependent transfer of energy from a donor fluorophore to an acceptor fluorophore. Recent advances have led to both qualitative and quantitative improvements, including increased spatial resolution, distance range, and sensitivity [62]. This progress has been facilitated by the use of new fluorescent dyes and by novel optical methods and instrumentation.

In FRET, a donor fluorophore is excited by incident light, and if an acceptor is in its close proximity, the excited state energy from the donor can be transferred to the acceptor. This event leads to a reduction in the fluorescence intensity and in the excited state lifetime of the donor, and to a concomitant increase in the emission intensity of the acceptor. The efficiency of the process depends on the inverse sixth-distance between the donor and the acceptor [Eq. (1)], where R_0 is the distance at which half of the energy is transferred, and depends on the spectral characteristics of the dyes and their relative orientation.

$$\text{efficiency} = 1/\{1 + (R/R_0)^6\} \qquad (1)$$

This would allow one to measure: (a) interactions between molecules, e.g., between proteins, between a protein and a ligand or (b) distances between two sites in a macromolecule.

A crucial aspect of FRET biosensor development involves optimizing the relative positions of energy levels of the donor and acceptor dyes to make them operate in concert with the desired recognition elements. Although organic dyes are available to accomplish this, many of them have functional limitations, such as pH dependence, susceptibility to photobleaching, and narrow excitation bands coupled with broad emission peaks, which can compromise sensor performance. The photophysical properties of nanocrystals, such as size-tunable photoluminescence spectra, broad absorption, narrow emission lines, and high quantum yields, make them good candidates as energy donors able to overcome some of the problems associated with conventional molecular-based FRET systems. Indeed, it has been recently demonstrated [63] that the use of single-NC-incorporated nanosensors for DNA detection can reduce significantly, or even eliminate, the complication of background fluorescence encountered by conventional molecular FRET probes (Fig. 5.13). Further, the target signal is amplified through enhanced energy-transfer efficiency by increasing the number of acceptors linked to a donor. These features allow NC-incorporated nanosensors to generate a very distinct target signal that is easily distinguishable from the background in the presence of low target abundance and a large excess of the probe.

5.5.4 Fluorescence Lifetime

The Fluorescence Lifetime (FLT) is a measure of the time a fluorophore spends in the excited state before returning to the ground state by emitting a photon. If a fluorophore population is excited, the lifetime τ is the time it takes for the excited molecules to decay to 1/e or 36.8% of the original population. The decay of the intensity as a function of time is given by Eq. (2).

$$I_t = \alpha e^{-t/\tau} \qquad (2)$$

where I_t is the intensity at time t, α is a normalization term (pre-exponential factor) and τ is the lifetime of the excited state. Knowledge of the excited state lifetime of a fluorophore is crucial for quantitative interpretations of fluorescence measurements, such as quenching, polarization, and FRET. Excited state lifetimes are measured in the time domain by illuminating the sample with a short pulse of light and recording the temporal behavior of the emission intensity. In pioneering work, these short light pulses having widths on the order of several nanoseconds were

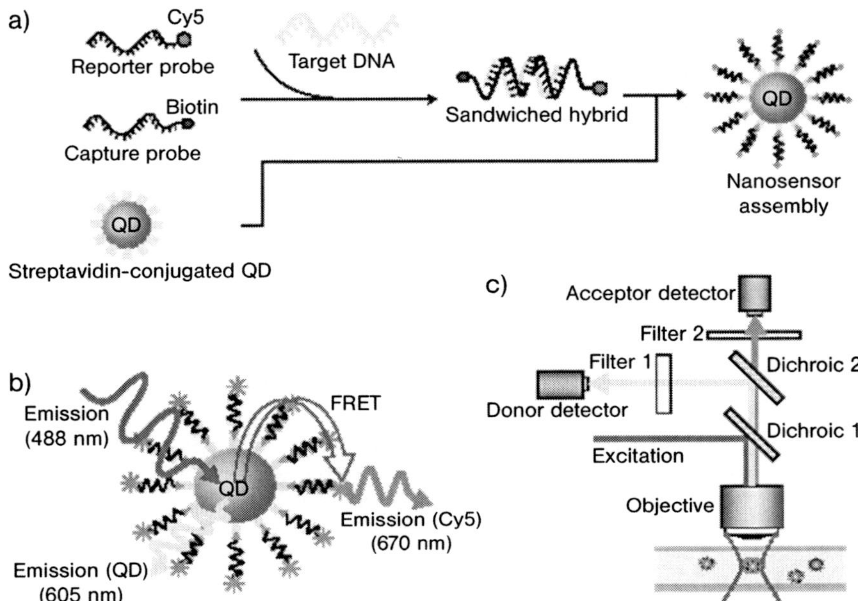

Fig. 5.13. Operative principle of single-NC-based DNA nanosensors. (a) Conceptual scheme showing the formation of a nanosensor assembly in the presence of targets. (b) Fluorescence emission from Cy5 caused by FRET between Cy5 acceptors and a NC donor in a nanosensor assembly. (c) Experimental setup. (From Ref. [63].)

generated using flash lamps. Modern laser sources can now routinely provide pulses with widths on the order of picoseconds or shorter. If the decay is a single exponential and the lifetime is long compared with the exciting light, then the lifetime can be determined directly from the slope of the curve. If the lifetime and the excitation pulse width are comparable, deconvolution methods taking in account the effect of the exciting pulse shape must be used to extract the lifetime.

Fluorescence lifetime is a parameter that is mostly unaffected by inner filter effects, static quenching and variations in the fluorophore concentration. For this reason, FLT can be considered as one of the most robust fluorescence parameters. Therefore, FLT measurements are advantageous in clinical and high-throughput screening applications, where it is necessary to discriminate against the high background fluorescence from biological samples. Also, the decay studies offer a route to multiplexing. The ability to distinguish two fluorophores with similar spectra but different lifetimes is another way to allow an increased number of parameters to be measured (Fig. 5.14).

Several approaches have been used for the development of FLT-based biosensors. For example, in simple binding assays, the binding of two components (one being fluorescently labeled) is accompanied by a change in the fluorescence lifetime. In quench-release type assays, a quenched species, exhibiting low but detectable fluo-

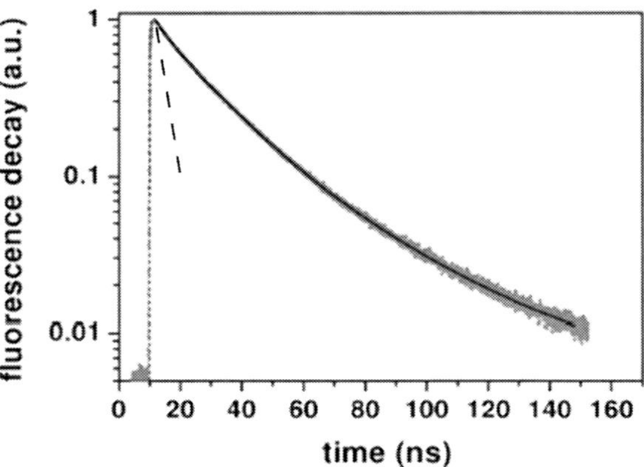

Fig. 5.14. Normalized ensemble fluorescence decay of CdSe@ZnS nanocrystals (1.8-nm radius, 575-nm peak emission). The solid curve is a triple exponential fit to the data with components at 3.4, 16.1, and 35.6 ns. The measured fluorescence decay of Rhodamine molecules in water is also displayed (dashed curve) and is well described by a single exponential decay with time constant 4.3 ns. (From Ref. [65].)

rescence, is initially present in large excess. If a fluorescence compound is released (e.g., by binding to a complementary DNA strand or by an enzymatic reaction) the FLT of the system increases. Finally, the study of lifetime is a powerful tool to measure the energy transfer efficiency in FRET assays by using a non-fluorescent acceptor, which helps to circumvent the problem of spectral cross talk between the donor and the acceptor.

FLT measurements offer additional opportunities in fluorescence microscopy, where the local probe concentration can not be controlled. Lifetime microscopy allows image contrast to be created from the fluorescence lifetime of a probe at each point of the image. Typical examples are the mapping of cell parameters, such as pH, ion concentrations or oxygen saturation. To enhance contrast in cellular imaging and analysis by FLT imaging, labeling agents with long lifetime fluorescence are required. Usually, organic dyes undergo radiative emission from allowed singlet–singlet electronic transitions, with typical decay rates of few nanoseconds. Unfortunately, such rapid emission merges with the time scale of short-lived self-fluorescence background from many naturally occurring species in biological specimens. In fact, self-fluorescence of several proteins often precludes observation of small details, unless the target is heavily loaded with dyes (which, in turn, results in phototoxicity problems). Lanthanide chelates with lifetimes in the sub-microsecond to millisecond range have also been successfully used [62]. However, because their FLT is too long, these probes have a limited photon turnover rate and, therefore, low sensitivity. In contrast, nanocrystals are more advantageous, in that they emit light slowly enough to avoid serious interference by self-fluorescence background, but also fast enough to maintain a high photon turnover rate. Indeed, ensemble measurements show a multi-exponential decay after pulsed excitation, which yields decay times between 10 and 40 ns. Benefiting from the long lifetime of NCs and using fast pulsed laser excitation, it is possible to reject all photons emitted within the first few ns after the pulse, while revealing only those arriving after the decay of fluorescence from all dyes or self-fluorescent proteins (time-gated imaging). Also, the allowed repetition rate of a few megahertz is compatible with rapid cell imaging.

5.5.5 Multiphoton Techniques

Another class of widely used optical methods for biological detection makes use of multiphoton-based approaches [65]. Such nonlinear optical techniques are based on the fact that two or more photons of low energy can be simultaneously absorbed by a fluorophore, producing the emission of a photon with a shorter wavelength. Since multiphoton absorption is a nonlinear effect, it strictly requires a high source intensity and a tightly focused incident beam. To illuminate the sample with the necessary intensity, while preserving it from thermal damage, a femtosecond pulsed laser source is commonly used. The laser beam is then focused through an objective onto a sample treated with a fluorescent dye. Although single photons do not possess enough energy to induce an absorption transition, the photon density is sufficiently high in the centre of the tightly focused laser beam to cause direct absorption of two photons, so that an up-converted emission can result in. The fluorescence signal can be thus only detected in the focus of the laser beam. The focus can be also rasterized through the sample to create a 2D mapping. A typical advantage of this technique is that it suffers little from photobleaching effects. Indeed, irradiation in the far-red and near-infrared (NIR) spectral regions is characterized by low scattering and absorption by thee biological environment [66]. More recently, NIR radiation has been used to excite visible CdSe@ZnS core–shell NCs in mouse capillaries using multiphoton fluorescence imaging [67].

5.5.6 Metal-enhanced Fluorescence

The use of metallic materials to favorably alter the intrinsic photo-physical characteristics of fluorophores in a controllable manner is a fairly new concept that has received much attention only in the past few years.

Although fluorescence is a highly sensitive technique, allowing for the detection of single molecules, it exhibits intrinsic detection limits. These latter are typically related to the fact that the detection of a fluorophore is usually limited by its quantum yield, self-fluorescence of the sample, and/or by its photostability. In typical fluorescence measurements, the experimental set-up is such that the sampled area is large relative to the size of the fluorophores and relative to the absorption and emission wavelengths. In this arrangement, the fluorophores irradiate freely in space. Interestingly, for fluorophores close to metallic surfaces or nanoparticles the free-space condition is altered, resulting in dramatic spectral changes. The so-called Metal Enhanced Fluorescence (MEF) [68] arises from the interaction of the dipole moment of the fluorophore and the surface plasmon field of the metal, resulting in an increase in the radiative decay rate and in stronger fluorescence emission. Indeed, the quantum yield (Q_0) and lifetime of the fluorophore (τ_0) in the free-space condition are given by Eqs. (3) and (4), respectively.

$$Q_0 = \Gamma/(\Gamma + k_{nr}) \tag{3}$$

$$\tau_0 = 1/(\Gamma + k_{nr}) \tag{4}$$

where Γ and k_{nr} are the radiative and non-radiative decay rate, respectively. Due to the close proximity to a metal surface, the radiative decay rate increases by a term Γ_m. In this case, the quantum yield (Q_m) and lifetime of the fluorophore (τ_m) near the metal surface are given by Eqs. (5) and (6) [69]:

$$Q_m = (\Gamma + \Gamma_m)/(\Gamma + \Gamma_m + k_{nr}) \tag{5}$$

$$\tau_m = 1/(\Gamma + \Gamma_m + k_{nr}) \tag{6}$$

These equations allow the prediction of unusual behavior of a fluorophore near to a metallic surface. Notably, from Eqs. (3)–(6), as Γ_m increases, the quantum yield increases, while the lifetime decreases. These effects have also been justified theoretically [70], being analogous to the increase in Raman signals observed in surface-enhanced Raman scattering. However, a remarkable difference is that fluorescence enhancement is activated by interactions through space and does not require direct contact between the fluorophores and the metal surface.

Remarkably, metal surfaces can affect the radiative decay rates of fluorophores and can also be beneficial to resonance energy transfer. In practice, this means that weakly emitting materials (dyes, proteins, or DNA) with low quantum yields can be transformed into more efficient fluorophores with shorter fluorescence lifetimes. The reduction in fluorescence lifetime associated to MEF implies that molecules spend less time in the excited state, thus reducing photobleaching effects. The dramatic increase in the fluorescence emission occurs for fluorophores located at well-defined distances (50–200 Å) above metal surfaces with nanoscale roughness. This enables the use of photosensitive fluorophores that can hardly be employed in conventional fluorescence studies because of their high susceptibility to photobleaching. Another interesting effect is observed when fluorescence originates from nanomaterials with periodic patterns. In this case, the fluorescence emission can be spatially directed over a small viewing angle [71], which leads to emission sufficiently intense to be revealed by detectors without extensive optics. Such high directional emission combined with the increase in quantum yield offers the potential for efficient MEF-based sensors and microarrays [72].

5.5.7
Surface Plasmon Resonance

In the past decade, sensing by Surface Plasmon Resonance (SPR) has been demonstrated to be a powerful tool to probe of protein–ligand, protein–protein, and protein–DNA interactions. More interestingly, this technique provides a means not only to quantify the equilibrium and kinetic constants related to such events, but also for setting up very sensitive, label-free biochemical assays.

This optical technique is based on the detection of change in reflectivity of a laser beam, incident to a metallic thin layer, due to the interaction with surface plasmons propagating on metallic surface. A surface plasmon is a collective surface excitation of conduction electrons in a metal that can be excited by means of photon irradiation [73]. This concept originates in the plasma approach of Maxwell's theory, in which the free electrons of a metal are treated as an electron liquid of high density (plasma) and density fluctuations occurring on the surface of such a liquid are called plasmons, surface plasmons (SP), or surface polaritons. According to Maxwell's theory, surface plasmons can propagate along a metallic surface and are characterized by a spectrum of eigenfrequencies related to the wave-vector by a dispersion relation. Since the dispersion relation of SPs never intersects the dispersion relation of light in air, which can be represented by a straight line, the resonances cannot be excited directly by a freely propagating beam of light incident upon a metal surface. However, if the exciting beam travels through an optically denser medium (e.g., glass) of high dielectric constant, it is possible to decrease the slope of the straight line corresponding to the photon dispersion. If this is the case, plasmons can be excited by p-polarized light undergoing total internal reflection on the glass surface. More precisely, the plasmons are excited by the evanescent wave associated with the internal reflection, which can penetrate deeply even to the metal/air interface (Fig. 5.15). The fields associated with the surface plasmon (SP) extend into the media adjacent to the interface and decay exponentially away from it. Consequently, the SP is sensitive to changes in the environment near the interface and therefore is a sensing probe.

Whenever the momentum of the incoming photon matches with that of the plasmon, surface plasmon resonance is excited. This phenomenon can be inferred

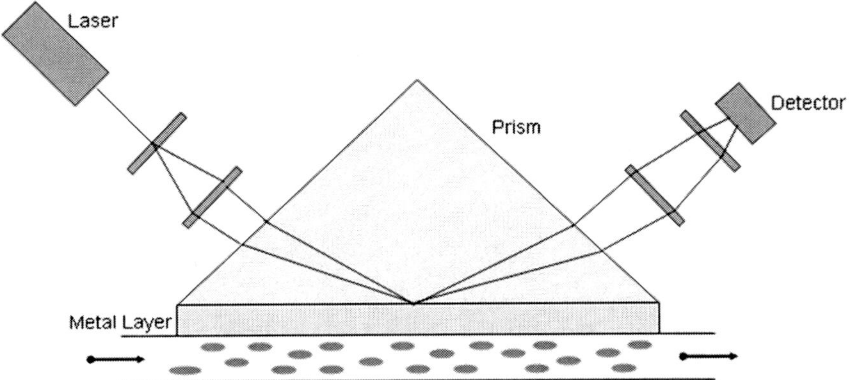

Fig. 5.15. Kretschmann configuration for surface plasmon resonance.

from the drop in intensity of reflected light when a suitable angle for resonance is approached at which an energy transfer between evanescent wave and surface plasmon occurs [74]. The value of the resonant angle is a sensitive function of the dielectric constants of the two contacting media.

The resonance conditions are influenced by the material adsorbed onto the thin metal film. Owing to this property, the surface plasmon resonance can be used to monitor surface reactions, as any organic or inorganic layer deposited on the metal surface causes changes in dielectric function of the medium, determining the new conditions for resonance (i.e., at a different resonance angle). The shape of the whole resonance curve, i.e., its depth and width depends on optical absorption within the metal and on radiation losses resulting from surface roughness. Therefore, optical systems based on SPR can provide a safe, remote, non-destructive means of sensing. SPR has already been used for gas sensing, biosensing, immunosensing and electrochemical studies. A satisfactory linear relationship is found between the resonance energy and the concentration of biochemically relevant molecules, such as proteins, sugars, and DNA [75]. The SPR signal, which is expressed in resonance units, therefore represents a measure of mass concentration at the sensor chip surface. This means that the analyte and ligand association and dissociation can be revealed, while rate constants as well as equilibrium constants can be then calculated.

A fraction of the light energy incident at a sharply defined angle can interact with the delocalized electrons in the metal film (plasmon), thus reducing the reflected light intensity. The precise value of incidence angle at which this occurs is determined by several factors, among which the most relevant is the refractive index of the backside layer of the metal film into which the evanescent wave propagates. This layer is the non-illuminated side of the metal film on which target molecules are immobilized and addressed by ligands in a mobile phase running along a flow cell. When a binding event occurs, the local refractive index is modified, leading to a change in SPR angle, which can be monitored in real-time by detecting changes in the intensity of the reflected light (Fig. 5.16).

The change rates for the SPR signal can be analyzed to extract apparent rate constants for the association and dissociation steps of the monitored reaction. The ratio of these values gives the apparent equilibrium constant (affinity). The magnitude of the change in the SPR signal is directly proportional to the immobilized mass, and can thus be roughly interpreted in terms of the reaction stoichiometry. Signals are easily obtained from sub-microgram quantities of material.

Since the SPR signal depends only on binding to the immobilized template, it is also possible to study binding events from molecules in extracts, i.e., in the presence of poorly purified components. The light source for SPR is a near-infrared light emitting diode that has a fixed range of incident angles. The SPR response is monitored by an array of light sensitive diodes covering the whole wedge of the reflected light. The angle at which minimum reflection occurs is detected and converted into resonance units.

Many biomolecular binding events cause changes in the refractive index at the surface layer, which are detected as changes in the SPR signal. In general, the re-

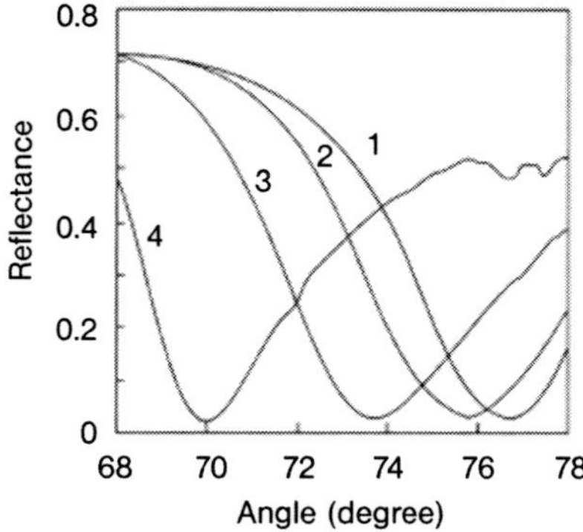

Fig. 5.16. Typical SPR curves of surfaces modified with Au-MIP/MIP gel (1), with Au-MIP/MIP of lower nanoparticle density (2), with MIP/MIP (3), and with 11-mercaptoundecanoic acid (4) in water at 30 °C. (From Ref. [76].)

fractive index change for a given change of mass concentration at the surface layer is practically of comparable magnitude for all proteins and peptides, and is similar for glycoproteins, lipids and nucleic acids.

5.5.8 Surface-enhanced Resonance Spectroscopy

Raman spectroscopy provides highly resolved vibrational spectra at room temperature without suffering from rapid photobleaching of the investigated molecule. However, Raman scattering is an intrinsically inefficient process with scattering cross sections ($\sim 10^{-30}$ cm^2 per molecule) approximately 14 orders of magnitude smaller than the absorption cross sections ($\sim 10^{-16}$ cm^2 per molecule) of fluorescent dye molecules. Therefore, to achieve sensitivity at the single molecule level, the normal Raman scattering efficiency must be enhanced 10^{14}-fold or more. Surface-enhanced Raman Spectroscopy (SERS) is a Raman Spectroscopic technique that provides a greatly enhanced Raman signal from Raman-active molecules that are adsorbed onto suitably prepared metal surfaces. An increase in the intensity of Raman signal of the order of 10^4–10^6 has been routinely observed, and can be as high as 10^8–10^{14} for some systems [77]. SERS was first discovered for analytes adsorbed onto Au, Ag, Cu or alkali (Li, Na, K) metal surfaces, with useful excitation wavelength near or in the visible region. Interestingly, a giant signal enhancement has been found using silver and gold faceted nanocrystals, allowing both detection and identification of single, non-fluorescent molecules. Therefore SERS is not only highly sensitive, but it is also selective as a result of enhancement mechanisms operating only at surfaces.

The importance of SERS is that the surface selectivity and sensitivity is extended to various molecular systems that are inaccessible by conventional Raman scattering, such as to *in situ* and ambient analyses of electrochemical, catalytic, biological, and organic systems. SERS measurements can be carried out under ambient conditions in a broad spectral range. Both electromagnetic-field and chemical-based models have been used to explain the SERS phenomenon. Electromagnetic effects dominate over the chemical effects, which in fact contribute to enhancement by one–two orders of magnitude [78]. Chemical enhancement (CE) involves changes in the electronic states of the analyte following its surface chemisorption [79], while electromagnetic enhancement (EME) is related to the metal surface roughness. The latter can be developed in several ways, such as, for example, by means of (a) oxidation–reduction cycles on electrode surfaces; (b) vapor deposition of metal particles onto substrates; (c) lithographically produced metal spheroid assemblies; (d) metal deposition over a mask of polystyrene nanospheres; (e) metal colloids [80]. All of these methods basically aim to provide rough metal surface for which roughness is actually due to metal nanoparticles or aggregates.

The intensity of the Raman scattered radiation is proportional to the square of the magnitude of the electromagnetic field incident to the analyte [Eq. (7)]:

$$I_{\mathrm{R}} = E^2 \tag{7}$$

where I_{R} is the intensity of the Raman field, and E is the total electromagnetic field associated with the analyte. Moreover, E is given by Eq. (8):

$$E = E_{\mathrm{a}} + E_{\mathrm{p}} \tag{8}$$

where E_{a} is the electromagnetic field on the analyte in the absence of any roughness features and E_{p} is the electromagnetic field emitted from the particulate metal roughness feature [81]. Of course, the electromagnetic field of the excitation radiation is incident to the analyte. In normal Raman scattering ($E_{\mathrm{p}} = 0$), E_{a} is relatively small since the source of the electromagnetic radiation (the laser) is relatively far away. However, in SERS the roughness features provide an additional field, E_{p}, in close proximity to the adsorbate, that is associated with the metal particles. Because of this proximity, any electromagnetic field from the particle will strongly contribute to E. Thus, the magnitude of the particle field is of crucial importance in determining the overall intensity of the Raman signal. Since the plasmon excitation is also confined in a small material volume (related to the characteristic features of the surface roughness), the field E_{p} associated with it is also very intense. The large electromagnetic energy density on the particle effects the scattering I_{R}, as described by Eqs. (7) and (8). It follows that analyte molecules adsorbed between two SERS-active particles will yield a higher signal than those molecules being close to only one particle.

Since the surface roughness features are so relevant to SERS, much attention has been devoted to their study. Normal roughening procedures produce heterogeneous surfaces where the particles have various sizes, shapes, and orientations [82]. The frequency of incident light that is most effective in producing an intense field on the particle severely depends on the degree of surface roughness as well as on the nature of metal used. The specific region on the particle where analyte adsorption occurs affects the magnitude of the SERS enhancement, as the EME is strongest where the particle has the highest curvature.

The CE mechanism is less understood than the EME enhancement, but brings about some interesting considerations. The molecule adsorbed onto the surface has necessarily to interact with the surface. CE exists because of this interaction, which has been described in several ways. The proximity of metal adsorbate may allow pathways of electronic coupling from which novel charge-transfer intermediates are formed that have a higher Raman scattering cross-section than that of a non-adsorbed analyte. This fact resembles a resonance Raman effect. Another explanation is that the molecular orbitals of the adsorbate overlap with the conduction band of the metal, thus modifying the analyte chemical properties [83]. Interestingly, the CE effect results in an alteration in the scattering cross-section, whereas the EME effect generally leads to change in the scattering intensity without affecting the scattering cross-section.

The distance dependence of SERS is used in the overlayer technique. An ultrathin solid or liquid film is adsorbed onto the SERS-active substrate, with the analyte within the film taking advantage of the EME of the substrate. It follows that the Raman spectrum of the analyte is greatly enhanced. This makes SERS applicable to various targets, such as organic films, semiconductors, insulators, etc. Spacer layers can be adsorbed onto the substrate, and can alter the chemistry of either the substrate or that of the overlayer, extending the utility of SERS even further. An advantage to the overlayer technique is that it suffers less from the inhibitory effect of metal surface defects, such as pinholes in silver [84].

As most of signal enhancement is due to electromagnetic fields whose intensities decrease with increasing distance from the point source, it follows that an analyte would benefit from SERS substrates even if it were located at some distance away from the active surface. Theoretical calculations have been performed, and the SERS enhancement, G, has been found to scale with distance according to Eq. (9):

$$G = [r/(r + d)] \tag{9}$$

where r is the radius of the spherical metal roughness feature, and d is the distance of the analyte from that feature. Experimentally, G decreases by a factor of ten at a distance of 2–3 nm, or equivalent to one–two monolayers.

Traditionally, SERS-based sensing has employed roughened metal surfaces or colloidal particles immobilized on a substrate to create relatively large active areas for sensing [85]. More recently, efforts have been focused on the potential of using individual nanoparticles or small clusters of nanoparticles as the sensing elements [86]. This approach has paved the way to a wide range of applications that so far have been difficult to address with conventional chemical sensors. A remarkable example is the possibility of using these probes inside single living cells. The small size of the nanoparticles (50–100 nm in diameter) combined with the highly localized probe volume makes these particles work as localized sensors suitable for monitoring biological processes in living cells.

5.6 Advantages and Disadvantages of Nanocrystals in Optical Detection

Colloidal nanocrystals have several peculiar advantages over organic dyes that are conventionally used in optical bio-detection [29d–87]. Indeed, their wide absorption spectrum, coupled with their extinction cross-section, makes it possible to excite different luminescent nanoparticles with different sizes by one single excitation wavelength. Organic fluorophores, however, exhibit discrete and well-spaced absorption bands with cross-sections nearly an order of magnitude smaller. In addition, most semiconductor nanocrystals have narrow (few tenths of nanometers) symmetric emission peaks that can be finely tuned from blue to red up to the near-infrared region [88]. The high color purity achievable by precise size control represents the main advantage offered by colloidal nanocrystals, especially in multiplexing applications. In contrast, organic dyes are often characterized by asymmetric spectra with distinct phonon progression towards the longer wavelength, which leads to undesirable spectral overlap among different channels in a detecting system for the measurements. Moreover, the typical fluorescence quantum yields of NCs, defined as the ratio of the number of photon emitted in radiative channel to the number of photon absorbed, can be as high as that of organic dyes, approaching values close to 30%. As a further advantage, semiconductor NCs are much less susceptible to photobleaching than organic dyes, which are rapidly and irreversibly photodegraded [52a].

The sensitivity of luminescence of NCs to their surface status (defects and/or dangling bonds) and to the local environment has typically represented the major limitation to technological applications. Now, core–shell heterostructures, where the nanocrystal is coated with a thin inorganic shell of a wider band gap semiconductor, provide increased fluorescence quantum efficiencies (>50%) and greatly improved photochemical stability (Fig. 5.17).

The sensitive detection in thick specimen requires wavelengths such that excitation light can penetrate up to the desired depth, and the emitted light must be able to travel back to a photodetector. Several semiconductor materials have been used to generate bright nanocrystals that emit between 650 and 2000 nm [89]. NCs with emission maxima between 750 and 860 nm have been employed to image the coronary vasculature in a rat model, while NCs emitting in the 840–860 nm region have been used to sentinel lymph-node mapping in cancer [90, 31c]. This use of near-IR emitting NCs can be thus profitably extended for "*in vivo*" biosensing applications, as NCs have been demonstrated to retain their emission in tissues *in vivo* for months.

Colloidal semiconductor nanocrystals represent, therefore, a valid alternative to organic fluorophores and their superior fluorescence properties account for the large success illustrated by several applications as biosensors.

Historical inconveniences of colloidal nanocrystals, such as the loss of emission in aqueous environment or the poor biocompatibility and bioconjugability, have been effectively overcome in recent years by suitable functionalization methods that have been reported in the Section 5.3. A serious difficulty with the large-scale

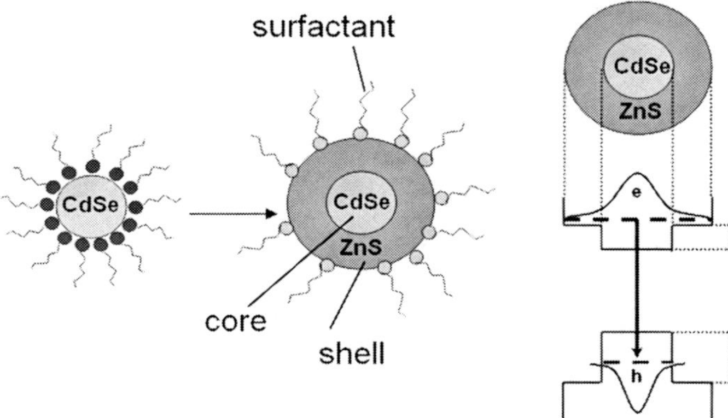

Fig. 5.17. Schematic diagram of core–shell nanocrystals of CdSe@ZnS. The growth of a thin layer of ZnS on CdSe nanocrystal cores causes an increased optical confinement.

use of colloidal nanocrystals, especially those subjected to bioconjugation treatments, lies in the still expensive fabrication costs. However, it is expected that commercial expansion of chemical manufacturers will lead to a more competitive environment and, possibly, to lower production costs.

5.7 Applications

The impact of nanotechnology on biosensing has been tremendous owing to the availability of the more robust, versatile and efficient materials for sensing. The following subsections review several successful examples of the use of colloidal nanoparticles in optical biosensing.

5.7.1 Biosensing with Semiconductor Nanocrystals

The broad spectrum of "colors", which can be obtained by exciting semiconductor NCs of different size with a single excitation laser wavelength, poses the basis for the simultaneous detection of several markers in biosensing and assay applications. NCs can be used as fluorescent labels for immunosensing and DNA sensing, providing tunable wavelength, narrow emission peaks, and 100-fold higher stability than molecular fluorescent dyes [91].

The novel applications of NCs have highlighted their wide potential in present technology. Several examples can be found on the use of luminescent CdSe@ZnS core–shell nanocrystal bioconjugates in quantum dot-based sensors. Biomolecules can be labeled with luminescent NCs. They have been also demonstrated to be involved in efficient fluorescence resonance energy transfer between neighboring NCs of different sizes, while their emission is readily quenched by bound fluorescent dyes. Sensor assemblies have been reported that employ NCs linked to dye-labeled biological receptors to exploit donor–acceptor energy transfer between the NCs and the receptors for detection purposes based on molecular recognition events. The effect of variation in concentration of the energy acceptors bound to a nanocrystal surface on the luminescence quenching of either in soluble and solid phase conditions has been investigated. The obtained results suggest the occurrence of fluorescence quenching between NCs and surface anchored dye–biological receptor [36b].

An application of quantum dots is the multiplexed optical encoding and the high-throughput analysis of genes and proteins. This is a very important issue, because most complex human diseases (such as cancer) involve many genes and proteins. Tracking a panel of molecular markers at the same time will allow scientists to understand, classify and to differentiate such complex human diseases.

The advantageous core–shell nanocrystals have also been exploited in combination with polymers [29c] (Fig. 5.18).

First, polystyrene microspheres doped with single-size CdSe@ZnS nanocrystals in known amounts are prepared and analyzed. Fluorescence intensity analysis of the beads clearly discriminates ten different intensity levels and indicates a high degree of bead uniformity and bead identification accuracy. Next, the beads with three colors of quantum dots are prepared. Significantly, fluorescent resonant energy transfer from the small nanocrystals to the large nanocrystals is not observed. The triple-coded bead is then used to demonstrate a DNA hybridization assay. Overall, these results indicate that a new spectral coding technology with potentially wide application to various biological assays that can be developed using fluorescent nanocrystals, embedded in microspheres. In this regard, NC probes are particularly attractive, because of the simultaneous excitation and continuous tuning by variation of particle size and chemical composition. The use of six colors and ten intensity levels can, theoretically, encode one million protein or nucleic acid sequences. Specific capturing molecules, such as peptides, proteins, and oligonucleotides can be covalently linked to the beads, which encode a specific spectroscopic signature. To determine whether an unknown analyte is captured or not, conventional assay methodologies (similar to direct or sandwich immunoassay) can be applied.

Bioconjugated NCs have also been used successfully in time-gated fluorescence detection of tissue sections. This optical technique is based on lifetime fluorescence microscopy employed for the selective and sensitive characterization of fluorescence species and in biophysical studies of proteins. Silanized nanocrystals were added to a 3T3 mouse fibroblasts cells grown on fibronectin treated coverslips, without signs of toxic response to the presence of NCs in the medium being observed [64].

The use of luminescent colloidal semiconductor nanocrystals as fluorophores have expanded to the range in FRET-based sensing applications [92]. If fully exploited, the unique NC properties should allow development of FRET-based nano-

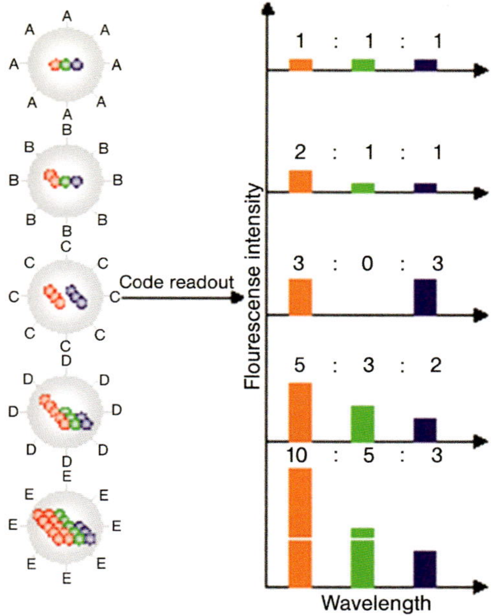

Fig. 5.18. Schematic illustration of optical coding based on wavelength and intensity multiplexing. Large spheres represent polymer microbeads, in which small colored spheres (multicolor quantum dots) are embedded according to predetermined intensity ratios. Molecular probes (A–E) are attached to the bead surface for biological binding and recognition, such as DNA–DNA hybridization and antibody–antigen/ligand–receptor interactions. The numbers of colored spheres (red, green, and blue) do not represent individual NCs, but are used to illustrate the fluorescence intensity levels. Optical readout is accomplished by measuring the fluorescence spectra of single beads. Both absolute intensities and relative intensity ratios at different wavelengths are used for coding purposes; for example, (1:1:1) (2:2:2), and (2:1:1) are distinguishable codes. (From Ref. [29c].)

scale assemblies capable of continuously monitoring target (bio)chemical species in diverse environments [93].

It has been confirmed that semiconductor nanocrystals can be FRET donors, quenchable with efficiencies up to 99%, when using organic fluorophores, non-emissive dyes, gold nanoparticles, or other nanocrystals as acceptors (Fig. 5.19) [94].

NCs conjugated to maltose binding proteins have been used as *in situ* biosensors for carbohydrate detection. A fluorescence quenching is observed by adding a maltose derivative covalently bound to a FRET acceptor dye, while PL emission is recovered upon addition of native maltose, as the sugar–dye compound is displaced. In this work, the recovery of nanocrystal fluorescence upon maltose addition can be directly related to maltose concentration, because of both the physical orientation and stoichiometry of the maltose receptors on the nanocrystals are considered as

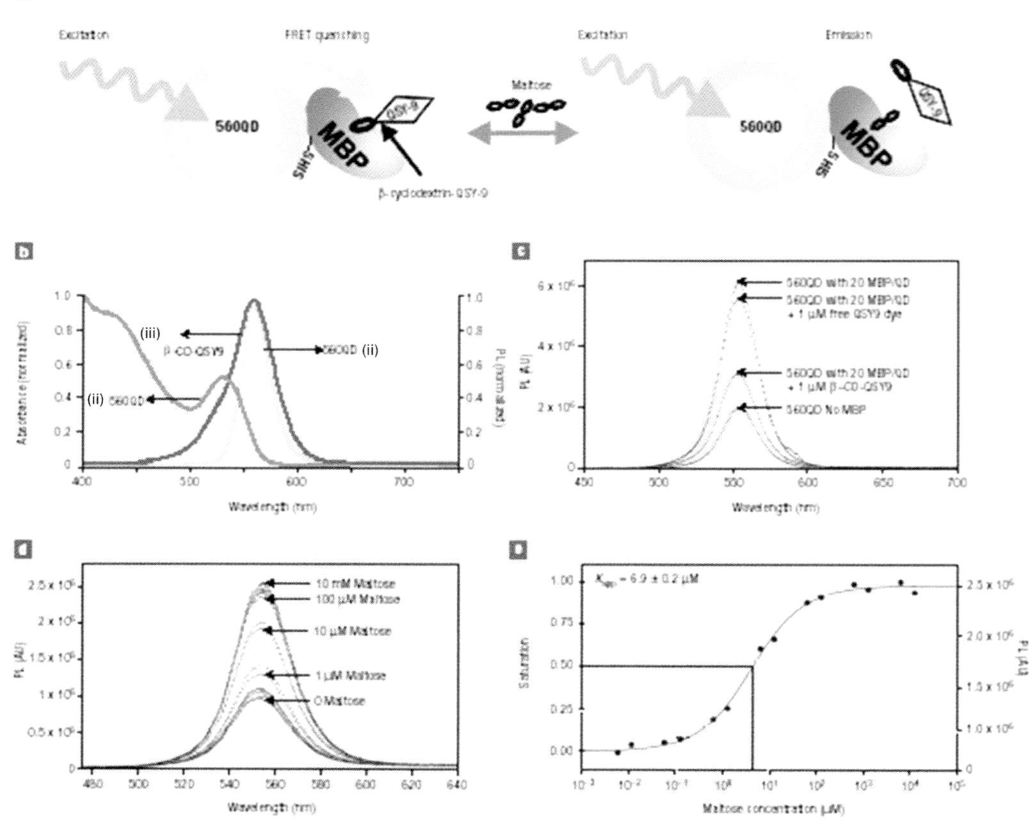

Fig. 5.19. Function and properties of a maltose nanosensor. (a) Nanocrystal-MBP nanosensor operation principle. Each NC is surrounded by MBP. The formation of complex NC-MBP with CD-QSY9 (maximum absorption ~565 nm) results in quenching of NC emission. Added maltose displaces CD-QSY9 from the sensor assembly, resulting in an increase in direct NC emission. (b) Spectral properties of the nanosensor. Absorption (i) and emission spectra (ii) of MBP-conjugated NC are displayed along with the absorption spectra (iii) of CD-QSY9. (c) Demonstration of NC-MBP FRET quenching. PL spectra (AU = arbitrary units) were collected from NCs and NCs mixed with an average of 20 MBP/NC, obtaining an increase in PL (~300%). The same NC-MBP conjugates were then mixed with either free QSY9 dye or CD-QSY9. (d) NC-10MBP maltose sensing. Titration of a NC-10MBP/NC conjugate (quantum yield ~39%) preassembled with CD-QSY9 with increasing concentrations of maltose. (e) Transformation of titration data. The right-hand axis shows PL at 560 nm, and fractional saturation is shown on the left-hand axis. (From Ref. [94].)

parameters. Although the FRET quenching efficiency was low, this work demonstrates the potential of NC-based *in situ* biosensing.

Other important applications of fluorescent nanocrystals as FRET-based biosensors have been very recently developed for ultrahigh sensitive detection of DNA [63]. The DNA nanosensor can consist of two target-specific oligonucleotide probes, which include a reporter probe labeled with a fluorophore and a capture probe labeled with biotin, in addition to a NC conjugated with several streptavidins. The nanocrystal works as both a FRET energy donor and a target concentrator, producing an extremely low level of background fluorescence that is difficult to achieve with conventional organic fluorophores. When a target DNA is present in solution, it is sandwiched by the two probes. Several sandwiched hybrids are then captured by a single NC through biotin–streptavidin binding, resulting in a local concentration of targets in a nanoscale domain. The resulting assembly brings the fluorophore acceptors and the nanocrystal donor into close proximity, leading to fluorescence emission by the acceptors by means of FRET upon illumination of the donor, indicating the presence of targets.

5.7.2 Biosensing with Metallic Nanoparticles

The use of gold colloids in biodetection dates back to the early 1970s when the immunogold staining procedure was invented. Since that time, the labeling of targeting molecules, such as antibodies, with gold NCs has deeply modified the approach to detection of biologically relevant components.

The optical transduction of receptor–ligand interactions at the surface of noble metal NCs via changes in their extinction spectrum is attractive for the development of biosensors, because a colorimetric transduction scheme is considerably simpler to implement than conventional, planar SPR.

It has been shown [95] that, for oligonucleotide-functionalized 30-nm Au NCs, the amount of optical absorbance recorded for an array surface of bound particles is directly related to the number of bound particles and to the concentration of hybridizable targets in solution. This study also demonstrates that, using microstructured substrates, DNA-modified areas with dimensions in the micrometer range specifically recognize complementary NC-bound DNA strands, with the signal background from unfunctionalized areas of the chip remaining rather low.

The environmental sensitivity of the color of nanosized gold has been exploited in a first prototype of biosensor in [96]. This study demonstrated the possibility of colorimetrically monitoring the binding of antibody to the surface of Au NCs functionalized with protein antigens.

Highly selective, colorimetric detection methods based on polynucleotide-modified gold particle probes have been also proposed (Fig. 5.20) [97]. Specifically, in the presence of 13-nm Au NCs, the color of the solution changes from red to blue, as a consequence of analyte-induced formation of Au NC aggregates in which the individual surface plasmon resonances are coupled. This simple phenomenon points to the use of metal NCs as DNA detection agents for nucleic acid targets. Actually, it has been found that spotting the solution onto a white support amplifies the colorimetric change and provides a permanent record for each test. Such system can detect about 10 fmol of oligonucleotides.

This type of colorimetric biosensor based on Au NCs has been extensively used in the analysis of biomolecules.

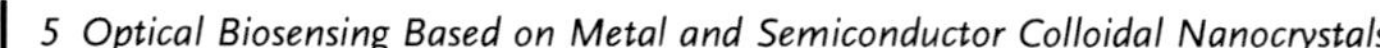

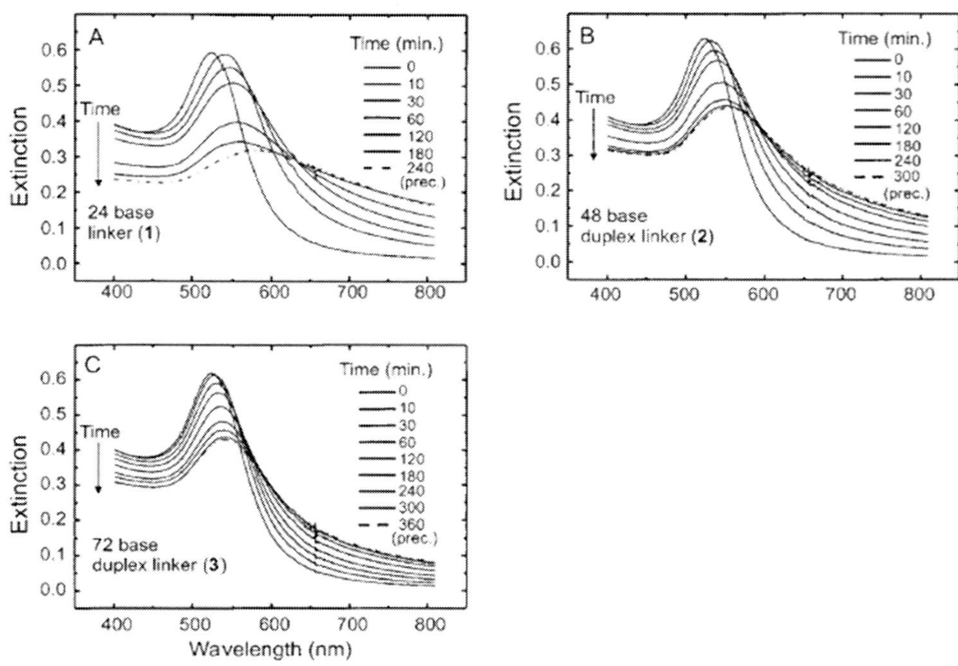

Fig. 5.20. Temporal changes in the UV/Vis spectra of nanoparticle aggregates grown from DNA linkers **1–3** (defined in parts A–C, respectively). The y-axes are labeled "extinction" since the larger aggregate structures will contain scattering as well as absorbance components. (From Ref. [97d].)

Gold NCs have been modified with an aptamer, which is specific to a defined growth factor, to allow monitoring of the change in color and extinction coefficient due to NC aggregation [44c]. The color of the aptamer-modified Au NCs significantly changes at low concentrations (<400 nM), while only minor changes are detected at higher concentrations (>400 nM). In addition, the ionic strength of the medium and the NC surface status surface are also important parameters in determining the ultimate sensitivity and specificity of the aptamer–nanoparticle-based probes. Efforts are now focused on the development of aptamer modified Au nanorods, to achieve further sensitivity enhancement. Special precautions must, however, be taken to prevent aspecific adsorption of biomolecules.

The performances of a label-free optical biosensor [98] have been tested in the real-time quantification of biomolecular interactions on a surface in a commercially available UV–visible spectrophotometer and in colorimetric end-point assays with an optical scanner. Such a sensor shows a concentration-dependent binding and a detection limit of 16 nM for streptavidin.

In this perspective, coupling the optical properties of noble metal colloids with improved bioconjugation protocols has the potential to provide excellent detection capabilities. SPR has been widely used to monitor a broad range of analyte–surface binding reactions, including the adsorption of small molecules, ligand–receptor binding, protein adsorption on self-assembled monolayers, antibody–antigen binding, DNA and RNA hybridization, and protein–DNA interactions [99]. The SPR sensing mechanism relies on the measurement of small changes in refractive index that occur in response to analyte binding at or near the surface of a noble metal (Au, Ag, Cu) thin film. Chemosensors and biosensors based on SPR spectroscopy possess many desirable characteristics: (a) a refractive index sensitivity of the order of 10^{-1} pg mm^{-2}; (b) a large sensing length scale (~200 nm) dictated by the exponential decay of the evanescent electromagnetic field; (c) multiple instrumental modes of detection (viz., angle shift, wavelength shift, and imaging); (d) real-time detection on the 10^{-1}–10^{-3} s time scale for measurement of binding kinetics; and (e) lateral spatial resolution of the order of 10 μm allows for multiplexing and miniaturization, especially using the SPR imaging mode of detection [100]. Although SPR spectroscopy is an intrinsically nonselective sensor platform, a high degree of analyte selectivity can be conferred by means of attaching high specificity ligands to the surface and then passivating the sensor surface with nonspecific binding molecules. Other practical advantages of SPR can be recognized, such as that of being label-free [101] useful for probing unpurified complex mixtures, such as clinical samples, easy to integrate with the available commercial instrumentation and with advanced microfluidic sample handling [100]. The development of large-scale biosensor arrays composed of highly miniaturized signal transducer elements that enable the real-time, parallel monitoring of multiple species is an important driving force in biosensor research.

Au NCs linked to bioreceptors provide labeled conjugates that can be used to follow the biorecognition event at a biosensor surface. Resonance enhancement of the absorption properties of nanosized metals binding to a surface by biorecognition interactions has also been used as an effective approach for biosensor devices [102]. Au nanoparticles have been extensively applied in bioaffinity binding events based on enhanced SPR. SPR biosensing has been improved dramatically in DNA immunosensing in the presence of Au NCs [103]. A sensor interface has been modified with antibody units, with the surface being functionalized with the complementary antigen component, and, finally, making this affinity assembly react with a secondary antibody labeled with Au NCs. The association of an antigen with its antibody-functionalized surface can be detected by SPR. However, the change in the SPR spectrum is larger if the secondary antibody is bound to the surface. The secondary antibody is, notably, not able to associate with the modified surface, unless the antigen is already bound there, thus providing an amplification route for the primary recognition event. Amplification is dramatically increased when the secondary antibody is labeled with gold. The binding of the Au NCs to the immunosensing interface leads to a large shift in plasmon angle, to a broadening of the plasmon resonance, and to an increase in the minimum reflectance, thereby allowing for picomolar detection of the antigen. Similarly, an enhancement in sensitivity by about 3 orders of magnitude has been obtained in DNA analysis when Au NC-functionalized DNA molecules have been used as probes [76].

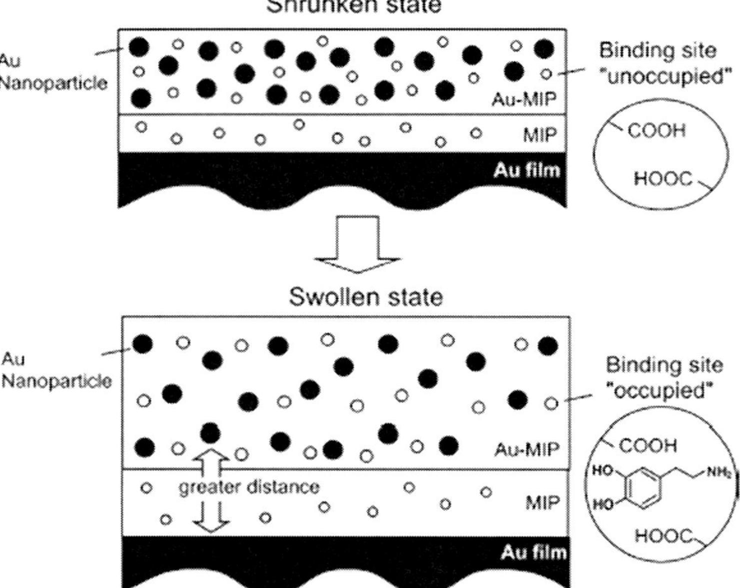

Fig. 5.21. Schematic representation of Au-MIP/MIP-coated SPR sensor chip for detection of an analyte, dopamine. (From Ref. [76].)

An SPR sensing device sensitive to a low molecular weight analyte has been fabricated starting from molecularly imprinted polymer gel with embedded Au NCs onto a gold substrate of a chip sensor. The sensing mechanism is based on swelling of the imprinted polymer gel that is triggered by an analyte binding event within the polymer gel (Fig. 5.21). The swelling increases the distance between the Au NCs and substrate, shifting the SPR resonance angle to higher values.

The modified sensor chip shows changes in the SPR angle in response to dopamine concentration, which agrees with the supposed sensing mechanism. The presence of Au NCs has been proven to be essential for enhancing the signal intensity. The analyte-binding process and the consequent swelling appears to be reversible, thus allowing re-use of the sensor chip.

There has been a recent explosion in the use of metallic nanostructures to favorably modify the spectral properties of fluorophores and to alleviate some of the photophysical constraints associated with these fluorophores. The use of fluorophore–metal interactions can lead to ultrabright over-labeled proteins, generating a new class of probes based on MEF proteins covalently labeled with fluorophores to be used as reagents, e.g., in immunoassays or for the immunostaining of biological specimens with specific antibodies. For these applications, the commonly used fluorescein is unfortunately self-quenching, thus resulting in a homo-transfer process [68]. Self-quenching can be largely avoided by the close proximity to Ag islands [68]. Such a decrease in self-quenching has been ascribed to an increase in the rate of radiative decay. These results suggest the possibility of exploiting ultrabright labeled proteins based on high labeling ratios, and the release of self-quenching through metal enhanced fluorescence.

An approach has also been reported that should provide a readily measurable change in fluorescence intensity in DNA hybridization assays [104] due to the presence of Ag particles. This approach should ensure an intensity enhancement relative to the background with an increase in the number of detected photons per fluorophore molecule by a factor of ten or more. In the presence of thiolated oligonucleotide single-stranded DNA as the capturing sequence bound to Ag particles, an increase in the emission intensity has been observed upon addition of a single strand fluorescein-labeled DNA. This effect can be attributed to the localization of the fluorescein labeled DNA near to the particles by hybridization with the capturing DNA.

A metal-enhanced multiphoton excitation has also been studied. For multiphoton excitation of fluorescence, most of excitation occurs at the focal point of the excitation beam, where the local intensity is the highest. For a two-photon absorption process, the rate of excitation is proportional to the square of the incident intensity. This suggests that two-photon excitation can be greatly enhanced [69–105]. Such an enhancement in the excitation rate is expected to provide selective excitation of fluorophores near to metal islands of colloids, even if the solution contains a considerable concentration of other fluorophores that could undergo two-photon excitation at the same wavelength, but which are more distant from the metal surface. Recently, the enhanced and localized multiphoton excitation of fluorophores adjacent to metallic silver islands has been observed to be accompanied by a reduction in lifetime, as compared with that detected using one-photon excitation. At present, following the widespread use of multiphoton excitation in microscopy and medical imaging, the use of metallic nanostructures appears promising to both enhance and localize fluorescence.

Recent works have demonstrated the effectiveness of MEF in solution-based biosensing applications. Silica-coated Ag colloids can represent enhanced-fluorescence sensing platform as a three- to fivefold enhancement has been observed [106]. It has also be found that oligonucleotide-modified Ag particles and fluorophore-labeled complementary oligonucleotides can yield an increased emission from the fluorophore-labeled complementary oligonucleotide after hybridization, which could allow access to DNA detection based on the aggregation of metallic NCs binding to the fluorophore label [3b].

The unusual optical properties of noble metal NCs have been also used to design a label-free biosensor in a chip format. The NC size affects the sensitivity of the biosensor significantly. The sensor has been fabricated by making Au NCs chemisorbed nanoparticles on amine-functionalized glass. In the investigate size range (12–48 nm), sensors fabricated from 39-nm-diameter Au NCs exhibited the highest sensitivity to the change in the bulk refractive index as well as the largest "analytical volume", defined as the region around the nanocrystal within which a change in refractive index induces a change in the optical properties of the immobilized NCs. The detection limit for streptavidin–biotin binding of a sensor fabricated from 39-nm diameter nanoparticles is 20× better than that of a sensor fabricated from 13-nm-diameter gold nanoparticles by another method [101].

Surface-enhanced Raman scattering (SERS) of substrates bound to nanoparticles allows molecular vibrational spectra to be amplified by a factor of 10^5 [107]. Although this technique has not been widely applied for detecting biorecognition events on surfaces, a cytochrome *c*–Au NC conjugate associated with silver surface has been revealed by a SERS spectrum [108].

The use of a versatile format based on gold nanoparticles-based microarray demonstrated high selectivity and sensitivity for protein binding and kinase functionality, indicating that the method is based on the fact that specific antibody–protein binding or peptide phosphorylation events can be marked by attaching gold nanoparticles and then depositing silver to enhance the signal. The detection principle is based on the resonance light scattering. The attachment of gold nanoparticles is achieved by standard avidin–biotin chemistry. The obtained results show that a low detection limit for protein binding has been achieved together with a large dynamic range. Similarly, for kinase functionality a high sensitivity has been reached [109]. Recently the SERS response of individual nanoscale pH sensors based on functionalized silver nanoparticles has been tested *in vitro* [110].

The 4-MBA-capped nanoparticles (4-MBA, 4-mercaptobenzoic acid, is a nonresonant molecule) have shown a pH-dependent response in a pH range relevant to biological systems, showing promise for application in intracellular chemical measurements (Fig. 5.22). The nanoparticle sensors can be incorporated into living cells while retaining their functionality and robust signal.

5.8 Towards Marketing

Generally, the term "technological applications" strictly refers to scientific products that can be potentially put on the road towards markets. The combination of biology, material science, and modern optical techniques has certainly great potential in offering a new generation of devices able to solve biosensing problems in a sensitive, selective, fast and high-performance way. Nevertheless, present applications of nanocrystals in the biosensing field are still far from being recognized as commercializable.

Presently, in detection devices relying on chip-based biosensing, the capturing molecules on the chip surfaces are patterned on the microscale. A large parallel screening of diverse analytes can be thus achieved in a small area, which greatly facilitates simultaneous sensing. Further miniaturization in the form of nanoarrays would allow for orders of magnitude larger multiplexed detection, eventually improving detection limits due to the smaller analyte capture area. Several estab-

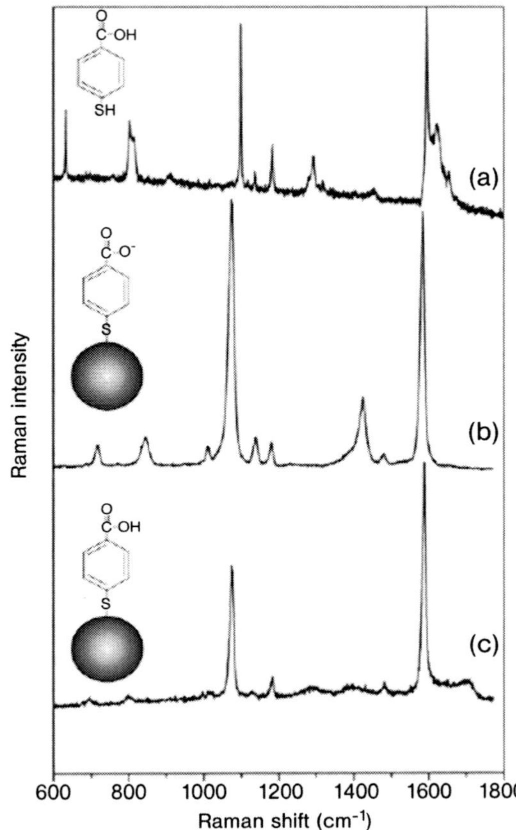

Fig. 5.22. Raman spectrum of solid 4-mercaptobenzoic acid (4-MBA) (a) and the SERS spectra of 4-MBA attached to silver nanoparticles at pH 12.3 (b) and at pH 5.0 (c). Insets to the left of each spectra illustrate the dominant state of the molecule under the conditions described. (From Ref. [110].)

lished techniques, as well as emerging technologies, such as dip-pen nanolithography, nanografting, and focused ion beam lithography are possible candidate tools to fabricate nanoscale patterns of biomolecules or hybrid systems.

Lithographic techniques can also be used to fabricate nanoscopic patterns or containers on surfaces that should significantly reduce the sample volume required and, possibly, lower detection limits even more. The development of such nanopatterning techniques can be expected to represent the next step that has to be made towards achieving further miniaturization [3b]. The commercial exploration of nanomaterials for biosensing deals also with the development, the production and the commercialization of nanocrystals suitably engineered. In this regard, a large progress has been realized, as authenticated by the first related patents [111].

Industrial interest in this direction is indicated by the number of companies that have worked in synergy with academic institutions, where the seed ideas have been generated. Quantum size effects in semiconductor NCs for tagging biomolecules or the use of bio-conjugated noble metal nanoparticles to label and sense biomolecules or bio-related analytes are among the most explored application fields.

Other key issues for the progress of nanomaterial based biosensing are related to material toxicity and safety issues. The properties of colloidal NCs should be better exploited in *in vivo* biosensing devices. In this perspective, the issue of toxicity of the used nanosized material became crucial. The toxicity suspects of many semiconductor nanocrystals have recently alerted the scientific community, becoming also a societal topic of considerable importance and discussion. Indeed, *in vivo* toxicity is likely to be a key factor in determining whether nanoparticles would be approved by regulatory agencies as biosensing probes for clinical use on human beings. For example, a recent study [31c] indicates that CdSe NCs are highly toxic to cultured cells under UV illumination for extended periods of time. This is not surprising because the UV irradiation can break covalent chemical bonds and induce NC photodissolution, ultimately releasing toxic cadmium ions into the culture medium. In the dark, however, NCs protected by a polymer coating have been found to be essentially non-toxic to cells and animals, as no effects on cell division or ATP production have been ascertained. For polymer-encapsulated NCs, chemical or enzymatic degradation of the inorganic cores is unlikely.

Finally, possible toxicity problems associated with long-term exposure of technical human staff devoted to routine production, manipulation, and "*in vitro*" use of NCs for fundamental research purposes must be carefully considered [112].

5.9 Conclusions

The potential of nanomaterial-based optical biosensing has been assessed by recent research developments and implemented into various novel and/or improved optical techniques, competing strongly with traditional methods. The merits of nanocrystals can be recognized in the breakthrough, compared with conventional assays, in terms of both selectivity and sensitivity. Optimization of the physical, chemical and fabrication parameters is necessary to ensure extension of these devices from experimental prototypes to large-scale industrial production.

Future design and engineering of biosensing devices able to detect analytes in complex matrices and environments will, essentially, need close interdisciplinary exchange and stimulating collaborations among material scientists, engineers, analytical, (bio)chemical and biomedical researchers. This can, hopefully, be expected to guarantee the realization of efficient sensors at relatively low cost for easy commercialization.

References

1 (a) Bartl, M. H., Boettcher, S. W., Frindell, K. L., Stucky, G. D., Molecular assembly of function in titania-based composite material systems, *Acc. Chem. Res.* **2005**, 38, 263–271. (b) Caminade, A.-M., Majoral, J.-P., Nanomaterials based on phosphorus dendrimers, *Acc. Chem. Res.* **2004**, 37, 341–348.

2 (a) Shipway, A. N., Katz, E., Willner, I., Nanoparticle arrays on surfaces for electronic, optical and sensor applications, *Chem. Phys. Chem.* **2000**, 1, 18–52. (b) Huynh, W. U., Dittmer, J., Alivisatos, A. P., Hybrid nanorod-polymer solar cells, *Science* **2002**, 295, 2425–2427. (c) Narayanan, R., El-Sayed, M. A., Catalysis with transition metal nanoparticles in colloidal solution: Nanoparticle shape dependence and stability, *J. Phys. Chem. B* **2005**, 109, 12663–12676.

3 (a) Alivisatos, A. P., Colloidal quantum dots. From scaling laws to biological applications, *Pure Appl. Chem.* **2000**, 72, 3–9. (b) Rosi, N. L., Mirkin, C. A., Nanostructures in biodiagnostics, *Chem. Rev.* **2005**, 105, 1547–1562. (c) Hu, J., Li, L.-S., Yang, W., Manna, L., Wang, L.-W., Alivisatos, A. P., Linearly polarized emission from colloidal semiconductor quantum rods, *Science* **2001**, 292, 2060–2063. (d) Bakalova, R., Ohba, H., Zhelev, Z., Ishikawa, M., Baba, Y., Quantum dots as photosensitizers?, *Nat. Biotechnol.* **2004**, 22, 1360–1361.

4 (a) Rao, C. N. R., Kulkarni, G. U., Thomas, P. J., Edwards, P. P., Size-dependent chemistry: Properties of nanocrystals, *Chem. Eur. J.* **2002**, 8, 28–35. (b) Kamat, P. V., Photophysical, photochemical and photocatalytic aspects of metal nanoparticles, *J. Phys. Chem. B* **2002**, 106, 7729–7744. (c) El-Sayed, M. A., *Acc. Chem. Res.* **2004**, 37, 326–333. (d) El-Sayed, M. A., *Acc. Chem. Res.* **2001**, 34, 257–264.

5 (a) Alivisatos, A. P., Perspectives on the physical chemistry of semiconductor nanocrystals, *J. Phys. Chem.* **1996**, 100, 13 226–13 239. (b) Buffat, P., Borel, J. P., Size effect on the melting temperature of gold particles, *Phys. Rev. A* **1976**, 13, 2287–2298. (c) Peters, K. F., Cohen, J. B., Chung, Y. W., Melting of Pb nanocrystals, *Phys. Rev. B – Condensed Matter* **1998**, 57, 13 430–13 438. (d) Besson, J. M., Itie, J. P., Polian, A., Weill, G., Mansot, J. L., Gonzalez, High-pressure phase transition and phase diagram of gallium arsenide, *Phys. Rev. B – Condensed Matter* **1991**, 44, 4214–4234. (e) Chen, C. C., Herhold, A. B., Johnson, C. S., Alivisatos, A. P., Dependence of structural metastability in semiconductor nanocrystals, *Science* **1997**, 276, 398–401. (f) Herhold, A. B., Chen, C. C., Johnson, C. S., Tolbert, S. H., Alivisatos, A. P., Klein, D. L., Bowen Katari, J. E., Roth, R. A., Alivisatos, P., McEuen P. L., A new approach to electrical studies of single nanocrystals, *Appl. Phys. Lett.* **1996**, 68, 2574.

6 Comparelli, R., Tamborra, M., Striccoli, M., Curri, M. L., Agostiano A., unpublished results.

7 Gaponenko, S. V., *Optical Properties of Semiconductor Nanocrystals*, Cambridge University Press, Cambridge, UK, **1998**.

8 Cozzoli, P. D., Manna, L., Curri, M. L., Kudera, S., Giannini, C., Striccoli, M., Agostiano, A., Shape and phase control of colloidal ZnSe nanocrystals, *Chem. Mater.* **2005**, 17, 1296–1306.

9 (a) Hodak, J. H., Henglein, A., Hartland, G. V., Photophysics and spectroscopy of metal particles, *Pure Appl. Chem.* **2000**, 72, 189–197. (b) Mulvaney, P., Surface plasmon spectroscopy of nanosized metal particles, *Langmuir* **1996**, 12, 788–800. (c) Link, S., El-Sayed, M. A., Spectral

properties and relaxation dynamics of surface plasmon electronic oscillations in gold and silver nanodots and nanorods, *J. Phys. Chem. B* **1999**, 103, 8410–8426. (d) LINK, S., EL-SAYED, M. A., Size and temperature dependence of the plasmon absorption of colloidal gold nanoparticles, *J. Phys. Chem. B* **1999**, 103, 4212–4217.

10 MURPHY, C. J., SAU, C. J., GOLE, T. K., ORENDORFF, A. M., GAO, C. J., GOU, HUNYADI, J., LI, T., Anisotropic metal nanoparticles: Synthesis, assembly, and optical applications, *J. Phys. Chem. B* **2005**, 109, 13 857–13 870.

11 COZZOLI, P. D., FANIZZA, E., CURRI, M. L., LAUB, D., AGOSTIANO, A., Low-dimensional chainlike assemblies of TiO_2 nanorod-stabilized Au nanoparticles, *Chem. Comm.* **2005**, 942–944.

12 COZZOLI, P. D., CURRI, M. L., AGOSTIANO, A., Efficient charge storage in photoexcited TiO_2 nanorod-noble metal nanoparticle composite systems, *Chem. Comm.* **2005**, 3186–3188.

13 (a) HENGLEIN, A., Small-particle research: Physicochemical properties of extremely small colloidal metal and semiconductor particles, *Chem. Rev.* **1989**, 89, 1861–1873. (b) HENGLEIN, A., Physicochemical properties of small metal particles in solution: "Microelectrode" reactions, chemisorption, composite metal particles, and the atom-to-metal transition, *J. Phys. Chem.* **1993**, 97, 5457–5471.

14 (a) ROTHLISBERGER, U., ANDREONI, W., PARRINELLO, M., Structure of nanoscale silicon clusters, *Phys. Rev. Lett.* **1994**, 72, 665–668. (b) MURRAY, C. B., NORRIS, D. B., BAWENDI, M. G., Synthesis and characterization of nearly monodisperse CdE (E = sulfur, selenium, tellurium) semiconductor nanocrystallites, *J. Am. Chem. Soc.* **1993**, 115, 8706–8715.

15 (a) IDRISS, H., PIERCE, K., BARTEAU, M. A., Carbonyl coupling on the titanium dioxide TiO2(001) surface, *J. Am. Chem. Soc.* **1991**, 113, 715–716. (b) IDRISS, H., PIERCE, K., BARTEAU, M. A., Synthesis of stilbene from benzaldehyde by reductive coupling on TiO2 (001) surfaces, *J. Am. Chem. Soc.* **1994**, 116, 3063–3074. (c) LEWIS, L. N., Chemical catalysis by colloids and clusters, *Chem. Rev.* **1993**, 93, 2693–2730. (d) SCHMID, G., Large clusters and colloids. Metals in the embryonic state, *Chem. Rev.* **1992**, 92, 1709–1727. (e) GATES, B. C., Rotational currents as a measure of excited-state dipole moments, *Chem. Rev.* **1995**, 95, 511–525. (f) MORENO-MANAS, M., PLEIXATS, R., Formation of carbon-carbon bonds under catalysis by transition-metal nanoparticles, *Acc. Chem. Res.* **2003**, 36, 638–643. (g) ROUCOUX, A., SCHLUZ, J., PATIN, H., Reduced transition metal colloids: A novel family of reusable catalysts? *Chem. Rev.* **2002**, 102, 3757–3778.

16 SHIOJIRI, S., HIRAI, T., KOMASAWA, I., *J. Chem. Eng. Jpn.* **1997**, 30, 86. (b) LADE, M., MAYS, H., SCHMIDT, J., WILLUMEIT, R., SCHOMAKER, R., On the nanoparticle synthesis in microemulsions: Detailed characterization of an applied reaction mixture, *Colloids Surf. A* **2000**, 163, 3–15.

17 LISIECKI, I., Size, shape and structural controls of metallic nanocrystals, *J. Phys. Chem. B* **2005**, 109, 12 231–12 244.

18 (a) WONG, E. M., BONEVISH, J. E., SEARSON, P. C., Ultrafast studies of photoexcited electron dynamics in γ- and α-Fe_2O_3 semiconductor nanoparticles, *J. Phys. Chem. B* **1998**, 102, 7770–7775. (b) CURRI, M. L., AGOSTIANO, A., MANNA, L., DELLA MONICA, M., CATALANO, M., CHIAVARONE, L., SPAGNOLO, V., LUGARA, M., Synthesis and characterization of CdS nanoclusters in a quaternary microemulsion: The role of the cosurfactant, *J. Phys. Chem. B* **2000**, 104, 8391–8397. (c) ADAIR, J. H., LI, T., KIDO, T., HAVEY, K., MOON, J., MECHOLSKY, J., MORRONE, A., TALHAM, D. R., LUDWIG, M. H., WANG, L., Recent developments in the preparation and properties of nanometer-size spherical and platelet-shaped particles and composite particles, *Mater. Sci. Eng. R* **1998**, 23, 139–242. (d) MACDOUGALL, J. E.,

Eckert, H., Stucky, G. D., Herron, N., Wang, Y., Moller, K., Bein, T., Cox, D., Synthesis and characterization of group III-V semiconductor clusters: Gallium phosphide GaP in zeolite Y, *J. Am. Chem. Soc.* **1989**, 111, 8006–8007.

19 (a) Manna, L., Scher, E. C., Alivisatos, A. P., Shape control of colloidal semiconductor nanocrystals, *J. Clusters Sci.* **2002**, 13, 521–532. (b) Lee, S.-M., Cho, S.-N., Cheon, J., Anisotropic shape control of colloidal inorganic nanocrystals, *Adv. Mater.* **2003**, 15, 441–444. (c) Peng, X., Green chemical approaches toward high-quality semiconductor nanocrystals, *Chem. Eur. J.* **2002**, 8, 334–339. (d) Burda, C., Chen, X., Narayanan, R., El-Sayed, M. A., Chemistry and properties of nanocrystals of different shapes, *Chem. Rev.* **2005**, 105, 1025–1102.

20 Yin, Y., Alivisatos, A. P., Colloidal nanocrystal synthesis and the organic–inorganic interface, *Nature*, **2005**, 437, 664–670.

21 (a) Milliron, D. J., Hughes, S. M., Cui, Y., Manna, L., Li, J., Wang, L.-W., Alivisatos, A. P., Colloidal nanocrystal heterostructures with linear and branched topology, *Nature* **2004**, 430, 190–195. (b) Shieh, F., Saunders, A. E., Korgel, B. A., General shape control of colloidal CdS, CdSe, CdTe quantum rods and quantum rod heterostructures, *J. Phys. Chem. B* **2005**, 109, 8538–8542. (c) Cozzoli, P. D., Kornowski, A., Weller, H., Low-temperature synthesis of soluble and processable organic-capped anatase TiO_2 nanorods, *J. Am. Chem. Soc.* **2003**, 125, 14 539–14 548.

22 (a) Mokari, T., Rothenberg, E., Popov, I., Costi, R., Banin, U., Selective growth of metal tips onto semiconductor quantum rods and tetrapods, *Science* **2004**, 304, 1787–1790. (b) Kudera, S., Carbone, L., Casula, M. F., Cingolani, R., Falqui, A., Snoeck, E., Parak, W. J., Manna, L., Selective growth of PbSe on one or both tips of colloidal semiconductor nanorods, *Nano Lett.* **2005**, 5, 445–449. (c) Pacholski, C., Kornowski, A., Weller, H., *Angew. Chem.* **2004**, 116, 4878–4881; *Angew. Chem. Int. Ed. Engl.* **2004**, 43, 4774–4777.

23 (a) Gu, H., Yang, Z., Gao, J., Chang, C. K., Xu, B., Heterodimers of nanoparticles: Formation at a liquid-liquid interface and particle-specific surface modification by functional molecules, *J. Am. Chem. Soc.* **2005**, 127, 34–35. (b) Gu, H., Zheng, R., Zhang, X., Xu, B., Facile one-pot synthesis of bifunctional heterodimers of nanoparticles: A conjugate of quantum dot and magnetic nanoparticles, *J. Am. Chem. Soc.* **2004**, 126, 5664–5665. (c) Teranishi, T., Inoue, Y., Nakaya, M., Oumi, Y., Sano, T., Nanoacorns: Anisotropically phase-segregated CoPd sulfide nanoparticles, *J. Am. Chem. Soc.* **2004**, 126, 9914–9915. (d) Yu, H., Chen, M., Rice, P. M., Wang, S. X., White, R. L., Sun, S., Dumbbell-like bifunctional Au-Fe_3O_4 nanoparticles, *Nano Lett.* **2005**, 5, 379–382. (e) Gao, X., Yu, L., MacCuspie, R. I., Matsui, H., Controlled growth of Se nanoparticles on Ag nanoparticles in different ratios, *Adv. Mater.* **2005**, 17, 426–429.

24 (a) Colvin, V. L., Shlamp, M. C., Alivisatos, A. P., Light-emitting diodes made from cadmium selenide nanocrystals and a semiconducting polymer, *Nature* **1994**, 370, 354–357. (b) Dabbousi, B. O., Bawendi, M. G., Onotsuka, O., Rubner, M. F., Electroluminescence from CdSe quantum-dot/polymer composites, *Appl. Phys. Lett.* **1995**, 66, 1316–1318.

25 Huynh, W. U., Peng, X., Alivisatos, A. P., CdSe nanocrystal rods/poly(3-hexylthiophene) composite photovoltaic devices, *Adv. Mater.* **1999**, 11, 923–927.

26 (a) Bruchez, M. J., Moronne, M., Gin, P., Weiss, S. Alivisatos, A. P., Semiconductor nanocrystals as fluorescent biological labels, *Science* **1998**, 281, 2013–2016. (b) Chan, W. C. W., Nie, S., Quantum dot bioconjugates for ultrasensitive nonisotopic detection, *Science* **1998**, 281, 2016–2018.

27 (a) Mirkin, C. A., Letsinger, R. L., Mucic, R. C., Storhoff, J. J., A DNA-based method for rationally assembling nanoparticles into macroscopic materials, *Nature* **1996**, 382, 607–609. (b) Loweth, C. J., Caldwell, W. B., Peng, X. G., Alivisatos, A. P., DNA-based assembly of gold nanocrystals, *Angew. Chem. Int. Ed.* **1999**, 38, 1808–1812.

28 (a) Zanchet, D., Micheel, C. M., Parak, W. J., Gerion, D., Williams, S. C., Alivisatos, A. P., Electrophoretic and structural studies of DNA-directed Au nanoparticle groupings, *J. Phys. Chem. B* **2002**, 106, 11758–11763. (b) Shenhar, R., Rotello, V. M., Nanoparticles: Scaffolds and building blocks, *Acc. Chem. Res.* **2003**, 36, 549–561. (c) Dubertret, B., Skourides, P., Norris, D. J., Noitreaux, V., Brivan-Lou, A. H., Libchaber, A., In vivo imaging of quantum dots encapsulated in phospholipid micelles, *Science* **2002**, 298, 1759–1762.

29 (a) Mitchell, G. P., Mirkin, C. A., Letsinger, L. R., Programmed assembly of DNA functionalized quantum dots, *J. Am. Chem. Soc.* **1999**, 121, 8122–8123. (b) Rosenthal S. J., Bar-coding biomolecules with fluorescent nanocrystals, *Nat. Biotechnol.* **2001**, 19, 621–622. (c) Han, M., Gao, X., Su, Z. J., Nie, S., Quantum-dot-tagged microbeads for multiplexed optical coding of biomolecules, *Nat. Biotechnol.* **2001**, 19, 631–635. (d) Michalet, X., Pinaud, F. F., Bentolila, L. A., Tsay, J. M., Doose, S., Li, J. J., Sundaresan, S. G., Wu, A. M., Gambhir, S. S., Weiss, S., Quantum dots for live cells, In Vivo Imag. Diagnostics, *Science* **2005**, 307, 538–544.

30 (a) Goldman, E. R., Balighian, E. D., Mattoussi, H., Kuno, K. M., Mauro, J. M., Tran, P. T., Anderson, G. P., Avidin: A natural bridge for quantum dot-antibody conjugates, *J. Am. Chem. Soc.* **2002**, 124, 6378–6382. (b) Goldman, E. R., Anderson, G. P., Tran, P. T., Mattoussi, H., Charles, P. T., Mauro, J. M., Conjugation of luminescent quantum dots with antibodies using an engineered adaptor protein to provide new reagents for fluoroimmunoassays, *Anal. Chem.* **2002**, 74, 841–847. (c) Wang, Z. Lee, J., Cossins, A. R., Brust, M., Microarray-based detection of protein binding and functionality by gold nanoparticle probes, *Anal. Chem.* **2005**, 77, 5770–5774.

31 (a) Jaiswal, J. K., Mattoussi, H., Mauro, J. M., Simon, S. M., Long-term multiple color imaging of live cells using quantum dot bioconjugates, *Nat. Biotechnol.* **2003**, 21, 47–51. (b) Fan, H., Leve, E. H., Scullin, C., Gabaldon, J., Tallant, D., Bunge, S., Boyle, T., Wilson, M. C., Brinker, C. J., Surfactant-assisted synthesis of water-soluble and biocompatible semiconductor quantum dot micelles, *Nano Lett.* **2005**, 5, 645–648. (c) Gao, X., Yang, L., Petros, J. A., Marshall, F. F., Simons, J. W., Nie, S., In vivo molecular and cellular imaging with quantum dots, *Curr. Opin. Biotechnol.* **2005**, 16, 63–72.

32 Meziani, M. J., Pathak, P., Harruff, B. A., Hurezeanu, R., Sun, Y. P., Direct conjugation of semiconductor nanoparticles with proteins, *Langmuir* **2005**, 21, 2008–2011.

33 Bailey, R. E., Nie, S., in *The Chemistry of Nanomaterials: Synthesis, Properties and Applications*, Ed. C. N. R. Rao, A. Muller, A. K. Cheetnam, Wiley-VCH Verlag, Weinheim, **2004**, pp. 405–417.

34 Pathak, S., Choi, S. K. Arnheim, N., Thompson, M. E., Hydroxylated quantum dots as luminescent probes for in situ hybridization, *J. Am. Chem. Soc.* **2001**, 123, 4103–4104.

35 (a) Sutherland, A. J., Quantum dots as luminescent probes in biological systems, *Curr. Opin. Solid State Mater Sci.* **2002**, 6, 365–370. (b) Medintz, I. G., Uyeda, H. T., Goldman, E. R., Mattoussi, H., Quantum dot bioconjugates for imaging, labelling and sensing, *Nat. Biotechnol.* **2005**, 4, 435–446.

36 (a) Mattoussi, H., Mauro, J. M., Goldman, E. R., Anderson, G. P., Sundar, V. C., Mikulec, F. V., Bawendi, M. G., Self-assembly of CdSe-ZnS quantum dot bioconjugates using an engineered recombinant protein, *J. Am. Chem. Soc.* **2000**, 122, 12 142–12 150. (b) Tran, P. T., Goldman, E. R., Anderson, G. P., Mauro, J. M., Mattoussi, H., Use of luminescent CdSe-ZnS nanocrystal bioconjugate in quantum dot based nanosensor, *Phys. Stat. Sol.* **2002**, 229, 427–432.

37 Parak, W. J., Gerion, D., Pellegrino, T., Zanchet, D., Micheel, C., Williams, S. C., Boudreau, R., Le Gros, M. A., Larabell, C. A., Alivisatos, A. P., Biological applications of colloidal nanocrystals, *Nanotechnology* **2003**, 14, R15–R27.

38 Uyeda, H. T., Medintz, I. L., Jaiswal, J. K. Simon, S. M., Mattoussi, H., Synthesis of compact multidentate ligands to prepare stable hydrophilic quantum dot fluorophores, *J. Am. Chem. Soc.* **2005**, 127, 3870–3878.

39 Willard, D. M., Carillo, L. L., Jung, J., Van Orden, A., CdSe-ZnS quantum dots as resonance energy transfer donors in a model protein-protein binding assay, *Nano Lett.* **2001**, 1, 469–474.

40 Shenton, W., Davis, S. A., Mann, S., Direct self-assembly of nanoparticles into macroscopic materials using antibody-antigen recognition, *Adv. Mater.* **1999**, 11, 449–452.

41 Rosenthal, S. J., Tomlinson, I., Adkins, E. M., Schroeter, E. M., Adams, S., Swafford, L., McBride, J., Wang, J. DeFelice, L. J., Blakely, R. D., Targeting cell surface receptors with ligand-conjugated nanocrystals, *J. Am. Chem. Soc.* **2002**, 124, 4586–4594.

42 Pinaud, F., King, D., Moore, H.-P., Weiss, S., Bioactivation and cell targeting of semiconductor CdSe/ZnS nanocrystals with phytochelatin-related peptides, *J. Am. Chem. Soc.* **2004**, 126, 6115–6123.

43 Gestwicki, J. E., Strong, L. E., Kiessling, L. L., Visualization of single multivalent receptor-ligand complexes by transmission electron microscopy, *Angew. Chem. Int. Ed.* **2000**, 39, 4567–4570.

44 (a) Park, S. J., Lazarides, A. A., Mirkin, C. A., Brazis, P. W., Kannewurf, C. R., Letsinger, R. L., The electrical properties of gold nanoparticle assemblies linked by DNA, *Angew. Chem.* **2000**, 112, 4003–4006. (b) Bardea, A., Dagan, A., Ben-Dov, I., Amit, B., Willner, I., Amplified microgravimetric quartz-crystal-microbalance analyses of oligonucleotide complexes: A route to a Tay–Sachs biosensor device, *Chem. Commun.* **1998**, 839–840. (c) Patolsky, F., Ranjit, K. T., Lichtenstein, A., Willner, I., Dendritic amplification of DNA analysis by oligonucleotide-functionalized Au-nanoparticles, *Chem. Commun.*, **2000**, 1025–1026. (d) Huang, C. C., Huang, Y. F., Cao, Z., Tan, W., Chang, H. T., Aptamer-modified gold nanoparticles for colorimetric determination of platelet-derived growth factors and their receptors, *Anal. Chem.* **2005**, 77, 5735–5741. (e) Brewer, S. H., Glomm, W. R., Johnson, M. C., Knag, M. K., Franzen, S., Probing BSA binding to citrate-coated gold nanoparticles and surfaces, *Langmuir* **2005**, 21, 9303–9307.

45 (a) Elechiguerra, J. L., Burt, J. L., Morones, J. R., Camacho-Bragado, A., Gao, X., Lara, H. H., Yacaman, M. J., Interaction of silver nanoparticles with HIV-1, *J. Nanobiotechnol.* **2005**, 3, 6. (b) MacDonald, I. D. G., Smith, W. E., Orientation of cytochrome *c* adsorbed on a citrate-reduced silver colloid surface, *Langmuir* **1996**, 12, 706–713.

46 (a) Mahtab, R., Rogers, J. P., Murphy, C. J., Protein-sized quantum dot luminescence can distinguish between "straight", "bent", and "kinked" oligonucleotides, *J. Am. Chem. Soc.* **1995**, 117, 9099–9100. (b) Mahtab, R., Rogers, J. P., Singleton, C. P., Murphy, C. J., Preferential adsorption of a "kinked"

DNA to a neutral curved surface: Comparisons to and implications for nonspecific DNA-protein interactions, *J. Am. Chem. Soc.* **1996**, 118, 7028–7032. (c) MAHTAB, R., HARDEN, H. H., MURPHY, C. J., Temperature- and salt-dependent binding of long DNA to protein-sized quantum dots: Thermodynamics of "inorganic protein"-DNA interactions, *J. Am. Chem. Soc.* **2000**, 122, 14–17.

47 WANG, Y. A., LI, J. J., CHEN, H., PENG, X., Stabilization of inorganic nanocrystals by organic dendrons, *J. Am. Chem. Soc.* **2002**, 124, 2293–2298.

48 (a) GERION, D., PINAUD, F., WILLIAMS, S. C., PARAK, W. J., ZANCHET, D., WEISS, S., ALIVISATOS, A. P., Synthesis and properties of biocompatible water-soluble silica-coated CdSe/ZnS semiconductor quantum dots, *J. Phys. Chem. B* **2001**, 105, 8861–8871. (b) CORREA-DUARTE, M. A., GIERSIG, M., LIZ-MARZAN, L. M., Stabilization of CdS semiconductor nanoparticles against photodegradation by a silica coating procedure, *Chem. Phys. Lett.* **1998**, 286, 497–501. (c) YANG, Y., GAO, M., Preparation of fluorescent SiO_2 particles with single CdTe nanocrystal cores by the reverse microemulsion method, *Adv. Mater.* **2005**, 17, 2354–2357.

49 PARAK, W. J., PELLEGRINO, T., PLANK, C., Labelling of cells with quantum dots, *Nanotechnology* **2005**, 16, R9–R25.

50 (a) PARAK, W. J., GERION, D., ZANCHET, D., WOERZ, A. S., PELLEGRINO, T., MICHEEL, C., WILLIAMS, S. C., SEITZ, M., BRUEHL, R. E., BRYANT, Z., BUSTAMANTE, C., BERTOZZI, C. R., ALIVISATOS, A. P., Conjugation of DNA to silanized colloidal semiconductor nanocrystalline quantum dots, *Chem. Mater.* **2002**, 14, 2113–2119. (b) KIRCHNER, C., LIEDL, T., KUDERA, S., PELLEGRINO, T., MUNOZ JAVIER, A., GAUB, H. E., STOLZLE, S., FERTIG, N., PARAK, W. J., Cytotoxicity of colloidal CdSe and CdSe/ZnS nanoparticles, *Nano Lett.* **2005**, 5, 331–338.

51 DEPALO, N., MALLARDI, A., FANIZZA, E., COMPARELLI, R., STRICCOLI, M., CURRI, M. L., AGOSTIANO, A., unpublished results.

52 (a) WU, X., LIU, H., LIU, J., HALEY, K. N., TREADWAY, J. A., LARSON, J. P., GE, N., PEALE, F., BRUCHEZ, M. P., Immunofluorescent labelling of cancer marker Her2 and other cellular targets with semiconductor quantum dots, *Nat. Biotechnol.* **2003**, 21, 41–46. (b) GAO, X., CUI, Y., LEVENSON, R. M., CHUNG, L. W. K., NIE, S., In vivo cancer targeting and imaging with semiconductor quantum dots, *Nat. Biotechnol.* **2004**, 22, 969–976. (c) PELLEGRINO, T., MANNA, L., KUDERA, S., LIEDL, T., KOKTYSH, D., ROGACH, A. L., KELLER, S., RADLER, J. NATILE, G., PARAK, W. J., Hydrophobic nanocrystals coated with an amphiphilic polymer shell: A general route to water soluble nanocrystals, *Nano Lett.*, **2004**, 4, 703–707.

53 (a) WANG, Y., WONG, J. F., TENG, X., LIN, X. Z., YANG, H., "Pulling" nanoparticles into water: Phase transfer of oleic acid stabilized monodisperse nanoparticles into aqueous solutions of α-cyclodextrin, *Nano Lett.* **2003**, 3, 1555–1559. (b) DEPALO, N., COMPARELLI, R., CURRI, M. L., STRICCOLI, M., AGOSTIANO, A., Cyclodextrin mediated phase transfer in water of organic capped CdS nanocrystals, *Synth. Met.* **2005**, 148, 43–46. (c) DEPALO, N., COMPARELLI, R., STRICCOLI, M., CURRI, M. L., CATUCCI, L., AGOSTIANO, A., Water phase transfer of oleic capped semiconductor nanocrystals mediated by α-cyclodextrins, *Proc. SPIE*, **2005**, vol. 5838, 245–251, Nanotechnology II, ed. PAOLO LUGLI. (d) HOU, Y., KONDOH, H., SHIMOJO, M., SAKO, E. O., OZAKI, N., KOGURE, T., OHTA, T., Inorganic nanocrystal self-assembly via the inclusion interaction of α-cyclodextrins: Toward 3d spherical magnetite, *J. Phys. Chem. B* **2005**, 109, 4845–4852.

54 DEL VALLE, E. M. M., Cyclodextrins and their uses: A review, *Process*

Biochem. 2004, 39, 1033–1046 and references therein.

55 (a) Bonacchi, D., Caneschi, A., Gatteschi, D., Sangregorio, C., Sessoli, R., Falqui, A., Synthesis and characterisation of metal oxides nanoparticles entrapped in cyclodextrin, *J. Phys. Chem. Solids* **2004**, 65, 719–722. (b) Mahalingam, V., Onclin, S., Peter, M., Ravoo, B. J., Huskens, J., Reinhoudt, D. N., Directed self-assembly of functionalized silica nanoparticles on molecular printboards through multivalent supramolecular interactions, *Langmuir* **2004**, 20, 11 756–11 762. (c) Crespo-Biel, O., Dordi, B., Reinhoudt, D. N., Huskens, J., Supramolecular layer-by-layer assembly: Alternating adsorptions of guest- and host-functionalized molecules and particles using multivalent supramolecular interactions, *J. Am. Chem. Soc.* **2005**, 127, 7594–7600. (d) Feng, J., Miedaner, A., Ahrenkiel, P., Himmel, M. E., Curtis, C., Ginley, D., Self-assembly of photoactive TiO_2-cyclodextrin wires, *J. Am. Chem. Soc.* **2005**, 127, 14 968–14 969. (e) Feng, J., Ding, S. Y., Tucker, M. P., Himmela, M. E., Kim, Y. H., Zhang, S. B., Keyes, B. M., Rumbles, G., Cyclodextrin driven hydrophobic/hydrophilic transformation of semiconductor nanoparticles, *Appl. Phys. Lett.* **2005**, 86, 33 108–33 111.

56 Shi, J., Zhu, Y., Zhang, X., Bayens, W. P. G., Garcia-Compan, A. M., Recent developments in nanomaterial optical sensors, *Trends Anal. Chem.* **2004**, 23, 351–360.

57 Fritzsche, W., Taton, T. A., Metal nanoparticles as labels for heterogeneous, chip-based DNA detection, *Nanotechnology* **2003**, 14, R63–R73.

58 (a) Yguerabide, J., Yguerabide, E. E., Light-scattering submicroscopic particles as highly fluorescent analogs and their use as tracer labels in clinical and biological applications. I. Theory., *Anal. Biochem.* **1998**, 262, 137–156. (b) Kreibig, U., Vollmer, M., *Optical Properties of Metal Clusters*, Springer, Berlin, **1995**.

59 (a) Mie, G. Beitraege zur optik trueber medien, speziell-kolloidaler metalosungen, *Ann. Phys.* **1908**, 25, 377–445. (b) Kelly, K. L., Coronado, E., Zhao, L. L., Schatz, G. C., The optical properties of metal nanoparticles: the influence of size, shape, and dielectric environment, *J. Phys. Chem. B* **2003**, 107, 668–677.

60 Kanaras, A. G, Wang, Z., Bates, A. D., Cosstick, R., Brust, M., Towards multistep nanostructure synthesis: Programmed enzymatic self-assembly of DNA/gold systems, *Angew. Chem.* **2003**, 115, 201–204.

61 Jares-Eryman, E. A., John, T. M., FRET imaging, *Nat. Biotechnol.* **2003**, 21, 1387–1395.

62 Selvin, P. R., The renaissance of fluorescence resonance energy transfer, *Nat. Struct. Biol.* **2000**, 7, 730–734.

63 Zhang, Y., Yeh, H. C., Kuroki, M. T., Wang, T. H., Single-quantum-dot-based DNA nanosensor, *Nat. Mater.* **2005**, 4, 826–831.

64 Bewersdorf, J., Pick, R., Hell, S. W., Multifocal multiphoton microscopy, *Opt. Lett.* **1998**, 23, 655–657.

65 Dahan, M., Laurence, T., Pinaud, F., Chemla., D. S., Alivisatos, A. P., Sauer, M., Weiss, S., Time-gated biological imaging by use of colloidal quantum dots, *Opt. Lett.* **2001**, 26, 825–827.

66 Weissleder, R., A clearer vision for in vivo imaging, *Nat. Biotechnol.*, **2001**, 19, 316–317.

67 Larson, D. R., Zipfel, W. R., Williams, R. M., Clark, S. W., Bruchez, M. P., Wise, F. W., Webb, W. W., Water soluble quantum dots for multiphoton fluorescence imaging in vivo, *Science* **2003**, 300, 1434–1436.

68 Aslan, K., Gryczynski, I., Malicka, J., Matveeva, E., Lakowicz, J. R., Geddes, C. D., Metal-enhanced fluorescence: An emerging tool in biotechnology, *Curr. Opin. Biotechnol.* **2005**, 16, 55–62.

69 Geddes, C. D., Lakowicz, J. R., Metal enhanced fluorescence, *J. Fluoresc.* **2002**, 12, 121–129.

70 Weitz, D. A., Garoff, S., Gersten, J. I., Nitzan, A., The enhancement of Raman scattering, resonance Raman scattering and fluorescence from molecules absorbed on a rough silver surface, *J. Chem. Phys.* **1983**, 78, 5324–5338.

71 Kitson, S. C., Barnes, W. L., Sambles, J. R., Photoluminescence from dye molecules on silver gratings, *Opt. Commun.*, **1996**, 122, 147–154.

72 (a) Lakowicz, J. R., Radiative decay engineering: Biophysical and biomedical applications, *Anal. Biochem.*, **2001**, 298, 1–24. (b) Lakowicz, J. R., Shen, Y., D'Auria, S., Malicka, J., Fang, J., Gryczynski, Z., Gryczynski, I., Radiative decay engineering 2. Effects of silver island films on fluorescence intensity, lifetimes, and resonance energy transfer, *Anal. Biochem.* **2002**, 301, 261–277.

73 Raether, H., *Surface Plasmons*, Springler-Verlag, Berlin, **1988**.

74 (a) Homola, J., Yee, S. S., Gauglitz, G., Surface plasmon resonance sensors: Review, *Sens. Actuators B-Chem.*, **1999**, 54, 3–15. (b) Cullen, D. C., Lowe, R., A direct surface plasmon-polariton immunosensor: Preliminary investigation of the non-specific adsorption of serum components to the sensor interface, *Sens. Actuators B-Chem.* **1990**, 1, 576–579.

75 Jung, L. S., Jung, L. S., Campbell, C. T., Chinowsky, T. M., Mar, M. N., Yee, S. S., Quantitative interpretation of the response of surface plasmon resonance sensors to adsorbed films, *Langmuir* **1998**, 14, 5636–5648.

76 Matsui, J., Akamatsu, K., Hara, N., Miyoshi, D., Nawafune, H., Tamaki, K., Sugimoto, N., SPR sensor chip for detection of small molecules using molecularly imprinted polymer with embedded gold nanoparticles, *Anal. Chem.* **2005**, 77, 4282–4285.

77 (a) Kneipp, K., Kneipp, H., Itzkan, I., Dasar, R. R., Feld, M. S., Ultra-sensitive chemical analysis by Raman spectroscopy, *Chem. Rev.* **1999**, 99, 2957–2975. (b) Lecomte, S., Wackerbart, H., Soulimane, T., Buse, G., Hildebrandt, P., Time-resolved surface-enhanced resonance Raman spectroscopy for studying electron-transfer dynamics of heme proteins, *J. Am. Chem. Soc.* **1998**, 120, 7381–7382.

78 Kambhampati, P., Child, C. M., Foster, M. C., Campion, A., On the chemical mechanism of surface enhanced Raman scattering: Experiment and theory, *J. Chem. Phys.*, **1998**, 108, 5013–5026.

79 Weaver, M. J., Zou, S., Chan, H. Y. H., The new interfacial ubiquity of surface-enhanced Raman spectroscopy, *Anal. Chem.* **2000**, 72, 38A–47A.

80 Jensen, T. R., Malinsky, M. D., Haynes, C. L., Van Duyne, R. P., Nanosphere lithography: Tunable localized surface plasmon resonance spectra of silver nanoparticles, *J. Phys. Chem. B* **2000**, 104, 10 549–10 556.

81 Moskovits, M., Surface-enhanced spectroscopy, *Rev. Mod. Phys.* **1985**, 57, 783–826.

82 Emory, S. R., Haskins, W. E., Nie, S., Direct observation of size-dependent optical enhancement in single metal nanoparticles, *J. Am. Chem. Soc.* **1998**, 120, 8009–8010.

83 Campion, A., Kambhampati, P., Surface-enhanced Raman scattering, *Chem. Soc. Rev.* **1998**, 27, 241–250.

84 Zou, S., Williams, C. T., Chen, E. K.-Y., Weaver, M. J., Surface-enhanced Raman scattering as a ubiquitous vibrational probe of transition-metal interfaces: Benzene and related chemisorbates on palladium and rhodium in aqueous solution, *J. Phys. Chem. B* **1998**, 102, 9039–9049.

85 Tian, Z. Q., Ren, B., Wu, D. Y. Surface-enhanced Raman scattering: From noble to transition metals and from rough surfaces to ordered nanostructures, *J. Phys. Chem. B* **2002**, 106, 9463–9483.

86 (a) McFarland, A. D., Van Duyne, R. P., Single silver nanoparticles as real-time optical sensors with zeptomole sensitivity, *Nano Lett.* **2003**,

3, 1057–1062. (b) Priikulis, J., Svedberg, F., Käll, M., Enger, J., Ramser, K., Goksör, M., Hanstorp, D., Optical spectroscopy of single trapped metal nanoparticles in solution, *Nano Lett.* **2004**, 4, 115–118.

87 Alivisatos, A. P., The use of nanocrystals in biological detection, *Nat. Biotechnol.* **2004**, 22, 47–52.

88 West, J. W., Halas, N. J., Engineered nanomaterials for biophotonics applications: Improving sensing, imaging, and therapeutics, *Annu. Rev. Biomed. Eng.* **2003**, 5, 285–292.

89 Hines, M. A., Scholes, G. D., Colloidal PbS nanocrystals with size-tunable near-infrared emission: Observation of post-synthesis self-narrowing of the particle size distribution, *Adv. Mater.* **2003**, 15, 1844–1849.

90 Kim, S., Lim, Y. T., Soltesz, E. G., De Grand, A. M., Lee, J., Nakayama, A., Parker, J. A., Mihaljevic, T., Laurence, R. G., Dor, D. M., Cohn, L. H., Bawendi, M. G., Frangioni, J. V., Near-infrared fluorescent type II quantum dots for sentinel lymph node mapping, *Nat. Biotechnol.* **2004**, 22, 93–97.

91 Niemeyer, C. M., Nanoparticles, proteins, and nucleic acids: Biotechnology meets materials science, *Angew. Chem. Int. Ed.* **2001**, 40, 4128–4158.

92 Chen, Q., Ma, Q., Wan, Y., Su, X., Lin, Z., Jin, Q., *Luminescence* **2005**, 20, 251–255.

93 Clapp, A. R., Medintz, I. L., Mauro, J. M., Fisher, B. R., Bawendi, M. G., Mattoussi, H., Fluorescence resonance energy transfer between quantum dot donors and dye labeled protein acceptors, *J. Am. Chem. Soc.* **2004**, 126, 301–310.

94 Medintz, I. L., Clapp, R. A., Mattoussi, H., Goldmann, E. R., Fisher, B., Mauro, J. M., Self-assembled nanoscale biosensors based on quantum dot FRET donors, *Nat. Mater.* **2003**, 2, 630–638.

95 Reichert, J., Csaki, A., Kohler, J. M., Fritzsche, W., Chip-based optical detection of DNA hybridization by means of nanobead labeling, *Anal. Chem.* **2000**, 72, 6025–6029.

96 Englebienne, P., Use of colloidal gold surface plasmon resonance peak shift to infer affinity constants from the interactions between protein antigens and antibodies specific for single or mltiple epitopes, *Analyst* **1998**, 123, 1599–1603.

97 (a) Elghanian, R., Storhoff, J. J., Mucic, R. C., Letsinger, R. L., Mirkin, C. A., Selective colorimetric detection of polynucleotides based on the distance-dependent optical properties of gold nanoparticles, *Science* **1997**, 277, 1078–1081. (b) Storhoff, J. J., Elghanian, R., Mucic, R. C., Mirkin, C. A., Letsinger, R. L., One-pot colorimetric differentiation of polynucleotides with single base imperfections using gold nanoparticle probes, *J. Am. Chem. Soc.* **1998**, 120, 1959–1964. (c) Reynolds, R. A., Mirkin, C. A., Letsinger, R. L., Homogeneous, nanoparticle-based quantitative colorimetric detection of oligonucleotides, *J. Am. Chem. Soc.*, **2000**, 122, 3795–3796. (d) Storhoff, J. J., Lazarides, A. A., Mucic, R. C., Mirkin, C. A., Letsinger, R. L., Schatz, G. C., What controls the optical properties of DNA-linked gold nanoparticle assemblies? *J. Am. Chem. Soc.* **2000**, 122, 4640–4650.

98 Nath, N., Chilkoti, A., A colorimetric colloidal gold sensor to interrogate biomolecular interactions in real-time on a surface, *Anal. Chem.* **2002**, 74, 504–509.

99 Hutter, E., Fendler, J. H., Roy, D., Surface plasmon resonance studies of gold and silver nanoparticles linked to gold and silver substrates by 2-amino-ethanethiol and 1,6-hexanedithiol, *J. Phys. Chem. B* **2001**, 105, 11 159–11 168.

100 Haes, A. J., Van Duyne, R. P., A nanoscale optical biosensor: Sensitivity and selectivity of an approach based on the localized surface plasmon resonance spectroscopy of triangular silver nanoparticles, *J. Am. Chem. Soc.* **2002**, 124, 10 596–10 604.

101 Nath, N., Chilkoti, A., Label-free biosensing by surface plasmon resonance of nanoparticles on glass: Optimization of nanoparticle size, *Anal. Chem.* **2004**, 76, 5370–5378.

102 Mu, Y., Zhang, H., Zhao, X., Song1, D., Wang1, Z., Sun, J., Li, M., Q. Jin, Q., An optical biosensor for monitoring antigen recognition based on surface plasmon resonance using avidin-biotin system, *Sensors* **2001**, 1, 91–101.

103 Sarkar, D., Somasundaran, P., Overcoming contamination in surface plasmon resonance spectroscopy, *Langmuir* **2002**, 18, 8271–8277.

104 Malicka, J., Gryczynski, I., Lakowicz, J. R., DNA hybridization assays using metal-enhanced fluorescence, *Biochem. Biophys. Res. Commun.* **2003**, 306, 213–218.

105 Maliwal, B. P., Malicka, J., Gryczynski, I., Gryczynski, Z., Lakowicz, J. R., Fluorescence properties of labeled proteins near silver colloid surfaces, *Biopolymers* **2003**, 70, 585–594.

106 Nie, S., Emory, R., Probing single molecules and single nanoparticles by surface-enhanced Raman scattering, *Science* **1997**, 21, 1102–1106.

107 Stuart, D. A., Haes, A. J., Yonzon, C. R., Hicks, E. M., Van Duyne, R. P., Biological applications of localised surface plasmonic phenomenae, *IEE Proc.-Nanobiotechnol.* **2005**, 152, 13–32.

108 Keating, C. D., Kovaleski, K. K., Natan, M. J., Heightened electromagnetic fields between metal nanoparticles: Surface enhanced Raman scattering from metal-cytochrome *c*-metal sandwiches, *J. Phys. Chem. B* **1998**, 102, 9414–9425.

109 Whang, Z., Lee, J., Cossins, A. R., Brust, M., Microarray-based detection of protein binding and functionality by gold nanoparticles probes, *Anal. Chem.* **2005**, 77, 5770–5774.

110 Talley, C. E., Jusinski, L., Hollars, C. W., Lane, S. M., Huser, T., Intracellular pH sensors based on surface-enhanced Raman scattering, *Anal. Chem.* **2004**, 76, 7064–7068.

111 Mattoussi, H., Kuno, M. K., Goldman, E. R., Anderson, G. P., Mauro, J. M., in *Optical Biosensors: Present and Future*, eds. Ligler, F. S., Rowe C. A., Elsevier, Amsterdam, **2002**, pp. 537–569.

112 Aitken, R. J., Creely, K. S., Tran, C. L., *Nanoparticles: An Occupational Hygiene Review*, Institute of Occupational Medicine for the Health and Safety Executive Research Report 274, **2004**.

6
Quantum Dot-based Nanobiohybrids for Fluorescent Detection of Molecular and Cellular Biological Targets

Rumiana Bakalova, Zhivko Zhelev, Hideki Ohba, and Yoshinobu Baba

6.1
Introduction

Over the last several years, semiconductor quantum dots (QDs) have won recognition as a new generation of fluorophores in bioimaging and biosensing, because of their unique spectral properties compared with traditional organic dyes: high quantum yield (>50% versus 15–50% for classical organic dyes); high molar extinction coefficients (~10–100× that of traditional organic dyes); broad absorption spectra with narrow, symmetric fluorescence spectra (full-width at half-maximum ~25–40 nm) spanning the ultraviolet to near-infrared; large effective excitation/emission Stokes shifts; high resistance to photobleaching and exceptional resistance to photo- and chemical degradation. All these characteristics make QDs brighter fluorescent probes (10–20× brighter than classical organic dyes) under photon-limited *in vivo* and *in situ* conditions, where the light intensities are severely attenuated by scattering and absorption. These novel optical properties can be used to optimize the signal-to-noise (signal-to-background) ratio and to improve the sensitivity of the fluorescence detection devices, as well as to increase the quality of fluorescent cellular and molecular labeling, and deep-tissue *in situ* and *in vivo* fluorescent imaging. Moreover, the size-tunable fluorescent emission (as a function of core size for binary semiconductor materials) and the broad excitation spectra (which allow excitation of mixed QD populations at a single wavelength) give possibilities for application of QDs in multiplexed fluorescent analyses.

The present chapter overviews the current status and future trends of QD-based nanobiohybrids for ultrasensitive fluorescent detection of molecular and cellular biological targets. Section 6.2 outlines the basic principles of design and synthesis of highly fluorescent QDs, appropriate for life science research, with their advantages and drawbacks over classical fluorophores. Section 6.3 summarizes currently data on applications of QD-based nanobioprobes in cellular and deep-tissue imaging *in situ* and *in vivo*, using the following fluorescent detection methods: fluorescent confocal microscopy, two-photon microscopy, fluorescence correlation spectroscopy, single-molecule microscopy. It describes the potential of multifunctional

Nanotechnologies for the Life Sciences Vol. 8
Nanomaterials for Biosensors. Edited by Challa S. S. R. Kumar

ISBN: 978-3-527-31388-4

QDs for positron emission tomography and functional magnetic resonance imaging. Sections 6.4 and 6.5 focus on the application of QDs in several basic fluorescent biotechnologies *in vitro* (e.g., FRET analyses, time-resolved photoluminescence spectroscopy, immunoblotting). These sections describe, for the first time, the great capacity of QDs for improvement and quantification of immunoblot analysis of protein expression, using a fluorescent detection, as well as for development of new generation optical recognition-based biosensors. A new type of QDs possessing a long fluorescence life-time (approximately 90 ns) is also reported. Finally, Section 6.6 briefly describes future trends for QD-based composite materials as novel fluorescent standards for thin calibration of fluorescent instrumentation.

6.2
Quantum Dots – Basic Principles of Design and Synthesis, Optical Properties, and Advantages over Classical Fluorophores

6.2.1
Basic Principles of Design and Synthesis of Quantum Dots

QDs are made from inorganic colloidal semiconductors. They are single crystals a few nanometers in diameter whose size and shape can be precisely controlled by the duration, temperature, and ligand molecules used in the synthesis [1]. This process yields QDs that have unique composition- and size-dependent absorption and emission (Fig. 6.1).

Absorption of a photon with energy above the semiconductor band gap energy results in the creation of an exciton. The absorption has an increased probability at higher energies (i.e., shorter wavelengths) and results in a broadband absorption spectrum, in marked contrast to standard fluorophores. For nanocrystals smaller than the so-called Bohr exciton radius (a few nanometers), energy levels are quantized, with values directly related to the QD size – an effect called quantum confinement that gives the name "quantum dots". The relative recombination of an exciton (characterized by a long life-time, >10 ns) [2] leads to the emission of a photon in a narrow, symmetric energy band.

Surface defects in the QD crystal structure act as temporary "traps" for the electron or hole, preventing their radiative recombination. The alternation of trapping and untrapping events results in intermittent fluorescence (blinking) visible at the single-molecule level [3–5]. This reduces the overall quantum yield (the ratio of emitted to absorbed photons). The way to overcome these problems and to protect surface atoms from oxidation and other chemical reactions is to grow a shell of a few atomic layers of a material with a large band gap on top of the nanocrystal core. The shell can be designed carefully to obtain quantum yields close to 90%. This step also enhances QD's photostability by several orders of magnitude in comparison with conventional dyes.

For application of QDs in life science experiments, it is necessary to dissolve the core–shell nanocrystals in aqueous solutions at physiological conditions (pH $\sim$ 7–7.4 and physiologically normal ion strength), avoiding aggregation completely.

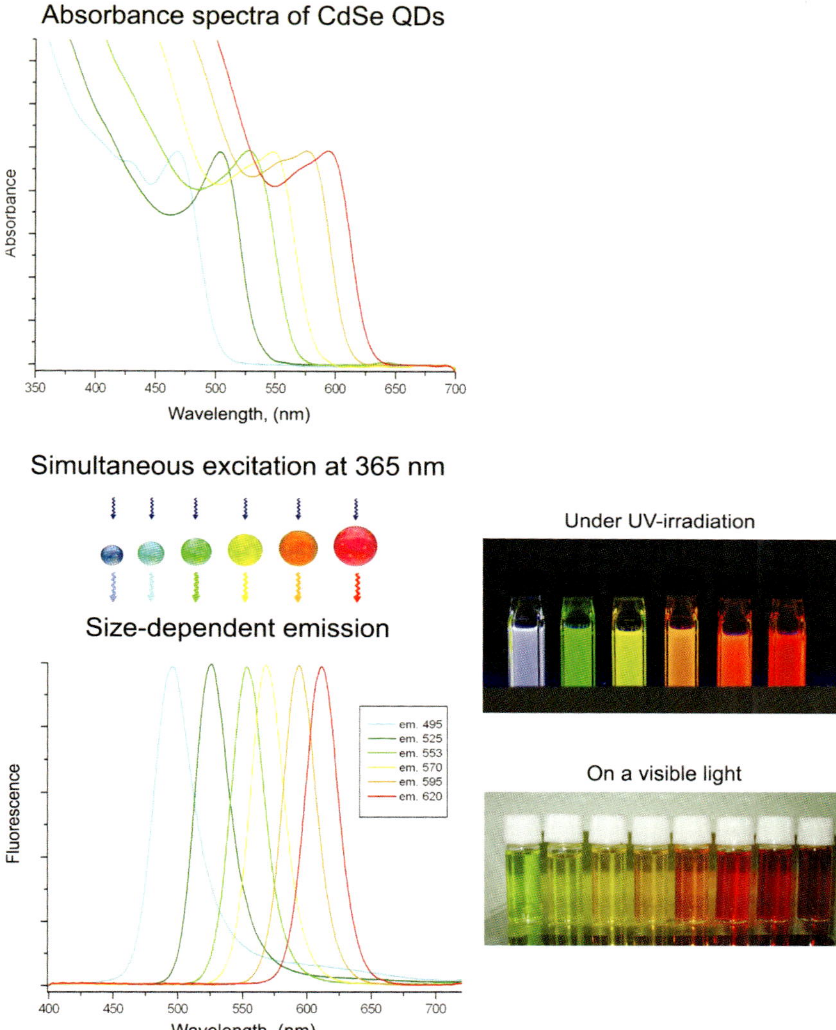

Fig. 6.1. Size-dependent absorption and emission of CdSe quantum dots (from ~2 to ~6 nm in diameter, starting from the blue color).

Presently, the best available QDs for biological applications are made of CdSe cores overcoated with a layer of ZnS. The size of CdSe/ZnS QDs varies from 3 to 10 nm. Their quantum yield is very high in organic solvents (~85%); however, the direct phase-transfer to aqueous solutions using appropriate surface-modifying agents results in a significant decrease or complete loss of their fluorescent properties (as a result of direct contact between the metal surface and the polar environment). Thus, additional organic coating of CdSe/ZnS QDs is necessary to preserve (at least partially) their high quantum yield in aqueous solutions. The organic (usually polymer) coating is required not only to preserve the quantum yield but, mostly, for isolation of the metal surface, which is sensitive to oxidation. For example, QDs that are covered only with water-soluble surface ligand can be easily oxidized by the oxygen dissolved in aqueous solution. As a result, QDs aggregate, form about 100–500 nm clusters, and lose completely their unique properties. The polymer coat forms a box around single QDs and thus avoids the aggregation and preserves the unique properties of the semiconductor core–shell nanocrystal.

Several excellent reviews summarize the advantages and drawbacks of the synthetic strategies for water-soluble QDs [6–11]. QDs have been synthesized using both two-element systems (binary dots) and three-element systems (ternary alloy dots). The most successful strategy for additional coating of CdSe/ZnS QDs before their subsequent water-solubilization is an organic coating using "diblock" or "triblock" copolymers [8, 12], amine box dendrimers [13], dihydrolipoic acid derivatives [14, 15], modified acrylic acid polymer [16–18]. However, current organic coats have one important shortcoming – the obtained QD particles have a non-defined heterogeneous size after the organic coating. Moreover, the total size of the organic-coated CdSe/ZnS QDs usually varies in the range 15–50 nm – a size that is commensurable with or larger than that of biomolecules (proteins, oligonucleotides, etc.) that the particles have to be conjugated with. Despite their relatively large size, recent life science experiments have shown that bioconjugated QD probes behave like fluorescent proteins and do not suffer from serious binding kinetic or steric hindrance problems and can be used in fluorescent imaging of molecular and cellular targets [8, 9, 12, 19–32]. In this "mesoscopic" size range, QDs also have a greater surface area and a lot of functionalities to develop multifunctional nanoparticles that can be used for linking to multiple diagnostic and therapeutic agents. However, the comparatively large sizes of water-soluble QDs make them inappropriate for several approaches, e.g., FRET-based biosensing technologies (the Forster-radius for effective FRET is calculated as 5–10 nm). Since the organic coating is necessary to avoid aggregation of the metal nanoparticles and to preserve their high quantum yield in aqueous solutions, future expectations have been directed to the development of methods for overcoating of QDs with a maximal size of organic coat of $\sim$8 nm.

6.2.2 Optical and Chemical Properties – Advantages Compared with Classical Fluorophores

Semiconductor QDs have attracted much interest for bioimaging and biosensing, because of their unique spectral properties over traditional organic fluorophores. Briefly, the following characteristics distinguish QDs from the commonly used fluorophores: high quantum yield (more 50% versus 15–50% for standard organic dyes); high molar extinction coefficients (in the order of $0.5–5 \times 10^6$ M^{-1} cm^{-1}, which is $\sim$10–100× larger than those of traditional organic dyes – $5–10 \times 10^4$

M^{-1} cm^{-1}) [7, 33]; broad absorption spectra with narrow, symmetric photoluminescence spectra (full-width at half-maximum ~25–40 nm) spanning the UV to near-infrared; low life-time-limited emission rates for single QDs (~5–10× lower than those of single organic dyes), because of their longer excited state life-times (20–50 ns); large effective Stokes shifts; high resistance to photobleaching (several thousand times more stable than organic dyes) and exceptional resistance to photo- and chemical degradation [34–36], which makes them well-suited for continuous tracking studies over a long period of time. All these characteristics make QDs brighter fluorescent probes (10–20× brighter than organic dyes) [37, 38] under photon-limited *in vivo* and *in situ* conditions, where light intensities are severely attenuated by scattering and absorption. These novel optical properties can be used to optimize the signal-to-background (signal-to-noise) ratio. For example, the comparatively long fluorescence life-time of QDs enables the use of time-gated detection to separate their signal from that of shorter lived species, such as background autofluorescence encountered in viable cells and animals. Moreover, the longer excited state life-times of QDs provide a means to separate the QD fluorescence from background fluorescence using a time-domain imaging technique [39, 40]. Figure 6.2 shows a comparison of the excited state decay curves of QDs and organic dyes.

Assuming that the initial fluorescence intensities of QDs and dyes after a pulse excitation are the same and that the fluorescence life-time of QDs is one order of magnitude longer, one can estimate that the QD and dye intensity ratio (I_{QD}/I_{dye}) will increase rapidly from 1 at time $t = 0$ to ~100 in only 10 ns ($t = 10$ ns). Thus,

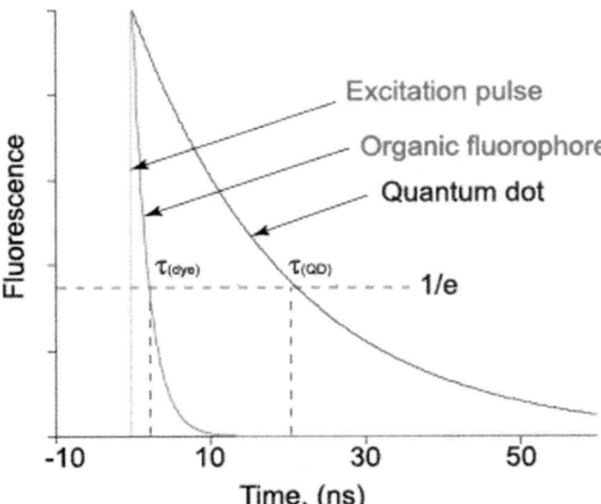

Fig. 6.2. Comparison of the excited state decay curves (monoexponential model) of QDs and classical organic dyes. The longer excited state life-time of QDs allows the use of time-domain imaging to discriminate against the background fluorescence (short life-times); $\tau_{(dye)}$ and $\tau_{(QD)}$ are the delay times for the fluorescence signals to decrease to 1/e of their original values, where e is the natural log constant (= 2.718). (Kindly provided from X. Gao [8]).

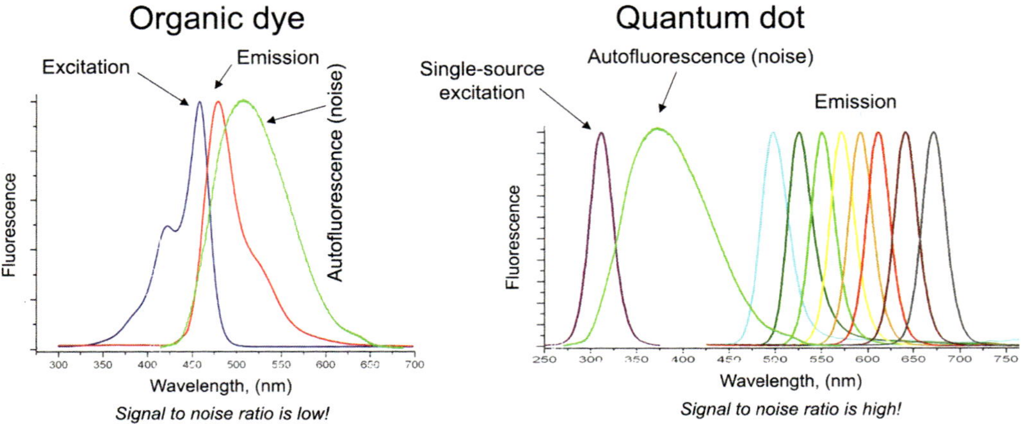

Fig. 6.3. Comparison of excitation/emission Stokes shifts between QDs and classical organic dyes – an important issue for optimization of the signal-to-noise (signal-to-background) ratio.

the image contrast (measured by signal-to-noise or signal-to-background ratio) can be markedly improved by time-delayed data acquisition.

The large Stokes shifts of QDs (measured by the distance between the excitation and emission peaks) can be used to further improve detection sensitivity (Fig. 6.3). As shown in Fig. 6.3, the Stokes shifts of QDs can be as large as 300–400 nm, depending on the wavelength of the excitation light. Organic dye signals with a small Stokes shift are often mimicked by strong tissue and cell autofluorescence. In contrast, QD signals with a large Stokes shift are clearly recognizable above the background. This "color contrast" is only available to QD probes, as the signals and background can be separated by wavelength-resolved or spectral imaging [12].

The most attractive features for life science research are the size- and composition-tunable emission of QDs from visible to infrared wavelengths, which give a possibility for their use to image and track multiple molecular targets simultaneously, e.g., combinatorial optical encoding, in which multiple colors and intensities are combined to encode thousands of genes, proteins or small-molecule compounds (a), and simultaneous sensing and imaging of several molecular and more complex targets (b) [17, 41–48].

Single QDs can be observed and tracked over an extended period of time (up to a few hours) with fluorescent confocal microscopy [49], total internal reflection microscopy [50, 51], or basic wide-field epifluorescence microscopy [19, 52]. Single-molecule microscopy is possibly one of the most exciting new capabilities offered to biologists. A related technique, fluorescence correlation spectroscopy, has allowed determination of the brightness per particle and also provides measurement of the average QD size [53]. QDs are also excellent probes for two-photon confocal microscopy [51, 54, 55] because they are characterized by a large absorption cross section. They can be used simultaneously with standard dyes. In particular, QDs have a largely untapped potential as novel tracers in PET and MRI imaging [9], as customizable donors of a fluorescence resonance energy transfer (FRET) pair [23, 24, 28, 56–61], and as photosensitizers [31, 32, 62]. In the next several subsections we discuss the application of QD-based probes in most usable imaging and sensing methodologies in life science experiments.

6.3 Quantum Dots for Fluorescent Labeling and Imaging

The unique optical properties of QDs make them very attractive as novel fluorophores in various life science investigations, in which traditional fluorescent organic dyes fall short of providing long-term stability and simultaneous detection of multiple fluorescent signals. The ability to make QDs water soluble and target them to specific biomolecules has led to promising applications in cellular labeling, deep-tissue imaging, *in vitro* fluorescent assays, and FRET-based techniques.

6.3.1 Structure of Quantum Dot Nanobiohybrids for Fluorescent Microscopic Imaging

Since water-soluble QDs have many functional groups (NH_2–, COOH–, SH–, etc.) on their surface, they are easily conjugated with different biomolecules (proteins, DNAs, RNAs, small ligands, etc.), as well as with appropriate chemical substances (e.g., drugs). The resulting conjugates combine the properties of both materials, i.e., the spectroscopic characteristics of the nanocrystal and the biomolecular function of the surface-attached entities.

There are two types of conjugates, depending on the QD size. Using QDs with size commensurable with or slightly larger than that of many biomolecules, it is possible to obtain QD-based nanobiohybrids that consist of a single QD with several biomolecules attached on its surface. Thus, the large number (10 to 100) of potential surface attachment groups can be used to "graft" different functionalities to individual QDs, resulting in multifunctional probes. For instance, in addition to a recognition moiety, QDs can be equipped with a membrane-crossing or cell-internalization capability, and/or and enzymatic function.

In contrast, using small QDs (approximately 2–3 nm in diameter) it is possible to obtain nanobiohybrids that consist of several QD particles attached to a single biomolecule (usually protein) [26]. In this case, QDs have an important privilege for fluorescence detection methods in comparison with classical organic dyes – the attachment of several small size QDs on the surface of one protein molecule does not result in fluorescence self-quenching, in contrast to classical dyes (e.g., cyanines, fluorescein, etc.) where the conjugation of several dye molecules with one protein molecule results in a strong fluorescence self-quenching. This assumption is based on our comparative study of the fluorescence intensity of multifunctional PMAM dendrimers conjugated with different numbers of QD particles (~2 nm in diameter) or cyanine 3/cyanine 5 molecules (unpublished data).

Several strategies can be used to manipulate the molecular orientation of the ligands attached on the QD surface as well as their molar ratios with respect to QDs. However, QD probes with precisely controlled ligand orientations and molar ratios are still not available.

Two problems associated with fluorescence microscopy – cell and tissue autofluorescence in the visible spectrum (which can mask the signals from labeled molecules) and the requirement of long observation times – have created a need for new QD probes that emit in the near-infrared (NIR) region (wavelength > 700 nm) and are more photostable than current organic fluorophores [29].

6.3.2
Quantum Dots for Fluorescent Cell Imaging

QDs have made the most progress and attracted the greatest interest in cellular labeling and imaging. Within the last two-three years, numerous reports have described the ability of one or more "color/size" of biofunctionalized QDs to label cells [20–22, 26, 30, 32, 37, 50, 63–68]. Many of these reports show that QD labeling permits extended visualization of cells under continuous laser illumination as well as multicolor imaging, highlighting the advantages offered by these fluorophores (Fig. 6.4) [20, 22, 26, 65–68].

A clear differentiation can be made between labeling of live and fixed cells (dead with crosslinked components to maintain cellular architecture). Fixed cells can be treated "harshly" to facilitate entry of the QDs by chemically creating pores. To label live cells, the process must be handled softly to maintain cellular viability. Live-cell experiments introduce a few extra levels of complexity, depending on the application: whole-cell QD labeling, labeling of membrane-bound proteins, and cytoplasmic, subcellular or nuclear target labeling. The major impediment is the entry of the relatively large QDs into the cell across the cellular membrane lipid bilayers. For small size water-soluble QDs, the positive charge on the surface can facilitate this process, but for negatively charged QDs this is usually extremely difficult. Such QDs can be delivered into the cells only through some specific (receptor-mediated) transport based on the functional molecules conjugated on the QD surface. The results indicate that large amounts of QDs can be delivered into live mammalian cells via three different mechanisms: non-specific pinocytosis, microinjection, and peptide-induced transport [16, 17, 22, 69–71]. For example, transferrin has been used to facilitate endocytosis of QDs by mammalian cells and to label pathogenic bacteria and yeast cells [38], whereas lectin-conjugated QDs were used to detect specifically Gram-negative bacteria [61, 72]. Different types of QD functionalization have also been explored as a way to target QDs to cell surface proteins. Some examples include streptavidin, secondary or primary antibodies, receptor ligands such as epidermal growth factor or serotonin, recognition peptides, and affinity pairs such as biotin–avidin after engineering of the target protein [19–22, 25, 70, 73, 74]. To reduce the size of QD probes, researchers have used ligands of cell surface receptors bound to QDs via a biotin–streptavidin link or by direct crosslinking [21, 25, 70]. Some cell surface proteins can be recognized by small

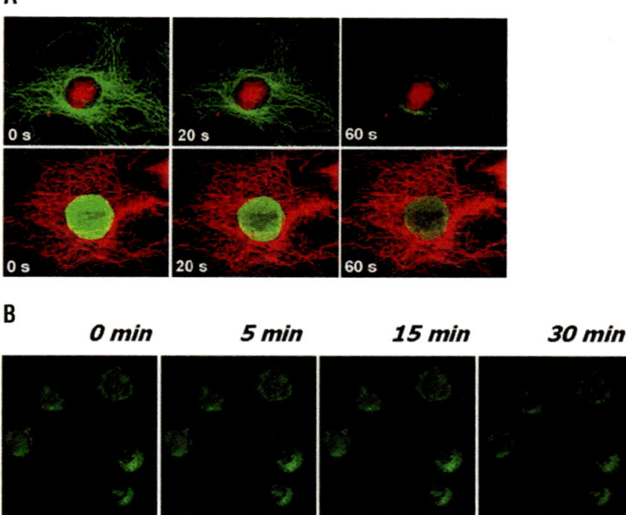

Fig. 6.4. Stability of QD fluorescence during continuous laser scanning on a confocal fluorescent microscope. (A) Top row: Nuclear antigens labeled with QD630-streptavidin (red), and microtubules labeled with Alexa-488 conjugated with anti-mouse IgG (green) simultaneously in 3T3 cells. Bottom row: Microtubules labeled with QD630-streptavidin (red), and nuclear antigens stained with Alexa-488 conjugated with anti-human IgG (green). Specimens were continuously illuminated for 1 min with light from a 100 W mercury lamp under a 100×1.30 oil-immersion objective. An excitation filter ($\lambda_{ex} = 485$ nm) was used to excite both Alexa-488 and QD630. Emission filters at $\lambda_{em} = 535$ nm and $\lambda_{em} = 635$ nm were used to collect Alexa-488 and QD630 signals, respectively. Images were captured with a CCD camera at 10 s intervals for each color automatically. (According to Wu et al. [20]). (B) Fluorescent microscopic imaging of interaction of QD-lectin conjugates with leukemia cells (green): dynamic of the signal during 30 min scanning. (According to Zhelev et al. [26]).

peptides (screened by phage-display technology) [74], so it is attractive to use peptides for QD functionalization. The same concepts could also be used to label subcellular or nuclear targets. However, QDs need to enter the cell cytoplasm and to reach their target without being trapped in the endocytosis. A surprising finding is that two billion QDs could be delivered into the nucleus of a single cell, without compromising its viability, proliferation or migration [8, 16, 71, 75]. The ability to image single-cell migration and differentiation in real time is expected to be important to several research areas such as embryogenesis, cancer metastasis, stem-cell therapeutics and lymphocyte immunology.

The main advantage of QDs in cellular imaging resides in their resistance to photobleaching over long periods of time (minutes to hours), allowing the acquisition of images that are crisp and well contrasted. The increased photostability of QDs is especially useful for acquisition of many consecutive focal-plane images and their reconstruction into a high-resolution three-dimensional (3D) projection, e.g., for 3D optical sectioning, where a major issue is bleaching of fluorophores during acquisition of successive z-sections. This compromises the correct reconstruction of 3D structures [76]. The very small number of QDs is necessary to produce a fluorescent signal. In addition, QDs are available in a virtually unlimited number of well-separated colors, all excitable by a single wavelength. This property could be used in fluorescent confocal microscopy to perform nanometer-resolution colocalization of multiple-color individual QDs.

6.3.3
Quantum Dots for Fluorescent Deep-tissue Imaging *In Vivo*

The long-term stability and brightness of QDs make them ideal candidates for *in vivo* fluorescent targeting and imaging. At present, such imaging uses two groups of organic fluorophores: the fluorescent proteins expressed by the cells themselves and fluorescent dyes that are exogenously loaded into the cells. Organic fluorophores have numerous limitations, restricting their usefulness for *in vivo* fluorescent microscopy. A first limitation is the difficulty in simultaneous imaging of multiple independent organic fluorophores, based on two characteristics of these fluorophores. First, they require distinct excitation wavelengths and have a very small Stokes shift between excitation and emission wavelengths (Fig. 6.3). Thus, multiple excitation lines are needed for imaging multiple organic fluorophores and each additional excitation line limits the spectra available for emission collection. Second, they emit over a broad region of the visible spectra. Thus the emission of different fluorophores overlaps with each other and with much of the tissue autofluorescence. A second limitation is the susceptibility of organic fluorophores to photodamage and metabolic degradation, which restricts their use in long-term *in vivo* imaging [53]. Moreover, some organic fluorophores (e.g., fluorescein) have a limited response to multiphoton excitation, which is a particular liability when trying to image into tissues.

QDs completely overcome these limitations. They can be excited by a wide spectrum of single and multiphoton excitation light, which is well separated from their emission spectra. The large Stokes shifts allow the possibility to overcome, or at least to minimize, tissue autofluorescence that appears near the excitation region (Fig. 6.5). Thus, only one excitation wavelength is needed to simultaneously excite several different QD probes. QDs also have narrow emission spectra, which are tunable to any desired wavelength from blue to infrared. Thus from their emission spectra several different QD probes can be easily distinguished. QDs are also virtually resistant to photobleaching and are as bright as the best-known organic fluorophores. Further, they have a good multiphoton absorption between 700 and 1000 nm. These features make QDs desirable for long-term multicolor *in vivo* imaging.

Theoretical modeling studies have indicated that two spectral windows are available for *in vivo* QD imaging – one at 700–900 nm and other at 1200–1600 nm [77–

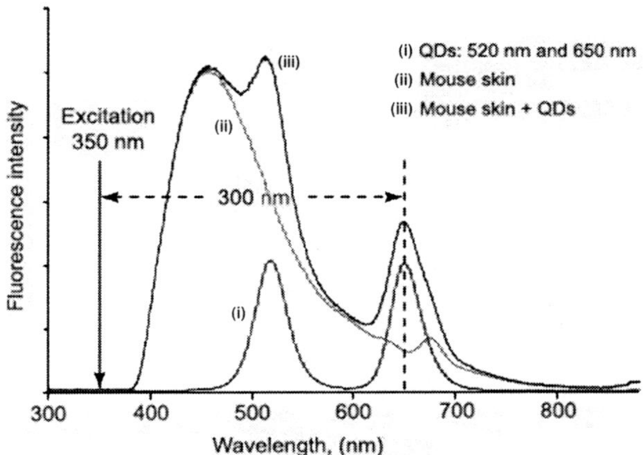

Fig. 6.5. Comparison of mouse skin autofluorescence and QD emission spectra, obtained under excitation at 350 nm. The results demonstrate that QD signals can be shifted to a spectral region where autofluorescence is reduced. (This figure was kindly provided from X. Gao [8].)

79]. There are great expectations for QDs emitting in the near-IR region that are ideal for *in vivo* imaging. In this region of the electromagnetic spectrum there is low tissue scattering and absorption, yielding the greatest tissue penetration depth and optical signal. In this case, QDs offer an excellent alternative to organic fluorophores for *in vivo* animal imaging. However, biocompatible near-IR-emitting organic fluorophores suffer from low QY (quantum yield), broad emission spectra, and an inability to multiplex. Near-IR QDs can, potentially, be designed to have high QYs and molar absorption coefficients, leading to a highly luminescent and useful *in vivo* contrast agent. In fact, high QY near-IR organic soluble QDs have already been designed [29, 80–82] – a major hurdle is the preservation of the optical properties of organic-soluble near-IR QDs after surface modification with biocompatible coatings and water solubilization. The highest reported QY for biocompatible near-IR-emitting QDs made from CdSe/CdTe alloy is 17%.

Another important issue for application of QDs *in vivo* is their overcoating with high molecular weight poly(ethylene glycol) (PEG) molecules that gives a possibility to reduce their accumulation in the liver and bone marrow, and to ensure effective targeting and imaging of the desired tissue *in vivo* (e.g., blood vessels, solid tumors, tumor metastases, lymph nodes, etc.) [12, 16, 18]. In contrast to small organic dyes, which are eliminated from the circulation within minutes after injection, PEG-coated QDs remained in the blood circulation for an extended period of time (half-life more than 3 h) (Fig. 6.6A) [18, 29]. This long-circulating feature can be explained by the unique structural properties of QD nanoparticles. PEG-coated QDs fall within an intermediate size range: they are small enough and sufficiently

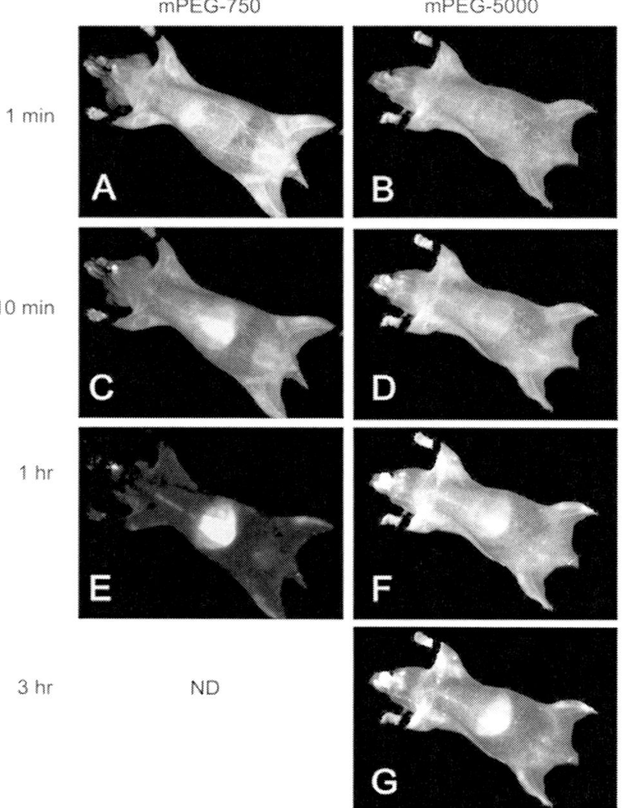

Fig. 6.6. (A) Noninvasive imaging of 645 nm-emitting mPEG-750 QDs (A, C, E) and 655 nm-emitting mPEG-5000 QDs (B, D, F, G.). Nude mice are imaged at different post-injection times. Significant liver uptake is detected using mPEG-750 QDs even at 10 min, while using mPEG-5000 QDs such uptake is visible after 1 h. The same QDs are very stable and can be also detected after 3 h. (Reprinted from Ref. [18] with the permission of ACS Publications).

hydrophilic to slow down opsonization and reticuloendothelial uptake, but are large enough to avoid renal filtration. In this context, near-IR QDs also have great potential to be used in diagnostic blood vessel (e.g., stenosis, aneurism, stroke, etc.).

To guarantee high binding affinities of QD probes to the respective cellular or molecular target *in vivo*, the nanocrystals are usually conjugated with antibodies or small peptides [12, 63]. Antibodies offer higher binding affinity of QDs than peptides, but they add size to the nanoparticles (~5–30 nm). Furthermore, antibodies could limit the co-coating of other molecules (e.g., polymers) onto the surface of QDs. By contrast, peptides are smaller than antibodies – large libraries of targeting peptides for specific diseases can be identified using screening tech-

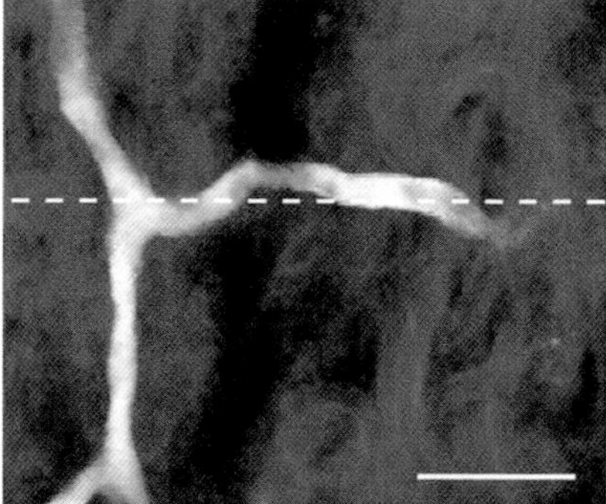

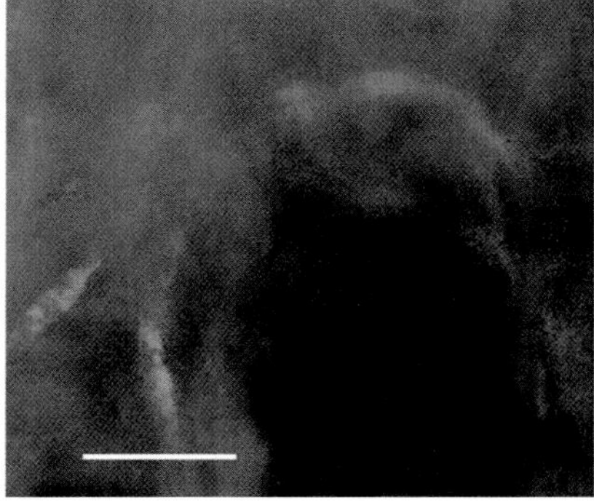

Fig. 6.6. (B1) In vivo imaging of vasculature labeled by a tail vein injection of water-soluble QDs (fluorescence wavelength, 550 nm). Fluorescent capillaries, containing ~1 μM QDs, were clearly visible trough the skin at the base of the dermis (~100 μm deep). Blue pseudocolor is collagen imaged via its second harmonic signal at 450 nm. (B2) Comparison image at the same depth as in (B1), acquired by injecting FITC-dextran at its solubility limit. Scale bars in (B1) and (B2): 20 μm. Excitation is at 880 nm for (B1) and 780 nm for (B2). (Reprinted from ref. [84] with the permission of Science Publishing Group.)

niques, such as phage-display – and provide greater flexibility in QD surface engineering.

Currently, there are several excellent examples for application of QDs in *in vivo* cellular targeting and imaging, including fluorescent deep-tissue imaging. Two-photon laser exciton confocal microscopy has been used to visualize blood vessels in mice after intravenous injection of QDs, showing that higher contrast and imaging depth can be obtained at a lower excitation power than with organic dyes [53]. Larson and coworkers found that the two-photon absorption cross-sections of QDs are two to three orders of magnitude larger than those of traditional organic fluorophores (Fig. 6.6.(B)) [81].

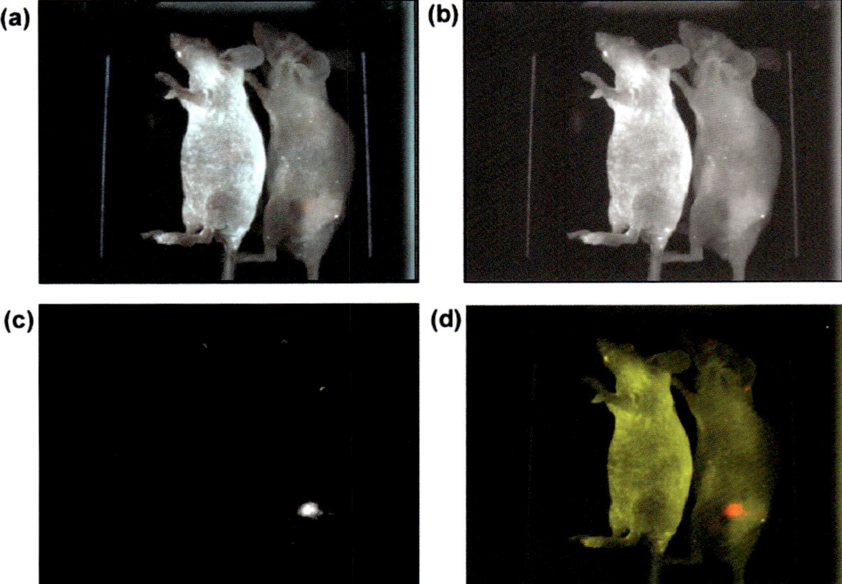

Fig. 6.7. *In vivo* fluorescence imaging of QD-PSMA antibody conjugates in mouse bearing C4-2 human prostate tumor. Orange-red fluorescence signals indicate a prostate tumor growing in a live mouse (right-hand mouse). Control studies using a healthy mouse (without tumor) and the same amount of QD injected showed no localized fluorescence signals (left-hand mouse). (a) Original image; (b) unmixed autofluorescence image; (c) unmixed QD image; (d) super-imposed image. (This figure was kindly provided from X. Gao [12].)

QD–peptide conjugates have been also used to target tissue-specific vascular markers (lung blood vessels and cancer cells) by intravenous injection in mice [18]. Histological sections of different organs after 5 or 20 min of circulation showed that QD-peptide conjugates reach their targets and are internalized by endocytosis in target cells but not in surrounding tissues, probably because of their large size relative to dye molecules (which stain surrounding tissues). *In vivo* imaging of targeted QD delivery has recently been achieved in mice by Gao et al., who intravenously injected PEG-coated QDs functionalized with antibodies to prostate-specific membrane antigen (PSMA) (Fig. 6.7) [12]. However, since their QDs emit in the visible spectrum, the authors use a spectral demixing algorithm to separate tissue autofluorescence from QD signal in grafted tumors.

This problem has been eliminated by Kim et al. [29], who have injected near-IR-emitting QDs intradermally into mice and pigs and demonstrated the visualization of sentinel lymph nodes in these animals via optical imaging (Fig. 6.8).

To improve tissue penetration, Kim et al. have prepared a novel core–shell nanostructure called type II QDs with fairly broad emission at 850 nm and a moderate QY of ~13% [29]. In contrast to conventional QDs (type I), the shell materials in type II QDs have valence and coordination band energies both lower than those of the core materials. As a result, the electrons and holes are physically separated and the nanoparticles emit light at reduced energies (longer wavelengths). Injection of only picomolar amounts of the QDs enables visualization of sentinel lymph nodes 1 cm below the skin using excitation rates of only 5 mW cm^{-2}. QDs rapidly migrate to nearby lymph nodes and can be imaged virtually background-free, allowing image-guided resection of a lymph node in a pig. The sentinel nodes have been confirmed by a second intradermal injection of blue dye (the current "gold standard"), which also flowed to the same nodes. The advantage of QD over blue dye is the ability to "see" through several centimeters of tissue so that there would be real-time visual guidance of the surgery, with resolution being limited only by the visual acuity of the surgeon. Visual inspection would also allow the surgeon to confirm that all sentinel nodes had been removed from the node field. There is also exciting potential for histopathologists to focus exactly on the part of the sentinel node containing the QDs by using a fluorescent microscope.

Sentinel lymph node biopsy is elegantly simple in concept but quite often very difficult to perform accurately in individual patents. The sentinel nodes can be very deep, up to 10 cm from the skin, and sometimes the nodes are inside the body cavities; situations such as these are considered as problematic for optical imaging using near-IR fluorescence. Therefore, anything that potentially improves the technique is welcome. The work of Kim et al. [29] points to the possibility that QD probes could be used for real-time intra-operative optical imaging, providing an *in situ* visual guide so that a surgeon could locate and remove small lesions (e.g., metastatic tumors) quickly and accurately. At present, however, appropriate high-quality QDs with near-IR-emitting properties are not yet available. Most materials (e.g., PdS, PdSe, CdHgTe and CdSeTe) are either not bright enough or not stable enough for biomedical imaging applications. There is an urgent need to develop bright and stable near-IR-emitting QDs that are broadly tunable in the far-red and IR spectral regions.

Fig. 6.8. Near-IR QD sentinel lymph node mapping in mouse and pig. (a) Images of mouse injected intradermally with 10 pmol of near-IR QDs in the left paw. Left-hand side – pre-injected near-IR autofluorescence image; middle – 5 min post-injection white light color video image; right – 5 min post-injection near-IR fluorescence image. An arrow indicates the putative axillary sentinel lymph node. Fluorescence images have identical exposure times and normalization. (b) Images of the mouse obtained 5 min after re-injection with 1% isosulfan blue (standard) and exposure of the actual sentinel lymph node. Left-hand side – color video; right – near-IR fluorescence images. Isosulfan blue and near-IR QDs are localized in the same lymph node (arrows). (c) Images of the surgical field in a pig injected intradermally with 400 pmol of near-IR QDs in the right groin. Four time points are shown from top to bottom: before injection (autofluorescence), 30 s after injection, 4 min after injection, and during image-guided resection. For each time-point: Left – color video; middle – near-IR fluorescence; and right – color-near-IR merge images. (This figure was kindly provided from J. Frangioni [29].)

a
Pre-injection autofluorescence
Color video 5 min post-injection
NIR fluorescence 5 min post-injection
1 cm
b
Color video
NIR fluorescence
c
Color video
NIR fluorescence
Color-NIR merge
Pre-injection (autofluorescence)
30 sec post-injection
4 min post-injection
Image-guided resection
1 cm
N

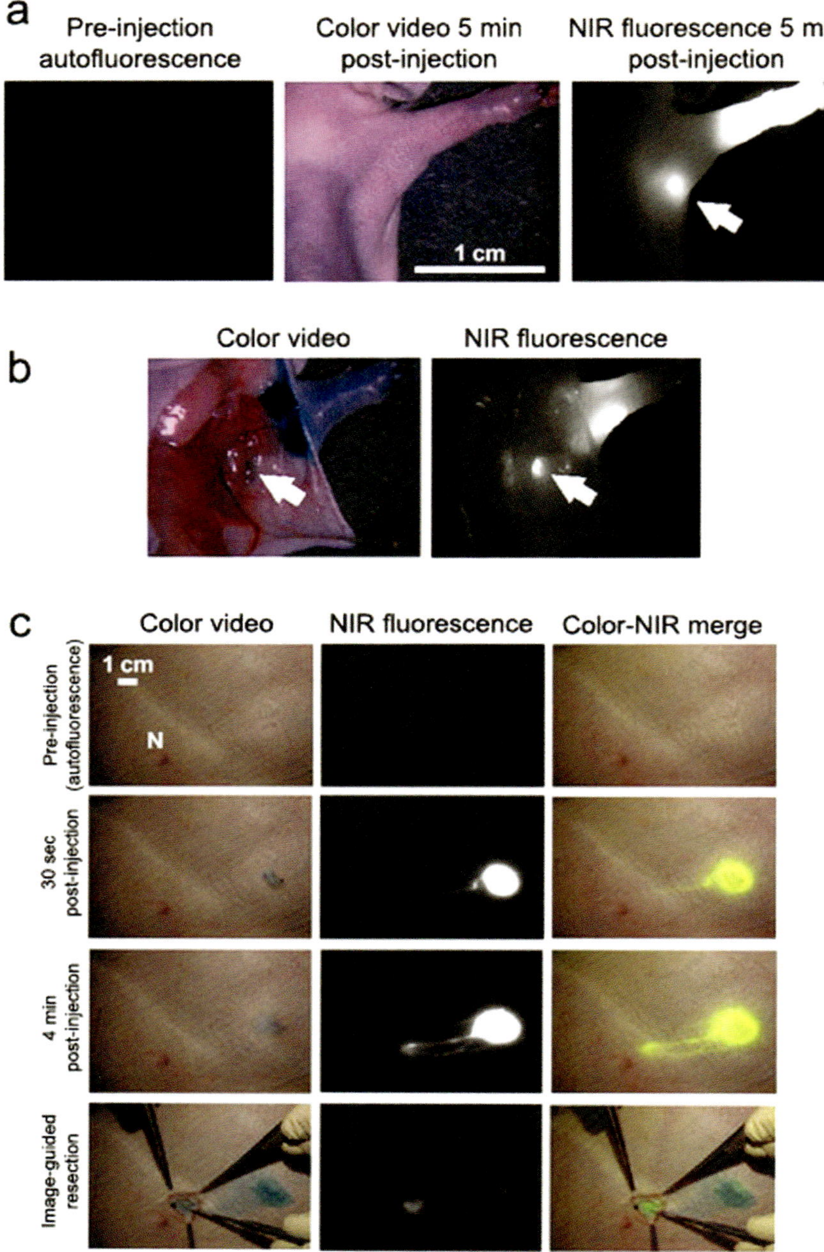

Fig. 6.8

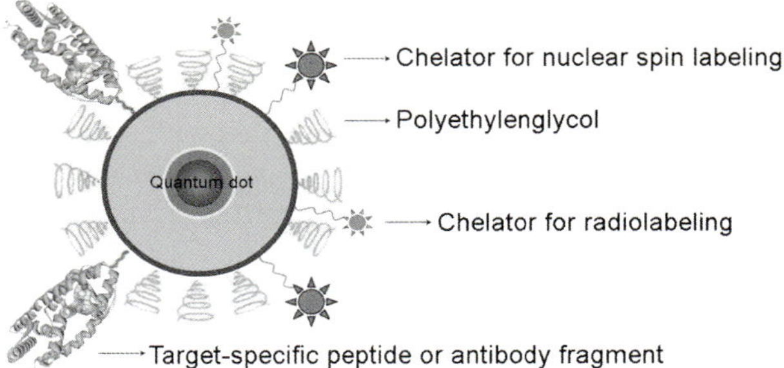

Fig. 6.9. Multifunctional QD-based probes for simultaneous application in different imaging techniques (e.g., fluorescent imaging, MRI, PET).

6.3.4
Potential of Quantum Dots for Positron Emission Tomography (PET) and functional Magnetic Resonance Imaging (fMRI)

The ability to design a wide variety of QDs and targeting molecules provides a new set of tools for engineering novel contrast probes for non-invasive imaging techniques such as PET, MRI, and CT. As a research tool, these probes could have many uses, from monitoring tissue implants to studying the real-time dynamics of tumor metastasis.

Despite their relatively large hydrodynamic radii (10–15 nm), bioconjugated QD probes do not suffer from serious binding kinetic or steric-hindrance problems. In this "mesoscopic" size range, QDs also have more surface areas and functionalities that can be used for linking to multiple diagnostic (e.g., radioisotopic or magnetic) and therapeutic (e.g., anticancer) agents. Multimodality imaging probes could be created by integrating QDs with paramagnetic or superparamagnetic agents, as well as with chelators for radioactive isotopes (Figs. 6.9 and 6.10).

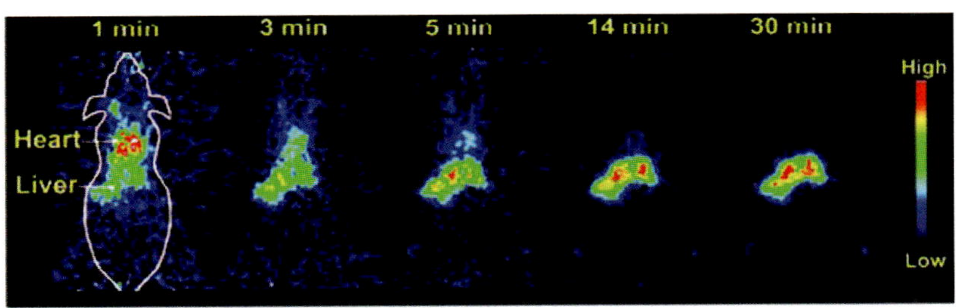

Fig. 6.10. MicroPET imaging of QD uptake in live animals. QDs having DOTA (a chelator for radiolabeling) and PEG on their surface were radiolabeled with ^{64}Cu (positron-emitted isotope with a half-life of 12.7 h). These QDs were then injected via the tail vein into nude mice (~80 μCi per animal) and imaged in a small animal PET-scanner. Rapid and marked accumulation of QDs in the liver was detected. This could be avoided by functionalizing QDs with higher molecular weight PEG chains. (Reprinted from Ref. [9] with the permission of Science Publishing Group.)

Researchers have recently attached QDs to Fe_2O_3 and FePt nanoparticles and even to paramagnetic gadolinium chelators [83, 84]. By correlating the deep imaging capabilities of MRI with ultrasensitive optical imaging, a clinician could visually identify tiny tumors or other small lesions during an operation and remove the diseased cells and tissue completely. Medical imaging modalities such as MRI and PET can identify diseases non-invasively, but they do not provide a visual guide during surgery. The development of magnetic or radioactive QD probes has the potential to solve this problem.

6.4 Quantum Dots for Immunoblot Analysis with Fluorescent Detection

6.4.1 Basic Principles of Classical and QD-based Immunoblot Analyses

Although immunoblotting technology is about 20 years old it is still a major analytical approach for detection of protein expression in cells and tissues, used in molecular and cellular biology and molecular medicine. Currently, it is one of the basic methodologies (together with northern blot analysis and Light Cycler technology) for microarray data validation and verification in functional genomics and proteomics projects.

Figure 6.11(A) shows the basic principle of classical immunoblot analysis (Western blot).

The protein fractions are separated from cell lysate using gel-electrophoresis and are subsequently transferred to a PVDF membrane. The protein of interest is detected using two-step immunoblotting procedure: incubation of PVDF membrane with primary monoclonal antibody that interacts specifically with the target protein and does not interact with other protein fractions (a); washing of membrane and incubation with secondary antibody, conjugated with enzyme (usually horse radish peroxidase, HRP) (b). The blotted antigen is detected by the chemiluminescence of the HRP-catalyzed reaction of appropriate substrate, exposed to X-ray or Polaroid films.

Notably, despite its widespread and long-standing use, immunoblotting has not been much improved from its initial state 20 years ago. The standard immunoblot analysis suffers from several shortcomings: a semi-quantitative nature; a long duration (usually more 24 h); a poor reproducibility of the obtained data; a risk to reach saturation of the enzyme reaction and to mimic the concentration-dependent difference of the blotted antigens; a low sensitivity for detection of "tracer" proteins – including the impossibility of using directly cell lysates for protein separation and immunoblotting, as well as a necessity to use preliminary procedures for immunoprecipitation and concentration of these proteins (each step holds a risk for accumulation of analytical errors). Every researcher in the genomics/proteomics field has been faced with the difficulties in obtaining reproducible and high quality immunoblot images.

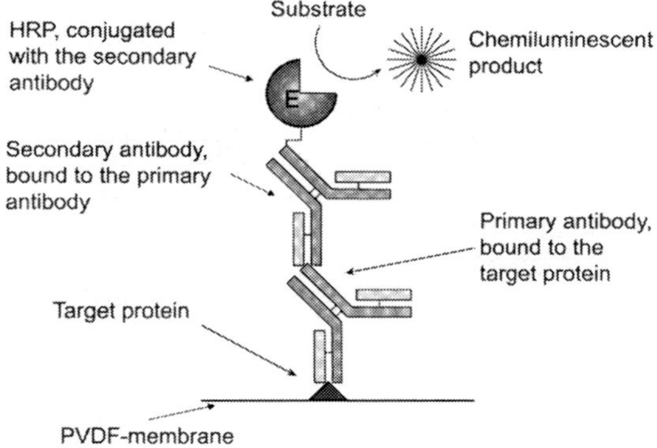

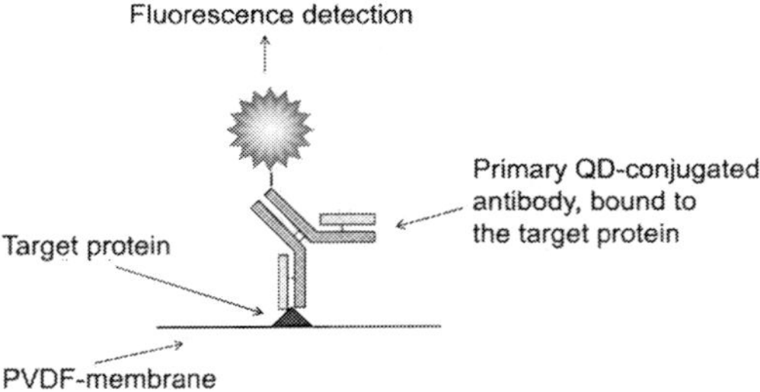

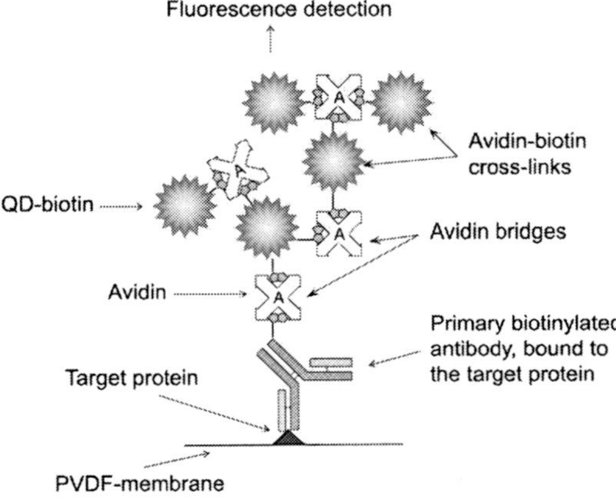

Fig. 6.11. Principle schemes of classical and QD-based immunoblot analyses.

In the middle of August 2004 the Quantum Dot Corporation (USA) offered on the market two western blotting kits, and in the January 2005 issue of *Nature Methods* [85] they undoubtedly demonstrated the potential of QD-based immunoblotting technology in multiplex fluorescent detection of proteins or protein states from a single immunoblot. Their protocol is similar to the classical one; however, the secondary antibody is conjugated with QD instead of enzyme, and the fluorescence of blot images instead of chemiluminescence is detected. The described procedure takes the same time as standard western blot analysis with chemiluminometric detection. However, the narrow emission of QD-labeled protein blots enables simplified image acquisition and quantification that gave a possibility to overcome the semiquantitative nature of western blot analysis. It is possible to store QD-labeled blots in a buffer at 4–8 °C with minimal loss of the fluorescent signal for imaging at a later date.

We have tried to improve QD-based western blotting technology, to simplify and to make it quantitative and applicable for detection of "tracer" proteins directly in cell lysates [27, 86].

Below, we present two immunoblotting protocols, named as "mono type" and "sandwich type" and show the privilege of QD-based western blot analysis in the detection of "tracer" proteins through classical technology with colorimetric detection. Figure 6.11(B, C) presents the principle schemes of both procedures.

6.4.2
QD-based Immunoblot Analysis of "tracer" Proteins – Privileges over Classical Immunoblot Analysis

QDs with emission maximum at 535 nm were used in this study. Telomere associated proteins, telomeric repeat binding factor (TRF1, a 56 kDa protein) and TRF1-interacting nuclear protein 2 (Tin2, ~40 kDa protein), were selected as proteins of interest. TRF1 and Tin2 are known to be poorly expressed in Philadelphia-positive (Ph+) cells derived from patients with chronic myelogenous leukemia (e.g., K-562 cells) and several modifications of the standard immunoblotting protocol have to be performed to identify these two proteins in K-562 cells. Details are described in Refs. [27, 86]. The blotted QD-labeled antigens (proteins) were detected directly by a fluorescence gel imaging system supplied with an appropriate emission filter. The experimental results in Fig. 6.12 represent typical blots of TRF1 and Tin2, analyzed in cell lysates by classical (C) and QD-based western blotting protocols (A, B).

Using "mono-type" QD-based western blotting technology, it was possible to detect both proteins without preliminary immunoprecipitation and concentration procedures (Fig. 6.12A). However, the fluorescence of QD-labeled TRF1 and Tin2 blots was weak. It was practically impossible to detect these "tracer" proteins directly in cell lysate, without their preliminary immunoprecipitation and concentration (Fig. 6.12C). In contrast, TRF1 and Tin2, detected in K-562 cell lysates by "sandwich type" QD-based western blotting technology, possessed a bright fluorescence as a result of biotin–avidin crosslinks (Fig. 6.12B). The images in Fig.

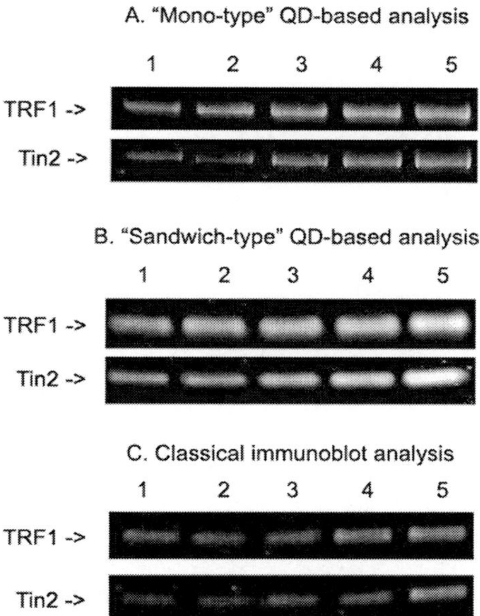

Fig. 6.12. Representative blots of "tracer" proteins TRF1 and Tin2, analyzed in K-562 cell lysate using classical (C) and QD-based (A, B) immunoblot analysis. Numbers 1–5 correspond, respectively, to protein concentrations of cell lysates (in µg), applied to each gel patch, of 1–10, 2–20, 3–30, 4–40, and 5–50. In (A) and (B) blot images were obtained by fluorescent detection, and the images in (C) were obtained by chemiluminescent detection.

6.12(A, B) were generated using an identical exposure time (15 min) and excitation/emission settings. Since avidin can interact non-specifically with other non-biotinylated proteins on the PVDF membrane, we recommend the use of neutravidin or streptavidin in "sandwich type" QD-based western blot analysis. Notably, the fluorescent signal was stable during continuous scanning in gel imager. No changes in the blot fluorescence intensity were registered between 5 and 30 min scanning – time enough for membrane imaging and data acquisition.

The higher sensitivity of QD-based methodology in comparison with classical analysis can be explained, at least partially, by the inherent sensitivity of fluorescence compared with chemiluminescence. The QD-labeled membranes can be kept in buffer at 4 °C for a week without loss of image quality. It was also possible to improve markedly the brightness of blots with serial incubations of PVDF-membrane in QD-labeled neutravidin and QD-labeled biotin. However, concentration-dependent saturation of QD-signal was detected at high protein concentrations of cell lysate (applied to each gel patch) and this procedure can be recommended for work with samples containing less than 20 µg protein per gel patch.

The proposed QD-based immunoblotting technologies with fluorescent detection allow the possibility of quantification of the blot fluorescence, and they also

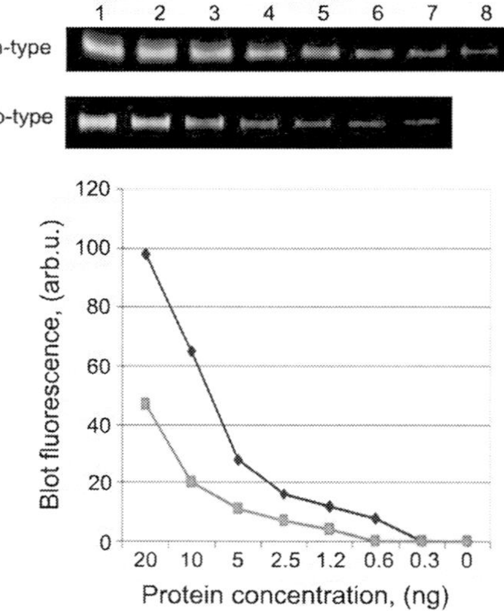

Fig. 6.13. Sensitivity of "sandwich type" and "mono-type" QD-based immunoblot analyses, as determined by pure soybean agglutinin (two-serial dilution starting from 20 ng per gel patch – number 1). Upper curve – "sandwich type"; lower curve – "mono type".

shortened the time of analysis of "tracer" proteins in comparison with the standard methodology. The sensitivity of our protocols was up to 1 and 5 ng of pure protein per gel patch for "sandwich" and "mono" type, respectively (Fig. 6.13). The detection of QD-labeled immunoblotted proteins is simple, in contrast to the detection procedures of the standard western blot analysis. Standard technology requires a transfer of the chemiluminescent signal on Hyperfilms, which is a time- and cost-consuming process. Both technologies in our study were developed for detection of TRF1 and Tin2 in K-562 cells, but they can also be applied for detection of other proteins in different cellular and tissue samples.

6.5 Quantum Dots for FRET Analyses, Time-resolved Fluorimetry, and Development of Optical Recognition-based Biosensors

6.5.1 Quantum Dots for FRET-based Bioanalyses

In the past several years, FRET has been involved in many biochemical analyses and applied to the development of simple fluorescence detection techniques, including a PCR with real-time FRET measurements; DNA hybridization analyses and formation and dissociation of hairpin structures; an elucidation of the dynamics of telomerization or DNA replication; and an investigation of interactions between proteins, nucleic acids and small molecules [23, 87–90].

The essential requirement for FRET is that the emission spectrum of the donor has to overlap with the absorption spectrum of the acceptor, and distance between donor and acceptor has to be within 1–10 nm. Reports on QDs as FRET-donors in a biological context appeared from 2001 [56, 57]; however, the full potential has not been demonstrated yet. Several elegant studies in this field have been designed and provided recently by the U.S. Naval Research Laboratory [23, 24, 58, 59]. By self-assembling acceptor dye-labeled proteins onto QD donor surfaces, two unique advantages over organic fluorophores for FRET became apparent: QD donor emission could be size-tuned to improve spectral overlap with an acceptor dye (using a size selection, it is possible to obtain QDs with emission maximum corresponding exactly to the excitation maximum of the FRET-acceptor) (a), and several acceptor dyes interacting with a single QD-donor substantially improved FRET efficiency (b). Obviously, the unique optical properties of QDs make them the most appropriate FRET-donors. However, currently there is a serious limitation in the widespread use of QDs in FRET-based bioanalyses related to the size of highly luminescent water-soluble QDs, which is usually beyond the Forster radius. Efforts are being directed to the development of small (up to 10 nm in diameter) highly luminescent water-soluble QDs with optimal FRET capacity.

6.5.2
Quantum Dots for Time-resolved Fluorimetry

Some of the presently described QDs possess a comparatively long fluorescence half-life and can be appropriate for time-resolved fluorimetric analyses. Figure 6.14 demonstrates the fluorescence half-life of water-soluble broad-fluorescent CdSe QDs [86]. The fluorescence half-life varied from 27 to 92 ns, depending of the emission wavelength. Lakowicz and colleagues have described sharp-fluorescent CdSe/ZnS QDs with a mean fluorescence half-life of ~17 ns and a heterogeneous intensity decay curve [91]. For comparison, the reported fluorescence half-life of commercial CdSe/ZnS QDs, dispersed in organic solvents, is approximately 10–12 ns [92–94].

6.5.3
Quantum Dots for development of New Generation Optical Recognition-based Biosensors

The FRET capacity and comparatively long fluorescence half-life of QDs are very promising in the evolution of optical recognition-based biosensors that can monitor rapidly the concentrations of target species in their physiological environment, in a continuous, simple and reliable manner. Presently, the most popular recognition-based optical sensors consist of surface immobilized functional

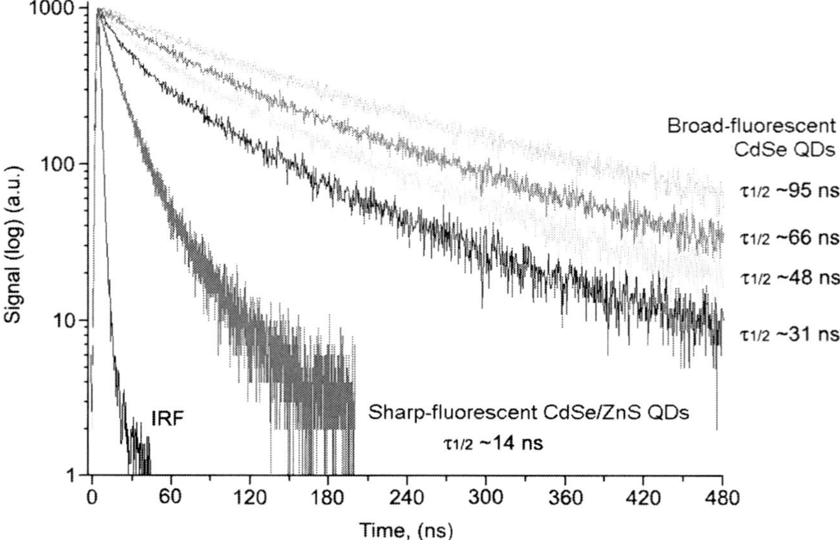

Fig. 6.14. Time-resolved fluorescence life-time spectra of water-soluble broad-fluorescent CdSe QDs synthesized at room temperature. For comparison, sharp fluorescent QDs synthesized at high temperature were recorded at an emission wavelength corresponding to their fluorescent maximum. All data were recorded at an excitation wavelength of 365 nm, frequency 40 kHz, using a Hamamatsu FLS920S spectrometer. IRF = instrumental response function of the nanosecond flash-lamp.

materials – usually a thin polymer film with incorporated hydrophobic organic dye [95, 96]. These sensors possess, predominantly, a pressure or temperature sensing ability, and an ability to detect the concentration of ions and small molecular weight molecules (e.g., oxygen, carbon dioxide, etc.). They do not possess an ability to detect the concentration of middle (100–1000 Da) or high (10–100 kDa) molecular weight molecules and more complex components, since the diffusion of such substances into the film and their accessibility to the dye are strongly restricted. All presently known recognition-based optical sensors work in two dimensions and in a non-aqueous environment. They have no possibility for analysis of complex chemical and biochemical substances, as well as enzyme activities.

The leading tendencies in the development of new generation optical recognition-based biosensors are: to increase the intensity of the specific optical signal; to decrease the intensity of the non-specific noise; to ensure the possibility for detection of water-soluble target compounds in their aqueous environment; and to increase the number of simultaneously analyzed targets in one sensor device.

What can QDs offer to biosensor evolution?

QDs have attracted much attention as one of the most promising nanotransducers that have several advantages over other known nanomaterials. Figures 6.15 and 6.16 represent two common examples for the design of QD-based sensors for detection of chemical or biochemical targets and for detection of enzyme activity.

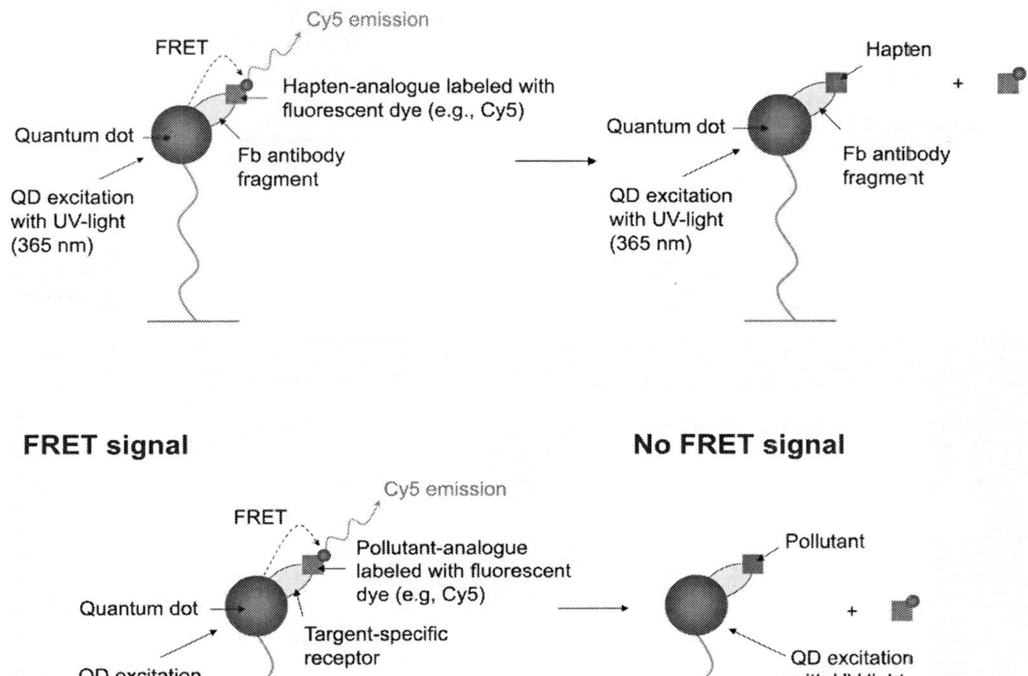

Fig. 6.15. Schemes of QD-based optical sensors for detection of more complex chemical and biochemical targets.

In the first sensor, a QD nanotransducer is conjugated with a receptor-specific for some chemical (or biochemical) target substance that has to be analyzed. The receptor is bound to target analogue, labeled with classical dye (e.g., cyanine). The analytical capacity of the sensor is based on the FRET efficiency between QD and dye. The target analogue has to be with a lower binding affinity to the receptor than the real target. Thus, it can be competitively replaced from the target, which will reflect the FRET signal in a concentration-dependent manner. In the second sensor (biosensor), a quantum dot nanotransducer is conjugated with some enzymatic substrate, e.g., double-stranded RNA (dsRNA) labeled with fluorescent quencher (e.g., black hall quencher, BHQ). The analytical capacity of this biosensor is based on the lack of fluorescence in the presence of quencher near the QD surface and the appearance of fluorescent signal if there is RNase activity leading to a

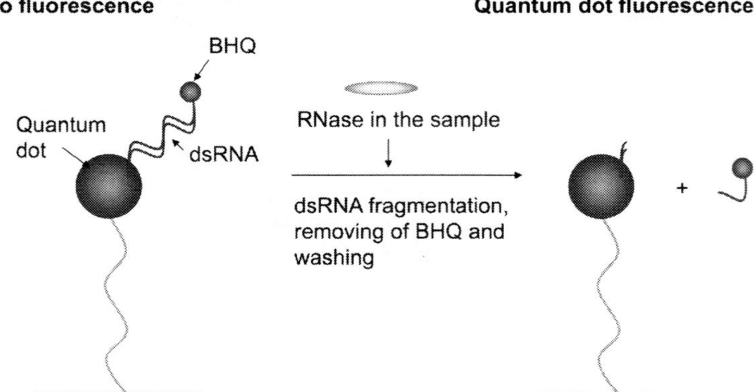

Fig. 6.16. Scheme of QD-based optical sensor for detection of enzyme activity.

fragmentation of dsRNA and removal of the quencher from the QD area. Fluorescence detection in both cases can be realized using fiber optic devices.

QD nanotransducers meet almost all requirements of the leading tendencies in the design and fabrication of new generation recognition-based optical biosensors. Currently, only QDs manifest a large excitation/emission Stokes shift, which gives an opportunity to overcome the autofluorescence of other components in the sensor device and to minimize the noise, using filter combinations that effectively isolate the desired fluorescent signal. Moreover, the autofluorescence usually appears rapidly, while the FRET-signal appears slower and, using a time-resolved fluorimetry, it is possible additionally to increase the signal-to-noise ratio. The size-dependent multicolor coding properties of QDs render these nanoparticles as an indisputable favorite in multiplex fluorescent analyses. Thus, size-distinguished QD nanotransducers immobilized on 2D or 3D matrices will give an opportunity to develop optical sensor devices for the simultaneous detection of several targets. Presumably, it will be not necessity to separate the size-distinguished QD nanotransducers in the area of matrix platform because the unique fluorescent properties of QDs give a possibility to detect more than six fluorescent signals at the same time. Because different colors QD nanotransducers can be conjugated with different reporters, they can be functionally separated on the matrix.

Finally, the significant progress in the synthesis of highly fluorescent water-soluble QDs makes possible the fabrication of nanobiosensors for detection of biological targets and enzyme activities in their physiological environment, as well as environmental pollutants and food ingredients in aqueous solutions.

Obviously, the bridge between QD technology and fiber optic technology can open up new trends in biosensor evolution, resulting in development of ultrasensitive 3D optical sensors for multiplex detection of low, middle and high molecular weight substances, and enzyme activities, in their natural environment.

6.6
Quantum Dots as New Fluorescent Standards for the Thin Calibration of Fluorescent Instrumentation

Fluorescence-based measurements are becoming the standard for high-throughput screening technologies, sensing technologies, imaging technologies, forensic determination, genomic research, clinical diagnostics, etc. One of the critical challenges in the measurement of fluorescence is to obtain wavelength calibration. A small discrepancy in wavelength can lead to a large discrepancy in the measured fluorescence intensity. Moreover, absolute fluorescent measurements are also difficult since few fluorescent standards are available. A fluorescence standard is necessary, especially where quantification is required. At present, it is nearly impossible to quantify the fluorescence from an assay. Traditional fluorescing materials, such as organic dyes, lose their fluorescence intensity due to photodegradation and have significant disadvantages as standards. Semiconductor QDs are very promising in this application since they are typically very stable – a characteristic required for a standard. Being composed of simple inorganic compounds they are chemically relatively inert and very resistant to photochemical damage. The size-tunable fluorescent emission of QDs and the large Stokes-shift between excitation and emission give a possibility for application of QD composite materials for the thin calibration of fluorescent instrumentation.

Future efforts are being directed to the development of QD composites (consisting of QDs incorporated in transparent materials such as polymers, silica, etc.) and to applying them in fluorescent measurements – to characterize these materials and to evaluate their possible use as novel fluorescence standards for calibration of biotech instrumentation. This is the priority of all leading institutes in the field of measurements and standardization.

Fluorescent standards based on QD composites will possess the following improved characteristics in comparison with conventional fluorescent standards (based on organic dyes incorporated into transparent composite materials): a higher quantum yield (70–80% vs. 15–50% for classical standards); a higher chemical and photochemical stability; a long fluorescence life-time; a possibility for fabrication of fluorescent standards that have Stokes' shifts of several nanometers for thin calibration of the fluorescent instrumentation, which is impossible with traditional fluorescent standards.

References

1 Alivisatos, A. P., Semiconductor clasters, nanocrystals, and quantum dots, *Science* **1996**, 271, 933–937.

2 Dahan, M., Laurence, F., Pinaud, D. S., Alivisatos, A. P., Sauer, M., Weiss, S., Time-gated biological imaging by use of colloidal quantum dots, *Opt. Lett.* **2001**, 26, 825–827.

3 Nirmal, M., Dabbousi, B. O., Bawendi, M. G., Macklin, J. J., Trautman, J. K., Harris, T. D., Brus, L. I., Fluorescence intermittency in

single cadmium selenide nanocrystals, *Nature* **1996**, 383, 802–804.
4 Tang, J., Marcus, R. A., Mechanisms of fluorescence blinking in semiconductor nanocrystal quantum dots, *J. Chem. Phys.* **2005**, 123(5), 054 704.
5 Shimizu, K. T., Woo, W. K., Fisher, B. R., Eisler, H. J., Bawendi, M. G., Surface-enhanced emission from single semiconductor nanocrystals, *Phys. Rev. Lett.* **2002**, 89(11), 117 401.
6 Medintz, I. L., Uyeda, H. T., Goldman, E. R., Mattoussi, H., Quantum dot bioconjugates for imaging, labeling and sensing, *Nature Mater.* **2005**, 4(6), 435–446.
7 Alivisatos, A. P., Gu, W., Larabell, C., Quantum dots as cellular probes, *Annu. Rev. Biomed. Eng.* **2005**, 7, 55–76.
8 Gao, X., Yang, L., Petros, J. A., Marshall, F. F., Simons, J. W., Nie, S., *In vivo* molecular and cellular imaging with quantum dots, *Curr. Opin. Biotechnol.* **2005**, 16(1), 63–72.
9 Michalet, X., Pinaud, F. F., Bentolila, L. A., Tsay, J. M., Doose, S., Li, J. J., Sundaresan, G., Wu, A. M., Gambhir, S. S., Weiss, S., Quantum dots for live cells, *in vivo* imaging, and diagnostics, *Science* **2005**, 307(5709), 538–544.
10 Jaiswal, J. K., Simon, S. M., Potentials and pitfalls of fluorescent quantum dots for biological imaging, *Trends Cell Biol.* **2004**, 14(9), 497–504.
11 Penn, S. G., He, L., Natan, M. J., Nanoparticles for bioanalyses, *Curr. Opin. Chem. Biol.* **2003**, 7(5), 609–615.
12 Gao, X. H., Cui, Y. Y., Levenson, R. M., Chung, L. W. K., Nie, S. M., In vivo cancer targeting and imaging with semiconductor quantum dots, *Nat. Biotechnol.* **2004**, 22, 969–976.
13 Gao, W., Li, J. J., Wang, Y. A., Peng, X., Layered structural heme protein magadiite nanocomposites with high enzyme-like peroxidase activity, *Chem. Mater.* **2003**, 15, 3125–3133.
14 Mattoussi, H., Mauro, J. M., Goldman, E. R., Anderson, G. P., Sundar, V. C., Miculec, F. V., Bawendi, M. G., Self-assembly of CdSe-ZnS quantum dot bioconjugates using an engineered recombinant protein, *J. Am. Chem. Soc.* **2000**, 122, 12 142–12 150.
15 Uyeda, H. T., Medintz, I. L., Jaiswal, J. K., Simon, S. M., Mattoussi, H., Synthesis of compact multidentate ligands to prepare stable hydrophilic quantum dot fluorophores, *J. Am. Chem. Soc.* **2005**, 127, 3870–3878.
16 Dubertret, B., Skourides, P., Norris, D. J., Noireaux, V., Brivanlou, A. H., Libchaber, A., *In vivo* imaging of quantum dots encapsulated in phospholipids micelles, *Science* **2002**, 298, 1759–1762.
17 Mattheakis, L. C., Dias, J. M., Choi, Y. J., Gong, J., Bruchez, M. P., Liu, J., Wang, E., Optical coding of mammalian cells using semiconductor quantum dots, *Anal. Biochem.* **2004**, 327, 200–208.
18 Ballou, B., Lagerholm, B. C., Ernst, L. A., Bruchez, M. P., Waggoner, A. S., Noninvasive imaging of quantum dots in mice, *Bioconj. Chem.* **2004**, 15, 79–86.
19 Dahan, M., Levi, S., Luccardini, C., Rostaing, P., Riveau, B., Triller, A., Diffusion dynamics of glycine receptors revealed by single-quantum dot tracking, *Science* **2003**, 302, 442–445.
20 Wu, X. Y., Liu, H. J., Liu, J. Q., Haley, K. N., Treadway, J. A., Larson, J. P., Ge, N. F., Peale, F., Bruchez, M. P., Immunofluorescent labeling of cancer marker Her2 and other cellular targets with semiconductor quantum dots, *Nat. Biotechnol.* **2003**, 21, 41–46.
21 Lidke, D. S., Nagy, P., Heintzmann, R., Arndt-Jovin, D. J., Post, J. N., Grecco, H. E., Jares-Erjiman, E. A., Jovin, T. M., Quantum dot ligands provide new insites into erbB/HER receptor-mediated signal transduction, *Nat. Biotechnol.* **2004**, 22, 198–203.
22 Jaiswal, J. K., Mattoussi, H., Mauro, J. M., Simon, S. M., Long-term multiple color imaging of live cells using quantum dot bioconjugates, *Nat. Biotechnol.* **2003**, 21, 47–51.

23 Medintz, I. L., Claap, A. R., Mattoussi, H., Goldman, E. R., Fisher, B., Mauro, J. M., Self-assembled nanoscale biosensors based on quantum dot FRET donors, *Nature Mater.* **2003**, 2, 630–638.

24 Medintz, I. L., Konnert, J. H., Claap, A. R., Stanish, I., Twigg, M. E., Mattoussi, H., Mauro, J. M., Deschamps, J. R., A fluorescence resonance energy transfer-derived structure of a quantum dot-protein bioconjugate nanoassembly, *Proc. Natl. Acad. Sci. U.S.A.* **2004**, 101, 9612–9617.

25 Rosenthal, S. J., Tomlinson, I., Adkins, E. M., Schroeter, S., Adams, S., Swafford, L., McBride, J., Wang, Y., DeFelice, L. J., Blakely, R. D., Targeting cell surface receptors with ligand-conjugated nanocrystals, *J. Am. Chem. Soc.* **2002**, 124, 4586–4594.

26 Zhelev, Z., Ohba, H., Bakalova, R., Jose, R., Fukuoka, S., Nagase, T., Ishikawa, M., Baba, Y., Fabrication of quantum dot-lectin conjugates as novel fluorescent probes for microscopic and flow cytometric identification of leukemia cells from normal lymphocytes, *Chem. Commun.* **2005**, 1980–1982.

27 Bakalova, R., Zhelev, Z., Ohba, H., Baba, Y., Quantum dot-based western blot technology for ultrasensitive detection of tracer proteins, *J. Am. Chem. Soc.* **2005**, 127(26), 9328–9329.

28 Bakalova, R., Zhelev, Z., Ohba, H., Baba, Y., Quantum dot-conjugated hybridization probes for preliminary screening of siRNA sequences, *J. Am. Chem. Soc.* **2005**, 127(32), 11 328–11 335.

29 Kim, S., Lim, Y. T., Soltesz, E. G., DeGrand, A. M., Lee, J., Nakayama, A., Parker, J. A., Mihaljevic, T., Laurence, R. G., Dor, D. M., Cohn, L. H., Bawendi, M. G., Frangioni, J. V., Near-infrared fluorescent type II quantum dots for sentinel lymph node mapping, *Nat. Biotechnol.* **2004**, 22(1), 93–96.

30 Ozkan, M., Quantum dots and other nanoparticles: What can they offer to drug discovery? *Drug Discovery Today* **2004**, 9(24), 1065–1071.

31 Bakalova, R., Ohba, H., Zhelev, Z., Ishikawa, M., Baba, Y., Quantum dots as photosensitizers?, *Nat. Biotechnol.* **2004**, 22(11), 1360–1361.

32 Bakalova, R., Ohba, H., Zhelev, Z., Nagase, T., Jose, R., Ishikawa, M., Baba, Y., Quantum dot anti-CD conjugates: Are they potential photosensitizers or potentiators of classical photosensitizing agents in photodynamic therapy of cancer? *Nano Lett.* **2004**, 4(9), 1567–1573.

33 Leatherdale, C. A., Woo, W. K., Mikulec, F. V., Bawendi, M. G., On the absorption cross section of CdSe nanocrystal quantum dots, *J. Phys. Chem. B* **2002**, 106, 7619–7622.

34 Murphy, C. J., Optical sensing with quantum dots, *Anal. Chem.* **2002**, 74, 520A–526A.

35 Alivisatos, A. P., The use of nanocrystals in biological detection, *Nat. Biotechnol.* **2004**, 22, 47–52.

36 Mattoussi, H., Kuno, M. K., Goldman, E. R., Anderson, G. P., Mauro, J. M., in *Optical Biosensors: Present and Future*, E. S. Ligler, C. A. Rowe (Eds.), Elsevier, Amsterdam, **2002**, pp. 537–569.

37 Bruchez, J. M., Moronne, M., Gin, P., Weiss, S., Alivisatos, A. P., Semiconductor nanocrystals as fluorescence biological labels, *Science* **1998**, 281, 2013–2015.

38 Chan, W. C. W., Nie, S. M., Quantum dot bioconjugates for ultrasensitive nonisotopic detection, *Science* **1998**, 281, 2016–2018.

39 Jakobs, S., Subramaniam, V., Schonle, A., Jovin, T. M., Hell, S. W., EFGP and DsRed expressing cultures of Escherichia coli imaged by confocal, two-photon and fluorescence lifetime microscopy, *FEBS Lett.* **2000**, 479, 131–135.

40 Pepperkok, R., Squire, A., Geley, S., Bastiaens, P. I. H., Simultaneous detection of multiple green fluorescent proteins in live cells by fluorescence lifetime imaging microscopy, *Curr. Biol.* **1999**, 9, 269–272.

41 Gao, X., Chan, W. C., Nie, S., Quantum dot nanocrystals for ultrasensitive biological labeling and multicolor optical encoding, *J. Biomed. Opt.* **2002**, 7(4), 532–537.

42 Cunin, F., Schmedake, T. A., Link, J. R., Li, Y. Y., Koh, J., Bhatia, S. N., Sailor, M. J., Biomolecular screening with encoded porous-silicon photonic crystals, *Nat. Mater.* **2002**, 1(1), 39–41.

43 Chan, P., Yuen, T., Ruf, F., Gonzalez-Maeso, J., Sealfon, S. C., Method for multiplex cellular detection of mRNAs using quantum dot fluorescent in situ hybridization, *Nucleic Acids Res.* **2005**, 33(18), e161.

44 Liang, R. Q., Li, W., Li, Y., Tan, C. Y., Li, J. X., Jin, Y. X., Ruan, K. C., An oligonucleotide microarray for microRNA expression analysis based on labeling RNA with quantum dot and nanogold probe, *Nucleic Acids Res.* **2005**, 33(2), e17.

45 Robelek, R., Niu, L., Schmid, E. L., Knoll, W., Multiplexed hybridization detection of quantum dot-conjugated DNA sequences using surface plasmon enhanced fluorescence microscopy and spectrometry, *Anal. Chem.* **2004**, 76(20), 6160–6165.

46 Wang, J., Liu, G., Merkoci, A., Electrochemical coding technology for simultaneous detection of multiple DNA targets, *J. Am. Chem. Soc.* **2003**, 125(11), 3214–3215.

47 Ho, Y. P., Kung, M. C., Yang, S., Wang, T. H., Multiplexed hybridization detection with multicolor colocalization of quantum dot nanoprobes, *Nano Lett.* **2005**, 5(9), 1693–1697.

48 Gao, X., Nie, S., Quantum dot-encoded mesoporous beads with high brightness and uniformity: Rapid readout using flow cytometry, *Anal. Chem.* **2004**, 76(8), 2406–2410.

49 Lacoste, T. D., Michalet, X., Pinaud, F., Chemla, D. S., Alivisatos, A. P., Weiss, S., Ultrahigh-resolution multicolor colocalization of single fluorescent probes, *Proc. Natl. Acad. Sci. U.S.A.* **2000**, 97, 9461–9466.

50 Michalet, X., Pinaud, F., Lacoste, T. D., Dahan, M., Bruchez, M. P., Alivisatos, A. P., Weiss, S., Properties of fluorescent semiconductor nanocrystals and their application to biological labeling, *Single Mol.* **2001**, 2, 261–276.

51 Powe, A. M., Fletcher, K. A., St. Luce, N. N., Lowry, M., Neal, S., McCarroll, M. E., Oldham, P. B., McGown, L. B., Warner, I. M., Molecular fluorescence, phosphorescence, and chemiluminescence spectrometry, *Anal. Chem.* **2004**, 76(16), 4614–4634.

52 Hohng, S., Ha, T., Near-complete suppression of quantum dot blinking in ambient conditions, *J. Am. Chem. Soc.* **2004**, 126, 1324–1325.

53 Larson, D., Zipfel, W. R., Williams, R. M., Clark, S. W., Bruchez, M. P., Wise, F. W., Webb, W. W., Water-soluble quantum dots for multiphoton fluorescence imaging in vivo, *Science* **2003**, 300, 1434–1436.

54 Lee, S., Thomas, K. R. J., Thayumanavan, S., Bardeen, C. J., Dependence of the two-photon absorption cross section on the conjugation of the phenylacetylene linker in dipolar donor-bridge-acceptor chromophores, *J. Phys. Chem. A* **2005**, 109(43), 9767–9774.

55 Bharali, D. J., Lucey, D. W., Jayakumar, H., Pudavar, H. E., Prasad, P. N., Folate-receptor-mediated delivery of InP quantum dots for bioimaging using confocal and two-photon microscopy, *J. Am. Chem. Soc.* **2005**, 127(32), 11 364–11 371.

56 Willard, D. M., Carillo, I. I., Jung, J., Van Orden, A., CdSe-ZnS quantum dots as resonance energy transfer donors in a model protein-protein binding assay, *Nano Lett.* **2001**, 1, 469–474.

57 Mamedova, N. N., Kotov, N. A., Rogach, A. L., Studer, J., Albumin-CdTe nanoparticle bioconjugates: Preparation, structure, and interunit energy transfer with antenna effect, *Nano Lett.* **2001**, 1, 281–286.

58 Claap, A. R., Medintz, I. L., Mauro, J. M., Fisher, B. R., Bawendi, M. G., Mattoussi, H., Fluorescence

resonance energy transfer between quantum dot donors and dye-labeled protein acceptors, *J. Am. Chem. Soc.* **2004**, 126, 301–310.

59 Claap, A. R., Medintz, I. L., Fisher, B. R., Anderson, G. P., Mattoussi, H., Can luminescent quantum dots be efficient energy acceptors with organic dye donors? *J. Am. Chem. Soc.* **2005**, 127(4), 1242–1250.

60 Ebenstein, Y., Mokari, T., Banin, U., Quantum dot-functionalized scanning probes for fluorescence-energy-transfer-based microscopy., *J. Phys. Chem. B* **2004**, 108(1), 93–99.

61 Kloepfer, J. A., Cohen, N., Nadeau, J. L., FRET between CdSe quantum dots in lipid vesicles and water- and lipid-soluble dyes, *J. Phys. Chem. B* **2004**, 108(44), 17 042–17 049.

62 Samia, A. C. S., Chen, X., Burda, C. Semiconductor quantum dots for photodynamic therapy, *J. Am. Chem. Soc.* **2003**, 125(51), 15 736–15 737.

63 Akerman, M. E., Chan, W. C. W., Laakkonen, P., Bhatia, S. N., Rouslahti, E., Nanocrystal targeting *in vivo*, *Proc. Natl. Acad. Sci. U.S.A.* **2002**, 99, 12 617–12 621.

64 Xiao, Y., Barker, P. E., Semiconductor nanocrystal probes for human metaphase chromosomes, *Nucleic Acids Res.* **2004**, 32, e28.

65 Hanaki, K., Momo, A., Oku, T., Komoto, A., Maenosono, S., Yamaguchi, Y., Yamamoto, K., Semiconductor quantum dot/albumin complex is a long-life and highly photostable endosome marker, *Biochem. Biophys. Res. Commun.* **2003**, 302, 496–501.

66 Sukhanova, A., Devy, J., Venteo, L., Kaplan, H., Artemyev, M., Oleinikov, V., Klinov, D., Pluot, M., Cohen, J. H., Nabiev, I., Biocompatible fluorescent nanocrystals for immunolabeling of membrane proteins and cells, *Anal. Biochem.* **2004**, 324, 60–67.

67 Kaul, Z., Yaguchi, T., Kaul, S. C., Hirano, T., Wadhwa, R., Taira, K., Mortalin imaging in normal and cancer cells with quantum dot immuno-conjugates, *Cell Res.* **2003**, 13, 503–507.

68 Hoshino, A., Hanaki, K.-I., Suzuki, K., Yamamoto, K., Applications of T lymphoma labeled with fluorescent quantum dots to cell tracing markers in mouse body, *Biochem. Biophys. Res. Commun.* **2004**, 314, 46–53.

69 Chen, F., Gerion, D., Fluorescent CdSe/ZnS nanocrystal-peptide conjugates for long-term, nontoxic imaging and nuclear targeting in living cells, *Nano Lett.* **2004**, 4, 1827–1832.

70 Derfus, A. M., Chan, W. C. W., Bhatia, S. N., Intracellular delivery of quantum dots for live cell labeling and organelle tracking, *Adv. Mater.* **2004**, 16, 961–966.

71 Voura, E. B., Jaiswal, J. K., Mattoussi, H., Simon, S. M., Tracking early metastatic progression with quantum dots and emission scanning microscopy, *Nat. Med.* **2004**, 10, 993–998.

72 Dwarakanath, S., Bruno, J. G., Shastry, A., Phillips, T., John, A. A., Kumar, A., Stephenson, L. D., Quantum dot-antibody and aptamer conjugates shift fluorescence upon binding bacteria, *Biochem. Biophys. Res. Commun.* **2005**, 325(3), 739–743.

73 Pinaud, F., King, D., Moore, H.-P., Weiss, S., Bioactivation and cell targeting of semiconductor CdSe/ZnS nanocrystals with phytochelatin-related peptides, *J. Am. Chem. Soc.* **2004**, 126, 6115–6123.

74 Winter, J. O., Liu, T. Y., Korgel, B. A., Schmidt, C. E., Recognition molecule directed interfacing between semiconductor quantum dots and nerve cells, *Adv. Mater.* **2001**, 13, 1673–1677.

75 Lewin, M., Carlesso, N., Tung, C. H., Tang, X. W., Cory, D., Scadden, D. T., Weissleder, R., Tat peptide-derivatized magnetic nanoparticles allow *in vivo* tracking and recovery of progenitor cells, *Nat. Biotechnol.* **2000**, 18, 410–414.

76 Tokumasu, F., Dvorak, J., Development and application of quantum dots

for immunocytochemistry of human erythrocytes, *J. Microsc.* **2003**, 211, 256–261.

77 Smith, A. M., Ruan, G., Rhyner, M. N., Nie, S., Engineering luminescent quantum dots for in vivo molecular and cellular imaging, *Ann. Biomed. Eng.* **2006**, 34, 3–14.

78 Pinaud, F., Michalet, X., Bentolila, L. A., Tsay, J. M., Doose, S., Li, J. J., Iyer, G., Weiss, S., Advances in fluorescence imaging with quantum dot bio-brobes, *Biomaterials* **2006**, 27, 1679–1687.

79 Lim, Y. T., Kim, S., Nakayama, A., Stott, N. E., Bawendi, M. G., Frangioni, J. V., Selection of quantum dot wavelengths for biomedical assays and imaging, *Mol. Imaging* **2003**, 2, 50–64.

80 Kim, S., Bawendi, M. G., Oligomeric ligands for luminescent and stable nanocrystal quantum dots, *J. Am. Chem. Soc.* **2003**, 125(48), 14 652–14 653.

81 Larson, D. R., Zipfel, W. R., Williams, R. M., Clark, S. W., Bruchez, M. P., Wise, F. W., Webb, W. W., Water-soluble quantum dots for multiphoton fluoresence imaging in vivo, *Science* **2003**, 300, 1434–1436.

82 Rosenthal, S. J., Tomlinson, I., Adkins, E. M., Schroeter, S., Adams, S., Swafford, L., McBride, J., Wang, Y., DeFelice, L. J., Blakely, R. D., Targeting cell surface receptors with ligand-conjugated nanocrystals, *J. Am. Chem. Soc.* **2002**, 124, 4586–4594.

83 Wang, D. S., He, J. B., Rosenzweig, N., Rosenzweig, Z., Superparamagnetic Fe_2O_3 beads-CdSe/ZnS quantum dots core-shell nanocomposite particles for cell separation, *Nano Lett.* **2004**, 4, 409–413.

84 Bakalova, R., Zhelev, Z., Aoki, I., Ohba, H., Imai, Y., Kanno, I., Silica-shelled single quantum dot micelles as imaging probes with dual and multimodality, *Anal. Chem.* **2006**, 78(16), 5925–5932.

85 Ornberg, R. L., Harper, T. F., Liu, H., Western blot analysis with quantum dot fluorescence technology: A sensitive and quantitative method for multiplexed proteomics, *Nat. Methods* **2005**, 2, 79–81.

86 Zhelev, Z., Bakalova, R., Ohba, H., Jose, R., Imai, Y., Baba, Y., Uncoated, broad-fluorescent, and size-homogenous CdSe quantum dots for bioanalyses, *Anal. Chem.* **2006**, 78(1), 321–330.

87 Gudmundsson, B., Bjarnadottir, H., Kristjansdottir, S., Jonsson, J. J., Quantitative assays for maedi-visna virus genetic sequences and mRNA's based on RT-PCR with real-time FRET measurements, *Virology* **2003**, 307, 135–142.

88 Yurke, B., Turberfield, A. J., Mills, A. P., Simmel, F. C., Neumann, J. L., A DNA-fuelled molecular machine made of DNA, *Nature* **2000**, 406, 605–608.

89 Dubertret, B., Calame, M., Libchaber, A. J., Single-mismatch detection using gold-quenched fluorescent oligonucleotides, *Nat. Biotechnol.* **2001**, 19, 365–370.

90 Patolsky, F., Gill, R., Weizmann, Y., Mokari, T., Banin, U., Willnerm, I., Lighting-up the dynamics of telomerization and DNA replication by CdSE-ZnS quantum dots, *J. Am. Chem. Soc.* **2003**, 125, 13 918–13 919.

91 Gryczynski, I., Malicka, J., Jiang, W., Fisher, H., Chan, W. C. W., Gryczynski, Z., Grudzinski, W., Lakowicz, J. R., Surface-plasmon-coupled emission of quantum dots, *J. Phys. Chem. B* **2005**, 109, 1088–1093.

92 Warner, J. H., Thomsen, E., Watt, A. R., Heckenberg, N. R., Rubinsztein-Dunlop, H., Time-resolved photoluminescence spectroscopy of ligand-capped PbS nanocrystals, *Nanotechnology* **2005**, 16, 175–179.

93 Lodahl, P., Van Driel, A. F., Nikolaev, I. S., Irman, A., Overgaag, K., Vanmaekelbergh, D., Vos, W. L., Controlling the dynamics of spontaneous emission from quantum dots by photonic crystals, *Nature* **2004**, 430, 654–657.

94 Woggon, U., Herz, E., Schops, O., Artemyev, M. V., Arens, Ch., Rousseau, N., Schikora, D., Lischka, K., Litvinov, D., Gerthsen, D., Hybrid epitaxial-colloidal semiconductor nanostructures, *Nano Lett.* **2005**, 5, 483–490.

95 Orellana, G., Moreno-Bondi, M. C. (Eds.), *Frontiers in Chemical Sensors*, Springer Verlag, Berlin, **2005**, Vol. 3.

96 Narayanaswamy, R., Wolfbeis, O. S. (Eds.), *Optical Sensors*, Springer Verlag, Berlin, **2004**, Vol. 1.

7
Detection of Biological Materials by Gold Nano-biosensor-based Electrochemical Method

Juan Jiang, Manju Basu, Sara Seggerson, Albert Miller, Michael Pugia, and Subhash Basu

7.1
Introduction

Escherichia coli (*E. coli*) is a fecal coliform bacteria commonly found in the intestine of healthy humans and animals, but disease-causing pathogenic forms such as *E. coli* O157:H7 are responsible for food-borne and water-borne illness because they can rapidly produce a powerful toxin, resulting in hemorrhagic colitis or hemolytic uremic syndrome, which may lead to death, especially in children. Not only *E. coli* but also other pathogenic bacteria produce toxins (such as *Cholera*, *Salmonella*, *Pseudomonas*, etc.) and they divide in human bodies quite rapidly (doubling times are 10 to 15 min). Time, sensitivity, and accuracy of analysis are the most important limitations affecting the usefulness of microbiological methods of detection for pathogenic bacteria. Most of these microbiological detection techniques involve the binding of a non-fluorescent or fluorescent dye to the bacteria and the culturing and screening in the presence of several antibiotics for 2 to 3 days. Thus, while a specific strain is being identified, a patient may die helplessly with kidney infection from *E. coli* or *Salmonella* (from food). A quicker detection method for pathogenic strains of such bacteria is urgently needed to maintain human health around the world.

Gold-based amperometric sensors are now recognized as very promising and powerful tools in bio-fluid or biomaterial investigations and their associated clinical studies. It is expected that such sensors will not only cut down the time of detection but also guide doctors to prescribe specific drug or antibiotic therapy to save a human being without contributing unduly to the growing problem of drug resistance. In addition, subtle changes in adsorbates at electrolyte/electrode interfaces affect the AC capacitance measured in two electrode cell systems by Electrochemical Impedance Spectroscopy (EIS) and could be used to sensitively detect and distinguish any strain of bacteria present on gold surface based biosensors, provided binding specificity can be provided by a surface bound specific antibody. An understanding of the electrical polarization mechanisms in bacteria leading to the AC capacitive response and the database that catalogues this determined capac-

Nanotechnologies for the Life Sciences Vol. 8
Nanomaterials for Biosensors. Edited by Challa S. S. R. Kumar

ISBN: 978-3-527-31388-4

itance behavior of selective antibody-bound bacteria are not available. Until such knowledge is developed, EIS biosensor development will continue to proceed on a case-by-case basis, limiting progress in the next few years.

Section 7.2 of this chapter gives a detailed method for the preparation of various kinds of gold nanowires arrays (GNW), while Section 7.3 describes the method for synthesis and attachment of a linker, dithiobissuccinimidylundecanoate (DSU), to these GNW, which significantly increases the binding capacity of a specific antibody. Section 7.4 details the method of bacteria detection utilizing GNW and Electrochemical Impedance Spectroscopy (EIS). Preliminary results demonstrate the potential applications of this kind of biosensor in clinical laboratories, environmental monitoring, and the food industry to achieve rapid and sensitive detections.

7.2 Template Synthesis of Gold Nano-wire Arrays for Biosensor Applications

7.2.1 General Template Synthesis

Template synthesis is one of the ways to form nanostructured materials. The nanostructured materials are formed by confining the nucleation and growth of the material inside the ordered templates. The advantage of this method is that it can be very generally applied. Various nanostructured materials, such as conductive polymer [1–5], metals [6–11], and semiconductors [12–14], can be prepared using the same type of template. Moreover, nanostructured materials with different shapes, such as nano sized wires, rods, tubules and particles, can be prepared by controlling the growth morphology of the materials inside a template cavity [15–18]. With a two-step replication, the template nanostructure can be duplicated in most materials [19]. Template synthesis has been able to produce materials with the dimensions of only several nanometers [20, 21], which are very difficult to make by lithographic methods. Template synthesis is also suitable for mass production.

Most template syntheses reported in the literature have been accomplished by the use of a nanoporous film. To date, there are mainly two types of films: "track-etch" polymeric films and porous alumina films [22, 23]. Track-etch films are made by bombarding a polymeric material to create damaged tracks and then chemically etching these tracks into pores. These porous films are commercially available as microporous or nanoporous polymeric filtration membranes, with pore diameters ranging from 10 nm to several microns [22, 23]. The drawbacks of "track-etch" membrane are randomly distributed pores and a wide distribution of pore diameters. Porous alumina films are produced electrochemically. Masuda et al. [24–26] have reported that, under certain conditions, the pores are self-ordered. All the pores have the same diameter and are uniformly distributed across the surface. The pores are vertically aligned, perpendicular to the film surface, and hexagonally arranged. In addition, the pore diameter is adjustable by controlling the conditions

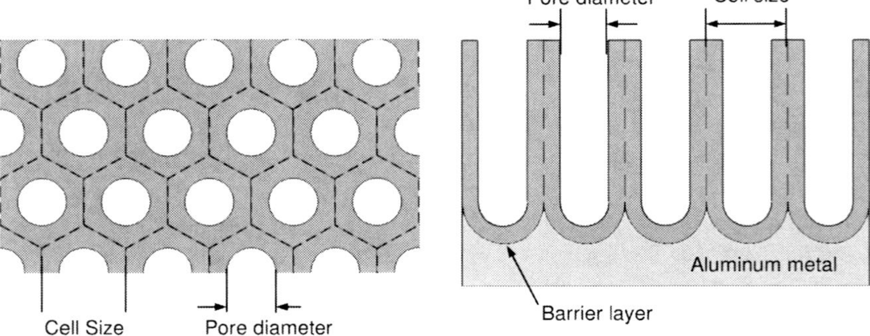

Fig. 7.1. Schematic illustration of the porous oxide film that grows on aluminum when it is anodized in acidic electrolytes.

of the electrochemical preparation process. A very high aspect ratio (pore length divided by the pore diameter) can be easily achieved, which is difficult to obtain by conventional lithographic techniques. Because of the regularity of the pore structure and ease of formation, the porous alumina films with pore diameters from 5 nm to more than 200 nm have been used as the template to prepare various nano-materials for applications in optoelectronic, electronic, and magnetic devices, in addition to catalysis.

The ideal structure of an anodized porous aluminum oxide film is a two-dimensional (2D) close-packed array of columnar hexagonal cells, each containing a central pore normal to the underlying metal surface and separated from it by a convex barrier-type film (Fig. 7.1). This porous Al_2O_3 film is made by electrochemical anodic oxidation of aluminum in a chilled acid bath. Traditionally, Al_2O_3 film is prepared to protect the metal against corrosion and to improve its abrasion and adsorption properties. The more recent and rapidly growing applications of porous anodic alumina in nano material synthesis are due to the self-assembled and well-controllable pore structure, high throughput, and low cost compared with conventional lithographic (electron beam, focused ion beam, X-ray, STM/AFM, etc.) techniques.

Anodic processes consist of the oxidation of metals and the formation of a hard oxide film, the partial dissolution of metals in the electrolyte, and the evolution of gaseous oxygen at the anode. The applied electric field sustains a continued growth of the film by causing metal ions and oxygen ions to be pulled through the growing film in opposite directions.

More than 40 years ago, Hoar and Mott [27] first proposed their field-assisted dissolution model to explain the pore initiation and growth phenomena. This electric field-assisted dissolution theory has been widely accepted since then to explain the steady-state growth of the pores [1, 28, 29]. This theory showed that pore growth is balanced by field-assisted dissolution at the bottom of the advancing pore, whose rate is determined by the local field and by the radius of curvature of the pore base. Therefore, the applied voltage is the most important factor that affects the formation and growth of the pores. Both pore diameter and pore spacing depend on the cell voltage. Other factors, such as pH and acid aggressiveness, also affect the oxide formation. The ratio of pore diameter to pore separation is found to be independent of voltage but varies with the pH of the electrolyte and at pH higher than 1.77 a nonporous oxide film forms [26]. Relatively non-aggressive electrolytes promote thicker barrier layers, larger cells and larger pores than aggressive electrolytes under constant current density conditions. The acid aggressiveness decreases in the order of sulfuric acid > oxalic acid > phosphoric acid. In sulfuric acid, pore sizes of 15–20 nm are usually formed; while in oxalic acid and phosphoric acid 40–60 nm pores and 120–180 nm pores respectively, are more commonly found.

The reason for the self-organization of the pores is still not very clear. The driving force could be the lateral component of the electrical field at the pore bottom curvature and also, could be the mechanical stress or thermal effects [26, 28, 30]. It is generally agreed that long anodization time and low anodization temperature improve the ordering of the pores. For different acids, the best ordering occurs in different potential regions. The experimental results of Masuda et al. showed that a well-ordered hexagonally arranged pore structure was accomplished with anodization in 0.5 M H_2SO_4 at 25 V, or in 0.3 M oxalic acid at 40 V, or in 0.3 M H_3PO_4 at 195 V [24, 25, 31]. However, very long-period anodizations, from 16 to 160 h, were used to achieve the best ordering.

The initial surface structure of the metal plays an important role in the pore initiation mechanism [28, 32]. Yue has shown that pores would preferentially nucleate at the bottom of depressions in electropolished aluminum surfaces or at defective sites, such as grain boundaries or cracks [33]. Therefore, improved ordering of the pore structure can be, possibly, obtained in a shorter time by introducing designed "preferred sites" to facilitate the nucleation of pores in a desired way. For example, an almost defect-free porous structure has been achieved by using a SiC stamp with a hexagonally ordered array of convexes to pre-texture the aluminum surface [24]. Two-step anodization can also achieve a better ordering and form well-ordered pores with a very short aspect ratio [19]. A long-term anodization in the first step generates a well-organized porous structure at the pore bottom. After complete removal of the anodic oxide film prepared in the first step by dissolution in the mixture of chromic acid and phosphoric acid, a highly ordered array of dimples remains on the surface of aluminum metal that becomes the pore initiation sites in the second anodization step. However, the overall preparation time is still long.

Previous work has determined that the electrochemically polished aluminum surface can exhibit a very regular pattern of aligned stripes or hexagonal arranged dots, depending on the time of the electropolishing and the voltage applied [33–39]. The patterns are attributed to preferential adsorption of organic molecules on the convex portion of the electrode due to its locally enhanced electric field [39]. The wavelength of the pattern increases monotonically with the applied voltage and also depends on the effective electrolyte polarizability. When the electrochemically polished patterned surface with the right wavelength has been used for anodization, a well-organized nano-porous oxide structure was formed in a very short time. Therefore, highly ordered nano-porous aluminum oxide template can be fabricated rapidly.

7.2.2
Template Formation

The porous alumina template was fabricated using 99.999% 0.1 mm thick aluminum foil (Alfa Aesar, Ward Hill, MA). The foil was annealed in nitrogen at 500 °C for 3 h before electropolishing at 40–42 V DC for 60 s in a LECO electropolishing tank EP-50 (LECO Corporation, St. Joseph, MI) with an electrolyte consisting of 70.0 vol.% of ethanol (CH_3CH_2OH), 13.8 vol.% distilled water, 10.0 vol.% butyl cellusolve ($CH_3(CH_2)_3OCH_2CH_2OH$), and 6.2 vol.% perchloric acid ($HClO_4$). The electropolishing temperature was 10 °C, controlled by a Julabo® FP30 refrigerated circulating bath (Julabo USA Inc., Allentown, PA). Anodization was performed in 0.3 M oxalic acid (pH 1.6) at a constant voltage of 40 V and 3 °C for 10 min. The reaction was executed in a jacketed glass reactor with vigorous agitation. The anodization process was controlled by a digital acquisition system running a custom LabVIEW© program that logged the anodization current and controlled the applied voltage at a preset level. After anodization, the foil was immersed in 5% (by weight) phosphoric acid (pH 1) with moderate stirring for 45 min to round the pores.

Field emission scanning electron microscopy (SEM, Hitachi, Model S4500) was used to observe the anodized alumina membranes and the nano-structured gold arrays. Atomic force microscope images were taken by a tapping mode atomic force microscope (AFM, Model Dimension 3100, Veeco Instruments, Woodbury, NY). X-Ray diffraction patterns were obtained using Cu $K\alpha$ radiation (wavelength = 1.5405 Å) on a Scintag X-ray diffractometer (Scintag Inc., Cupertino, CA).

Figure 7.2 shows a typical anodization curve of aluminum in 0.3 M oxalic acid at 40 V, 3 °C. The curve can be divided into three regions. In region I, the current density surges to a very high value in a very short time, representing the barrier oxide layer formation. Once the barrier oxide layer forms, the current starts to drop because of the lower mass transport through the oxide layer. After the current reaches a minimum, it increases again in region II, where pore nucleation occurs. Region III is the pore growth and self-adjustment region in which the dissolution of the aluminum metal and the oxide formation have the same rate.

The ordering near the pore bottom (pore advancing front) was observed to be better than that near the pore opening because of a longer self-adjustment time. Figure 7.3(a) and (b) are SEM images viewed from the top and bottom of a porous alumina film anodized in 0.3 M oxalic acid at 3 °C, 40 V for 16 h. At the pore bottom, the cells have an almost perfect hexagonal shape and uniform size. The average distance from the center of the cell to the center of the neighboring cell is 100 nm, which is also shown in the magnified cross-section image (Fig. 7.3(d)). Al-

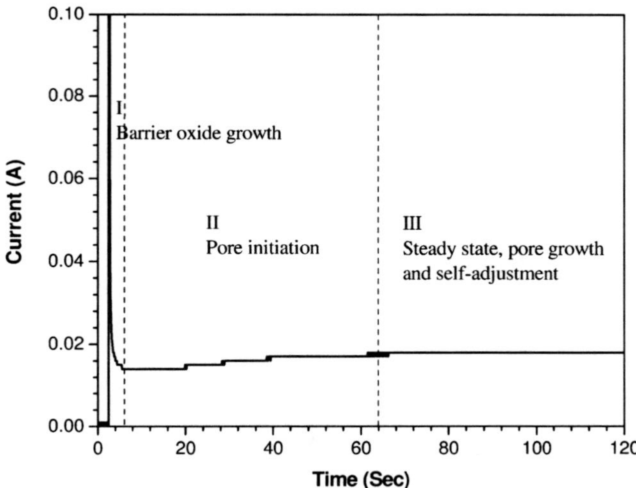

Fig. 7.2. Dependence of anodization current on time for aluminum anodized in 0.3 M oxalic acid at 40 V and 3 °C.

though there are still a couple of defect sites, the ordering at the pore advancing front is much better than that at the pore opening.

Figure 7.4 demonstrates the impact of the initial surface structure of the aluminum metal on the pore ordering of the template. One of the electropolishing patterns containing regular stripes of wavelength of 102 nm was observed when the aluminum foil was electropolished at 41 V, 10 °C for 60 s (Fig. 7.4a). Actually, at this voltage the electropolishing pattern is at the transition from regular stripes to regular hexagons [39]. The onset of the breakdown of the stripes to dots was observed. When this pre-patterned sample was anodized, the prior electropolishing pattern impacted the initial pore nucleation and hence the final pore ordering. Figure 7.4(b) shows one sample anodized in 0.3 M oxalic acid at 40 V, 3 °C for 5 min. The pores are aligned along parallel lines, and the average distance from line to line is about 100 nm, indicating that the pores are preferentially initiated along the valleys of the stripe pattern obtained from electropolishing. For comparison, Fig. 7.4(c) is the porous structure obtained when the aluminum was not electropolished.

The templates discussed here were produced without electropolishing since array order was not important. The aluminum foil was anodized for 10 min in 0.3 M oxalic acid at 3 °C. The pores were then widened and rounded in 5 wt.% H_3PO_4 for 45 min. Figure 7.4(c) shows a typical SEM image of an anodized oxide film. The average pore diameter is 65 nm with a standard deviation of 4.3 nm and the separation is 95 nm with a standard deviation of 6.2 nm. The porosity is 43% and the pore density is 1.4×10^{10} cm^{-2}. To determine the anodization rate, the thicknesses of the oxide layers were measured by SEM after the aluminum metal was removed in saturated $HgCl_2$ solution. The anodized aluminum oxide film was

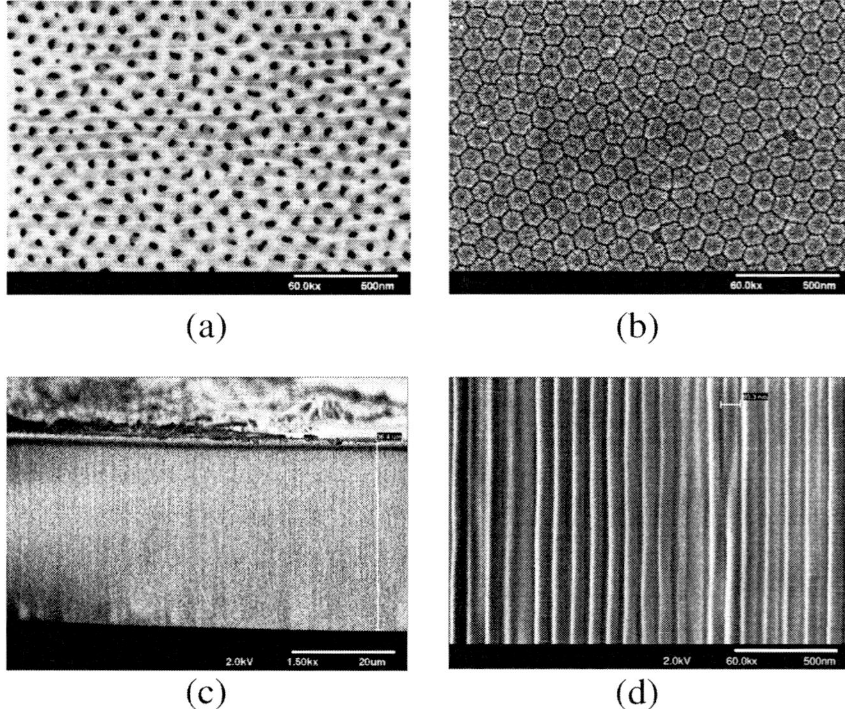

Fig. 7.3. SEM images of an alumina oxide film, anodized in 0.3 M oxalic acid at 40 V, 3 °C for 16 h. (a) View from the top of the oxide film; (b) from the bottom side of the oxide film; (c) a cross-section image that shows that the oxide layer is 37 μm thick; and (d) a higher magnification image of the cross-section, showing the parallel columnar pores.

about 2.5 μm thick for 60-min anodization, 5.3 μm thick for 120-min anodization and 37 μm for 16-h anodization. The calculated average growth rate is about 42 nm min^{-1}. Therefore, the estimated pore depth for 10-min anodization is around 420 nm.

7.2.3 Fabrication of Gold Nano-wire Arrays (GNW)

Gold nanowire arrays were formed by a two-step deposition procedure, which consisted of electrochemical deposition and electroless plating. Electrodeposition was carried out in a CEC/TH electrochemical cell (Radiometer Analytical, Loveland, CO) with a platinum disc counter electrode. The electrolyte was $HAuCl_4 \cdot 3H_2O$ (1 g L^{-1}; ACS reagent grade, ICN Biomedicals Inc., Aurora, OH) + H_2SO_4 (7 g L^{-1}). An AC voltage of frequency 60 Hz and RMS amplitude of 15 V was applied at room temperature for 1 min. The electrochemical deposition was used to drive gold ions down to the bottom of the pores and form gold metal nuclei. Electroless

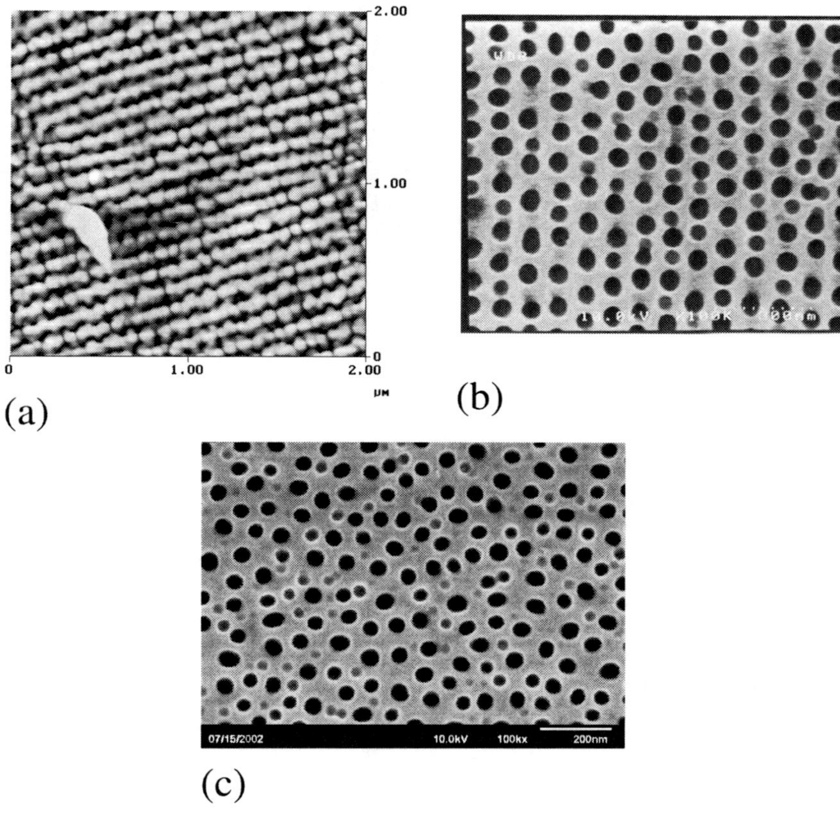

Fig. 7.4. (a) AFM image of an aluminum surface electropolished for 60 s at 41 V, 10 °C. A regular striped structure developed. (b, c) Porous alumina oxide formed at 40 V, 3 °C in 0.3 M oxalic acid for 5 min: (b) after electropolishing and (c) without electropolishing.

deposition was used to grow gold nanowires. Electroless deposition used a commercial Neorum TWB gold plating solution (Uyemura International Corporation, Ontario, CA). For the best plating speed and quality, the deposition temperature was controlled in the range of 68–72 °C. No stirring of the solution was supplied.

Two types of gold nanowire arrays were prepared by this template synthesis method. The experimental procedure is described schematically in Fig. 7.5. The first type of structure was formed by filling the pores, so that gold was deposited just to the top of the porous alumina template surface (Fig. 7.5(a)). The gold nanowires were still embedded in the oxide template and separated from the Al metal by a 3-nm thick oxide barrier, with the tips of the wires sticking out to interact with biomolecules for later biosensor applications. The second type of surface was made by a 2.5–3 h electroless plating of gold. After the pores of the template were com-

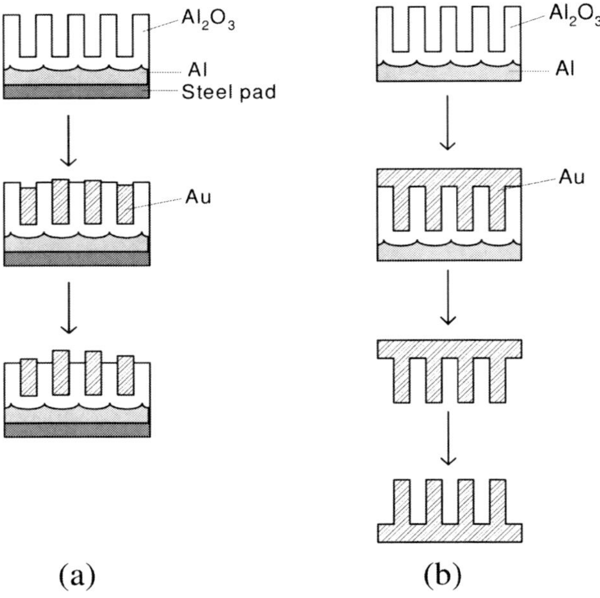

Fig. 7.5. Schematic view of the procedure for preparing gold nanowire (GNW) arrays, (a) embedded in the aluminum oxide template (GNW-Al_2O_3) and (b) free-standing (GNW).

pletely filled with gold, gold was plated onto the top surface of the oxide template. Before use, both the oxide template and the underlying unanodized aluminum metal were removed by immersing the sample in 1.5 M NaOH at 40 °C. After the dissolution, the overplated thin (~5 µm) gold film was left with gold nanowires protruding from the surface like the bristles of a brush (Fig. 7.5b).

For the application of biosensors, the substrate materials should be compatible with and have a good affinity for biological molecules and should also be stable in air or aqueous solution for reliable measurements. Gold is a good candidate for this purpose. Gold has good affinity to biomolecules – functional groups such as thiol (–SH) or amine ($-NH_2$) can chemically adsorb onto the gold surface. The bonds are strong enough to survive measurements in flowing liquids and even in a vacuum.

There are two common ways to electroplate metals inside the pores. The first is to electrodeposit metal from a metal ion solution by AC current. Alternating current was needed to deposit metal inside the pores because of the rectifying nature of the Al metal/oxide junction [9, 40, 41]. Another approach is to remove the aluminum metal and etch through the barrier layer of the oxide to get a through-hole oxide membrane. Then a thin layer of metal is evaporated onto one side of the oxide as the cathode, so electrodeposition by DC current can be carried out. Electroless plating has a distinct advantage over electroplating in its capability to coat a surface that is not electronically conductive, such as aluminum oxide. Martin et al.

also discovered that electroless plating allowed for more uniform gold deposition than electrochemical plating [18]. Therefore, a commercial electroless gold plating bath was chosen to grow the gold wires inside the aluminum oxide template. This plating solution is a non-cyanide solution and works at a neutral pH. Most of the conventional electroless gold plating baths are strongly alkaline [42]. However, alumina is only stable in the pH range between 4 and 9. When the pH is above 9, Al_2O_3 can be dissolved and form aluminate ion (AlO_2^-). Thereby the porous structure can be damaged.

The basic components of an electroless plating solution are a gold salt (Au^{z+}) and a reducing agent ($Red_{solution}$). This reducing agent will reduce gold ion to gold metal ($Au_{lattice}$) and form an oxidized species in the solution ($Ox_{solution}$) when a catalytic surface is present. The overall reaction is:

$$Au^{z+}_{solution} + Red_{solution} \xrightarrow{\text{catalytic surface}} Au_{lattice} + Ox_{solution}$$

Without a catalyst, this reaction is extremely slow and, therefore, prevents gold formation in the solution. A non-catalytic surface to be plated must be activated by generating catalytic metal nuclei on the surface to obtain a practical plating rate. The catalyst metal may be any of Pd, Pt, Au, Ag, or Cu. In this work, gold seeds were put down as the catalytic sites. The deposition of gold seeds was realized by AC electrodeposition. The locally enhanced electric field will drive the metal ions down to the pore bottom and form nuclei there [43]. Gold wires were then grown on these catalytic sites by electroless plating. The length of the wires can be simply controlled by the plating time.

Figure 7.6 is the top view SEM image of an alumina membrane with gold inside the pores. The time used for electrodeposition was 1 min and for electroless deposition it was 5 min. Since the electrodeposition solution was acidic (pH 2), containing H_2SO_4, prolonged immersion in this solution with an AC voltage applied may damage the template; consequently, the electrochemical deposition was performed

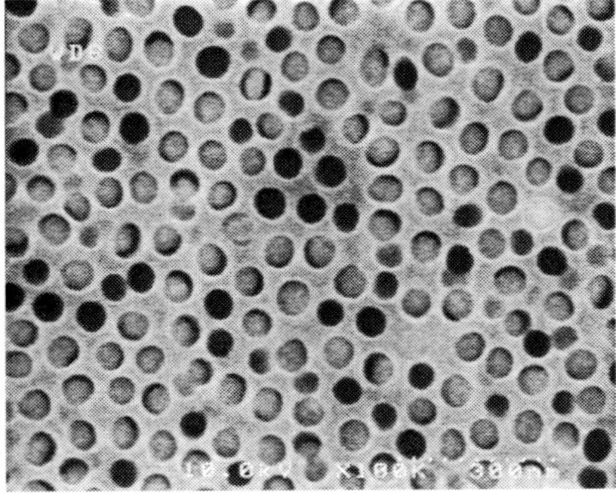

Fig. 7.6. SEM image of an aluminum oxide template with gold deposited inside the pores.

for no longer than 1 min. Notably, not all the pores are filled with gold. This is probably due to the low frequency (60 Hz) of the AC voltage applied. The filling percentage increases as the frequency increased from 100 to 750 Hz for electrodeposition of Ni and Bi inside the pores of alumina films [44]. This is because more nuclei are formed at higher frequencies, promoting the deposition better. This is similar to pulse plating, which has more "throwing" power to facilitate the filling of the pores. However, a lower frequency may be preferred if a uniform growth of the wires is required [45].

Because only the gold nuclei at the pore bottoms produced by electrodeposition were catalytic sites for the electroless plating, gold was only plated inside the pores and not on the top of the oxide layer. After the pores were completely filled with gold, gold began to plate on top of the oxide film. It was found experimentally that the time needed for electroless plating to fill the pore (the average length is ~420 nm) was about 5 min, when the temperature of the electroless plating bath was controlled between 68 and 72 °C. Owing to the special optical properties of nano-sized metal-oxide composite, the color of the gold-alumina composite changes depending on the aspect ratio of the gold wire inside the oxide [46–51]. This phenomenon was also observed in this research in that the gold-alumina changed from purple to dark blue, then green and yellow during the filling of the pores (65 nm in diameter, 420 nm long) of the oxide layer with gold.

Figure 7.7(a) and (b) shows top-view SEM images of the free-standing gold nanowires at lower and higher magnifications respectively, after the dissolution of both the alumina and the residual aluminum metal. The electroless plating was carried out for 2.5 to 3 h, resulting in an overplated gold film of about 5 μm thick. The wires stood on this overplated gold film. According to the SEM images, the average diameter of the wire is 62 nm with a standard deviation of 6.4 nm, and the distance from the center of the wire to the center of the neighboring wire is 93 nm, with a standard deviation of 7.5 nm. This is consistent with the pore size and the interpore distance of the template. Notably, in Fig. 7.7(a), some gold nanowires are

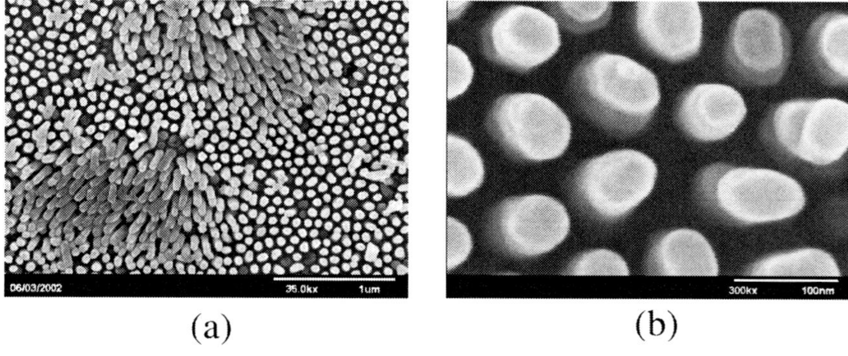

Fig. 7.7. SEM images of the top view of gold nanowires. Magnifications: (a) 60k×. (b) 300k×.

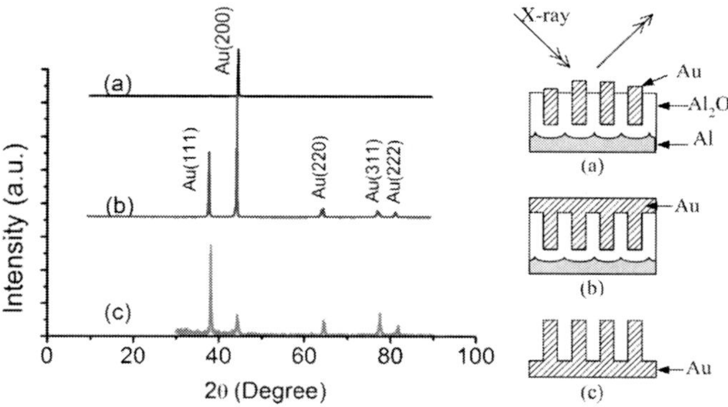

Fig. 7.8. X-ray diffraction patterns of (a) gold nanowires embedded in a porous alumina template; (b) an electrolessly plated gold film (~5 μm thick), and (c) free-standing gold nanowire arrays on top of the electroeless plated gold film. The average length of the wire is about 420 nm.

lying down in two regions. They were deformed intentionally with tweezers in an attempt to measure the length of the wires.

Figure 7.8(a) shows an X-ray diffraction pattern of the gold nanowire arrays inside the alumina template. There is only one peak, at about $2\theta = 44°$, which represents the Au(200) crystal plane. No Al_2O_3 peaks appeared at this scale, indicating the oxide was probably amorphous. The X-ray diffraction pattern of the electroless-plated gold film is shown as pattern (b) in Fig. 7.8. The peaks corresponding to Au(111), Au(200), Au(220), Au(311), and Au(222) crystal planes appeared, meaning that this electroless-plated gold film is polycrystalline with a face-centered cubic crystal structure. When the sample with free-standing gold nanowire arrays on top of the electrolessly plated gold film was scanned, again all the gold peaks appeared (pattern (c) in Fig. 7.8). This X-ray diffraction pattern is not only the diffraction pattern of the gold wires but may also have a contribution from the underlying electrolessly overplated gold film.

In this work, nanogold wire arrays were formed based on a template synthesis method. The template was a nano-porous aluminum oxide membrane produced by electrochemical anodization in 0.3 M oxalic acid. Gold nanowires were grown by filling the pores of the template by electrochemical plating and electroless plating. The gold nanowires could either be embedded in the oxide matrix or be free-standing on a thin gold film. The average diameter of the wire was 62 nm, which is consistent with the size of the pores. The wire could be as long as the depth of the pore or shorter, depending on the deposition time.

Gold has selective affinity to some functional groups, such as amine ($-NH_2$) and thiol ($-SH$). These functional groups exist extensively in bio-entities, such as enzymes and antibodies, which are common bio-sensing elements for the detection of a particular bio-reaction or bio-interaction. Both kinds of nano-structured gold surfaces fabricated in this work, including the free-standing gold nanowires and the gold nanowire arrays embedded inside the porous alumina template, could be the substrates for sensor applications. Electrochemical and optical transducers can be applied. The free-standing gold nanowire arrays have a large surface area, which improves the binding capacity of bio-sensing molecules or electrochemically active species; and enhances the reaction current and double-layer charging capacitance that can be detected with electrochemical methods, such as chronoamperometry or impedance spectroscopy. Complete sealing of the cylindrical surface of the gold nanowires inside the aluminum oxide template, with only the end disk-shaped electrode exposed to the electrolyte, enables detection by electrochemical methods, such as cyclic voltammetry, because of the reduced charging current and the enhanced signal-to-noise ratio [18, 52]. The special optical properties of this nanostructured metal-oxide composite may also find use in optical sensors [49, 51].

7.3 Synthesis of a Linker and its Attachment to Gold Posts of GNW followed by Binding to Specific Antibodies

Figure 7.9 gives the synthesis of DSU [53]. In the first reaction, 11-bromoundecanoic acid (**1**) was added to $Na_2S_2O_3$ in 50% aqueous 1,4-dioxane and

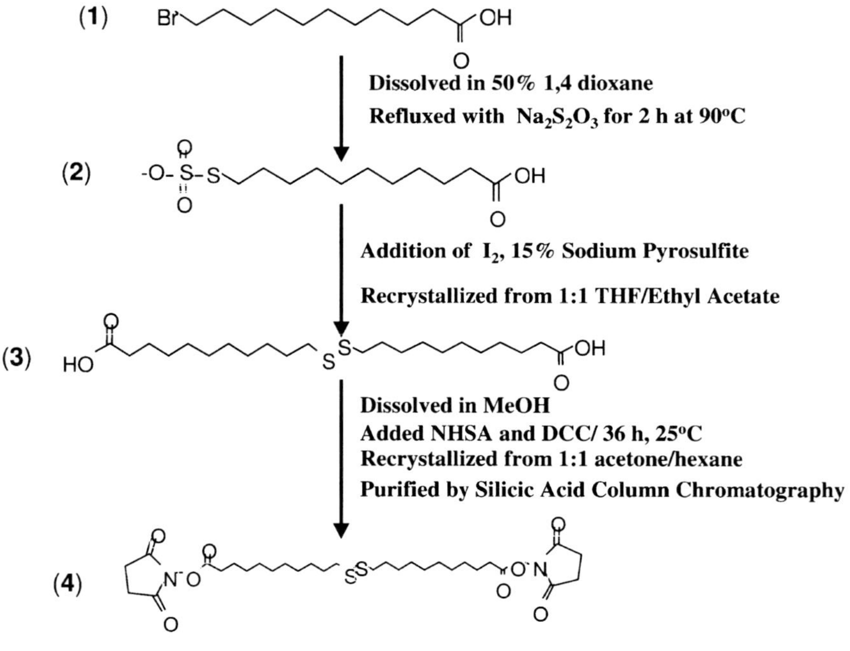

(**1**)
Dissolved in 50% 1,4 dioxane
Refluxed with $Na_2S_2O_3$ for 2 h at 90°C
(**2**)
Addition of I_2, 15% Sodium Pyrosulfite
Recrystallized from 1:1 THF/Ethyl Acetate
(**3**)
Dissolved in MeOH
Added NHSA and DCC/ 36 h, 25°C
Recrystallized from 1:1 acetone/hexane
Purified by Silicic Acid Column Chromatography
(**4**)

Fig. 7.9. Synthesis of the dithio-bissuccinimidylundecanoate (DSU) linker.

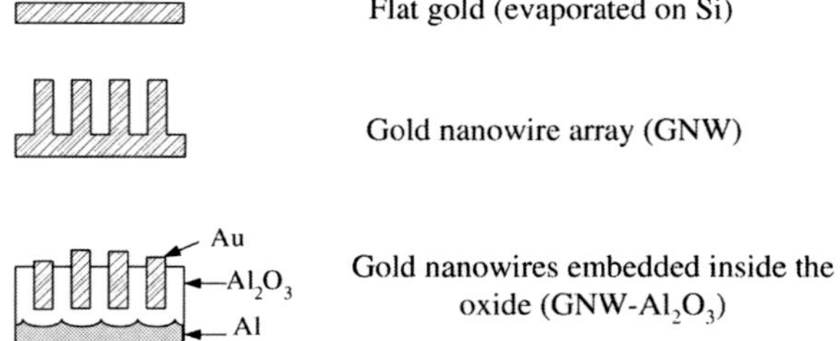

Fig. 7.10. Structures of the surfaces investigated.

refluxed for 2 h at 90 °C to yield the Bunte Salt (**2**). Oxidation to the disulfide was carried out by adding I_2. Surplus I_2 was retitrated with 15% aqueous sodium pyrosulfite. The 1,4-dioxane solvent was removed by rotary evaporation and the suspension was filtered to yield dithio-bis(undecanoic acid) (**3**). Purification was achieved by recrystallization from ethyl acetate/THF(1:1). *N*-Hydroxysuccinimide (NHSA) was added to a solution of (**3**) in MeOH, followed by the addition of dicyclohexylcarbodiimide (DCC) at 0 °C. The reaction mixture was cooled to 25 °C and stirred for 36 h. MeOH was removed by rotary evaporation, and the product was resuspended in THF and filtered off. The crude final product (**4**) was purified by recrystallization from 1:1 acetone/hexane, and by silicic acid column chromatography to yield DSU (**4**). All intermediates were characterized by IR spectroscopy.

Initial studies were carried out using various amounts of $(\text{Goat PAb})_{E.\ coli}$ to determine the degrees of direct binding to the gold plate. Competition studies were carried out among three kinds of testing surfaces (as shown in Fig. 7.10): (I) flat gold disc (FGD), (II) free-standing gold nanowire array (GNW), and (III) gold nanowire array surrounded by Al_2O_3 (GNW-Al_2O_3), to determine which had the higher affinity for antibody binding. The flat gold surface (structure I) was prepared by electron beam evaporation of gold on a Si(100) n-type wafer under a vacuum of about 1×10^{-6} Torr. The deposition speed was around 10 Å s^{-1}. The gold layer was 200 nm thick.

The DSU linker was applied to the surface to covalently immobilize the antibodies. Covalent immobilization of $(\text{Goat Pab})_{E.\ coli}$ was accomplished by immersing the gold discs in 1500 μL of a 1 mm DSU in MeOH solution for 30 min at room temperature [54]. After rinsing with MeOH (7 mL), the *N*-hydroxysuccinimidyl (NHS) terminated monolayers were dried under a stream of nitrogen gas and immediately used for immobilization of *E. coli* antibodies. Figure 7.11 depicts the self-assembly of DSU to form self-assembled monolayers (SAM) on the gold surface. Covalent immobilization takes place when a lysine residue of the antibody comes into contact with the DSU SAM to nucleophilically attack the carbonyl carbon, thereby displacing the *N*-hydroxysuccinimide to form an amide linkage (Fig. 7.12). Covalent immobilization of antibodies via SAM (Fig. 7.12) is desirable because of the tendency to increase the binding capacity of the gold surface for the antibody. This is predicted because the linker has a long carbon chain, protruding directly up from the plate, that can freely rotate in space. This creates a more steri-

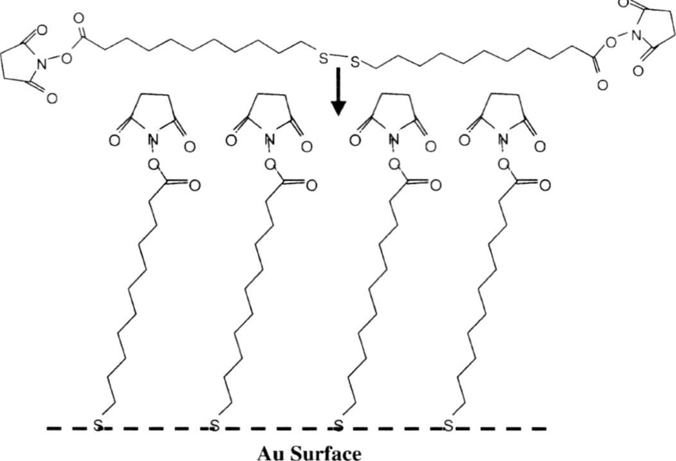

Fig. 7.11. Formation of self-assembled monolayers with the dithio-bissuccinimidylundecanoate (DSU) linker.

cally favorable condition for the antibodies to bind to monolayers because there is less chance of antibody crowding, whereas when antibodies are directly bound to the gold nanowire surfaces without the SAM there may be crowding on the gold surface that decreases the binding capacity.

Studies of the influence of the linker on antibody binding were also carried out among the same set of three surface types.

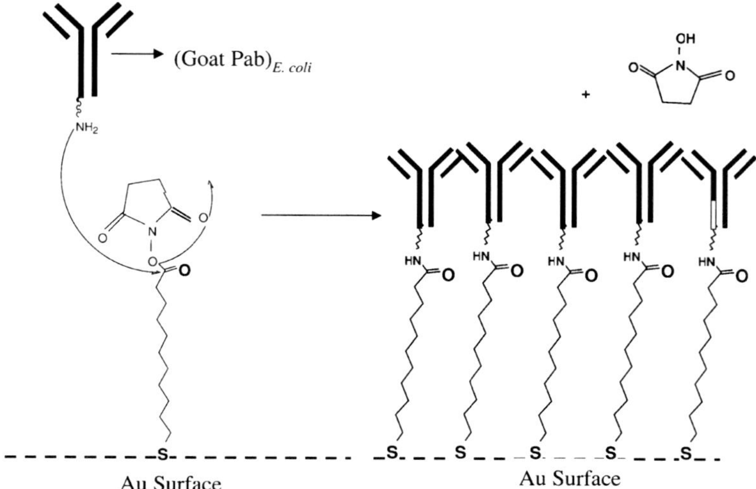

Fig. 7.12. Covalent immobilization of antibodies via a DSU self-assembled monolayer.

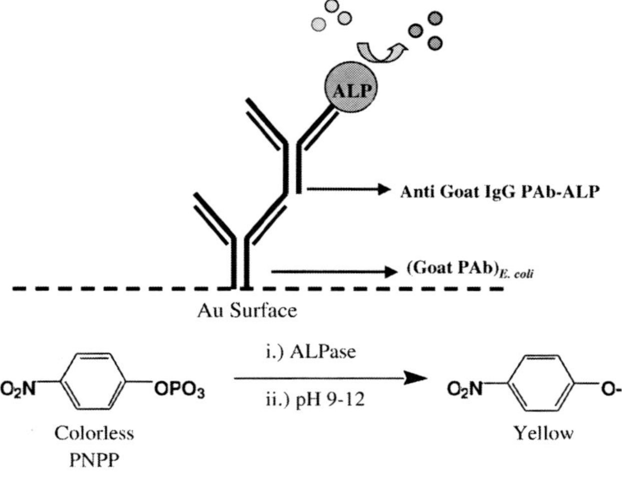

Fig. 7.13. Schematic representation of antibody binding detection by the ELISA sandwich assay.

The binding of anti-*E. coli* antibody was determined optically by enzyme-linked immunosorbent assay (ELISA). Alkaline-phosphatase-conjugated second antibody (rabbit anti-goat IgG) was added to the gold discs after washing off unbound first antibody. The anti-*E. coli* antibody binding to the gold substrates was then quantitated optically by measuring the hydrolysis of p-NO_2-phenylphosphates (Fig. 7.13). Tables 7.1–7.3 give the ELISA results.

From the data it was determined that, without linker, the gold surface with Al_2O_3 (GNW-Al_2O_3) bound the antibodies more efficiently than the gold nanowires (GNW), although gold nanowires have a huge surface area (e.g., GNW-1 vs. GNW-Al_2O_3-1). It was inferred that the antibodies are not able to fit completely inside the crevices or between the wires, thereby decreasing the binding capacity. For the surface with Al_2O_3, it was hypothesized that the antibodies may be binding to the Al_2O_3 as well as the Au.

Tab. 7.1. Relative affinity of (Goat PAb)$_{E.\ coli}$ to flat gold discs (FGD) with and without DSU linker.

	1 mM DSU (μL)	1st antibody (μg)	2nd antibody (μg)	O.D.$_{405\ nm}$
FGD-1	–	5	6	0.144
FGD-2	–	10	12	0.337
FGD-3	1500	5	6	0.224
FGD-4	1500	10	12	0.510
FGD-5	–	–	6	blank

Tab. 7.2. Relative affinity of (Goat PAb)$_{E. coli}$ to a gold nanowire array (GNW) with and without DSU linker.

	1 mM DSU (μL)	1st antibody (μg)	2nd antibody (μg)	O.D.$_{405\ nm}$
GNW-1	–	5	6	0.127
GNW-2	–	10	12	0.204
GNW-3	1500	5	6	0.648
GNW-4	1500	10	12	0.652
GNW-5	–	–	6	blank

Tab. 7.3. Relative affinity of (Goat PAb)$_{E. coli}$ to gold nanowire array embedded in Al_2O_3 (GNW-Al_2O_3) with and without DSU linker.

	DSU 1 mM (μL)	1st antibody (μg)	2nd antibody (μg)	O.D.$_{405\ nm}$
GNW-Al_2O_3-1	–	5	6	0.192
GNW-Al_2O_3-2	–	10	12	0.354
GNW-Al_2O_3-3	1500	5	6	0.257
GNW-Al_2O_3-4	1500	10	12	0.590
GNW-Al_2O_3-5	–	–	6	blank

It was determined that the linker greatly enhances antibody binding for the free-standing gold nanowire surfaces (GNW) (e.g., GNW-3 vs. GNW-1). The presence of the linker also enhances antibody binding for the GNW-Al_2O_3 surfaces (e.g., GNW-Al_2O_3-3 vs. GNW-Al_2O_3-1). However, the increase in antibody binding is not as great as it was for the gold nanowire surfaces (GNW-3 vs. GNW-Al_2O_3-3). For flat gold surfaces, DSU linker also enhances the antibody binding, but the effect is not as great as GNW and GNW-Al_2O_3 surfaces. This may be because the flat surface does not have the crevice characteristic and, therefore, exhibits a lower binding capacity for antibodies than other kinds of surfaces.

7.4 Development of Electrochemical Nano-biosensor for Bacteria Detection

7.4.1 General Detections for Biosensors

Biosensors have attracted considerable attention and have found extensive applications. However, although there is an extensive literature on various combinations of biological recognition elements and transducers in biosensor configurations, the conventional biosensors generally lack the combination of high-speed detection, high specific sensitivity and the ability to be integrated into a miniaturized sensor [55]. Therefore, in recent years, intensive research has been undertaken to develop the technology of portable, selective and sensitive biosensors capable of immediate results [56, 57]. The development of biosensors in this work was based on electrochemical impedance measurements on nanostructured substrates, with the aim of increasing the sensitivity, and decreasing the response time and the size of the device.

Among various transducers, electrochemical transduction strategies have distinct advantages in that they offer a much higher sensitivity and a simple, fast detection procedure. Owing to the capability of the measurement of nano- or picoampere current by conventional instruments, the pretreatment steps, such as separation and pre-enrichment, are no longer necessary. In addition, the equipment required for electrochemical analyses are simple and inexpensive as compared with most other analytical techniques. Furthermore, electrochemical techniques are more suitable for miniaturization, circuit integration and continuous electrical controls. With the development of microelectronics, the integrated circuit industry, nanotechnology and microfluidics, electrochemical transducers could be one of the most promising candidates for future sensor development.

Electrochemical impedance spectroscopy (EIS) was chosen as the main detection method because it is a very sensitive method that facilitates the direct measurement of the antigen–antibody interaction. This method does not require the presence of tags, such as redox probes, fluorophores, dyes or radioactive species, for the detection. This greatly simplifies the detection procedures. Moreover, EIS measurements require only a very small voltage (which is usually less than 20 mV) to be applied to the system. This small voltage applied is beneficial not only because it allows a complicated nonlinear problem to be simplified into a pseudo-linear problem during the analysis, but also because it is less destructive to the biological properties of the biomolecules, such as the activities of enzymes and antibodies, minimizing the impact on the sensitivity, stability and reliability of biosensors employing EIS.

Figure 7.14 depicts a model of the nano-biosensor. The nano-structured substrate, with biological recognition components immobilized, was the working electrode. The nano-structured substrates were gold nanowire arrays, which were fabricated as described in Section 7.2. To detect a specific type of bacteria, the biological recognition component was a bacteria-specific antibody. To enhance the binding strength and capacity, the specific antibody was covalently bound onto the gold surface through an organic "linker" molecule (Section 7.3). *Escherichia coli* (*E. coli*) was the target analyte for this sensor prototype. The reaction scheme at the electrode surface includes: (a) assembling the linker onto the gold substrates; (b) covalent binding of the anti *E. coli* antibody onto the linker; and (c) tethering of the target *E. coli* antigen. The voltage and current signals were recorded when the analyte interacted with the biological recognition component. This study was carried out using both the nano-structured gold substrates and the conventional

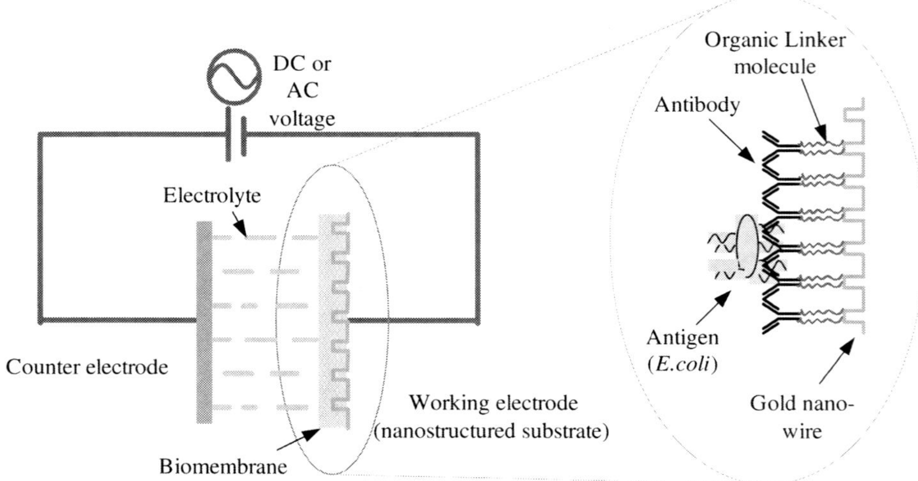

Fig. 7.14. Biosensor model for the detection of *E. coli* bacteria.

e-beam evaporated planar gold substrate. The effect of the nano-scale structure of the substrate on the detection sensitivity was investigated. Optimization of the dimension of the nano-structured gold substrates is expected in the future.

7.4.2
Experimental Conditions

E. coli antigen was cultured in the Department of Chemical and Biomolecular Engineering at the University of Notre Dame from the strain – ATCC #11775 (American Type Culture Collection, Manassas, VA). The bacteria were suspended in a 0.01 M phosphate saline buffer. The number of bacteria was counted during culturing after serial dilution using an optical microscope.

The antibody used was polyclonal goat anti *E. coli* antibody (catalog # 1091) from ViroStat (Portland, Maine). The concentration of the antibody was 0.4 $\mu g\ \mu L^{-1}$. Dithio-bissuccinimidylundecanoate (DSU) was synthesized as described in Section 7.3. It was used to covalently link the antibody onto the gold substrate. Thus, it is also called the "linker". The final concentration of DSU was 1 mM, dissolved in methanol.

All electrochemical measurements were performed using a Gamry FAS1 potentiostat (Gamry Instruments, Warminster, PA). Electrochemical impedance spectroscopy (EIS) was performed in the frequency range 10^{-2}–10^{5} Hz at an amplitude of 10 mV (RMS) around the open circuit potential. A three-electrode electrochemical cell was used. The reference electrode was a saturated calomel electrode (SCE) and the counter electrode was a platinum disk. The exposed geometric surface area of the working electrode was 0.2 cm^2, defined by a silicone rubber washer.

All measurements were performed in 0.01 M phosphate buffer saline (PBS buffer) solution (Sigma, St. Louis, MO), which has a pH of 7.4 at 25 °C. All dilutions were made using Millipore [Millipore, Bedford, MA (18 MΩ)] water.

Three kinds of surfaces with the same geometric area have been studied for comparison. They are (I) vacuum evaporated gold on Si wafer (FGD); (II) free-standing gold nanowire array (GNW), and (III) gold nanowire array distributed inside the pores of the anodized alumina template (GNW with Al_2O_3). Figure 7.10 depicts the structures of the substrates.

The general biomolecule immobilization steps were as follows. First 1 mM DSU (20 μL) was added to the substrate. The sample was rested at room temperature for 30 min, followed by washing in methanol. After the sample was dried under an air stream, antibody (10 μL, 0.4 μg μL^{-1}) was added and the sample was rested at room temperature for 2 h to let the antibody interact with the DSU linker. The sample was kept wet by sealing in a glass bottle. If needed, another 10 μL of PBS buffer was added to the sample after 1 h. The sample was then rinsed with PBS buffer to remove the loosely bound antibodies. *E. coli* bacteria were added in units from 10 to 25 μL, depending on the final surface concentration of bacteria desired and the *E. coli* concentration of the stock solution used. After adding the bacteria, the sample was rested for another 1.5 to 2 h at room temperature to let the antigens attach to the antibodies. Finally, the sample was rinsed with PBS buffer before the electrochemical measurements.

7.4.3
Electrochemical Impedance (EIS) Detection of *E. coli*

The detection of the *E. coli* antigen was investigated by sequentially adding a certain number of *E. coli* cell onto the antibody/linker coated substrates.

The EIS data can be analyzed by an equivalent circuit model [58, 59]. However, for this *E. coli* sensor device, the equivalent circuit model analysis could be very complicated because:

- There are three layers of molecules that have been immobilized on the surface: linker, antibody and antigen. Each layer may not be completely loaded. Also, in every layer, either the organic linker or the bioentities may be immobilized at different orientations, which also results a nonuniform biomembrane.
- The substrates studied have nanostructured surfaces. These nanostructures can cause an inhomogeneous current distribution along the surface and disturb the double-layer structure.
- There is an electrostatic force between the biomolecules, which changes the polarizability of the system and influences the impedance measurements.

Therefore, this biosensor study used a simplified circuit model that contained only an overall resistance and an overall capacitance connected serially.

It was assumed that the binding of the biomolecules causes the capacitance to change and the measured imaginary part of the impedance comes mainly from the capacitive behavior of the biomembrane. Therefore, the number of *E. coli* on the surface can be approximately determined from the system capacitance value. This overall capacitance can be calculated by the Eq. (1):

$$C = 1/[2\pi f|Z_{\text{imag}}|] \quad (1)$$

where $|Z_{\text{imag}}|$ is the magnitude of the imaginary part of the impedance, in ohms, f is frequency in Hz, and C is capacitance in farads. Equation (1) can also be rewritten as

$$|Z_{\text{imag}}| = \frac{1}{C}\omega^{-1} \quad (2)$$

where $\omega^{-1} = 1/2\pi f$, in rad^{-1} s. Therefore the capacitance C can be obtained as the inverse slope of a plot of the imaginary impedance Z_{imag} against the reciprocal frequency for the low frequency data.

7.4.3.1 EIS on Flat Gold Surfaces

Figures 6.15 and 6.16 are representative EIS results on an enzyme-coated flat gold electrode with different numbers of *E. coli* bacteria attached. Figure 6.15 is the Nyquist plot. Figure 7.16 is the same result plotted as the imaginary part vs. the frequency.

The capacitances can be calculated by the reciprocal slops of $|Z_{\text{imag}}|$ vs. ω^{-1} (Fig. 7.17a). If the capacitance of the electrode with linker and antibody, 4.3 μF, is

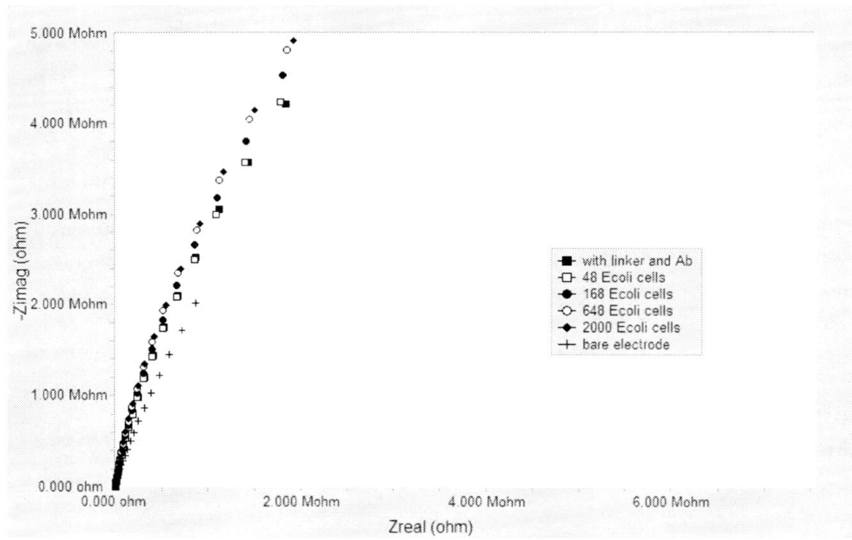

Fig. 7.15. Electrochemical impedance spectroscopy (EIS) on a flat gold surface with different amounts of *E. coli* bacteria attached (Nyquist plot).

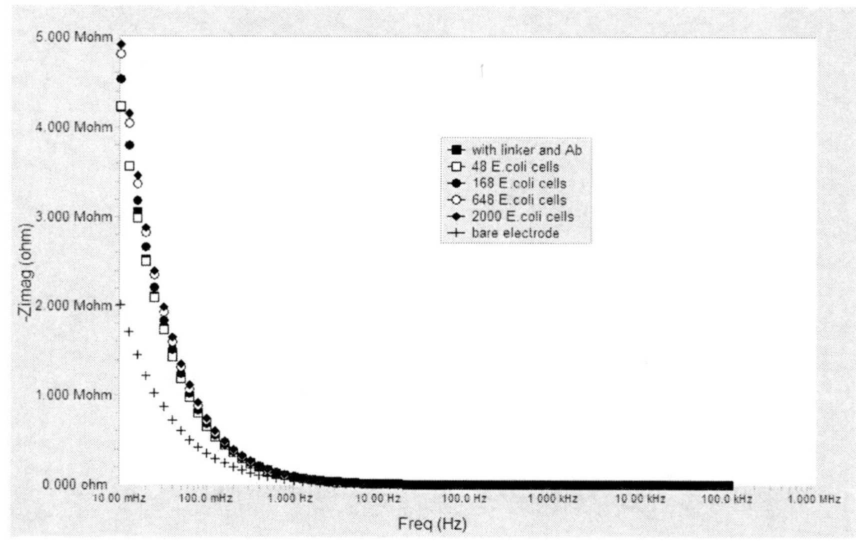

Fig. 7.16. EIS on a flat gold surface with different amounts of *E. coli* bacteria attached ($-Z_{imag}$ vs. frequency).

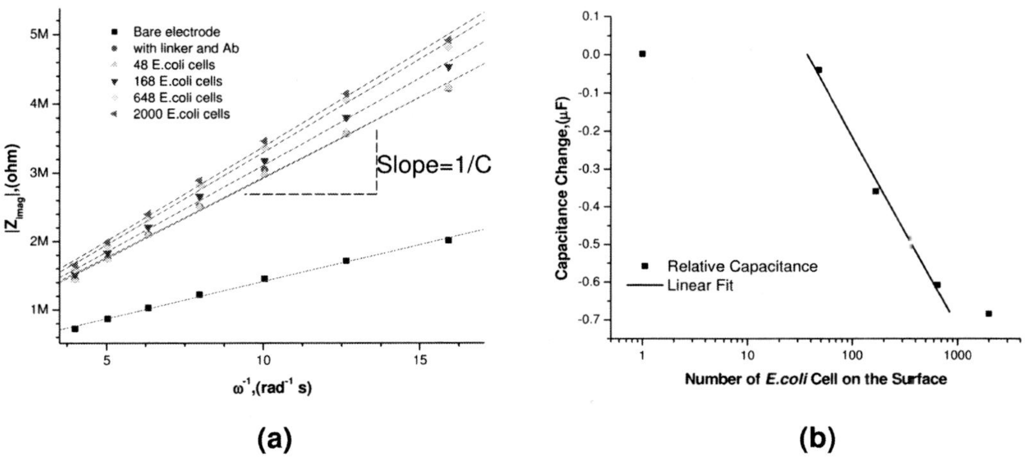

Fig. 7.17. (a) $|Z_{imag}|$ vs. ω^{-1}, and (b) relative capacitance change vs. the number of *E. coli* (log scale) on a flat gold electrode.

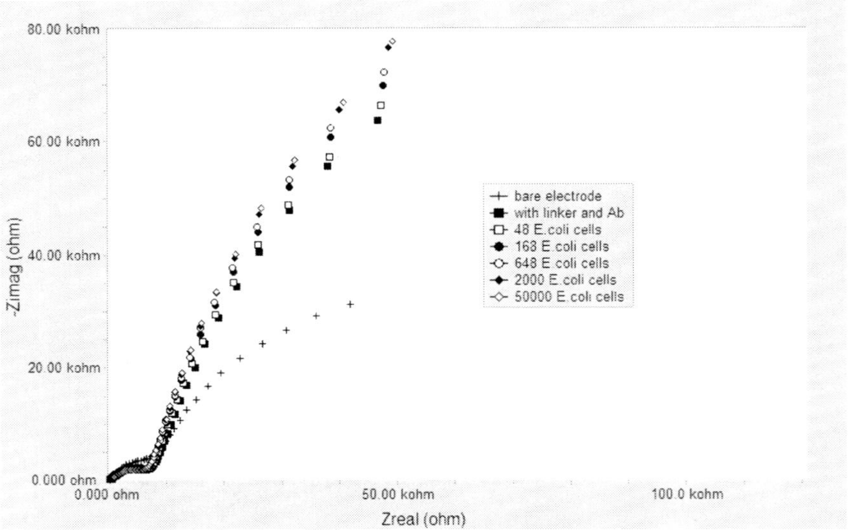

Fig. 7.18. EIS on GNW with different amounts of *E. coli* bacteria attached (Nyquist plot).

considered as a reference capacitance, Fig. 7.17(b) shows the relative capacitance change with the number of *E. coli* bacteria tethered. There is a linear relationship between the change of capacitance and the logarithmic number of *E. coli* cell on the surface when the number of *E. coli* cell is from 50 to 1000. The sensitivity of the capacitance change is about -0.5 μF per log(number of *E. coli* cells).

7.4.3.2 EIS on GNW

On a free standing gold nanowire array, typical EIS results are shown in Figs. 7.18 and 7.19. The same calculation procedure was applied as that for a flat gold. The imaginary impedance is also linear with the reciprocal radial frequency in the low frequency range (Fig. 7.20(a)). The relative capacitance change because of the binding of the *E. coli* cells is shown in Fig. 7.20(b). A similar linear relationship between the change of the capacitance and the log scale number of *E. coli* was obtained in the concentration range from 50 cells to 2000 cells. The slope was determined to be -29.82 μF per log(number of *E. coli* cells). The sensitivity saturates at a higher concentration than a flat gold electrode.

7.4.3.3 EIS on GNW with Al_2O_3

Figures 7.21 and 7.22 shows a typical set of the impedance results of *E. coli* attached GNW-Al_2O_3. The Nyquist plots of this sensor surface have very different shapes from the previous ones. This is because of the different structure of the sensor, in other words, the gold nanowires are embedded in an insulating alumina matrix. The capacitance of this insulating oxide layer also contributes to the system overall capacitance. However, the capacitance of the oxide layer could be assumed

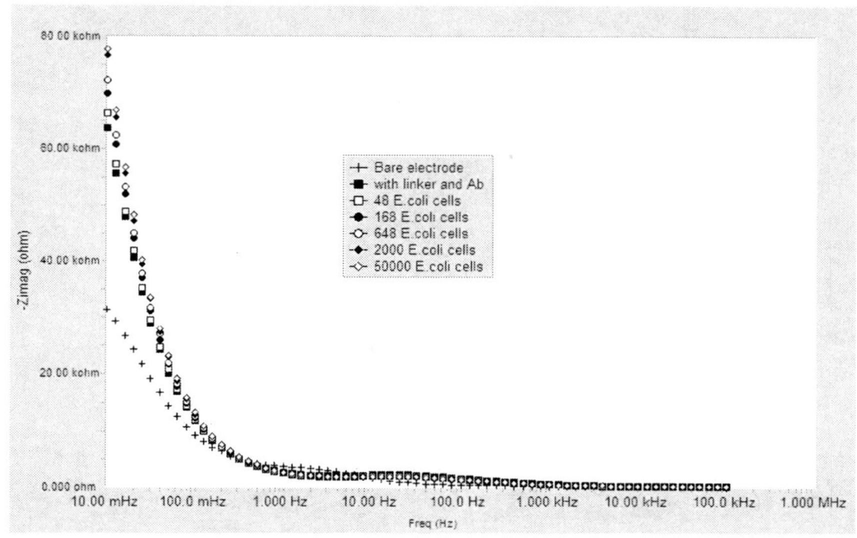

Fig. 7.19. EIS on GNW with different amounts of *E. coli* bacteria attached ($-Z_{imag}$ vs. frequency).

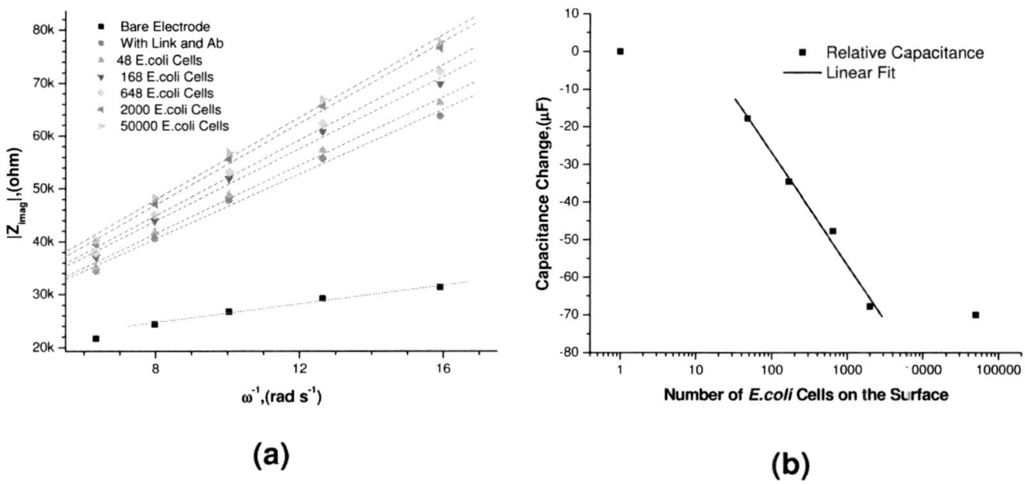

Fig. 7.20. (a) $|Z_{imag}|$ vs. ω^{-1} and (b) relative capacitance change vs. the number of *E. coli* (log scale) on a GNW electrode.

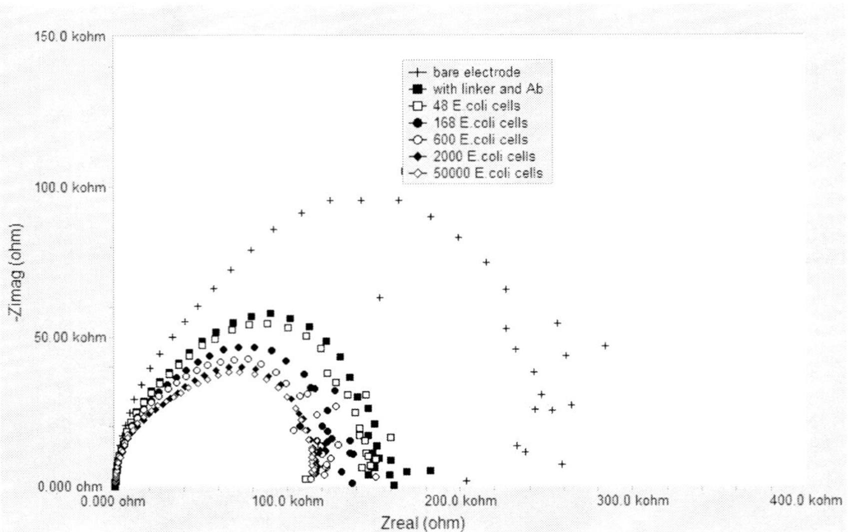

Fig. 7.21. EIS results for a GNW-Al_2O_3 sample with different amounts of *E. coli* cells attached (Nyquist plot).

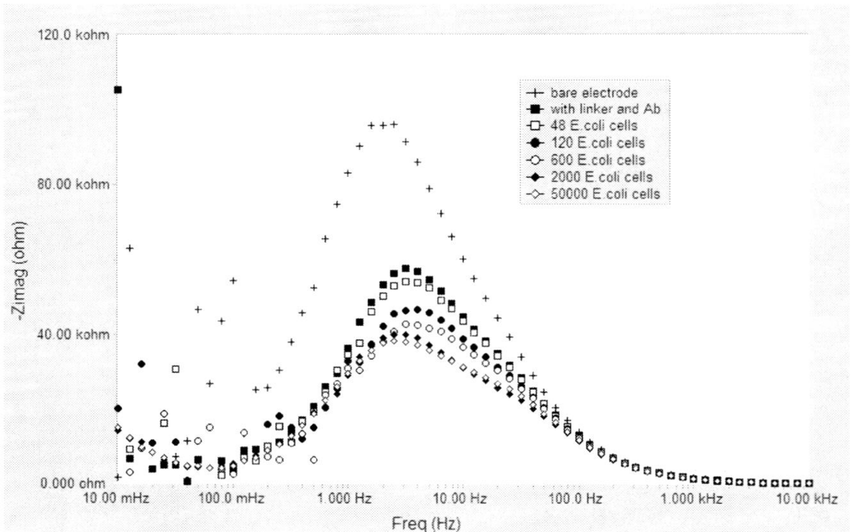

Fig. 7.22. Imaginary part of impedance vs. frequency of a GNW-Al_2O_3 sample with different amounts of *E. coli* cells attached.

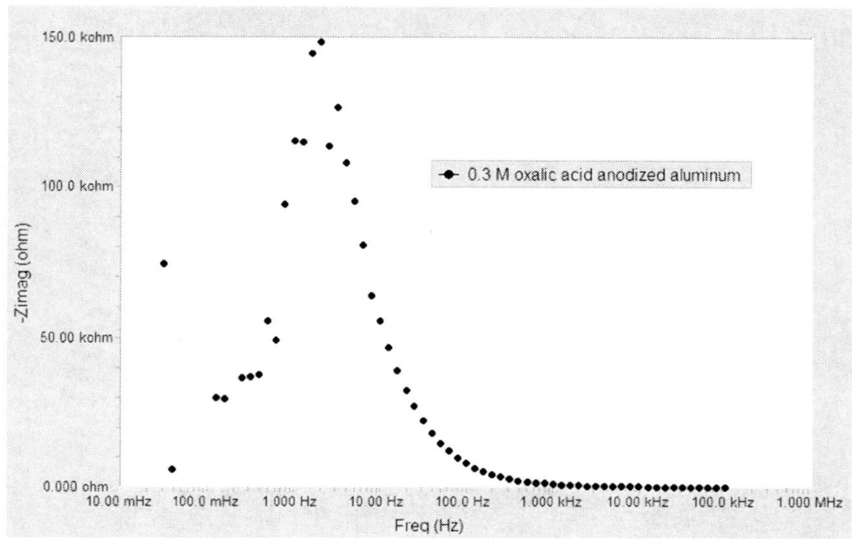

Fig. 7.23. Imaginary part of impedance vs. frequency of an anodized Al_2O_3/Al electrode in 0.01 M PBS buffer.

to be constant before and after the antigens interact with the antibodies, which means that the observed capacitance change is only due to the binding of *E. coli* cells. If the imaginary part of impedance is plotted vs. the frequency, a peak shape curve was obtained (Fig. 7.22). The imaginary impedance vs. frequency curve was a peak shape in many tests for this type of substrate. The peak frequencies for different samples were in the range 1 Hz–100 Hz. This peak is most likely due to the presence of Al_2O_3 on the substrate. The impedance results of an anodized Al_2O_3/Al electrode tested in 0.01 M PBS buffer shown in Fig. 7.23 confirmed this speculation.

The capacitance can be calculated for this type of substrate from the data in the region of the right-half of the peak in the figure. Figure 7.24(a) shows the fitting results. If the capacitance of the electrode with linker and antibody is set as the reference capacitance, the relative capacitance change vs. the number of *E. coli* cells can be plotted as shown in Fig. 7.24(b). An increase of the capacitance with the concentration of *E. coli* has been found. The sensitivity of the linear region is about 0.36 μF per log(number of *E. coli* cells). The reason why the capacitance increases instead of decreases after *E. coli* is bound is still not clear.

7.4.4
Summary of EIS Detection of *E. coli* Bacteria

Immunosensors for the detection of *E. coli* bacteria have been developed based on the gold nanowire arrays formed by the template synthesis technique. A specific

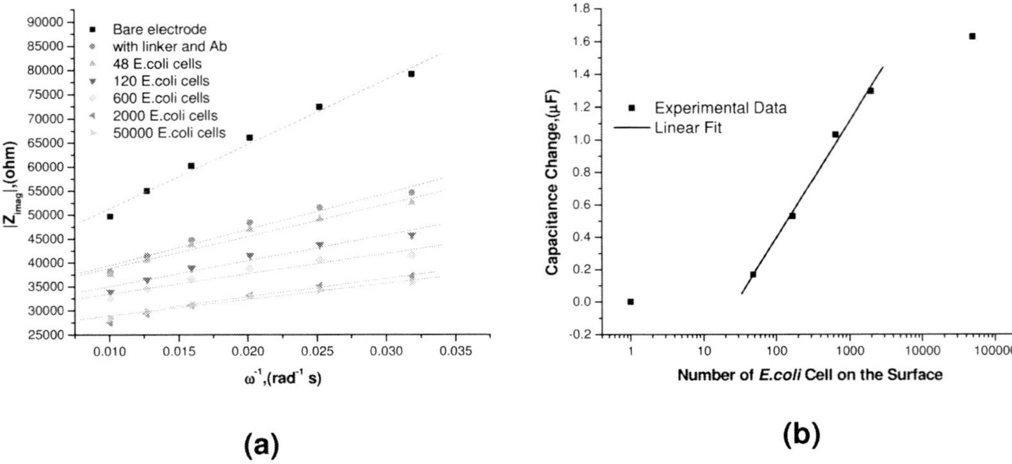

Fig. 7.24. (a) $|Z_{imag}|$ vs. ω^{-1}, and (b) relative capacitance change vs. the number of *E. coli* (log scale) on a GNW-Al_2O_3 sample.

type of antibody was immobilized on the active gold surface by covalent binding. EIS measurements were carried out to detect the complex formation between the antibodies and antigens. A thiol-containing linker molecule was synthesized to increase the binding capacity and strength of the specific antibody. The binding of *E. coli* results in a change in the overall capacitance. This capacitance change can be monitored by impedance measurements. Different substrate structures result in different working frequency ranges (compare Figs. 17(a), 20(a), and 24(a)). The preliminary results showed that this nano-gold biosensor is able to detect every 50 *E. coli* cells with a sensor area of 0.2 cm^2. For both planar and nanostructured gold substrates, the change of capacitance (ΔC) was linear with the logarithmic number of *E. coli* cell on the surface. However, the linear range was smaller on a flat gold surface. The detection sensitivities were determined to be the slopes of the linear portion of the curves for all substrates (Table 7.4). It can be seen that the gold

Tab. 7.4. Comparison of three substrates studied for *E. coli* sensor application.

Substrate	Detection sensitivity	Sensitive region
Flat gold	−0.5 μF/log(number of *E. coli* cell)	50–1000 cells
Gold nanowire array (GNW)	−29.8 μF/log(number of *E. coli* cell)	50–2000 cells
Gold nanowire array in the oxide (GNW-Al_2O_3)	0.36 μF/log(number of *E. coli* cell)	50–2000 cells

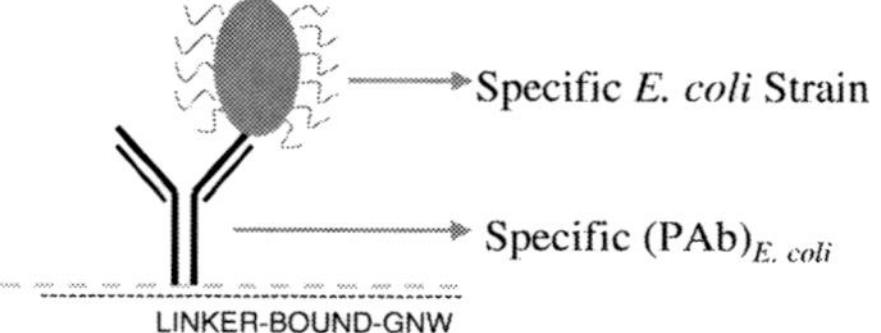

Fig. 7.25. Detection of pathogenic *E. coli* by antibody-linker-GNW biosensors.

nanowire array electrode is the most sensitive substrate according to the obtained capacitance change arising from the binding of the same amount of *E. coli* cells.

The study of the *E. coli* sensor has shown that an impedometric transduction strategy is suitable for detection of *E. coli* at low concentrations. The nanostructured materials can further improve the detection sensitivity. However, the specificity of this sensor device, as well as the long-term stability has not yet been established. The antibody used is not specific to a single strain of *E. coli* bacteria, various strains of *E. coli* can interact with this antibody.

7.5 Conclusions

This chapter has described a quick method for bacteria detection using a gold nanowire based biosensor and electrochemical measurements. Detection methods have been developed with the use of gold nanowire (GNW) substrates attached to a C_{11} linker arm in turn attached to the specific *E. coli* antibodies (Fig. 7.25). Preliminary results indicate that the GNW biosensor can detect as few as 50 *E. coli* cells with a sensor area of 0.2 cm^2. However, this report gives the initial results for this novel procedure to be used in wider applications. The specificity of detection of any cell depends on the availability of a specific antibody directed to the cell surface macromolecules or antigens (Fig. 7.25). This method would be applicable to detect any cancer cells based on the antigens present on cell surfaces, such as human colon and breast cancer cells containing Le^X and Sialo-Le^X [60]. Antibodies against these antigens are available commercially. Further studies on the detection of cancer cells by this method are in progress.

Acknowledgments

This chapter was written on research supported by an NSF-REU grant (Chemistry Department) and a grant-in-aid from the Bayer Corporation to Subhash Bash. Juan Jiang was the recipient of a Bayer Predoctoral Fellowship. Our thanks go to Dr. Patrick J. Boyle, Mr. Rui Ma, Mr. Joshua Henshaw, Ms. Rocio del A. Cordona, Mr. Kunal Saxena and Ms. Clare Lefave for their tireless technical help. Our special thanks to Mrs. Dorisanne Nielsen for her help in the preparation of the final manuscript.

References

1 Cai, Z. H., Martin, C. R., Electronically conductive polymer fibers with mesoscopic diameters show enhanced electronic conductivities. *J. Am. Chem. Soc.* **1989**, 111(11), 4138–4139.

2 Martin, C. R., Parthasarathy, R., Menon, V., Template synthesis of electronically conductive polymers – A new route for achieving higher electronic conductivities. *Synthetic Met.* **1993**, 55(2–3), 1165–1170.

3 Parthasarathy, R. V., Martin, C. R., Synthesis of polymeric microcapsule arrays and their use for enzyme immobilization. *Nature* **1994**, 369(6478), 298–301.

4 Penner, R. M., Martin, C. R., Controlling the morphology of electronically conductive polymers. *J. Electrochem. Soc.* **1986**, 133, 2206.

5 Vandyke, L. S., Martin, C. R., Electrochemical investigations of electronically conductive polymers. 4. Controlling the supermolecular structure allows charge transport rates to be enhanced. *Langmuir* **1990**, 6(6), 1118–1123.

6 Jiang, J., Electrochemical studies of nanostructured materials: The synthesis and size effect dependence of adsorption of organic molecules and applications to biosensors. Ph.D. dissertation, University of Notre Dame, Notre Dame, **2004**.

7 Gao, T., Meng, G. W., Zhang, J., Wang, Y. W., Liang, C. H., Fan, J. C., Zhang, L. D., Template synthesis of single-crystal Cu nanowire arrays by electrodeposition. *Appl. Phys. A* **2001**, 73(2), 251–254.

8 Li, L., AC anodization of aluminum, electrodeposition of nickel and optical property examination. *Sol. Energy Mater. Sol. Cells* **2000**, 64, 279–289.

9 Metzger, R. M., Konovalov, V. V., Sun, M., Xu, T., Zangari, G., Xu, B., Benakli, M, Doyle, W. D., Magnetic nanowires in hexagonally ordered pores of alumina. *IEEE Trans. Magn.* **2000**, 36(1), 30–35.

10 Peng, Y., Qin, D.-H., Zhou, R.-J., Li, H.-L., Bismuth quantum-wires arrays fabricated by electrodeposition in nanoporous anodic aluminum oxide and its structural properties. *Mater. Sci. Eng.* **2000**, B77, 246–249.

11 Shingubara, S., Okino, O., Sayama, Y., Sakaue, H., Takahagi, T., Two-dimensional nanowire array formation on Si substrate using self-organized nanoholes of anodically oxidized aluminum. *Solid State Electron.* **1999**, 43, 1143–1146.

12 Crouse, M. M., Miller, A. E., Jiang, J., Crouse, D. T., Basu, S. C., Enabling nanostructured materials via multilayer thin film precursor and applications to biosensors. *US Patent 6,869,671 B1*, March 22, **2005**.

13 Wang, Z., Li, H. L., Highly ordered zinc oxide nanotubules synthesized within the anodic aluminum oxide template. *Appl. Phys. A* **2002**, 74(2), 201–203.

14 Xu, D. S., Xu, Y. J., Chen, D. P., Guo, G. L., Gui, L. L., Tang, Y. Q., Preparation and characterization of CdS nanowire arrays by dc electrodeposit in porous anodic aluminum oxide templates. *Chem. Phys. Lett.* **2000**, 325, 340–344.

15 Brumlik, C. J., Martin, C. R., Template synthesis of metal microtubules. *J. Am. Chem. Soc.* **1991**, 113(8), 3174–3175.

16 Hulteen, J. C., Chen, H. X., Chambliss, C. K., Martin, C. R., Template synthesis of carbon nanotubule and nanofiber arrays. *Nanostructured Mater.* **1997**, 9(1–8), 133–136.

17 Martin, C. R., Vandyke, L. S., Cai, Z. H., Liang, W. B., Template synthesis of organic microtubules.

J. Am. Chem. Soc. **1990**, 112(24), 8976–8977.

18 Wirtz, M., Martin, C. R., Template-fabricated gold nanowires and nanotubes. *Adv. Mater.* **2003**, 15(5), 455–458.

19 Masuda, H., Watanabe, M., Yasui, K., Tryk, D., Rao, T., Fujishima, A., Fabrication of a nanostructured diamond honeycomb film. *Adv. Mater.* **2000**, 12(6), 444–447.

20 Wu, C. G., Bein, T., Conducting polyaniline filaments in a mesoporous channel host. *Science* **1994**, 264(5166), 1757–1759.

21 Wu, C. G., Bein, T., Conducting carbon wires in ordered, nanometer-sized channels. *Science* **1994**, 266(5187), 1013–1015.

22 Martin, C. R., Nanomaterials – A membrane-based synthetic approach. *Science* **1994**, 266(5193), 1961–1966.

23 Martin, C. R., Membrane-based synthesis of nanomaterials. *Chem. Mater.* **1996**, 8(8), 1739–1746.

24 Masuda, H., Hasegwa, F., Ono, S., Self-ordering of cell arrangement of anodic porous alumina formed in sulfuric acid solution. *J. Electrochem. Soc.* **1997**, 144(5), L127–L130.

25 Masuda, H., Yada, K., Osaka, A., Self-ordering of cell configuration of anodic porous alumina with large-size pores in phosphoric acid solution. *Jpn. J. Appl. Phys.* **1998**, 37(Pt.2)(11A), L1340–L1342.

26 Thamida, S. K., Chang, H. C., Nanoscale pore formation dynamics during aluminum anodization. *Chaos* **2002**, 12(1), 240–251.

27 Hoar, T. P., Mott, N. F., A mechanism for the formation of porous anodic oxide films on aluminum. *Phys. Chem. Solids* **1959**, 9, 97.

28 O'Sullivan, J. P., Wood, G. C., The morphology and mechanism of formation of porous anodic films on aluminum. *Proc. R. Soc. London A* **1970**, 317(1531), 511–543.

29 Thompson, G. E., Xu, Y., Skeldon, P., Shimizu, K., Han, S. H., Wood, G. C., Anodic-oxidation of aluminum. *Philos. Mag. B* **1987**, 55(6), 651–667.

30 Ba, L., Li, W. S., Influence of anodizing conditions on the ordered pore formation in anodic alumina. *J. Phys. D: Appl. Phys.* **2000**, 33, 2527–2531.

31 Masuda, H. A. F., Kenjiq, Ordered metal nanohole arrays made by a two-step replication of honeycomb structures of anodic alumina. *Science* **1995**, 268, 1466–1468.

32 Despie, A., Parkhutik, V. P., Electrochemistry of aluminum in aqueous solutions and physics of its anodic oxide, in *Modern Aspects of Electrochemistry*, Bockris, J. O. M., Conway, B. (Eds.), Plenum Press, New York, **1989**, Vol. 20.

33 Yue, D., Electrochemically Synthesized Nanostructures on Aluminum. Ph.D. dissertation, University of Notre Dame, Notre Dame, **1995**.

34 Bandyopadhyay, S., Miller, A. E., Chang, H. C., Banerjee, G., Yuzhakov, V., Yue, D. F., Ricker, R. E., Jones, S., Eastman, J. A., Baugher, E., Chandrasekhar, M., Electrochemically assembled quasi-periodic quantum dot arrays. *Nanotechnology* **1996**, 7(4), 360–371.

35 Crouse, D., Lo, Y. H., Miller, A. E., Crouse, M., Self-ordered pore structure of anodized aluminum on silicon and pattern transfer. *Appl. Phys. Lett.* **2000**, 76(1), 49–51.

36 Ricker, R. E., Miller, A. E., Yue, D. F., Banerjee, G., Bandyopadhyay, S., Nanofabrication of a quantum dot array: Atomic force microscopy of electropolished aluminum. *J. Electron. Mater.* **1996**, 25(10), 1585–1592.

37 Yue, D. F., Banerjee, G., Miller, A. E., Bandyopadhyay, S., Giant magnetoresistance in an electrochemically synthesized regimented array of nickel quantum dots. *Superlattices Microstruct.* **1996**, 19(3), 191–195.

38 Yuzhakov, V. V., Chang, H. C., Miller, A. E., Pattern formation during electropolishing. *Phys. Rev. B* **1997**, 56(19), 12 608–12 624.

39 Yuzhakov, V. V., Takhistov, P. V., Miller, A. E., Chang, H. C., Pattern selection during electropolishing due

to double-layer effects. *Chaos* **1999**, 9(1), 62–77.
40 Preston, C. K., Moskovites, M., Optical characterization of anodic aluminum oxide films containing electrochemically deposited metal particles. 1. Gold in phosphoric acid anodic aluminum oxide films. *J. Phys. Chem.* **1993**, 97, 8459–8503.
41 Wang, Y. H., Mo, J. M., Cai, W. L., Shi, G., Mechanism of metal deposited in porous alumina by alternating current. *Acta Phys. Chim. Sin.* **2001**, 17(2), 116–118.
42 Morimoto, K., Tsujimoto, M., Yaji, T., Murakami, T., Electroless gold plating bath. *US 5,364,460*, Nov. 15, **1994**.
43 Miller, A. E., Bandyopadhyay, S., Electrochemical synthesis of quasi-periodic quantum dot and nano-structrure arrays. *US 5,747,180*, **1998**.
44 Yin, A. J., Li, J., Jian, W., Bennett, A. J., Xu, J. M., Fabrication of highly ordered metallic nanowire arrays by electrodeposiiton. *Appl. Phys. Lett.* **2001**, 79(7), 1039–1041.
45 Wang, Z., Su, Y. K., Li, H. L., AFM study of gold nanowire array electrodeposited within anodic aluminum oxide template. *Appl. Phys. A* **2002**, 74, 563–565.
46 Foss, C. A., Hornyak, G. L., Stockert, J. A., Martin, C. R., Optical-properties of composite membranes containing arrays of nanoscopic gold cylinders. *J. Phys. Chem.* **1992**, 96(19), 7497–7499.
47 Foss, C. A., Hornyak, G. L., Stockert, J. A., Martin, C. R., Template-synthesized nanoscopic gold particles – Optical-spectra and the effects of particle-size and shape. *J. Phys. Chem.* **1994**, 98(11), 2963–2971.
48 Hornyak, G. L., Patrissi, C. J., Martin, C. R., Fabrication, characterization, and optical properties of gold nanoparticle/porous alumina composites: The nonscattering Maxwell-Garnett limit. *J. Phys. Chem. B* **1997**, 101(9), 1548–1555.
49 Hornyak, G. L., Patrissi, C. J., Martin, C. R., Valmalette, J. C., Dutta, J., Hofmann, H., Dynamical Maxwell-Garnett optical modeling of nanogold porous alumina composites: Mie and kappa influence on absorption maxima. *Nanostruct. Mater.* **1997**, 9(1–8), 575–578.
50 Hornyak, G. L., Phani, K. L. N., Kunkel, D. L., Menon, V. P., Martin, C. R., Fabrication, characterization and optical theory of aluminum nanometal nanoporous membrane thin film composites. *Nanostruct. Mater.* **1995**, 6(5–8), 839–842.
51 Hulteen, J. C., Patrissi, C. J., Miner, D. L., Crosthwait, E. R., Oberhauser, E. B., Martin, C. R., Changes in the shape and optical properties of gold nanoparticles contained within alumina membranes due to low-temperature annealing. *J. Phys. Chem. B* **1997**, 101(39), 7727–7731.
52 Martin, C. R., Mitchell, D. T., Template-synthesized nanomaterials in electrochemistry, in *Electroanalytical Chemistry*, **1999**, Vol. 21, pp. 1–74.
53 Wagner, P., Hegner, M., Kerman, P., Zuagg, F., Semenza, G., Covalent immobilization of biomolecules on Au via n-OH succinamide ester functionalized self-assembled monolayers for scanning probe microscopy. *Biophys. J.* **1996**, 70, 2052–2066.
54 Basu, M., Seggerson, S., Henshaw, J., Jiang, J., Cordona, R., Lefave, C., Boyle, P. J., Miller, A., Pugia, M., Basu, S., Nano-biosensor development for the bacterial detection during human kidney infection. *Glycoconjugate J.* **2004**, 21, 487–496.
55 Ghindilis, A. L., Atanasov, P., Wilkins, M., Wilkins, E., Immunosensors: Electrochemical sensing and other engineering approaches. *Biosens. Bioelectron.* **1998**, 13(1), 113–131.
56 Griffiths, D., Hall, G., Biosensors – What real progress is being made? *Trends Biotechnol.* **1993**, 11, 122–130.
57 Owen, V. M., Market requirements for advanced biosensors in healthcare. *Biosens. Bioelectron.* **1994**, 9, XXIX–XXXV.

58 Bard, A. J., Faulkner, L. R., *Electrochemical Methods: Fundamentals and Applications*, 2nd edn., John Wiley & Sons, Inc., New York, **2001**, 833.

59 Macdonald, J. R., *Impedance Spectroscopy*, John Wiley & Sons, New York, **1987**.

60 Basu, M., Hawes, J. W., Li, Z., Ghosh, S., Khan, F. A., Zhang, B. J., Basu, S., Biosynthesis in vitro of SA-Le^x and SA-diLe^x by 1-3 fucosyltransferases from colon carcinoma cells and embryonic brain tissues. *Glycobiology*, **1991**, 1, 527–535.

8
Dendrimer-based Electrochemical Detection Methods

Hak-Sung Kim and Hyun C. Yoon

8.1
Overview

The use of molecularly-organized nanoscale interfaces is of great interest in efforts to enhance the analytical capabilities of biosensors. In this regard, dendrimer-based biocomposite structures, including molecularly organized monolayers and multilayers, have been developed for catalytic and affinity biosensing. As a highly organized and plurifunctional macromolecule, dendrimers find wide applications in analytical sciences. The anticipated merits of using dendrimers as the layer-forming materials are based on their structural characteristics such as homogeneity, compatible size with biomolecules, internal porosity, and the high density of functional groups. These unique characteristics make them good candidates for the building units of films on biosensor surfaces, presenting both the advantages of using polymers (plurifunctionality) and molecularly controllable nanomaterials. This chapter focuses on the fabrication of dendrimer-based biocomposite mono-/multilayers and their biosensing applications. Implementations of biointerfaces for the bioelectrocatalytic enzyme sensors with the multilayer configuration and affinity sensors based on either bioelectrocatalytic signal amplification or immunoprecipitation-voltammetric detection principles with dendrimer monolayers are covered.

8.2
Introduction

8.2.1
Background

Molecularly organized nanostructures are of great interest in analytical sciences, molecular device technology, and biotechnology [1]. Especially, an enormous amount of research has been devoted to the development of molecularly organized

Nanotechnologies for the Life Sciences Vol. 8
Nanomaterials for Biosensors. Edited by Challa S. S. R. Kumar

ISBN: 978-3-527-31388-4

interfaces containing biomolecules as diagnostic tools (biosensors), biomimetic membranes and, recently, bioelectronic devices [2–4].

With the availability of highly controllable structuring and patterning techniques, especially the self-assembled monolayers (SAMs) methodology [5, 6], the pace of development in the related fields has greatly increased. The pioneering studies of Whitesides group regarding the mixed SAM technology and micropatterning such as microcontact printing have opened a new avenue for the technique [7–9]. By keeping in step with such developments, there have been numerous approaches for the well-organized and active biocomposite assemblies.

However, the quest continues for molecularly organized, structurally rigid, biochemically active and stable biocomposite superstructures. In this regard, the implementation of useful surface bio-functionalization methodologies as well as the introduction of novel materials having unique properties is one of the major subjects of research and development. As a promising molecule for biocomposite nanostructures, highly branched dendrimers draw much attention.

8.2.2
Dendrimers as a new Constituent of Biocomposite Structures

The past decade has seen expanding interest in the newly introduced synthetic dendritic polymers (dendrimers) [10, 11], which are highly branched, fractal-like macromolecules of defined molecular structure, size, and topology. The molecules are synthesized by an iterative sequence of reaction steps, or cascade synthesis, with sophisticated control in reaction conditions [12]. With the development in synthetic methodology, structurally perfect dendrimers with high purity and narrow polydispersity have been produced. The unique characteristics of dendrimers, such as structural homogeneity, molecular integrity, controlled composition, and the multiple homogeneous chain-ends available for consecutive conjugation reaction, have enabled their use as the material of choice in applications such as drug delivery, energy harvesting, ion sensing, catalysis, and information storage [13]. For these objectives, several approaches adopting dendrimers as the building block for the nanostructures have been conducted. These include deposition of dendritic multilayers via Pt-complexation, electrostatic interaction, and reaction with grafted copolymer [14–17]. Characteristics mentioned above, along with the recently recognized biocompatibility [13], merit the use of dendrimers for the fabrication of organized functional biocomposite nanostructures, comparable with those prepared with entangled linear and branched polymers. By us and other groups, the utility of the dendrimer has extended to bio-related fields [18–20].

Our recent reports demonstrate that dendrimers are advantageous as the building block for the construction of the multilayered biocomposite nanostructures and reagentless enzyme or affinity biosensors [18, 21–25]. As a building unit for organic thin films, dendrimers, as highly branched dendritic macromolecules, possess a unique surface of multiple chain-ends, and the number of surface groups can be precisely controlled as a function of synthetic generations (Fig. 8.1). For example,

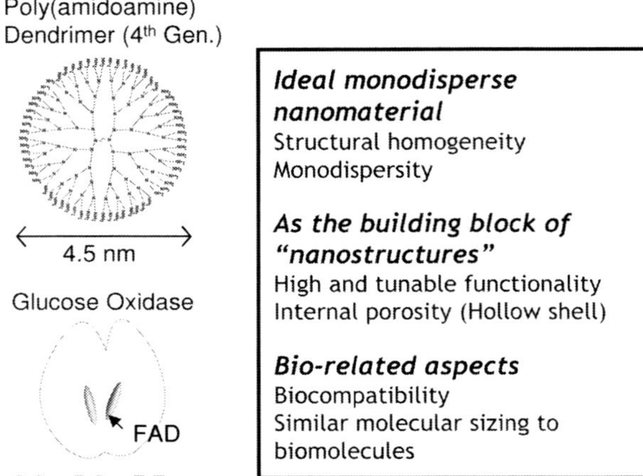

Fig. 8.1. Characteristics of fourth-generation (G4) poly(amidoamine) dendrimers. Molecular models for dendrimer and glucose oxidase are shown for size comparison.

fourth generation (G4) poly(amidoamine) dendrimers have 64 surface amine groups. The high concentration of functional end groups of dendrimers enables synthetic modifications of molecularly ordered nanostructures. Dendrimers provide multiple conjugation sites, and the remaining groups, after film formation, are accessible for further modification with functionalization groups for specific purposes. Dendrimers of high generation numbers offer some advantages over linear or partly branched polymers, e.g., structural homogeneity, controllable composition, comparable size to the participating biomolecules, and multiple homogeneous chain-ends groups, valuable for the conjugating reactions [21].

In this respect, we have employed the G4 poly(amidoamine) dendrimer as the underlying-layer for functionalizing or micropatterning ligands on the solid surface. Besides G4 poly(amidoamine) dendrimers, poly-L-lysine and poly(allylamine) and amine-terminated cystamine SAM, all presenting surface amine functionalities, have frequently been used, and we have compared their physicochemical and biochemical properties as thin films for arraying ligands on a solid surface [26].

8.3 Applications for Biosensors

In the past decade, self-assembled monolayers, silane-modified layers, Langmuir–Blodgett (LB) layers, and polymer grafting layers compatible with biomolecules have been commonly used for the presentation of ligands or proteins on a solid surface to modulate their interactions with reacting couple molecules. These organic thin films have an expanded application in biosensors (enzyme electrodes, immunoelectrodes, DNA analyses), artificial biomimetic membranes, and, recently, bioelectronic devices. In this regard, dendrimer-based biocomposite structures, organized mono- and multilayers, have been developed for catalytic and affinity biosensing.

8.3.1
Bioelectrocatalytic Enzyme Electrodes based on LBL (layer-by-layer) Assembly with Dendrimers

For a catalytic biosensing interface, a new approach to construct a multilayered enzyme film on the electrode surface has been developed [18, 21]. Figure 8.2 shows the schematic procedure to construct a film by alternate layer-by-layer (LBL) depositions of G4 poly(amidoamine) dendrimers and periodate-oxidized glucose oxidase (GOx). Cyclic voltammograms registered from gold electrodes modified with GOx/

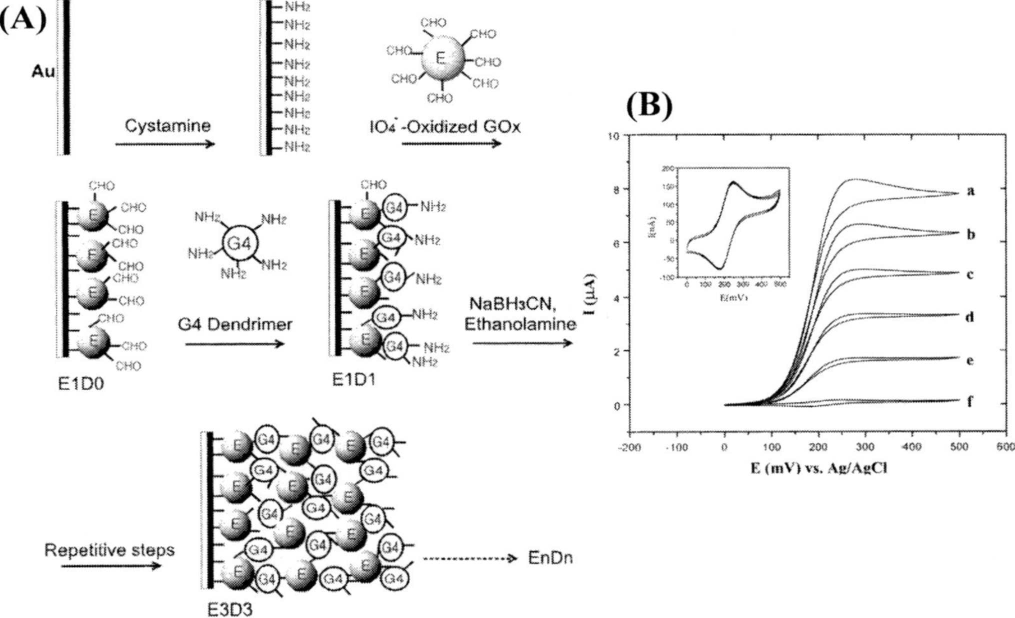

Fig. 8.2. (A) Schematic representation of a multilayered GOx/dendrimer network construction on a Au electrode surface. (B) Cyclic voltammograms of the GOx/dendrimer multilayered electrodes in the presence of 0.1 mM ferrocene-methanol as a diffusional electron-transferring mediator: (a) E5D5, (b) E4D4, (c) E3D3, (d) E2D2, and (e) E1D1 in the presence of 20 mM glucose; (f) E1D1 in the absence of glucose, unmediated. Inset: cyclic voltammograms for each layer numbers in the absence of glucose in solution. All curves were registered in 0.1 M phosphate buffer (pH 8.0) under Ar. Potential scan rate: 5 mV s^{-1}. (Modified after Ref. [18]).

dendrimer multilayers revealed that the bioelectrocatalytic response is directly proportional to the number of deposited bilayers, i.e., to the amount of active GOx immobilized on the gold electrode surface. From analysis of electrochemical signals, the amount of active enzyme per GOx/dendrimer bilayer during the multilayer forming steps was calculated, demonstrating that the multilayer is constructed in a spatially-ordered manner. Also, by ellipsometry measurement, a linear increase of the film thickness was observed, supporting formation of the desired multilayer. The five-bilayer associated electrode (E5D5) showed a sensitivity of 14.7 μA (mM glucose)$^{-1}$ cm^{-2} and remained stable over 20 days under daily calibrations. The proposed method is simple and applicable to the construction of thickness- and sensitivity-controllable biosensing [18].

As an extension of this research, poly(amidoamine) dendrimers having surface ferrocenyl functional groups were prepared and used for the fabrication of a reagentless bioelectrocatalyzed enzyme electrode [21]. Poly(amidoamine) dendrimers having various degrees of tethering with the ferrocenyls were prepared by controlling the molar ratio of ferrocenecarboxaldehyde to amine groups of dendrimers during synthesis (Fig. 8.3A). By LBL depositions of ferrocenyl-tethered dendrimers with periodate-oxidized GOx on gold surface, an electrochemically and enzymatically active multilayer was constructed. The resulting GOx/Fc-D multilayer-associated electrodes were electrochemically characterized, and the density of ferrocenyl groups, active enzyme content, and sensitivity were analyzed. Dendrimers with 32% modification level was found to be an optimum from the analyses in terms of GOx-dendrimer network formation, electrochemical connectivity of ferrocenyls, and electrode sensitivity (Fig. 8.3B, C). With the synthesized Fc(32%)-tethered dendrimers, mono- and multilayered electrodes were constructed, and their electrochemical and catalytic properties were characterized. The bioelectrocatalytic signals from the GOx/Fc-D electrodes were directly proportional to the bilayer numbers. From this result, it seems that the electrode sensitivity is directly controllable, and the multilayer-forming strategy with ferrocenyl-labeled dendrimers is useful for the development of reagentless biosensors.

8.3.2 Bioelectrocatalytic Immunosensors based on the Dendrimer-associated SAMs

8.3.2.1 Affinity Recognition Surface based on the Dendrimer-associated SAMs

All biological phenomena, such as generation and development, are regulated by functional interactions of protein–protein, receptor–ligand, antigen–antibody, enzyme–substrate, and protein–XNA. Therefore, the understanding of functions of biomolecules that specifically interact with particular proteins or ligands is the main theme in life science. As a tool for the functional study of proteins, affinity recognition chip technology is important. Therefore, we have focused on surface functionalization technology that will arrange and immobilize active proteins on affinity chip surfaces. Immobilization of active proteins on solid supports with an extended lifetime is the major concern of a protein chip.

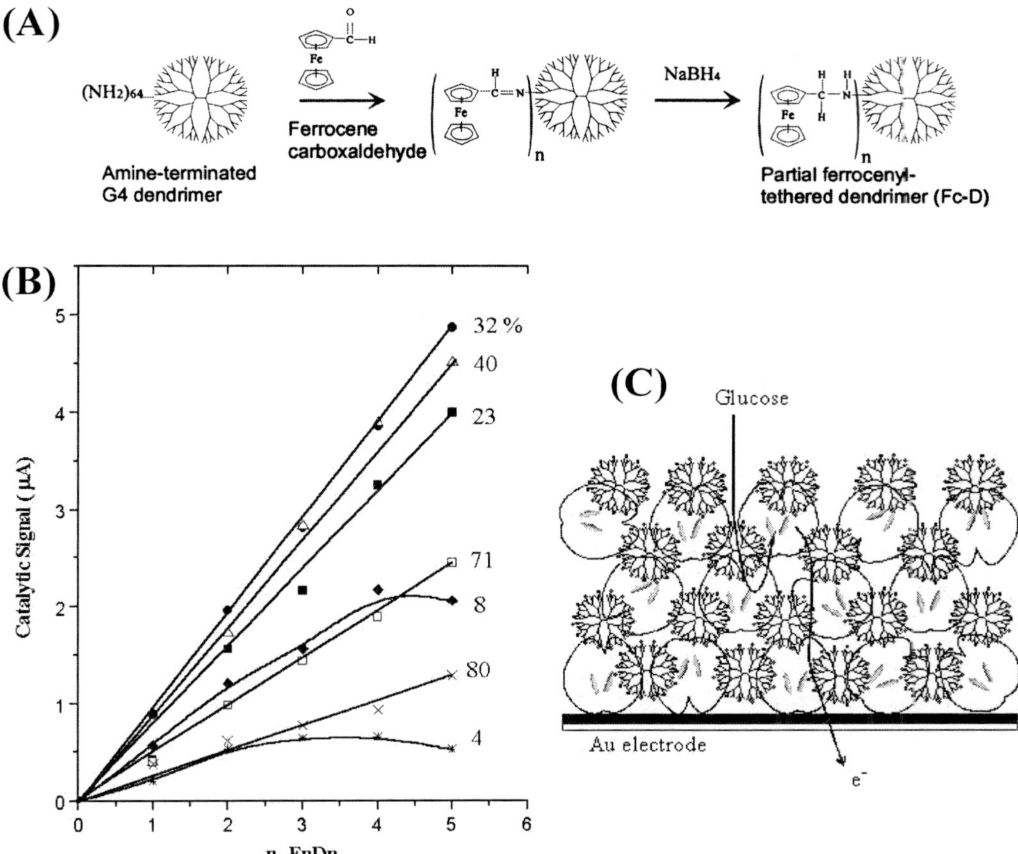

Fig. 8.3. (A) Schematic representation of a semi-synthetic preparation of partial ferrocenyl-tethered G4 poly(amidoamine) dendrimer (Fc-D). (B) Bioelectrocatalytic signal amplification for GOx/Fc-D electrodes containing dendrimers with various ferrocenyl functionalization levels as a function of deposited bilayer numbers. Signal values were sampled at +370 mV vs. Ag/AgCl reference electrode from cyclic voltammetric data in the presence of 20 mM glucose. (C) Schematic representation of the idealized multilayered GOx/Fc-D network on an Au electrode surface and the anticipated biochemical and charge-transfer reactions. (Modified after Ref. [21]).

In this respect, we emphasize three considerations to obtain sufficient sensitivity and accuracy from the affinity sensor. Aside from the signal generation efficacy, which will be covered later, the surface chemistry should be specially optimized to drive efficient immobilization of proteins or ligands. (a) Proteins must be immobilized on a chip surface in a stable and satisfactorily concentrated manner. (b) The orientation of protein should be controlled for favorable bio-specific interactions at the chip surface. (c) Nonspecific binding of molecules at the chip surface should be adequately circumvented. Additionally, and importantly, a highly sensitive detection system should be devised, by which the derived signal can be analyzed, because, in typical cases, the concentration of immobilized biomolecules on the chip is as low as 10^{-15}~10^{-14} mol/chip surface.

For a biospecific affinity-sensing interface, an affinity biosensor system based on avidin–biotin interaction has been developed [22]. The avidin–biotin interaction, a unique model of strong protein–ligand interaction, exhibits stronger affinity than any other antigen–antibody interactions. As the building block of an affinity sensing monolayer, a G4 poly(amidoamine) dendrimer having partial ferrocenyl-labeled surface groups was synthesized and used. The surface amine groups from dendrimers were also functionalized with biotin amidocaproate, and the biotinylated and electroactive monolayer was used for the affinity-sensing surface interacting with avidin. As shown in Fig. 8.4(A), an electrochemical signal from the affinity biosensor was generated by free enzyme GOx in electrolyte, depending on the degree of coverage of the sensing surface with avidin and subsequent surface shielding. The sensor signal decreased with increasing avidin concentration, and approached a minimum when the sensing surface was fully covered (Fig. 8.4B). The

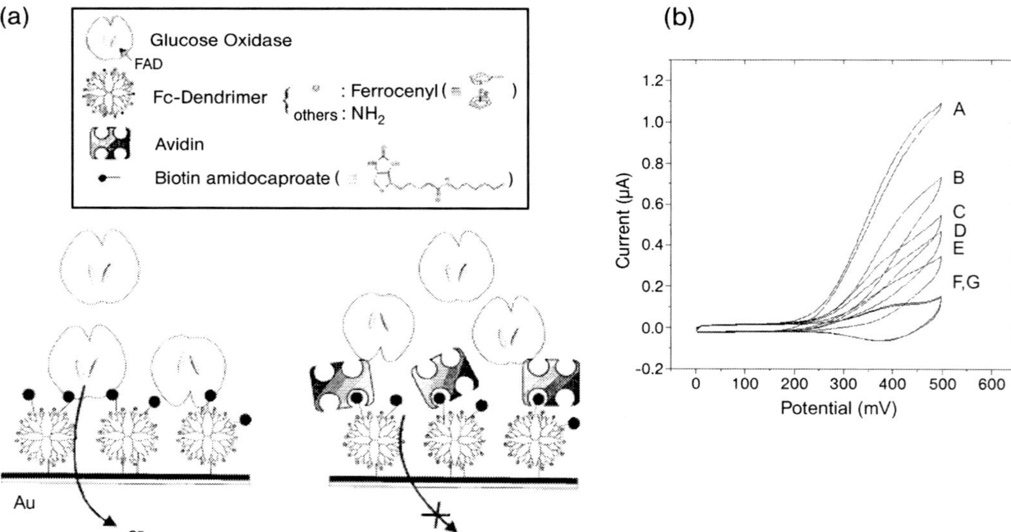

Fig. 8.4. (a) Construction and proposed operational principle for an affinity biosensor based on the avidin–biotin interaction on a gold electrode surface. Molecular models of the chemicals used for electrode construction are shown. (b) Cyclic voltammograms of the affinity biosensors as a function of reacted avidin concentration: (A) 0, (B) 1 ng mL^{-1}, (C) 10 ng mL^{-1}, (D) 100 ng mL^{-1}, (E) 1 mg mL^{-1}, and (F) 10 mg mL^{-1}. Cyclic voltammograms were obtained in the presence of 30 mg mL^{-1} of GOx as a signal generator and 10 mM glucose as a substrate; (G) background voltammogram in the absence of enzyme and substrate. Potential scan rate: 5 mV s^{-1}. (Modified after Ref. [22]).

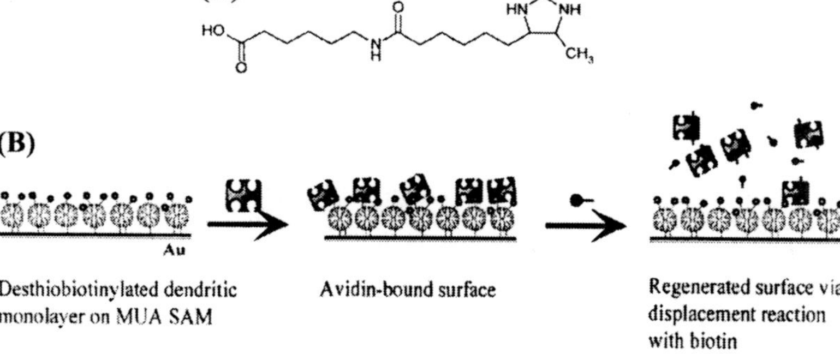

Fig. 8.5. (A) Molecular structures of d-desthiobiotin amidocaproate. (B) Schematic representation of the procedure employed for the biospecific association/dissociation of biomolecules at the affinity-sensing electrode surfaces. (Modified after Ref. [23]).

detection limit of avidin was about 4.5 pM, and the sensor signal was linear, ranging from 1.5 pM to 10 nM, under optimal conditions. Based on the kinetic analysis using the biotinylated glucose oxidase, an active enzyme coverage of 2.5×10^{-12} mol cm^{-2} on the avidin pretreated surface was calculated, demonstrating the formation of a spatially ordered and compact protein layer on the modified sensing interface.

As an extension of this research, a new approach regarding the development of a repeatedly renewable affinity-sensing surface was presented based on the reversible association/dissociation reactions between avidin and biotin analogues [23] (Fig. 8.5). The surfaces were constructed with dendrimer monolayers, whose surface chain-end groups have been functionalized with a biotin analog, desthiobiotin, which has a reduced affinity toward avidin (Fig. 8.5A). The functionalized monolayers provided an affinity recognition interface for avidin and further biospecific interactions with biotinylated molecules. The desthiobiotin–avidin associates underwent a dissociation (displacement) reaction with free biotin treatment, and this renewed the affinity surface and provided the possibility of repeated utilization of the affinity-sensing surface. Biotinylated glucose oxidase, as a model compound for signal generation, was used for the association reaction onto the avidin preincubated surface, and voltammetric measurements were performed to track the reaction steps by registering the activity of associated enzyme. Efficient association/dissociation reaction cycle traces were found, especially for the desthiobiotin amidocaproate modified electrodes, suggesting steric limitation regarding the ligand length for the biospecific interaction. With the optimized affinity-surface construction steps and reaction conditions, continuous association/dissociation reaction cycles were achieved, which is useful as a regenerable affinity surface.

Reversible affinity interactions at the functionalized electrode have also been extended to antibodies [24]. The surfaces were constructed with dendrimer mono-

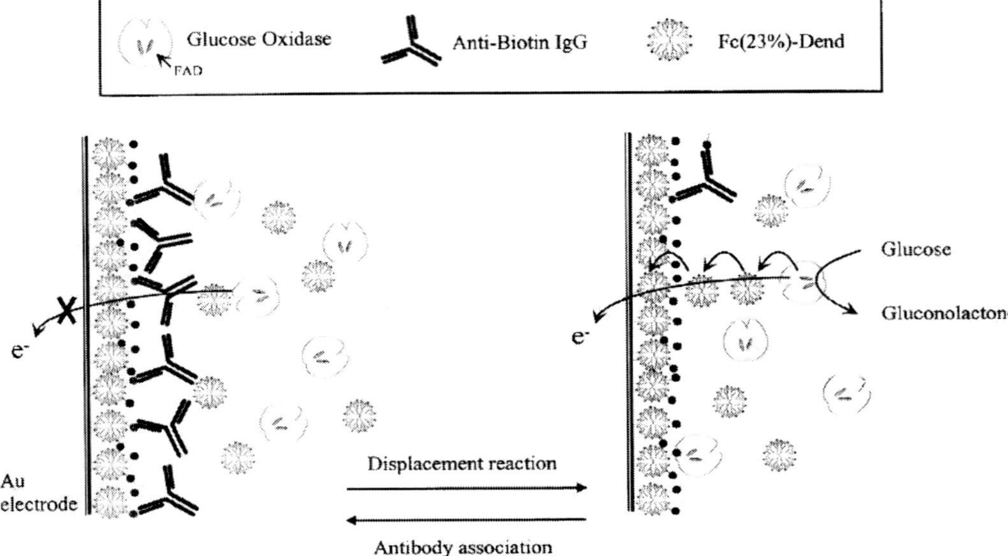

Fig. 8.6. Schematic representation of the idealized reactions in electrolyte and at the affinity-sensing electrodes for antibody-associated and regenerated surfaces: enzymatic catalysis, electron mediation with Fc-D in electrolyte, and charge transfer to Au electrodes. Dimensions of the components are not drawn to scale for simplicity. (From Ref. [24]).

layers, whose surface chain-end groups have been double-functionalized with biotins and ferrocenyls for the biospecific recognition and electron transfer. The functionalized monolayers provide a platform for biospecific recognition with monoclonal anti-biotin antibodies. The bound antibodies were dissociated with free biotin treatment, and the process renewed the affinity surfaces for repeated utilization. Figure 8.6 displays the electrochemical tracking of the association/dissociation reaction cycles, based on the shielding of the electrode surface with bound antibody molecules and subsequent hindrance in electron transfer, with free-diffusing signal generator and mediator. Factors influencing the biospecific interactions and measurements were considered. With the results, continuous association/dissociation reactions have been accomplished, holding great promise for reversible affinity biosensing.

8.3.2.2 Electrochemical Signaling from Affinity Recognition Reactions

Affinity sensing has been evolved from the viewpoints of detection of immune-related molecules with minute concentration range and signaling with acceptable quantification capability. Several detection principles for affinity/immuno sensing have been developed. A detection technique such as surface plasmon resonance (SPR) spectroscopy and a quartz crystal microbalance (QCM) can recognize the concentration of bound biomolecules by changes in mass and/or surface density (molecules/sensor area). However, notably, they are not suitable for application as portable biosensors. For SPR immunosensing, major progress has been made with bench-top scale instruments, enabling high-throughput and multiplexed analysis. In the concept of handheld biosensor/immunosensor, electrochemical methods have received the spotlight due to facile device-miniaturization and sensitive signal-quantification. Especially, recent efforts have focused on integrated devices, the so-called lab-on-a-chips [27, 28], combining the detection part with sample delivery parts based on microfluidics; thus a detection method that necessitates minimum instrumental part is highly desired. In that sense, we have focused on signaling methodologies that fulfills the needs of electrochemical detection and have developed two potential methods, including the immunoprecipitation-mediated signaling and the enzymatic back-filling immunoassay.

A signaling strategy for immunosensors that transduces biospecific affinity recognition reactions into electrochemical signals has been developed. This method combines the relatively well-known immunohistochemical reactions with electrochemistry [25]. As can be seen in Fig. 8.7, the cyclic voltammetric method, tracking the precipitation of insoluble products onto the sensing surface and the subsequent surface shielding and decrement in the electrode area, was employed for signal registration. Precipitation of insolubilities was induced by the catalytic reaction of peroxidase, which were labeled to the biospecifically associated protein or antibody molecules. We have investigated the functionalization of biotin ligand groups to the sensing monolayer and their biospecific interactions with anti-biotin antibody molecules as a model affinity recognition. The immunosensing interface was developed onto dendrimer-activated SAMs, as the base template for the functionalization of the antigen as well as generation. Additionally, the sensing system was applied for biotin/(strept)avidin couples, extending the usage of the developed strategy. With the affinity-sensing interface, a stepwise surface regeneration process has been developed, based on the combination of deposited product thin-film dissolution and bound-protein displacement reactions from the modified sensor surfaces [29]. The results exemplify the usefulness of immunoprecipitation-mediated signaling in terms of sensitivity and repeated use.

By using the developed immunosensing principle, a silicon-based immunosensor chip has been microfabricated and applied for the analysis of real samples such as blood serum [30]. Figure 8.8 shows an array-type gold electrode, which was fabricated on a silicon wafer, containing two electrode geometries of rectangular (100×500 μm) and circular (radius: 50 μm) types. These two types of electrodes showed distinct electrochemical characteristics of bulk and micro-electrodes, respectively, exhibiting different sensing parameters such as operation voltage, sensitivity, and dynamic detection range. Ferritin was employed as a model analyte for immunosensing because it has serological importance as a general marker protein for tumors and cancer recurrence. With the ferritin-functionalized immunosensor chips, biospecific interactions were performed with antiferritin antiserum and secondary antibody samples, followed by electrochemical signaling by an immunoprecipitation reaction with the label peroxidase enzyme. Under the optimized affinity-surface construction steps and signaling conditions, both types of microfabri-

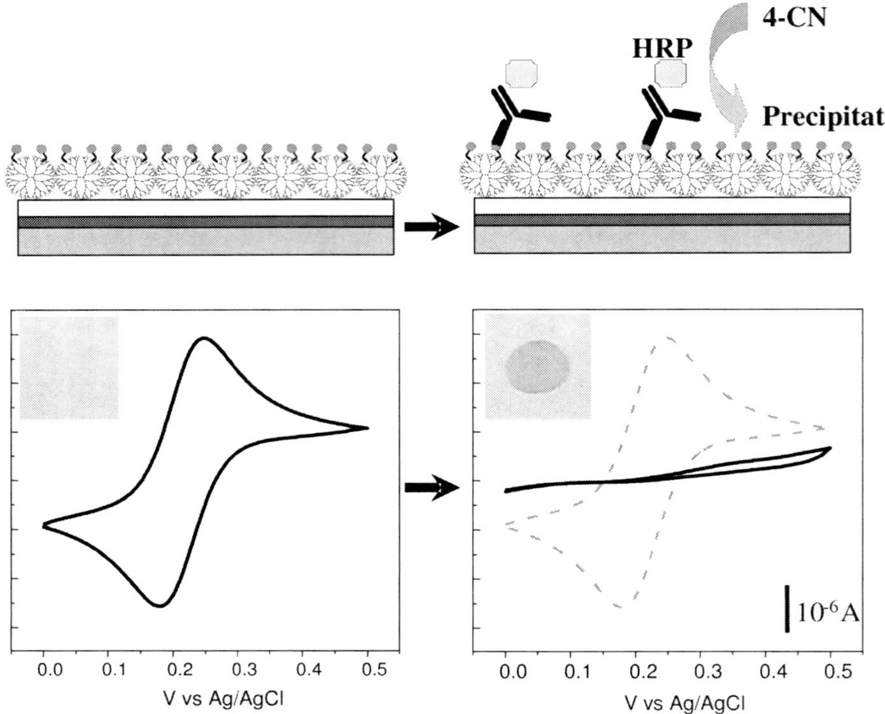

Fig. 8.7. (Top) Schematic representation of the affinity biosensor construction and proposed operational principle, and CCD camera images of a representative surface upon signaling reaction (inset). Component dimensions are not drawn to scale for simplicity. (Bottom) Cyclic voltammetric traces for sensor signaling at the dendrimer-assisted SAM surfaces. A freshly prepared and biotin-functionalized surface before (left) and after target protein association and precipitation reaction steps (right). Curves were registered in a 0.1 M phosphate buffer (pH 7) containing ferrocene-methanol (0.1 mM) as a signal tracer. Potential scan rate: 50 mV s^{-1}. (Modified after Ref. [25]).

cated electrodes, including rectangular (100 × 500 μm) and circular (radius: 50 μm) types, exhibited well-defined calibration results as a function of ferritin concentration in antiserum samples. Furthermore, circular-type micropatterned immunoelectrodes displayed the voltammetric characteristics of microelectrodes, which is advantageous in sensor operation under a fixed potential with low signal drift compared with the bulk-type electrodes. The results support the idea that the employed signaling method with the proposed immunosensor configuration is fit for sensor miniaturization and integration to biomicrosystems and lab-on-a-chips.

The analyte for affinity sensors, especially antibody, is usually bound to the surface-immobilized capture molecule and could cause a significant signal loss during subsequent handling procedures and signaling due to its limited affinity/

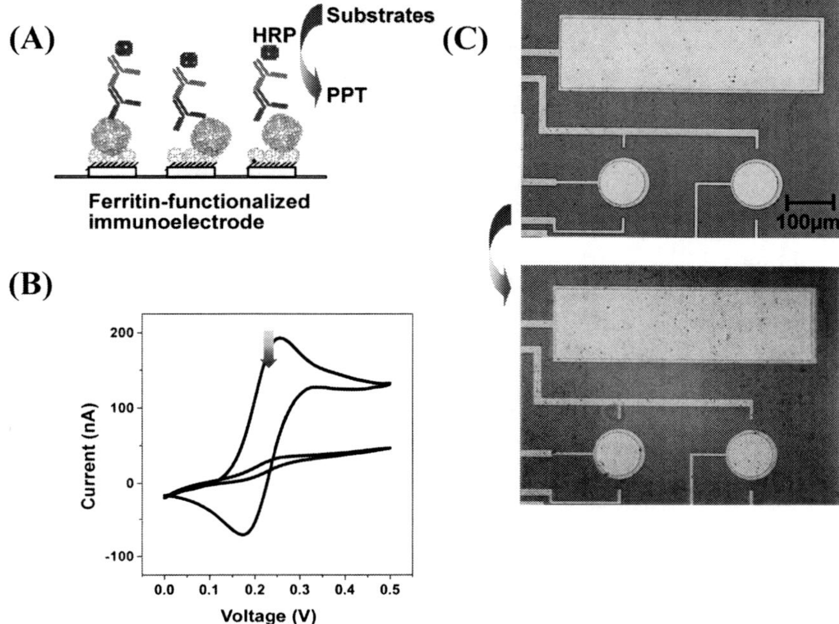

Fig. 8.8. (A) Schematic representation of the procedure for bioaffinity interface construction and the biocatalyzed precipitation reaction for signal generation from a ferritin immunosensor. Component dimensions are not drawn to scale for simplicity. (B) Representative voltammetric sensor signal. (C) Photographs of the signaling result along with the microfabricated immunosensors. Magnified views of an array micropatterned biochip were taken both before and after the signal generation reaction by the biocatalyzed immunoprecipitation. (Modified after Ref. [30]).

stability. Thus, we also focused on the signal stability of affinity sensors. The signal stability from immunosensors is usually not sufficient because the known binding constant between antigens and antibodies are several orders of magnitude lower than the well known and frequently used model of biotin–streptavidin ($K_a = 1 \times 10^{15}$ M^{-1}) [23, 24]. This suggests that the antibody-bound surface undergoes a gradual change, raising the possibility of antibody detachment and signal variation. Also, supplementary label molecules such as enzymes are applied to generate electrochemical signals from immunosensors. In a typical immunoassay, the detection process requires pre-treatment of ligands or expensive commercial reagent to activate reporter molecules such as secondary antibodies because the signaling mechanism is based on the proper enzymatic labeling. Therefore, a detection strategy for electrochemical immunosensors in which the signal is not dependent on the stability of bound antibody molecules to the sensing interface would be valuable. In this regard, we have developed a signal generation method from bioelectrocatalytic immunosensors that does not require routine and cumber-

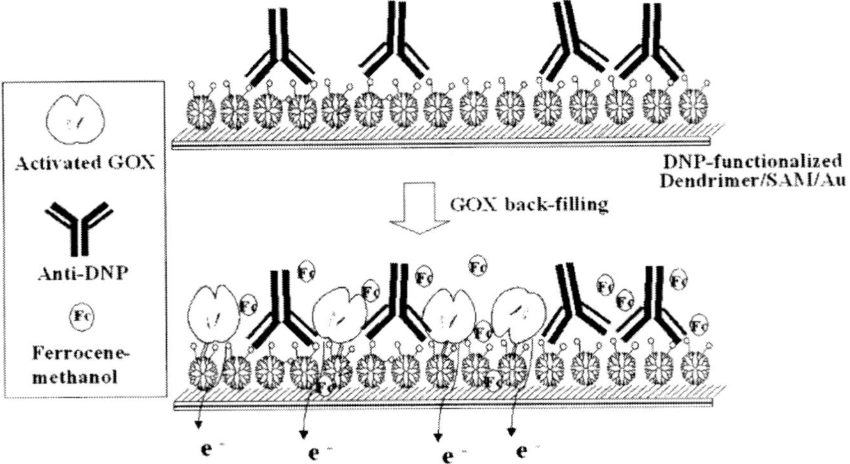

Fig. 8.9. Schematic representation of electrochemical immunosensing with the "back-filling" immobilization of enzyme (GOx) and bioelectrocatalysis (component dimensions not drawn to scale for simplicity). (From Ref. [31]).

some processes, including surface activation and antibody labeling or the use of labeled reporter molecules such as secondary antibodies [31]. The signaling method is based on the "back-filling" covalent immobilization of enzymes onto the immunosensor surface, circumventing the use of enzyme-tagged antibody and alleviating the signal instability from low-affinity binding (Fig. 8.9). In the back-filling assay, the electrochemical signal could be maintained despite the analyte dissociation by its low affinity with immobilized ligand because the signal generating enzyme is separated from the analyte.

As a model biorecognition reaction, a dinitrophenyl (DNP) antigen-functionalized immunosensor surface has been fabricated and the anti-DNP antibody was used as a target analyte [31]. For the construction of immunosensing surface, a poly(amidoamine) G4-dendrimer was employed not only as a building block for the electrode surface modification for antigen-functionalization but also as a matrix for binding of signaling enzyme (GOx). The non-labeled native antibody was biospecifically bound to the immobilized ligand, and the enzyme (periodate-activated GOx) reacted and back-filled the remaining surface amine groups on the dendrimer layer by an imine formation reaction. The DNP functionalization reaction was optimized to facilitate the antibody recognition and signaling; 65% displacement of surface amine to DNP was found to be an optimum (Fig. 8.10A). From quartz crystal microbalance measurements, the immunosensing reaction timing and the surface inertness to nonspecific biomolecular binding were investigated. By changing the DNP functionalization level in the calibration experiments, immunosensors exhibited distinct dynamic detection ranges and limits of detection, supporting the capability of parameter modulation (Fig. 8.10B).

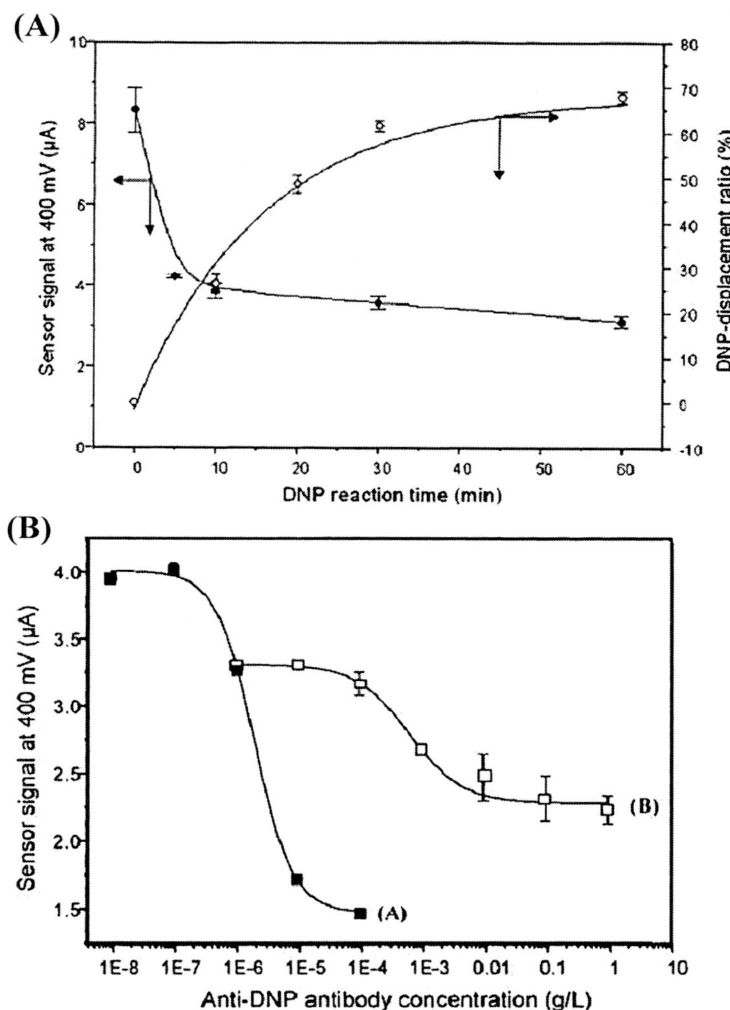

Fig. 8.10. (A) Comparison between the DNP displacement ratio (○) and bioelectro-catalyzed signals from the back-filled GOx (●) as a function of reaction time for DNP-functionalization. Error bars represent 95% confidence limits for three tests. (B) Calibration curves for the DNP/anti-DNP IgG affinity biosensors as a function of target protein concentration. Two types of immunosensors were tested, having different DNP-functionalization ratios of 13% (■) and 65% (□). (Modified after Ref. [31]).

8.3.3 Protein Micropatterning on Sensor Surfaces for Multiplexed Analysis

Micropatterning of biomolecules on solid surfaces has several applications, including the modulation of cell–substrate interactions in biomaterials and tissue engineering, the fabrication of array-type biosensors for multianalyte detection, and genomic/proteomic arrays. We have patterned molecularly organized films of dendrimers having a submicrometer edge resolution on solid chip surfaces such as evaporated gold and glass using microcontact printing technique [32]. Microcontact printing is a non-photolithographic technique that is broadly applicable for the generation of micrometer scale patterns on solid surfaces, e.g., gold, silver, copper, and silicon oxide [33]. In this process, an elastomeric stamp, typically made from poly(dimethylsiloxane) (PDMS), is prepared by casting a prepolymer against a master whose surface has been patterned with a complementary relief structure using conventional photolithography. Microcontact printing has been used to pattern alkanethiolates on metallic substrates, alkylamines on reactive SAM of alkanethiolates, alkyltrichlorosilanes on metal oxides, and proteins with submicrometer spatial resolution by area-selective printing [34, 35].

In our report [32], a patterned thin film of dendrimers was used as the underlying platform for ligand modification for the fabrication of protein microarray. Patterning of biological molecules was attempted on both gold and glass surfaces using fourth generation poly(amidoamine) dendrimer as an interfacing layer between solid surfaces and biomolecules to be patterned (Fig. 8.11). An alcoholic solution of dendrimer was employed as the inking material for the PDMS stamp. As for the patterning of avidin and anti-biotin antibody on gold, dendrimers representing amine groups were printed onto the preactivated 11-mercaptoundecanoic acid SAMs by microcontact printing, followed by biotinylation, and reacted with fluorescein-labeled avidin or anti-biotin antibody. Fluorescence analysis revealed that the patterns of avidin and anti-biotin antibody were successfully constructed within the resolution of less than a micrometer. The dendrimers were also printed onto an aldehyde-activated slide glass and reacted directly with anti-bovine serum albumin (BSA) antibodies that had been oxidized with sodium periodate. Also, distinct patterns of the anti-BSA antibodies were made with a comparable edge resolution to that of avidin patterns on gold (Fig. 8.11B). These results clearly show that dendrimers can be adopted as an interfacing layer for the patterning of biological molecules on solid surfaces with micrometer resolution, presenting a highly functionalizable microarray (plurifunctionality) made of a molecularly ordered nanomaterial.

The micropatterning technology of arranging dozens or hundreds of different biomolecules on a chip surface is required to analyze biospecific recognition events in a concurrent and massive manner. For example, 256 corrals of proteins dimensioned by 50×50 μm is made up on a 1×1 cm chip surface, and a large pool of biomolecules can be distributed at the chip surface by micro-spotting. By using the developed micropatterning technique with amine-terminated dendrimer, array-

Fig. 8.11. (A) Schematic representation of the microcontact printing and patterning of biomolecules on gold and glass. Component dimensions are not drawn to scale for simplicity. MUA = 11-mercaptoundecanoic acid, PF5 = pentafluorophenol, EDAC = 1-ethyl-3-(3-dimethylaminopropyl)carbodiimide hydrochloride, PAMAM = poly(amidoamine), PDMS = poly(dimethylsiloxane). (B) Fluorescence images of patterns constructed on the dendrimer-printed gold surface by using FITC-labeled avidin. Fluorescence profiles across the white lines on images are also registered. (Modified after Ref. [32]).

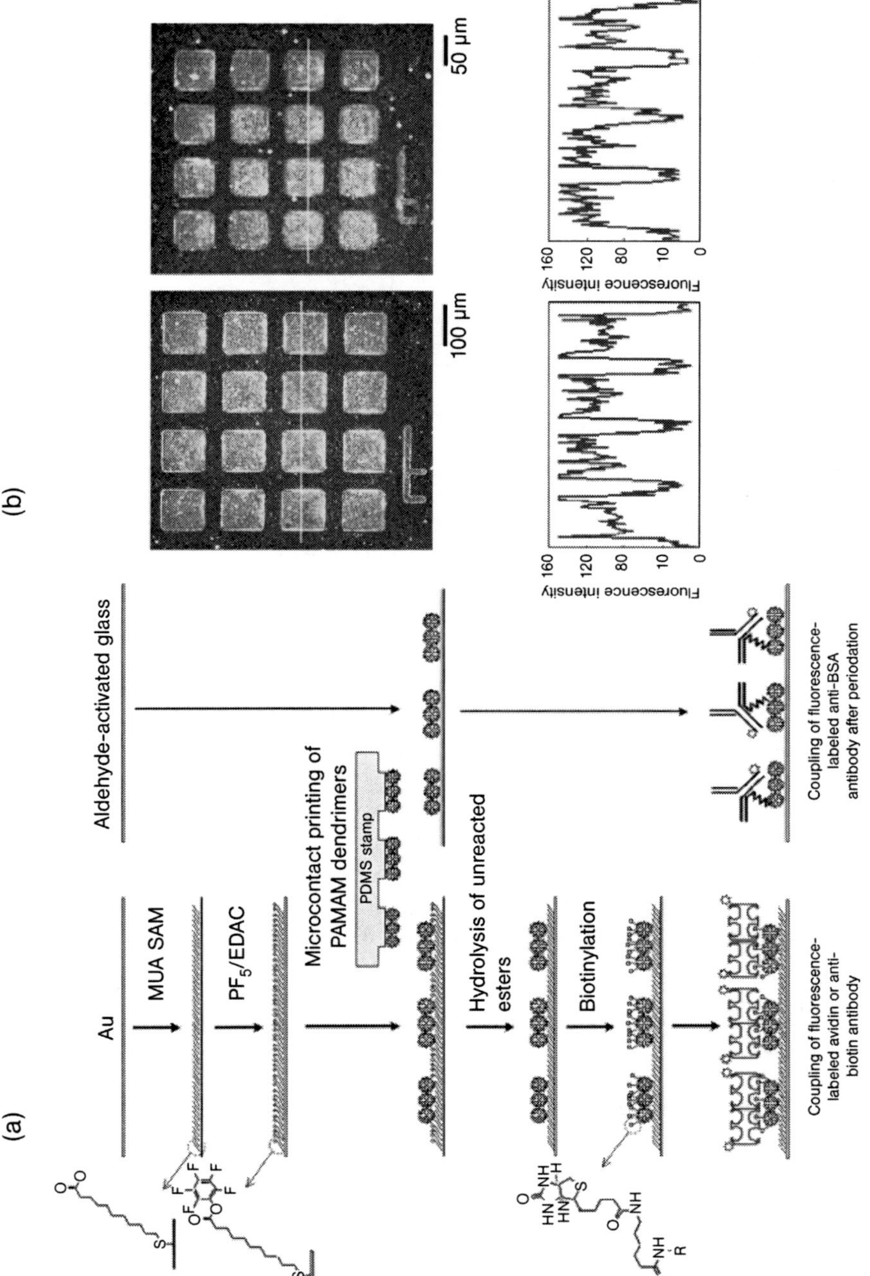

(a)
(b)
Au
MUA SAM
PF5/EDAC
Microcontact printing of PAMAM dendrimers
PDMS stamp
Hydrolysis of unreacted esters
Biotinylation
Coupling of fluorescence-labeled avidin or anti-biotin antibody
Aldehyde-activated glass
Coupling of fluorescence-labeled anti-BSA antibody after periodation
100 µm
50 µm
Fluorescence intensity
160
120
80
10
0

patterned chips could be made having different surface wettability or surface charge between the patterned area and background space, enabling efficient spotting and/or functionalization of target molecules. Multiplexed electrochemical affinity sensing with the micropatterned chips is being pursed, by adopting the aforementioned electrochemical signal generation strategies [25, 31].

8.4 Conclusions

There has been an increasing demand for efficient analytical tools for bioassays in the fields of clinical analysis and biochemical studies. Based on this requirement, a new scientific field in bioassay has been emerging, linking bioanalytical techniques with microelectronics technology. Especially, immuno- or affinity-sensing biochip technology, registering biospecific interactions such as antigen–antibody, ligand–receptor and protein–protein recognition reactions, is under great demand in terms of assay automation and throughput/output. Additionally, recent completion of the human genome project (HGP) and flourishing genomics/proteomics have opened new research fields of high throughput and user-friendly analysis, making the development of biosensing techniques more important.

Current research trends in protein biochips are, mainly, in two directions: the first is the design of biorecognition interfaces/surfaces presenting desired characteristics such as useful surface functionality, adequate immobilization density, biocompatibility, resistance to nonspecific binding, etc. Second is the development of novel transduction techniques, particularly stressing parallel sensing with array-type sensors. In this chapter, we have summarized our recent researches regarding the development of electrochemical biosensors and platform technologies for the effective immobilization of proteins or ligands on solid surfaces. We have developed technologies that use dendrimers as the building unit for biospecific recognition layers based on SAMs technology and accumulated signaling result by using the dendrimer-associated mono-/multilayers. Currently, we are attempting to integrate proteins on chip surfaces in a form of microarrays to prototype the protein chip microsystem.

Acknowledgments

H.C.Y acknowledges support from the Regional Technology Innovation Program of MOCIE (RTI04-03-05) and the ERC(BSEF)/KOSEF.

References

1 Niemeyer, C. M., Mirkin, C. A., *Nanobiotechnology: Concepts, Applications and Perspectivies*, Wiley-VCH, Weinheim, **2004**.

2 Sackmann, E., Tanaka, M., Supported membranes on soft polymer cushions: Fabrication, characterization and applications, *TIBTECH* **2000**, 18, 58–64.

3 Cornell, B. A., Braach-Maksvytis, V. L. B., King, L. G., Osman, P. D. J., Raguse, B., Wieczorek, L., Pace, R. J., A biosensor that uses ion-channel switches, *Nature* **1997**, 387, 580–583.

4 Liu, Q., Wang, L., Frutos, A. G., Condon, A. E., Corn, R. M., Smith, L. M., DNA computing on surfaces, *Nature* **2000**, 403, 175–179.

5 Ulman, A., *An Introduction to Ultrathin Organic Films: From Langmuir-Blodgett to Self-assembly*, Academic Press, Boston, **1991**.

6 Ulman, A., Formation and structure of self-assembled monolayers, *Chem. Rev.* **1996**, 96, 1533–1554.

7 Kumar, A., Abbott, N. L., Kim, E., Biebuyck, H. A., Whitesides, G. M., Patterned self-assembled monolayers and meso-scale phenomena, *Acc. Chem. Res.* **1995**, 28, 219–226.

8 Bain, C. D., Whitesides, G. M., Depth sensitivity of wetting: Monolayers of mercapto ethers on gold, *J. Am. Chem. Soc.* **1988**, 110, 5897–5898.

9 Lahiri, J., Isaacs, L., Grzybowski, B., Carbeck, J. D., Whitesides, G. M., Biospecific binding of carbonic anhydrase to mixed SAMs presenting benzenesulfonamide ligands: A model system for studying lateral steric effects, *Langmuir* **1999**, 15, 7186–7198.

10 Tomalia, D. A., Naylor, A. M., Goddard, W. A. III, Starburst dendrimers: Molecular-level control of size, shape, surface chemistry, topology, and flexibility from atoms to macroscopic matter, *Angew. Chem., Int. Ed. Engl.* **1990**, 29, 138–175.

11 Fischer, M., Vögtle, F., Dendrimers: From design to application – A progress report, *Angew. Chem. Int. Ed. Engl.* **1999**, 38, 884–905.

12 Chow, H.-F., Mong, T. K.-K., Nongrum, M. F., Wan, C.-W., The synthesis and properties of novel functional dendritic molecules, *Tetrahedron* **1998**, 54, 8543–8660.

13 Freemantle, M., Blossoming of dendrimers, *C&EN* **1999**, 77(44), 27–35.

14 Watanabe, S., Regen, S. L., Dendrimers as building blocks for multilayer construction, *J. Am. Chem. Soc.* **1994**, 116, 8855–8856.

15 Tsukruk, V. V., Rinderspacher, F., Bliznyuk, V. N., Self-assembled multilayer films from dendrimers, *Langmuir* **1997**, 13, 2171–2176.

16 Liu, Y., Bruening, M. L., Bergbreiter, D. E., Crooks, R. M., Multilayer dendrimer-polyanhydride composite films on glass, silicon, and gold wafers, *Angew. Chem., Int. Ed. Engl.* **1997**, 36, 2114–2116.

17 Abruna, H. D., Redox and photoactive dendrimers in solution and on surfaces, *Anal. Chem.* **2004**, 76, 310A–319A.

18 Yoon, H. C., Kim, H.-S., Multilayered assembly of dendrimers with enzymes on gold: Thickness-controlled biosensing interface, *Anal. Chem.* **2000**, 72, 922–926.

19 Anzai, J., Kobayashi, Y., Nakamura, N., Nishimura, M., Hoshi, T., Layer-by-layer construction of multilayer thin films composed of avidin and biotin-labeled poly(amine)s, *Langmuir* **1999**, 15, 221–226.

20 Franchina, J. G., Lackowski, W. M., Dermody, D. L., Crooks, R. M., Bergbreiter, D. E., Sirkar, K., Russell, R. J., Pishko, M. V., Electrostatic immobilization of glucose oxidase in a weak acid, polyelectrolyte hyperbranched ultrathin film on gold: Fabrication, characterization, and enzymatic activity, *Anal. Chem.* **1999**, 71, 3133–3139.

21 Yoon, H. C., Hong, M. Y., Kim, H. S., Functionalization of a poly(amidoamine) dendrimer with ferrocenyls and its application to the construction of a reagentless enzyme electrode, *Anal. Chem.* **2000**, 72, 4420–4427.

22 Yoon, H. C., Hong, M. Y., Kim, H. S., Affinity biosensor for avidin using a double functionalized dendrimer monolayer on a gold electrode, *Anal. Biochem.* **2000**, 282, 121–128.

23 Yoon, H. C., Hong, M. Y., Kim, H. S., Reversible association/dissociation reaction of avidin on the dendrimer

monolayer functionalized with a biotin analogue for a regenerable affinity-sensing surface, *Langmuir* **2001**, 17, 1234–1239.

24 Yoon, H. C., Lee, D., Kim, H. S., Reversible affinity interactions of antibody molecules at functionalized dendrimer monolayer: Affinity-sensing surface with reusability, *Anal. Chim. Acta* **2002**, 456, 209–218.

25 Yoon, H. C., Yang, H., Kim, Y. T., Biocatalytic precipitation induced by an affinity reaction on dendrimer-activated surfaces for the electrochemical signaling from immunosensors, *The Analyst* **2002**, 127, 1082–1087.

26 Hong, M. Y., Yoon, H. C., Kim, H. S., Protein-ligand interactions at poly(amidoamine) dendrimer monolayers on gold, *Langmuir* **2003**, 19, 416–421.

27 Bange, A., Halsall, H. B., Heineman, W. R., Microfluidic immunosensor systems, *Biosens. Bioelectron.* **2005**, 20, 2488–2503.

28 Ko, J. S., Yoon, H. C., Yang, H., Pyo, H. B., Chung, K. H., Kim, S. J., Kim, Y. T., A polymer-based microfluidic device for immunosensing biochips, *Lab Chip* **2003**, 3, 106–113.

29 Yoon, H. C., Ko, J. S., Yang, H., Kim, Y. T., Stepwise surface regeneration of electrochemical immunosensors working on biocatalyzed precipitation, *The Analyst* **2002**, 127, 1576–1579.

30 Yoon, H. C., Yang, H., Byun, S. Y., Ferritin immunosensing on microfabricated electrodes based on the integration of immunoprecipitation and electrochemical signaling reactions, *Anal. Sci.* **2004**, 20, 1249–1253.

31 Won, B. Y., Choi, H. G., Kim, K. H., Byun, S. Y., Kim, H. S., Yoon, H. C., Bioelectrocatalytic signaling from immunosensors with back-filling immobilization of glucose oxidase on biorecognition surfaces, *Biotechnol. Bioeng.* **2005**, 89, 815–821.

32 Hong, M. Y., Lee, D., Yoon, H. C., Kim, H. S., Patterning biological molecules onto poly(amidoamine) dendrimer on gold and glass, *Bull. Korean Chem. Soc.* **2003**, 24, 1197–1202.

33 Gates, B. D., Xu, Q., Stewart, M., Ryan, D., Wilson, C. G., Whitesides, G. M., New approaches to nanofabrication: Molding, printing, and other techniques, *Chem. Rev.* **2005**, 105, 1171–1196.

34 Kane, R. S., Takayama, S., Ostuni, E., Ingber, D. E., Whitesides, G. M., Patterning proteins and cells using soft lithography, *Biomaterials* **1999**, 20, 2363–2376.

35 Kung, L. A., Kam, L., Hovis, J. S., Boxer, S. G., Patterning hybrid surfaces of proteins and supported lipid bilayers, *Langmuir* **2000**, 16, 6773–6776.

9
Coordinated Biosensors: Integrated Systems for Ultrasensitive Detection of Biomarkers

Joanne I. Yeh

9.1
Overview

The ability to monitor biorecognition events and interactions on platforms offers pathways to the application of biological macromolecules as detectors. Coupled to the ability to precisely produce conductive elements on the nanoscale, biosensing offers unprecedented avenues for screening and detection at increasing sensitivities. Although biosensors have been an area of active investigation for several years, full realization of their potential has yet to be reached because the rates of reactions and sensitivities are significantly lower than in endogenous, biological systems. This is likely due to the random nature of how the various signal transducing units are placed relative to the electrode. Consequently, integrating the precise 3D information obtained from X-ray crystal structure analysis with nanotechnology platforms can result in a highly enhanced system. This gain is from optimization of the geometrical parameters that make up the various components of the biosensor. A nanobiosensor involves a biological molecule, linker or mediator, and nanoelectrodes; the various components can be equated with the electronic elements of a sensor as every component has to transduce the signal generated at the source (biomolecule) to the detector (electrode). Consequently, as in enzyme systems, rate improvements can occur from proximity and geometric effects, with potential enhancements of 10^2 to 10^3 at each junction. The additive consequence can be a gain of several orders of magnitude in rates, concomitantly improving sensitivities. In this chapter, the concept of coordinated biosensors is introduced, as an approach to align the signal transduction centers to enhance the kinetics of reactions. Additionally, nanoparticles are described that can further augment the systems by positioning the bioelement relative to the electrode surface. The various components of a bionanosensor and their role in enhancing the overall response of the system are described, from a structural perspective.

Nanotechnologies for the Life Sciences Vol. 8
Nanomaterials for Biosensors. Edited by Challa S. S. R. Kumar

ISBN: 978-3-527-31388-4

9.2 Introduction

Several recent reviews illustrates, compares and contrasts biosensor designs [1–5]. The main focus of this chapter is to provide a brief introduction of the biotechnologies behind these systems and to highlight the improvements that can be gained using an integrated approach, utilizing precise three-dimensional (3D) structural information.

The structure determination of deoxyribonucleic acid (DNA) double helix advanced the scientific world in numerous ways. This discovery lead to the understanding of how DNA replication is achieved, resulting in the birth of a new field, molecular biology. The foundation for these discoveries was the X-ray diffraction data from DNA fibers [6]. This scientifically defining result highlights the value structures can have on understanding biological reactions, functions, and states. The ability to explicitly observe the exact positions of atoms of a molecule provides vast insights into important questions, such as how a biomolecule functions, potential interactions to trigger cellular activity, and regulatory mechanisms to control activity. The link between structural biology – the study of the precise conformation and location of the atoms that comprise the molecule and how the molecule folds in three dimensions – and its link to bionanotechnology can be understood from how structural results are used in biotechnological applications. Among the more traditional uses are in structure-based drug design, mutational analysis, and bioengineering. However, a revolutionary approach is to integrate structural results with nanotechnology, as described in this chapter.

Proteins and nucleic acids have diverse roles in the body and studies of their 3D structures teaches us how these function in our bodies and help us understand diseases caused by abnormal forms of these biomolecules. Just as structural biology has provided a definitive fundamental understanding of functional states of biological macromolecules, nanotechnology has provided the platform and a means to bridge the gap between fundamental scientific understanding to enhanced applications [7–11]. A direct connection between the two disciplines can be seen in size: in X-ray crystallography, the resolutions at which molecules are determined are at the atomic dimensions of 10^{-10} m or angstroms (Å). When looking at protein–protein or oligomeric interactions, complexes, and organelles, the working functional "machinery" of a cell, these are on the order of hundreds of angstroms or in the nanometer range. Consequently, there exist complementarities of dimensionalities when linking structural results with nanotechnology applications.

As structural biology has had an immense impact on how scientists visualize biomolecules and their interactions, nanotechnology has changed our concept of working dimensions and applications at the single molecule level. Single eukaryotic cells are at least 250× larger than the nanodevices that are being developed. As a new field, nanotechnology already has had an immense impact on theoretical as well as empirical areas. In bioanalysis, nanoparticles can overcome many of the significant chemical and spectral limitations of more traditional reagents, leading to the ability to detect and monitor on the single molecule level. Fundamentally, as single molecule measurements become possible, profound changes to our understanding of biochemical reactions are occurring. With the removal of ensemble averaging, distributions and fluctuations of molecular properties can be characterized, transient intermediates identified, and catalytic mechanisms elucidated. Recently, it has become apparent that models based on data obtained from a population of molecules do not most accurately reflect the true biochemical nature of reactions [12–15]. This is particularly important when applying biomolecules to various applications because the most efficient means of catalyzing a reaction, for example, would not be one based on an averaged reaction but the most active and efficient one. Consequently, the ability to discriminate differences between single molecules allows us to identify the most effective states of a biomolecule, to obtain enhanced reactions and sensitivities.

The ability to precisely see the positions of atoms from crystallographic structural analysis allows for optimally aligning the various components of a nanobiosensor. Orientation and proximity can have immense effects on activity and this can be readily seen in native biological systems. The ultimate catalysts in nature are enzymes, which have the ability to enhance reactions by 10^5 to 10^{17}-fold. Enzymes hold reactive molecules in precise configurations one by one, orchestrating the formation and breakage of bonds to form products. Although enzyme biosensors have been under study for several years, their use have been limited by the inability to actively harness signals generated by a binding event, such as the electrons produced from a biochemical reaction. The rate constants in electrode-contacted enzymes are far lower than between enzymes and their natural electron acceptors. This decrease has been largely attributed to nonoptimal positioning of the bioelectrocatalyst with the electrode. This is because the active site of a redox enzyme is buried and electrons produced from a reaction cannot efficiently get out to the surface to be used or are not directed to the electrode so they are lost to solution. Consequently, the decreased rates of reactions and amount of enzyme required to obtain detectable signals make these enzyme-based sensors impractical and limited in their sensitivities. The application of structural insights to precisely align the biological components with nanoplatforms can more fully realize the vision of exploiting the innate specificities and enhancements of biological reactions in detection applications. This ability to direct the placement of various active centers (e.g., electron transducing sites or other active units), allows the production of systems with enhanced sensitivities and kinetic properties. This capability is firmly rooted in incorporating results from X-ray crystallography or NMR spectroscopy, techniques allowing for the atomic resolution determination of the 3D conformation of molecules. Consequently, detection of molecular events approximating single molecule sensitivity through integrating advances gained in structural biology and nanotechnology is feasible.

9.3 Elements of a Nanobiosensor

9.3.1 Biomolecular Components

Biosensors are chemical sensors capable of biorecognition through biochemical processes. Generally, bionanosensors contain a biological component that can be a protein (e.g. antibodies, enzymes) or nucleic acids or even whole cells [16, 17]. This bioelement is responsible for the binding and recognition of the target analyte, whether a small molecule or a large protein partner. The binding event is the basis for signal generation and a physical element, such as an electrode, captures the signal as the output. As the electrode, this component translates information from the biological element into a chemical or physical output with a defined sensitivity. The information that is detected can be chemical, energetic such as detection of light, or essentially any information that organisms innately process as all of these signals depend on biological molecules for their generation and/or signal detection and transduction. The array of potential sensory detection is vast, as combinatorial integration can potentially result in detection of multiple signals. The bioelement can consist of antibody/antigen; enzymes, particularly redox enzymes; nucleic acids including DNA, RNA; cellular components such as organelles; and synthetic or semi-synthetic materials. Some will be described briefly.

In antigen/antibody biosensing, interactions of immunogenic partners are the events being detected [19, 20]. The innate sensitivity of antibody–antigen complexes are exploited to detect the presence of antigens although conditions are usually manipulated to minimize nonspecific interactions. The structures of several antibodies have been determined to high resolution.

For enzyme-based biosensors, the detection process relies on the catalytic activity of the system. Enzymes are of great interest as they are natural catalysts, typically proteins but can also be catalytic ribonucleic acids (catalytic RNA). Enzymes can accelerate reactions 10^5- to 10^{17}-fold. Such magnitudes of kinetic amplification are possible through various mechanisms, including proximity and geometrical effects. The structures of enzymes show that they are mostly globular, with their active sites buried at the center or at oligomeric (multimeric) interfaces [21]. This serves to protect the active site from reacting with non-productive molecules, including solvent water molecules. This is achieved by protecting their labile active sites from bulk environment, forming a region with limited accessible and a local microenvironment. The catalytic site can thus have substantially perturbed pK_a and electronegativity values from the bulk solvent; this unique microenvironment is attained through the folded structure of the protein. Accordingly, the catalytic activity of enzymes depends on the integrity of their native protein conformations. Signals and products formed at the active site must somehow be released via a conformational change or through other means. This highlights the essential difficulties in using proteins as the source; immobilizing the protein onto an electrode surface to more directly establish electrical communication between the biocatalyst and electrode can alter the conformation of the protein. Not only can the biocatalyst not be optimally aligned to the electrode surface upon immobilization but the signals generated in a buried active site may not be efficiently routed to allow facile communication between the biocatalysts and electrodes. Consequently, these can limit applications of enzyme biosensors as the electrode-contacted electron transfer rate constants are far lower than those between the enzymes and their natural electron acceptors [22–24]. This highlights an area of optimization in bioelectrode development, where chemical immobilization method for the enzyme can be modified to maximize the enzyme's active conformation. In addition to immobilizing, the method would provide a means of directing the signals generated at the buried active site of the enzyme to the electrode surface, increasing the signals obtained. The 3D structures of numerous enzymes have been determined and these can used to enhance both of these important parameters.

Some of the most familiar systems involve the use of DNA-based sensors. Nucleic acid based biosensors (also called genosensors) are numerous and have been shown to detect nucleic acids at very low concentrations [25–27] (attomole levels). These types of sensors combine nucleic acid layers with electrochemical transducers for sensitivity. Several different approaches have been used to amplify signals, many based on electrochemical effects. Advantages of using nucleic acids for biosensing are their endogenous charge characteristics, allowing for electrochemical detection methods that allow for amplification. There are numerous structures of nucleic acids and most adopt canonical conformations that can be predicted based on sequence information.

9.3.2 Nanoparticles

Semiconductor nanoparticles coordinated to native biological macromolecules sensitive to external stimuli are attractive systems for various applications [28, 29]. As labeling reagents, undecagold particles have been used for several years in electron microscopy applications as phasing reagents. Their application in nanobiotech applications is an extension of their material properties, providing facile surface chemistries that can be exploited to further enhance optical, charge, and magnetic properties of the particles. Several types of nanoparticles have been used as biosensors components. Their function in a biosensor varies, from directing the binding of the biomolecular component to functioning as the sensor component itself. They have been used as probes, recognizing and differentiating an analyte of interest for diagnostic and screening purposes [30]. Detection of an interaction event via nanoparticle probes can be through color, mass, or other changes in physical properties [31–34].

Nanoparticles such as quantum dots, metallic nanobeads, and non-metallic particles such as those based on silica and natural materials such as chitosan and carbon nanotubes have been produced [35, 36]. In addition to their role in mediating biomolecular binding and linkage for electrochemical detection, these nanoparticles have been used in various bioanalytical formats. These include quantitation tags in optical detection of quantum dots, substrates for multiplexed bioassays, mediators of signaling events such as those using gold-based aggregation assays, and potentially as surface catalysts for biological reactions. The sizes of nanoparticles used for most of these applications are from 2 to 50 nm [2, 30].

Another type of nanomaterial is nanopores, molecular sieves that could be transport and possibly serve as inorganic channels. These nanopores have been demonstrated to allow permeation of charged molecules when an external electric field is applied. A protein capable of forming pore structures, α-hemolysin, has been inserted in lipid bilayers coating nanopores to form selective molecular sieves [37]. Other organic and synthetic nanopores have been fabricated in an attempt to form channels and detection systems capable of high throughput and specificity [38, 39]. These nanopores are typically over 5 nm in diameter and larger.

9.3.3 Nanoelectrodes

In nanoscale structures, electrons no longer behave like physical objects that flow in a continuous stream but take on wave mechanical and quantum properties and have the ability to tunnel through structures that would ordinarily be insulators. As single molecule measurements become more feasible with the advent of methods sensitive enough to study single molecule kinetics, thermodynamic, and electronics, significant deviations from ensemble measurements have been found. With the removal of ensemble averaging, distributions and fluctuations of molecular properties can be characterized, transient intermediates identified, and catalytic mechanisms elucidated. Towards facilitating single molecule measurements, nanoelectrode platforms have been investigated as nanosensors for enhancing signals.

A viable nanobiosensor must include an electrode surface that allows proteins to be immobilized yet retain their native structure. Proteins approaching a hydrophobic planar surface will denature upon contact. This phenomenon, called surface denaturation, occurs because the platform serves as a catalyst to the cooperative unfolding process whereby hydrophobic residues of a protein will interact with a surface to expose additional hydrophobic groups. Consequently, immobilization of a protein onto an electrode whose dimensions are significantly larger than the hydrodynamic radius of the macromolecule requires a spacer that will prevent or minimize surface denaturation effects. As described in the next section, peptides and DNA have been used for this purpose; furthermore, these have an additional role as a conductive bridge between the protein and electrode. However, nanoelectrode platforms with dimensions of 20–40 nm diameter and 80–100 nm center-to-center distance on an array platform mostly circumvents surface denaturation effects as their sizes and geometrical aspects results in 3D electrodes, bypassing complications that arise from planar surfaces as well as maximizing surface areas for binding proteins.

Nanoelectrodes formed from carbon nanotube and other materials can display altered properties simply by the nature of their dimensionality [40–43]. Reducing to smaller sizes can result in increased reactive surface areas as well as other effects, such as focusing of electric field due to their geometric configuration [36, 44, 45]. These are desirable effects, enhancing the system through their physical aspects. Carbon nanotube and metallic arrays have been used as electrodes, to promote electron transfer in redox reactions [36, 40, 46].

9.4 Coordinated Biosensors

Advantages in sensitivity and enhanced rates of reactions can be gained from integrating atomic resolution structural information with nanoelectrode arrays, to build detection devices of ultrahigh sensitivity and enhanced kinetics of reactions. In this section, a system utilizing a peroxidase consisting of the components described above will be described, to demonstrate how structural results can be applied to enhance nanobiosensors design. The enzyme NADH peroxidase has been used as the specific detector of hydrogen peroxide, converting a biological binding event into an electronic signal [47, 48]. Although this system used an oxidative metabolism enzyme, other redox proteins can be the bioelement, since detection is based on generation of electrons as the detectable signal. In addition to the traditional use of redox enzymes, other non-redox proteins can be used if the binding of a ligand triggers a conformational change that can be detected by an induced electronic event or via optical, thermal or other detectable physical changes. These interactions provide the basis for catalysis (in enzymes), recognition of antigen (with antibodies), in initiating cellular processes and signal transduction (in receptor systems). Alternatively, a virion or particles can theoretically be the bioelement of a sensor as structural information is available for many of these macromolecules. Our strategy integrates desirable properties of the individual components: the protein machinery for sensitivity and specificity of binding, peptide chemistry for aligning the various electron transducing units, and the nanoelectrodes for gain sensitivity in electronic detection.

9.4.1 Biomolecular Conduits: Signal Transducing Mediators

The use of redox enzymes as the source of electronic signals that can be propagated efficiently in response to a specific ligand or other physiological signal has many potential applications, such as amperometric biosensors [17, 18, 49]. A major limitation in using biomolecules in electrodes is that their incorporation is complicated by a lack of efficient pathways for the transport of electrons from their embedded redox sites to an electrode, as described above. Although electrical communication between redox proteins and electrode surfaces has been improved through various approaches [17, 19, 50–52], these methods involve contact of the enzyme at somewhat random conformations with respect to the electron relaying units, ultimately resulting in decreased rates of electron transfer. To produce electroactive biocatalysts in amperometric biosensors with sufficiently high rates of electron transfer, the relaying units must be linked in a controlled and mechanistically relevant manner [53–57]. The following section describes harnessing of electrons through a modularly formed, redox active assembly consisting of a redox enzyme, a metallized double-helical peptide, and a gold nanoparticle immobilized onto a carbon nanotubes electrode and a gold wire derivatized with a benzenedithiol compound. Each of these components and their role in biosensing has been described before. One unique component of this bioassembly is the use of metallized peptide as the conduit of electronic signals.

Peptides and DNA linkers can be made conductive through binding of metals and functionalized to incorporate specific reactive groups. The overall redox potential is driven by the coordination of the particular metals to the peptide. The reductive potentials reflect the ease or difficulty in transitioning between oxidized and reduced states for a particular metal ion. Imidazolium chemistry is well defined and allows for the binding of divalent metals. Depending on the metal, the potentials needed to drive the oxidative/reductive states of the metals can be shifted through precise ligation [58, 59]. Consequently, application of specific coordination chemistry through various ligation groups of a peptide or DNA allows for the more directed binding and stabilization of a subsequent redox state. This is an important consideration as the peptide or DNA linkers form a junction and functions in transducing the signals from the bioelement to the electrode. Functionalization of peptides and DNA allows for targeting, in principle, of virtually any protein, as long as unique groups (e.g., reactive amines, thiols, hydroxides) are present and assessable. In addition to their roles in linking and signal transduction, they have a physical function of spacing the protein away from the surface of the electrode, helping to minimizing surface denaturation effects.

Using the unique cysteine-sulfenic acid chemistry of the Npx redox center, described below, we designed a metallizable, multi-histidine peptide (MHP), with differentially functionalized termini that linked to a specific component of the signaling assembly (Fig. 9.1) and of sufficient length to penetrate the active site.

A 33-amino acid sequence from the leucine zipper region of the GCN4 transcription factor was used to design metal binding sites along a face of the helix by introducing histidines at i, $i+4$ positions [60]. These histidines bind divalent metals while still permitting the formation of a stable dimer with metals bound along each face of the double-stranded peptide, as confirmed through circular dichroism analysis. A reduced thiolate moiety at the carboxyl terminus (CT) reacts with the active site sulfenate of Npx to form a cysteine bond. Formation of the enzyme–peptide complex was confirmed by mass spectrometry (Fig. 9.2).

The peptide has a helical axis 47 Å long, sufficient to penetrate into the active site that is buried centrally in the enzyme. Before formation of the peptide–AuNP complex in the next step, cobalt metallization of the peptide was achieved by incubation of the MHP with $CoCl_2$, forming a substituted metallized peptide (CoMHP) (confirmed by mass spectrometry). The stoichiometry of Co to peptide duplex was determined by isothermal titration calorimetry, which resulted in a stoichiometry of 2 to 4.7, depending on conditions and design, Co per peptide strand. Further confirmation of binding to various divalent metals came through X-ray fluores-

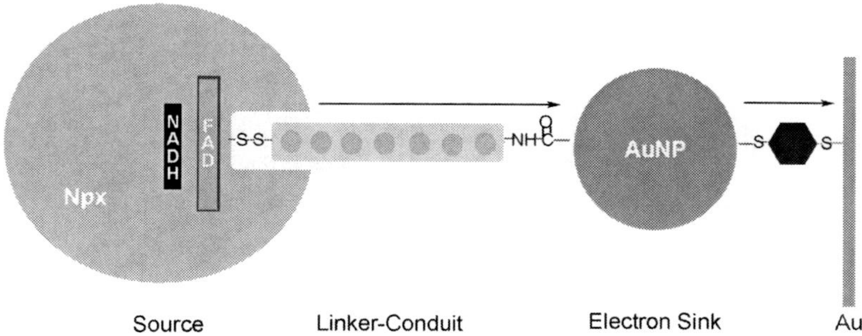

Fig. 9.1. Schematic of assembly of AuNP-Co-MHP-Npx electrode through stepwise linkage of each component with adsorption of the AuNP linked complex reconstituted onto a dithiol monolayer associated with an Au electrode. An electroactive monolayer assembly consisting of NADH peroxidase enzyme ("source"), CoMHP, ("linker-conduit"), and AuNP ("electron sink") was constructed by stepwise linkage of components and covalently connected to a monolayer on a macroscopic Au electrode.

cence spectroscopy (EXAF) measurements performed at beamline X8C at Brookhaven National Laboratory. EXAF results established that the peptide binds Ni^{2+}, Zn^{2+}, and Co^{2+}. For the assembly, Co was used as the Co^{3+}/Co^{2+} redox couple has reported values of 0.22–0.295 V vs. SCE for imidazolium complexes [58] and 0.37 to 0.4 V for conjugated ring complexes [59]. The mechanism by which electrons are transmitted through this assembly system likely requires the formation of the Co^{3+} state, which is transiently reduced back to Co^{2+} during the transduction of electrons upon NADH binding. The structure of the MHP peptide in the presence and absence of cobalt have been determined to high resolutions of 1.6 and 1.7 Å. These results show that the peptides not only maintain their structural integrity after mutation to histidines but that their elongated structure is even more pronounced (Fig. 9.3), confirming the design principles originating from the structural results of the intact GCN4 protein.

9.4.2
NADH Peroxidase: the Biocatalytic Element

For the peroxidase sensor [61], the assembly incorporated a bacterial NADH peroxidase (Npx), a 49 kDa flavoenzyme, that catalyzes the two-electron reduction of H_2O_2 into H_2O. This reduction is initiated upon binding of the cofactor NADH, which donates two electrons through the primary redox center, FAD. Under oxidizing conditions, the electrons are passed onto a molecule of peroxide, bound to the secondary redox center, a thiolate residue. This residue progresses from the reduced thiolate ($-S^-$) to the oxidized sulfenate ($-SO^-$) states during the catalytic cycle [48]. The 3D crystal structures of wild-type, oxidized and reduced forms of Npx [61] have been solved to atomic resolution so that the precise conformation of the active site residues and steric limitations within the active site are well-defined.

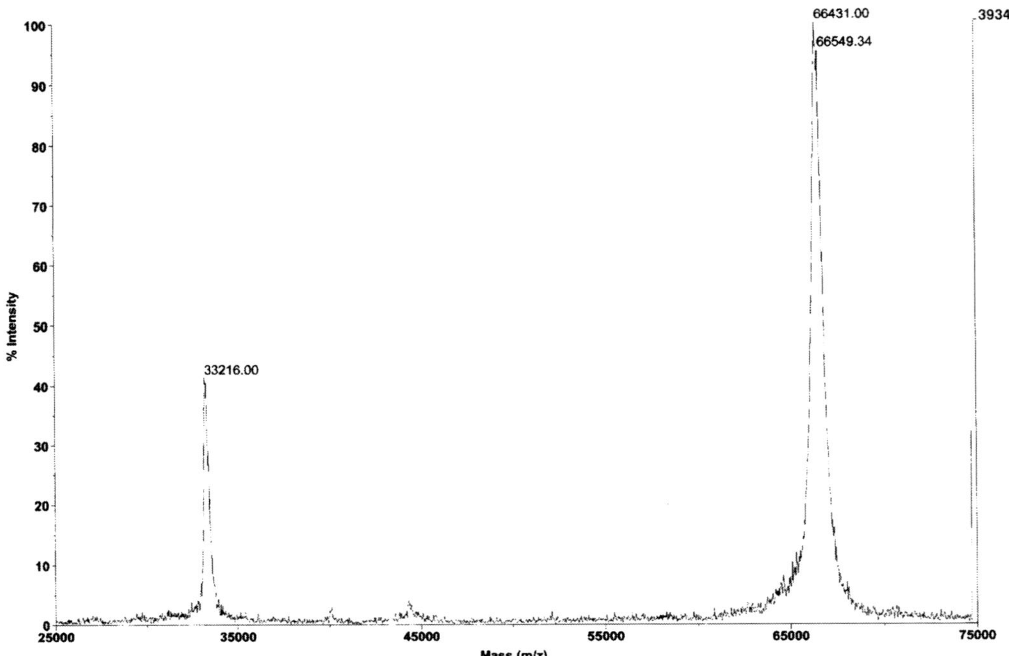

Fig. 9.2. Matrix-assisted laser desorption time-of-flight mass spectrometry (MALDI-TOFMS) analysis of the complex formed upon reaction of the peptide with the NADH peroxidase (Npx). A gold nanoparticle was also part of the complex and the resulting complex molecular weight is between 66431 and 66549 daltons (Da). The molecular weights of each component are Npx (49 500 Da), MHP (3934 Da), and gold nanoparticle (13 115 Da). MS analysis can be valuable in confirming the formation of various complexes as the bioassembly consists of multiple components.

The current resulting from NADH binding to Npx and the concomitant electron transfer through the assembly was measured using an electrochemical cell consisting of the biocatalytic molecular wire assembly ("working electrode") and an Au-wire reference electrode. Solution conditions for current measurements were the same as that used for enzyme activity measurements as these are optimized for NADH binding [47]. Before initiating the reactions, we applied a slight potential of 10 mV to the working electrode, to offset the difference in potential between the reference and working electrodes originating from immobilization of the assembly. It was expected that a charging potential may be necessary to overcome the tunneling barrier that may arise from the dithiol monolayer that bridges the AuNP to the Au bulk electrode, as reported previously [49]. However, other than the 10 mV offset potential applied, additional charging potential did not appear to be necessary in this system.

Once the system was equilibrated in the reaction buffer, an aliquot of H_2O_2 solution was added. This generated a signal in the positive Y-direction, indicating that

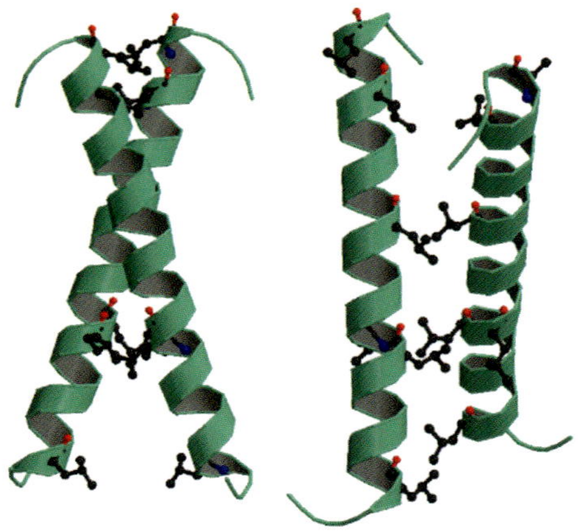

Fig. 9.3. Structure of the metallizable peptide (MHP), shown in ribbon depiction, in the absence of metals, refined to 1.6 Å resolution. These structural results confirm the conformational rigidity of the peptide, which is essential for efficient electron transduction. The leucines, which hold the peptide together, are shown in gray ball-and-stick, at the interface of the two strands. Two views are shown, differing by 90° along the helical axis.

electrons were flowing through the circuit toward the working electrode from the reference electrode. The role of hydrogen peroxide is two-fold – to oxidize unreacted free thiols prior to initiating the enzyme-mediated reaction and, more importantly, to generate the Co^{3+} oxidation state of the metal. As expected, the reaction of H_2O_2 has a time constant longer than that generated by NADH, whose binding initiates the electron transfer reaction mediated through the binding of the cofactor by the enzyme. Hence, H_2O_2 current generation represents reactions that are not mediated through the enzyme itself but rather through oxidation effects at the surface and on the metal centers. Oxidation of metals bound in the peptide primes the assembly for electron transduction. This is an important factor and aids in driving the overall reaction from the enzyme active site to the electrode. After the H_2O_2 signal returned to the starting baseline value, an aliquot of NADH was added to reach a final concentration of 0.3 μm. The current generated in response to NADH binding reached maximum values within 100 ms, indicating fast kinetics of electron transfer. Anodic current peak density values obtained for electrodes with three different surface areas were 50 nA for an electrode coated with 10^8 molecules of assembly and 85 μA for 10^{11} molecules of immobilized assembly. Two concurrent processes appeared to occur with the addition of NADH – a fast electron transfer signal that is transduced through the assembly to the electrode and a slower event that is the reduction of the disulfide bond linkage between the enzyme and the metallized peptide [61]. This reduction resulted in eventual liberation of the enzyme from the assembly and limited the current producing cycles of the assembly, providing additional verification of the route of electron transfer in this system.

9.4.3 Undecagold Nanoparticle: Role in Alignment and Directing Electron Flow

For the peroxidase sensor described, an undecagold nanoparticle was used to further help align Npx on the electrode surface. A sulfo-N-hydroxysuccinimido activated carboxyl group on the AuNP allows the formation of a covalent link to the MHP through a nucleophilic attack by the amino terminus primary amine group of MHP. The AuNP-CoMHP component is reacted with the Npx enzyme to form a disulfide bond between the carboxy terminal cysteine of the peptide and the redox active cysteine of Npx. For the final reconstitution of the Npx-CoMHP-AuNP assembly onto the Au electrode, a self-assembled monolayer was formed by adsorption of 1,4-benzenedithiol (BDT) onto Au wires. This provided the reaction surface for final immobilization by reaction of the AuNP with thiols of the BDT monolayer. The tri-modular assembly was formed on an Au electrode wire with a geometrical area of 0.25 cm^2 and a roughness factor of $\sim$1.2, functionalized with the 1,4-benzenedithiol monolayer. Surface coverage of the electrode was determined by cyclic voltammetry scans of the 1,4-benzenedithiol derivatized electrode, which indicated a number density of thiol groups of 3.2×10^{-10} mol cm^{-2}. Our first route for assembly formation prior to immobilization on the Au-electrode was non-optimal, which resulted in 10^8 molecules of assembly per 0.25 cm^2 electrode surface. This was found to be due largely to incubation times of 1–2 h for the initial CoMHP-AuNP formation. We have found that increasing the incubation times to overnight at 4 °C resulted in enhanced linkage between the peptide and AuNP, producing an electrode surface coverage of 10^{11} molecules cm^{-2}.

Using the gold nanoparticle, we labeled as described above then reacted the assembly through the MHP N-terminus to carbon nanotube arrays. The reaction between the peptide's N-terminus to the carboxyl groups at the tips of carbon nanotube arrays was catalyzed by EDC/NHS chemistry. Scanning electron microscopy of the resulting array shows that the bioassembly is localized to the tips of the electrodes and that this reaction can be quantitated (Fig. 9.4).

9.4.4 Integrated Signals

Based on reaction rate enhancements reported due to proximity and orientation effects in native enzyme and receptor systems, a coordinated assembly whose components are optimally tethered on an electrode would be expected to enhance rates of reactions by up to $\sim$100-fold [62, 63]. Furthermore, the tethering of a signaling assembly containing the redox enzyme onto an electrode could enhance electronic signal transduction rates by minimizing loss of signal to the solution. Particularly relevant to the formation of a bioelectrocatalytic system is the spatial orientation of the signaling unit and all its components, which must be reconstituted in a conformation that is conducive for charge transfer. Exhibition of enhanced electron transfer rates through the enzyme assembly, compared with the native rate of

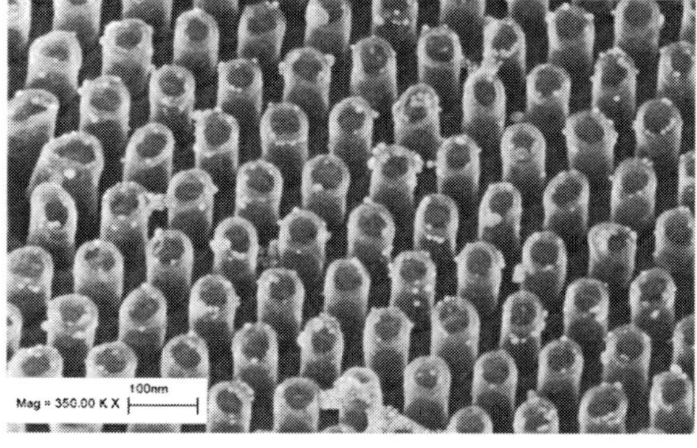

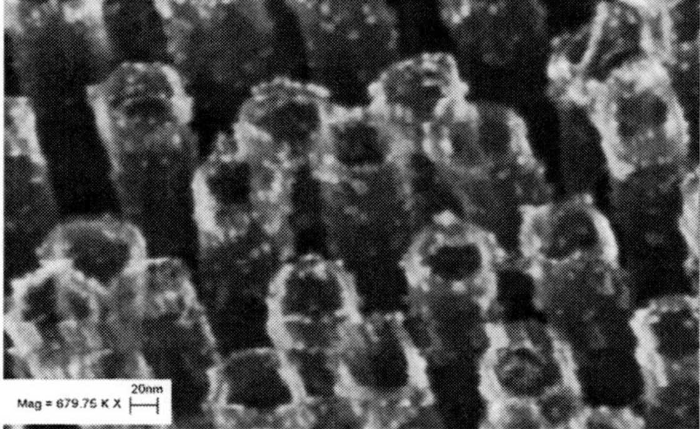

Fig. 9.4. Scanning electron microscopy (SEM) of an NADH peroxidase assembly conjugated at the tips of highly ordered carbon nanotubes (CNTs). In the array, the CNTs are 50 nm in diameter, have walls of 3-nm thick, and exhibit an exposed length of 60 nm, a total length of 10 μm, and a center-to-center spacing of 100 nm between adjacent tubes. The reaction conditions can be modified to link quantitatively the bioassembly onto the electrode platform.

unimolecular enzyme reduction by NADH, the electron donor, would suggest that the signaling units are aligned properly for electron transfer.

The currents measured correlate to an electron transfer rate constant on the order of 3000 s^{-1} within each assembly. This electron transfer rate is two orders of magnitude higher than the endogenous electron transfer rate from NADH to the native enzyme, 27 s^{-1}. This rate indicates that the metallized peptide is in an optimal conformation for electron transfer, and, in conjunction with the nanoparticle, forms effective conduits for the electrical signals.

An advantage to this system is that, in addition to the atomic resolution structures of Npx, the kinetics of this enzyme have been well characterized so that the rate constants of several steps in the reaction cycle are known. This information helps correlate the electron transfer rate constant derived from the assembly currents measured in this system with the native reaction rates and allows an assessment of the efficiency of the designed assembly in electronic signal transduction. Using the anodic currents and the number of assembly molecules on the electrode surface, we calculate a first-order electron transfer rate constant of $k_{et} = 3000\ s^{-1}$. This value is more than 100× larger than that of the reduction of the oxidized enzyme by NADH ($27\ s^{-1}$), the most comparable unimolecular kinetic step characterized in the native enzyme. We believe that the rate-determining step for electron transfer in this system is the initial binding of NADH, and that subsequent steps in electron transduction are fast compared with this event.

Metallized peptides appear to be an efficient mediator of electrons. The precise mechanism by which electrons are propagated through the peptide remains to be studied although, based on the 5.2 Å distance between cobalt ions modeled into the peptide, a hopping mechanism seems more likely than through-bond transduction. The use of metallized peptides as conduits for electronic signals can be extended to various systems, offering a convenient approach for the design of modular functional linkers. Linking of biomolecules to electronic circuitry can be essential to the design of efficient signal transduction assemblies in bioresponsive systems. The use of metal nanoparticles that act as electrical nanoplugs helps to align the enzyme on the conductive support. In our work, the AuNP appears to channel the current to the electrode to help maximize the efficiency of electron transfer.

This peroxidase biosensor initiates and conducts redox signals in the presence of H_2O_2 and NADH. The current generated by the binding of NADH, the electron donor, was transduced through the molecular assembly with high efficiency. A key component of the system is the metallized peptide and the formation of M^{3+}/M^{2+} states that allow for the transduction of the electronic signal. The gain in electron transfer rates and concomitant sensitivity is likely a result of positioning of the electron transducing units (Fig. 9.5), which provides an improvement at each junction, which can additively influence the overall rates of the biocatalytic element.

9.5 Conclusion

Important technological advances in the past few years have provided the tools needed to develop new technologies to monitor biorecognition and interaction events on solid devices and in solution [7, 64]. In conjunction with the ability to fabricate solid substrates with nanoscale features and precisions, biosensing offers unprecedented opportunities for screening and detection. In this chapter, the approach of integrating 3D structural information to align the various signal transducing elements in a nanobiosensor has been described. This can result in coordinated biosensing, where the elements of a biosensor are optimized geometrically to enhance kinetics of reactions.

To illustrate the concepts presented here, the NADH peroxidase biosensor was described as this was one of the first systems that demonstrated the gain that

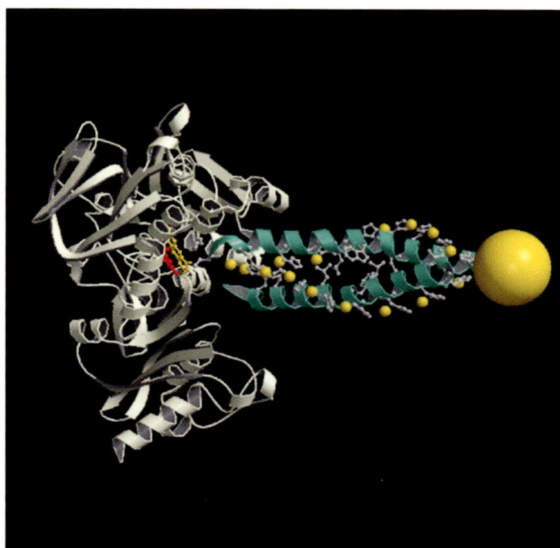

Fig. 9.5. Model of the assembly based on the atomic coordinates of the NADH peroxidase and MHP. The peptide coordinates cobalt atoms (small yellow spheres) through the histidine residues at every i, $i+4$ positions. AuNP was modeled in as a gold sphere, to scale with the biological molecules, with a diameter of 14 Å. The cofactors are shown in ball-and-stick, with nicotinamide adenine dinucleotide in red and flavin adenine dinucleotide in yellow.

can be obtained in by applying atomic resolution structural results to align transducing centers. These results highlight the feasibility of using these biosensors in various applications, including nanomedicine, where detecting markers of disease states requires high sensitivity to make them viable in clinical applications. This approach, applying detailed structural information at sub-nanometer resolutions, produces sensors at the nanometer dimensions that fully integrate the various biological and electronic elements to produce ultrasensitive detectors.

The efficiency, specificity, and modularity of biomolecules such as proteins and nucleic acids make them attractive material for designing highly sensitive and intelligent circuitry. Utilization of natural macromolecules as sensors exploits the inherent recognition and diversity of these molecules. Effective integration of biomolecules with nanoelectronic circuitry holds great promise for the design of compact and highly sensitive systems. Direct electronic coupling with high efficiencies between electrodes and enzymes can be achieved through several approaches and highlights the feasibility of fully utilizing biomolecules in nanobiosensors. The new field of nanomedicine, as with other studies on the nanoscale, can lead to substantially new breakthroughs in technologies, leading to new paradigms in medical detection and treatments. Many biomedical applications can benefit substantially from the use of coordinated biosensors.

Acknowledgments

We thank DARPA (F49620-03-1-0365) and NIH (GM066466) for funding.

References

1 Mosharov, E.V., Sulzer, D., Analysis of exocytotic events recorded by amperometry. *Nat. Methods* **2005**, 2, 651–658.

2 Medintz, I., Uyeda, H.T., Goldman, E.R., Mattoussi, H., Quantum dot bioconjugates for imaging, labeling, and sensing. *Nat. Mater.* **2005**, 4, 435–446.

3 Drummond, T.G., Hill, M.G., Barton, J.K., Electrochemical DNA sensors. *Nat. Biotechnol.* **2003**, 21, 1192–1199.

4 Ramsey, M., DNA chips: State-of-the-art. *Nat. Biotechnol.* **1998**, 16, 40–44.

5 Rosenthal, S.J., Bar-coding biomolecules with fluorescent nanocrystals. *Nature* **2001**, 7, 621–622.

6 Watson, J., Crick, F., A structure for deoxyribose nucleic acid. *Nature* **1953**, 171, 737–738.

7 Kubik, T., Bogunia-Kubik, K., Sugisaka, M., Nanotechnology on duty in medical applications. *Curr. Pharm. Biotechnol.* **2005**, 6, 17–33.

8 Storhoff, J.J., Elghanian, R., Mucic, R.C., Mirkin, C.A., Letsinger, R.L., One-pot colorimetric differentiation of polynucleotides with single base imperfections using gold nanoparticle probes. *J. Am. Chem. Soc.* **2000**, 120, 1959–1964.

9 Taton, T.A., Mirkin, C.A., Letsinger, R.L., Scanometric DNA array detection with nanoparticle probes. *Science* **2000**, 289, 1757–1760.

10 Katz, E., Willner, I., Integrated nanoparticle-biomolecule hybrid systems: Synthesis, properties, and applications. *Angew. Chem. Int. Ed.* **2004**, 43, 6042–6108.

11 Armstrong, F.A., Recent developments in dynamic electrochemical studies of adsorbed enzymes and their active sites. *Curr. Opin. Chem. Biol.* **2005**, 9, 110–117.

12 van Oijen, A.M., Blainey, P.C., Crampton, D.J., Richardson, C.C., Ellenberger, T., Xie, X.S., Single-molecule kinetics of l exonuclease reveal base dependence and dynamic disorder. *Science* **2003**, 301, 1235–1238.

13 Bustamante, C., Bryant, Z., Smith, S.B., Ten years of tension: Single-molecule DNA mechanics. *Nature* **2003**, 421, 423–427.

14 Yasuda, R., Noji, H., Yoshida, M., Kinosita Jr., K., Itoh, H., Resolution of distinct rotational substeps by submillisecond kinetic analysis of F1-ATPase. *Nature* **2001**, 410, 898–904.

15 Finer, J.T., Simmons, R.M., Spudich, J.A., Single myosin molecule mechanics: Piconewton forces and nanometre steps. *Nature* **1994**, 368, 113–119.

16 Armstrong, F.A., Wilson, G.S., *Electrochim. Acta* **2000**, 45, 263.

17 Willner, I., Katz, E., Integration of layered redox proteins and conductive supports for bioelectronic applications. *Angew. Chem. Int. Ed.* **2000**, 39, 1180–1218.

18 Heller, A., Amperometric biosensors. *Curr. Opin. Biotechnol.* **1996**, 7, 50–54.

19 Sleytr, U.B., Pum, D., Sara, M., Advances in S-layer nanotechnology and biomimetics. *Adv. Biophys.* **1997**, 34, 71–79.

20 Sleytr, U.B., Bayley, H., Sara, M., Breitwieser, A., Kupcu, S., Mader, C., Weigert, S., Unger, F.M., Messner, P., Jahn-Schmid, B., Schuster, B., Pum, D., Douglas, K., Clark, N.A., Moore, J.T., Winningham, T.A., Levy, S., Frithsen, I., Pankovc, J., Beale, P.,

GILLIS, H.P., CHOUTOV, D.A., MARTIN, K.P., Applications of S-layers. *FEMS Micobiol. Rev.* **1997**, 20, 151–157.

21 BRANDEN, C., TOOZE, J., An example of enzyme catalysis: Serine proteases, in *Introduction to Protein Structure*, 2nd edn., Garland Publishing, New York, 1999, pp. 205–220.

22 GERARD, M., CHAUBEY, A., MALHOTRA, D.B., Application of conducting polymers to biosensors. *Biosens. Bioelectron.* **2002**, 17, 345–359.

23 OHARA, T.J., RAJAGOPALAN, R., HELLER, A., "Wired" enzyme electrodes for amperometric determination of glucose or lactate in the presence of interfering substances. *Anal Chem.* **1994**, 66, 2451–2457.

24 BADIA, A., CARLINI, R., FERNANDEZ, A., BATTAGLINI, F., MIKKELSEN, S.R., ENGLISH, A.M., Intramolecular electron-transfer rates in ferrocene-derivatized glucose oxidase. *J. Am. Chem. Soc.* **1993**, 115, 7053.

25 WANG, J., POLSKY, R., XU, D.K., Silver-enhanced colloidal gold electrochemical stripping detection of DNA hybridization. *Langmuir* **2001**, 17, 5739–5741.

26 PARK, S.J., TATON, T.A., MIRKIN, C.A., Array-based electrical detection of DNA with nanoparticle probles. *Science*, **2002**, 295, 1503–1506.

27 WANG, J., POLSKY, R., MERKOCI, A., TURNER, K.L., "Electroactive beads" for ultrasensitive DNA detection. *Langmuir* **2003**, 19, 989–991.

28 ANDERSON, J.L., BOWDEN, E.F., PICKUP, P.G., Dynamic electrochemistry: Methodology and application. *Anal. Chem.* **1996**, 68, 379R–444R.

29 HAMAD-SCHIFFERLI, K., SCHWARTZ, J.L., SANTOS, A.T., ZHANG, S., JACOBSON, J.M., Remote electronic control of hybridization through inductive coupling to an attached metal nanocrystal antenna. *Nature* **2002**, 415, 152–155.

30 PENN, S.G., HE, L., NATAN, M.J., Nanoparticles for bioanalysis. *Curr. Opin. Chem. Biol.* **2003**, 7, 609–615.

31 SANTRA, S., ZHANG, P., WANG, K., TAPEC, R., TAN, W., Conjugation of biomolecules with luminophore-doped silica nanoparticles for photostable biomarkers. *Anal. Chem.* **2001**, 73, 4988–4993.

32 HAN, M., GAO, X., SU, J.Z., NIE, S., Quantum-dot-tagged microbeads for multiplexed optical coding of biomolecules. *Nat. Biotechnol.* **2001**, 19, 631–635.

33 VAN DE RIJKE, F., ZIJLMANS, H., LI, S., VAIL, T., RAAP, A.K., NIEDBALA, R.S., TANKE, H.J., Up-converting phosphor reporters for nucleic acid microarrays. *Nat. Biotechnol.* **2001**, 19, 273–276.

34 ZHENG, J., DICSON, R.M., Individual water-soluble dendrimer-encapsulated silver nanodot fluorescence. *J. Am. Chem. Soc.* **2002**, 124, 13 982–13 983.

35 YI, H., WU, L.-Q., BENTLY, W.E., GHODSSI, R., RUBLOFF, G.W., CULVER, J.N., PAYNE, G.F., Biofabrication with chitosan. *Biomacromolecules* **2005**, 6, 2881–2894.

36 GRUNER, G., Carbon nanotube transisitors for biosensing applications. *Anal. Bioanal. Chem.* **2006**, 384, 232–335.

37 ASTIER, Y., BRAHA, O., BAYLEY, H., Toward single molecule DNA sequencing: Direct identification of ribonucleoside and deoxyribonucleoside 5′-monophosphates by using an engineered protein nanopore equipped with a molecular adapter. *J. Am. Chem. Soc.* **2006**, 128, 1705–1710.

38 LI, N., YU, S., HARRELL, C.C., MARTIN, C.R., Conical nanopore membranes. Preparation and transport properties. *Anal Chem.* **2004**, 76, 2025–2030.

39 HARRELL, C.C., LEE, S.B., MARTIN, C.R., Synthetic single-nanopore and nanotube membranes. *Anal Chem.* **2003**, 75, 6861–6867.

40 WITHEY, G.D., LAZARECK, A.D., TZOLOV, M.B., YIN, A., AICH, P., YEH, J.I., XU, J.M., Ultra-high redox enzyme signal transduction using highly ordered carbon nanotube array electrodes. *Biosens. Bioelectron.* **2006**, 21, 1560–1565.

41 SHIM, M., SHI KAM, N.W., CHEN, R.J., LI, Y., DAI, H., Functionalization of carbon nanotubes for biocompati-

bility and biomolecular recognition. *Nano Lett.* **2002**, 2, 285.

42 Hyang, W., Taylor, S., Fu, K., Lin, Y., Zhang, D., Hanks, T.W., Rao, A.M., Sun, Y.-P., Attaching proteins to carbon nanotubes via diimide-activated amidation. *Nano Lett.* **2002**, 2, 311.

43 Chen, R.J., Zhang, Y., Wang, D., Dai, H., Noncovalent sidewall functionalization of single-walled carbon nanotubes for protein immobilization. *J. Am. Chem. Soc.* **2001**, 123, 3838.

44 Bradley, K., Briman, M., Star, A., Gruner, G., Charge transfer from adsorbed proteins. *Nano Lett.* **2004**, 4, 253.

45 Star, A., Bradley, K., Gabriel, J.-C.P., Gruner, G., Nanoelectronic sensors: Chemical detection using carbon nanotubes. *Pol. Mater: Sci. Eng.* **2003**, 89, 204.

46 Kohli, P., Harrell, C.C., Cao, Z., Gasparac, R., Tan, W., Martin, C.R., DNA-functionalized nanotube membranes with single-base mismatch selectivity. *Science* **2004**, 305, 984–986.

47 Parsonage, D., Miller, H., Ross, P.R., Claiborne, A., Purification and analysis of streptococcal NADH peroxidase expressed in E. coli. *J. Biol. Chem.* **1995**, 268, 3161–3167.

48 Poole, L.B., Claiborne, A., Interactions of pyridine nucleotides with redox forms of the flavin-containing NADH peroxidase from Streptococcus faecalis. *J. Biol. Chem.* **1986**, 261, 14 525–14 533.

49 Wang, J., Amperometric biosensors for clinical and therapeutic drug monitoring: A review. *J. Pharm. Biomed. Anal.* **1999**, 19, 47–53.

50 Schuhmann, W., Ohara, T.J., Schmidt, H.-L., Heller, A., Electron transfer between glucose oxidase and electrodes via redox mediators bound with flexible chains to the enzyme surface. *J. Am. Chem. Soc.* **1991**, 113, 1394.

51 Willner, I., Heleg-Shabtai, V., Blonder, R., Katz, E., Tao, G., Bückmann, A.F., Heller, A., Electrical wiring of glucose oxidase by reconstitution of FAD-modified monolayers assembled onto Au-electrodes, *J. Am. Chem. Soc.* **1996**, 118, 10 321.

52 Rajagopalan, R., Aoki, A., Heller, A., Effect of quaternization of the glucose oxidase "wiring" redox polymer on the maximum current densities of glucose electrodes, *J. Phys. Chem.* **1996**, 100, 3719.

53 Medvedev, D.M., Daizadeh, I., Stuchebrukhov, A.A., Electron transfer tunneling pathways in bovine heart cytochrome C oxidase. *J. Am. Chem. Soc.* **2000**, 122, 6571–6582.

54 Xiao, Y., Patolsky, F., Katz, E., Hainfeld, J.F., Willner, I., "Plugging into enzymes": Nanowiring of redox enzymes by a gold nanoparticle. *Science* **2003**, 299, 1877–1881.

55 Yeh, J.I., Zimmt, M.B., Zimmerman, A.L., Nanowiring of a redox enzyme by metallized peptides. *Biosens. Bioelectron.* **2005**, 21, 973–978.

56 Scott, D.L., Bowden, E.F., Enzyme-substrate kinetics of adsorbed cytochrome c peroxidase on pyrolytic graphite electrodes. *Anal. Chem.* **1994**, 66, 1217–1223.

57 Nahir, T.M., Clark, R.A., Bowden, E.F., Linear-sweep voltammetry of irreversible electron transfer in surface-confined species using the Marcus theory. *Anal. Chem.* **1994**, 66, 2595–2598.

58 Bottcher, A., Takeuchi, T., Hardcastle, K.I., Meade, T.J., Gray, H.B., Cwikel, D., Kapon, M., Dori, Z., Spectroscopy and electrochemistry of cobalt(III) Schiff base complexes. *Inorg. Chem.* **1997**, 36, 2498–2504.

59 Arounaguiri, S., Esawaramoorthy, D., Ashkkumar, A., Dattagupta, A., Maiya, B.G., Cobalt (III), nickel (II), and ruthenium (II) complexes of 1,10-phenanthroline family of ligands: DNA binding and photocleavage studies. *Proc. Indian Acad. Sci. (Chem Sci.)* **2000**, 112, 1–17.

60 Krantz, B.A., Sosnick, T.R., Engineered metal binding sites map the heterogeneous folding landscape of a coiled coil. *Nat. Struct. Biol.* **2001**, 12, 1042–1047.

61 Yeh, J.I., Claiborne, A., Hol, W.G.J., Structure of the native cysteine-sulfenic acid redox center of enterococcal NADH peroxidase refined at 2.8 Å resolution. *Biochemistry* **1996**, 31, 9951–9957.

62 Foubert, T.R., Burritt, J.B., Taylor, R.M., Jesaitis, A.J., Structural changes are induced in human neutrophil cytochrome b by NADPH oxidase activators, LDS, SDS, and arachidonate: Intermolecular resonance energy transfer between trisulfopyrenyl-wheat germ agglutinin and cytochrome b(558). *Biochim. Biophys. Acta* **2002**, 1567, 221–231.

63 Graef, I.A., Holsinger, L.J., Diver, S., Schreiber, S.L., Crabtree, G.R., Proximity and orientation underlie signaling by the non-receptor tyrosine kinase ZAP70. *EMBO J.* **1997**, 16, 5618–5628.

64 Wang, J., Amperometric biosensors for clinical and therapeutic drug monitoring: A review. *J. Pharm. Biomed. Anal.* **1999**, 19, 47–53.

10
Protein-based Biosensors using Nanomaterials

Genxi Li

10.1
Introduction

Protein-based biosensors are fabricated based on the electron transfer between proteins (enzymes) and electrodes [1, 2]. However, electron-transfer rates between redox proteins and electrode surfaces are usually prohibitively slow [3]. On the one hand, the electroactive prosthetic groups of most redox proteins are deeply buried in the electrically insulated peptide backbones. On the other hand, unfavorable orientations and adsorptive denaturations of proteins often occur at electrode surfaces. To achieve efficient electrical communication, one approach that has been proposed, and widely employed in previous years, is based on electrochemical mediators, both natural enzyme substrates or products, and artificial redox mediators (first-generation and second-generation biosensors) [1]. First-generation biosensors have many defects, such as possible interference because of the too high applied potential, systematic complexity caused by fluctuation of the concentration of dissolved oxygen, a relatively low detection limited since the tenuity of dissolved oxygen will significantly decrease the electrical currents, etc. [4–6]. Consequently, the idea of artificial mediators has been proposed, leading to the development of second-generation biosensors. Nevertheless, redox mediators used in conjunction with redox proteins are in no way selective but rather general redox catalysts, facilitating not only the electron transfer between electrode and protein but also various interfering reactions [7]. Therefore, third-generation biosensors, which are based on the direct electron transfer of proteins, i.e., the electrical communication between the proteins and signal transducers is accomplished directly without additional mediators, have received more and more attention [2]. Third-generation biosensors are superior in selectivity both because they should be operated in a potential window close to the redox potential of the protein itself, and therefore, are less prone to interfering reactions [4] and also because of the lack of yet another reagent in the reaction sequence.

Recent breakthroughs in nanotechnology have made various nanostructured materials more affordable for a broader range of applications, and nanotechnology is playing an increasingly important role in the development of biosensors [8–11].

Nanotechnologies for the Life Sciences Vol. 8
Nanomaterials for Biosensors. Edited by Challa S. S. R. Kumar

ISBN: 978-3-527-31388-4

Owing to the unique properties of nanomaterials, direct electrochemistry and catalytic activity of many proteins have been observed at electrodes modified with various nanomaterials such as TiO_2, ZrO_2, SiO_2, Fe_3O_4, metal nanoparticles, carbon nanotubes, etc. [12–17]. The sensitivity and performance of biosensors are being improved by using nanomaterials for their construction. Various nanostructures have been examined as hosts for protein immobilization via approaches including protein adsorption, covalent attachment, protein encapsulation, and sophisticated combinations of methods. Studies have shown that nanomaterials can not only provide a friendly platform for the assembly of protein molecules but also enhance the electron-transfer process between protein molecules and the electrode.

We have summarized the sensors fabricated with proteins (enzymes) in a recently published encyclopedia of sensors [1]. Later, we noticed that increasing numbers of protein-based biosensors were being fabricated by using nanomaterials. It has been necessary to focus on the application of nanomaterials in the development of protein-based biosensors. This chapter reviews protein-based biosensors according to the various sorts of nanomaterials used in their fabrication.

10.2
Metal Nanoparticles

Metal nanoparticles have many unique properties: large surface-to-volume ratio, high surface reaction activity, high catalytic efficiency, and strong adsorption ability. Metal nanoparticles have been used to facilitate the electron transfer in nanoelectronic devices owing to the following important functions: the roughening of the conductive sensing interface, the catalytic properties of the nanoparticles, and the conductivity properties of nanoparticles at nanoscale dimensions that allow the electrical contact of redox-centers in proteins with electrode surfaces [8, 18]. Gold nanoparticles and silver nanoparticles are the most intensively studied and applied metal nanoparticles in fabricating biosensors, which will be specially summarized later. Other metal nanoparticles, such as palladium [19] and platinum nanoparticles [20], etc. have also been used.

10.2.1
Gold Nanoparticles

Colloidal gold is a metallic colloid whose particles are formed by small octahedral units, called primary particles. The size of gold particles essentially depends on the way the colloid is formed [21]. To improve the stability of these particles, a kind of gold nanoparticle protected by lipid has also been invented [22]. Gold colloid has been widely used as a cytochemical label for the study of macromolecules, with transmission and scanning electron microscopy, light microscopy and freeze-etch electron microscopy, and to improve the signals of both surface-enhanced Raman spectroscopy and surface plasmon resonance [21].

In the last few years, electrochemical biosensors created by coupling biological elements with electrochemical transducers based on or modified with gold nanoparticles have played an increasingly important role in biosensor research. Gold colloid has many advantages for biosensor applications. First, gold nanoparticles can, very usefully, provide a stable surface for the immobilization of biomolecules, such that the molecules retain their biological activities. As is well known, the performance of enzyme biosensors depends not only on the nature of the enzyme molecules but also on the influences imposed on these molecules by immobilization. Modification of an electrode surface with gold nanoparticles provides a microenvironment similar to that of the redox proteins in native systems and offers the protein molecules more freedom in orientation, which can weaken the insulating property of the protein shell for the direct electron transfer and facilitate the electron transfer through the conducting tunnels of colloidal gold [21]. Second, gold nanoparticles can form conducting electrodes and are the site of electron transfer when anchored to the substrate surface, allowing direct electron transfer between redox proteins and electrode surfaces with no mediators required [23]. They can act as an electron-conducting pathway between prosthetic groups and electrode surface. They can provide useful interfaces at which redox processes of biological molecules, such as proteins or NADH, involved in biochemical reactions of analytical significance can be electrocatalyzed [23]. Xiao et al. have that reported the reconstitution of one apo-flavoenzyme, apo-glucose oxidase, on a 1.4-nm gold nanocrystal functionalized with the cofactor flavin adenine dinucleotide and integrated into a conductive film could yield a bioelectrocatalytic system with exceptional electrical contact with the electrode support [24]. Their work shows that electron transfer through the Au nanoparticles is much faster than electron transfer to O_2. Besides, various characteristics of gold colloid, such as their high surface-to-volume ratio, high surface energy, and ability to decrease the distance between proteins and metal particles, may also facilitate electron transfer between the redox sites of the proteins and electrode surface. The last advantage, but not the least, is that the size and surface morphology (important parameters when applied in biosensing) of gold colloid can be controlled easily experimentally. Colloidal gold surface morphology is vital in establishing the electrical contact between the proteins and the electrodes [25].

Taking horseradish peroxidase (HRP) as an example, although direct, nonmediated electron transfer is well known for some redox proteins and particularly for HRP [26, 27], it is usually very inefficient and is not generally used for practical biosensory devices. Zhao et al. found, in 1992, that a mixed layer of HRP and colloidal gold nanoparticles of 30 nm could be electrocatalytically active for hydrogen peroxide (H_2O_2) reduction without the need of electro-transfer mediators [28]. In addition, immobilized HRP was reduced directly on gold colloid through an amplification effect of the enzyme–substrate interaction. It was demonstrated that the utilization of gold nanoparticles in enzyme immobilization could increase the enzyme loading without compromising charge-transfer efficiency [21]. That was to say the small size of gold nanoparticles could allow the conductive material to come into close proximity of the active center of the enzyme, which might facilitate the electron-transfer process. The nanoparticles could also provide bioelectrocatalytic activity that can be utilized in biosensor devices. This work led to an extensive investigation of the direct electron transfer based on gold nanoparticles and immobilization of proteins, especially enzymes [15, 29–33]. Some related biosensors for H_2O_2 [28, 34–38], glucose [34, 39], xanthine and hypoxanthine [34, 40–42], dopamine [43], and cholesterol [44, 45] were also reported. These biosensors exhibited high sensitivity, good reproducibility, and long-term stability.

The catalytic activity of HRP is attributed to its iron heme group, which acts as the electroactive center. Since hemoglobin (Hb) contains four heme groups, it can be utilized as the HRP substitute in the detection of H_2O_2, by virtue of its low cost and stable properties in solution [46]. In the last few years, electrochemical sensors for H_2O_2 based on the peroxidase-like activity of Hb incorporated in many different kinds of materials, including colloidal gold, have been fabricated [22, 47–49]. Researchers have also made a novel Hb-based H_2O_2 sensor constructed on a gold nanoparticles-modified ITO electrode [50]. Gold nanoparticles, acting as bridges of electron transfer, can greatly promote the direct electron transfer between Hb and the electrode surface without the aid of any electron mediator [51]. Thus, the Hb-immobilized gold nanoparticles-modified electrode can be developed as a third-generation biosensor for the determination of H_2O_2.

Besides HRP and Hb, other heme proteins, such as cytochrome *c* (Cyt c), myoglobin (Mb), have also been investigated via being immobilized with gold nanoparticles [52]. Direct electron transfer between immobilized Mb and a colloidal gold modified carbon paste electrode, in addition to a glass carbon electrode, has also been studied [53]. Cyt c immobilized on colloidal gold modified carbon paste electrodes can also maintain its activity and electrocatalyze the reduction of H_2O_2 [54]. Meanwhile, a chitosan-stabilized gold nanoparticles-modified electrode can also be developed for biosensors [55, 56].

Fabricating a sensitive, stable glucose biosensor has long been of great interest. A novel method to fabricate a glucose biosensor has been proposed by immobilizing glucose oxidase (GOD) on gold nanoparticles, which have self-assembled on Au electrode modified with a thiol-containing three-dimensional (3D) network of silica gel [57] or with a cystasmine monolayer [58]. GOD can exhibit excellent bioelectrocatalytic response to the oxidation of glucose. Researchers have also found that aqueous colloidal gold nanoparticles can enhance the activity of aqueous enzymes [59]. Mena. et al. have compared several enzyme biosensor designs, prepared by immobilization of GOD onto different tailored gold nanoparticle-modified electrode surfaces [60].

The immobilization methods for connecting gold nanoparticles with electrodes and with proteins should both be discussed in detail. Generally, there are three ways to prepare gold nanoparticles-modified electrode: (a) by binding with functional groups of self-assembled monolayer (SAMs); (b) by direct deposition of gold colloid onto the electrode surface; (c) by co-modification of mixed gold colloid with other components in the composite electrode matrix [61].

Protein or enzyme can readily be immobilized on colloid gold by dipping a protein solution onto the colloid gold modified electrode surface. The electrostatic interaction between the negatively charged citrate surface of colloidal gold and positively charged groups of the protein leads to the adsorption of protein onto the electrode surface. The prepared electrodes can be characterized by several means such as cyclic voltammetry (CV), transmission electron microscope (TEM), atomic force microscopy (AFM), UV/visible spectroscopy (UV/Vis), surface-enhanced Raman spectroscopy (SERS) and electrochemical impedance spectroscopy (EIS), etc.

SAMs can provide a simple way to tailor surfaces with well-defined compositions, structures and thickness that can then be employed as specific functionalized surfaces for the immobilization of gold nanoparticles and enzymes [62]. Gold nanoparticle-modified electrode surfaces can be prepared by covalently binding gold nanoparticles with surface functional groups (–CN, $-NH_2$, or –SH) of SAMs modified solid surfaces [63–66]. Short-chain molecules, such as cysteamine (Cyst) and 3-mercaptopropionic acid (MPA), can be self-assembled on the gold disk electrode for further binding of gold nanoparticles. Biosensors based on immobilization of enzyme onto Cyst or MPA SAMs that had been previously bound onto gold nanoparticles have also been evaluated and compared [61].

An early work through fabrication of a monolayer of colloidal gold on an indium-doped tin oxide (ITO) support was performed in 1995 [67]. Either (2-aminopropyl)-trimethoxysilane or (3-mercaptopropyl)trimethoxysilane was first attached to the ITO support to yield an amine/thiol-functionalized thin film. The deposited gold thin film was quite stable and provided an active interface for the immobilization or redox-active self-assembled monolayers. The thiol functional group-derived carbon ceramic electrode (CCE) has been used to construct H_2O_2 biosensors [68]. Biosensors can also be constructed by immobilizing the proteins by adsorbing them onto the nanoparticles, by crosslinking them with bifunctional agents such as glutaraldehyde, or by mixing them with the other components of composite electrodes. When HRP is immobilized on a colloidal gold monolayer formed at the gold electrode by a long "cysteamine/glutaraldehyde/cysteamine" molecular bridge, it shows an excellent electrocatalytic response to the reduction of H_2O_2 [33]. This sensor shows a fast amperometric response, wide linear calibration range and low detection limit for H_2O_2 determination. In addition, HRP immobilized on colloidal gold has a high affinity to H_2O_2 with no loss of enzymatic activity.

Multilayers of conductive particles can give rise to porous, high-surface-area electrodes, where the local microenvironment can be controlled by the crosslinking elements. Superstructures of a controllable number of colloidal Au particle layers have been formed on a support surface by the alternate interaction of the assembly with a crosslinker and the negatively charged Au particles [69]. A two-dimensional (2D) double layer structure [70] and 3D superstructure consisting of gold nanoparticles for biosensors [71] have also been proposed. The utilization of gold nanoparticles as a building block in the construction of nanoscale functional devices has become a very promising aspect for electrochemical biosensing. The technology of derivatizing Au electrode surface by silica sol–gel and gold nanoparticles takes advantage of both self-assembly, nanoparticles, and the increased surface area of 3D electrodes. Ikeda and coworkers have demonstrated that gold nanoparticles can self-assemble both inside the network and on the surface of the silica gel [72]. These gold nanoparticles immobilized by a silica gel 3D network can act as tiny conducting centers and facilitate electron transfer between proteins and electrodes. Based on these investigations, a HRP-based H_2O_2 sensor and a GOD-based glucose sensor have been developed by embedding gold nanoparticles in thiol-containing silica sol–gel network [73]. An amperometric H_2O_2 biosensor has also been reported, based on immobilizing HRP to a nano-Au monolayer supported by a carbon sol–gel derived carbon ceramic electrode [68]. This sol–gel 3D network made it possible to control the effectiveness of the electrocatalytic process, the sensitivity and response rate of the nanoparticle biosensor by controlling the thickness of thiolated polymer or sol–gel, and the number of Au particle layers associated with the electrodes.

Another feasible method to fabricate protein-based biosensor involves covalent attachment of protein molecules to a gold nanoparticle monolayer modified Au electrode, such as GOD/gold nanoparticle systems [58]. Briefly, gold nanoparticles are first self-assembled on a gold electrode by dithiol via Au–S bonds. A cystamine monolayer is then chemisorbed onto those gold nanoparticles and exposed to an array of amino groups, which would further react with aldehyde groups of periodate oxidized GOD via the well-known Schiff base reaction. By this means, GOD could be covalently attached to the gold electrode, resulting in a stable biosensing interface.

Recently, a gold nanoparticle–$CaCO_3$ hybrid material (AuNP–$CaCO_3$) has been prepared and applied for sensor preparation [74]. AuNP–$CaCO_3$ can retain the porous structure and inherits the advantages from its parent materials, such as satisfying biocompatibility and good solubility and dispersibility in water; therefore, it can offer a promising template for enzyme immobilization and biosensor fabrication. HRP has been conjugated with AuNP–$CaCO_3$ to fabricate HRP-AuNP–$CaCO_3$ bioconjugates, which were then embedded into a silica sol–gel matrix to construct a novel biosensor.

There are also many reports on fabricating biosensors by mixing colloidal gold and carbon paste to incorporate colloidal gold into the electrode. A reagentless glucose biosensor based on the direct electron transfer of GOD [75], and a renewable tyrosinase biosensor [76], have been constructed by immobilizing the corresponding enzymes onto electrodes prepared by mixing a colloidal gold solution with the carbon paste components. A renewable reagentless H_2O_2 sensor has also been reported based on the direct electron transfer of HRP where HRP is mixed into colloidal gold-modified carbon paste electrodes [15]. Here, the nanoparticles are used to retain the enzymatic activity and facilitate direct electron transfer between HRP and carbon-sensing sites. In addition, nitrite sensors based on a Hb-colloidal gold nanoparticle modified screen-printed electrode and carbon paste electrode have also been developed [77, 78].

In addition to the proteins employed for the fabrication of the above-mentioned biosensors, some other proteins (enzymes) have been used. Xanthine oxidase is reported to be adsorbed onto colloidal gold that was previously evaporated on the surface of glassy carbon, which responds to xanthine or hypoxanthine in the absence of added mediator by electrochemical oxidation of the enzymatic oxidation product, uric acid [40]. A multi-enzyme biosensor for cholesterol [44] has been constructed to detect cholesterol in serum and whole blood, by making use of HRP and cholesterol oxidase. This sensor gives an electrochemical response that correlates well with the total cholesterol concentration. The biosensor operates at a sufficiently low potential to avoid interference from many sources, and the carrageenan hydrogel enables analysis of whole blood without electrode fouling. A more recent paper reports immobilization of tyrosinase by crosslinking it onto glassy carbon electrodes modified with electrodeposited gold nanoparticles [79]. The immobilized tyrosinase retained a high bioactivity on this electrode material, displaying stable, sensitive responses to various phenolic compounds.

The immobilization of proteins on colloidal gold has also been used for the development of immunological detection methods. An electrochemical method to monitor biotin–streptavidin interaction and determine the concentration of streptavidin has been established by the use of colloidal gold as an electrochemical label [80]. Velev and Kaler have created arrays of biosensors by *in situ* assembly of gold and silver colloidal particles onto micropatterned electrodes and immobilizing antibodies (IgG) on the colloidal particles [81]. Niemeyer and Ceyhan [82] have prepared biofunctionalized nanoparticles by DNA-directed immobilization of proteins at colloidal gold. Biometallic hybrid components can be used in an immunoassay for the detection of proteins. Thanh and Rosenzweig have developed a unique, sensitive and highly specific immunoassay system for antibodies using gold nanoparticles [83]. This assay is based on the aggregation of gold nanoparticles that are coated with protein antigens in the presence of their corresponding antibodies. Au nanoparticles have also been used as electrochemical labels in genetic diagnosis application [84]. *Escherichia coli* single-stranded DNA binding protein (SSB) has been attached onto a SAM of single-stranded oligonucleotide modified Au nanoparticle, and the resulting Au-tagged SSB was used as the hybridization label for the electrochemical detection of DNA hybridization. Changes in the Au oxidation signal were monitored upon binding of Au tagged SSB to probe and hybridization on the electrode surface. These works demonstrate the emergence of a new field of application for colloidal gold in protein immobilization and biosensing.

10.2.2 Silver Nanoparticles

Colloidal silver is another nanomaterial often used for biosensor fabrication. Its synthesis has been well demonstrated [85]. Much attention has also been given to the quantum characteristics of small granule diameter and large specific surface area as well as the ability to quickly transfer photoinduced electrons at the surfaces of colloidal particles [86]. It is also found that surface-assembled silver nanoparticles can act as an electrical bridge that "wire" the fast interfacial electron transfer between Cyt c and pyrolytic graphite electrodes, opening up new opportunities for the *in vitro* electron-transfer process of heme proteins by using silver nanoparticles and for fabricating bioelectronic devices. Furthermore, silver nanoparticles greatly enhance the electron-transfer reactivity of heme-proteins, such as HRP, Mb, Hb and their catalytic ability toward H_2O_2 and nitric oxide [87–89]. Glucose biosensors based on immobilization of GOD in silver nanoparticles have also been fabricated [90, 91]. The studies show that the colloidal silver nanoparticles function as electron-conducting pathways between the prosthetic groups and the electrode surface, therefore the electron-transfer rate between the enzyme and the electrode is increased significantly. In addition, by the vapor deposition method, Hb and colloidal silver nanoparticles can be entrapped in a titania sol–gel matrix on the surface of a glassy carbon electrode – thus a NO_2 biosensor is prepared [92].

10.2.3 Other Metal Nanomaterials

One promising metal for nanoparticles is platinum. Platinum nanoparticles with a diameter of 2–3 nm have been used with single-wall carbon nanotubes (SWCNTs) for fabricating electrochemical sensors with Nafion to form a network that connected Pt nanoparticles to the electrode surface [20]. The use of platinized carbon microelectrodes has also been reported for a glucose biosensor [93]. Nanocrystalline diamond exhibits several special properties that make it particularly suitable for biofunctionalization and biosensing [94]. As an improvement, platinum-modified boron-doped diamond microfiber electrodes were fabricated, which exhibited much higher sensitivity [95]. Electrodeposition of highly dispersed palladium nanoparticles on a glassy carbon electrode has also been reported for the construction of a glucose biosensor [96]. Another glucose biosensor has also been reported by co-depositing palladium nanoparticles and GOD onto a Nafion-solubilized carbon nanotube (CNT) film [19]. GOD can retain its biocatalytic activity and offer efficient oxidation and reduction of the enzymatically liberated H_2O_2, allowing for fast and sensitive glucose quantification.

Some glucose biosensors based on non-conducting polymer films have been reported; however, such biosensors always suffer from a low response current and a relatively high detection limit. To solve these problems, Cu nanoparticles are selected to increase the response current because it can electrochemically oxidize glucose [97]. A GOD and Cu nanoparticles-based sensor has a two times lower detection limit, a three times larger maximum current and 2.5× higher sensitivity than biosensors fabricated with no Cu nanoparticles.

10.3 Metallic Oxide Nanoparticles

Some researchers have used MnO_2 nanoparticles as eliminators of ascorbic acid interference to amperometric glucose and lactate biosensors [98, 99]. Others have constructed an enzyme field-effect transistor using a GOD membrane doped with MnO_2 powder [100]. In this configuration, MnO_2 acts as a catalyst to the decomposition of H_2O_2. Therefore, the produced O_2 from the decomposition of H_2O_2 can be replenished in the glucose oxidation reaction and the dynamic range of glucose determination is extended. Researchers have also fabricated some glucose biosensors based on the co-immobilization of GOD and MnO_2 nanoparticles [98, 101].

Nano titanium dioxide (TiO_2) is now an attractive biocompatible material widely used in toothpaste and cosmetics. Due to its unique physiochemical properties and its inclination to selectively combine with some groups of biomolecules, nanosized TiO_2 has been proposed as a promising interface for the immobilization of biomolecules. Attempts have been made to assemble heme proteins, including Cyt c, Mb and Hb, onto a nanocrystalline TiO_2 film; the TiO_2 film can not only offer a friendly platform to assemble the protein molecules, but also enhances the electron-transfer process between the protein molecules and the electrode [12]. The electrochemical characteristics of HRP entrapped in a TiO_2 nanoparticles film cast on pyrolytic graphite electrode has been examined [102]. The good biocompatibility of TiO_2 nanoparticles can make HRP retain its native state and show good electrocatalytic activity, which may have potential application in constructing a third-generation electrochemical biosensor.

Zirconium dioxide (ZrO_2) nanoparticles have also been used in protein-based biosensors. As an example, ZrO_2 nanoparticles 35 nm in diameter have been cast on a pyrolytic graphite electrode by dispersing them in dimethyl sulfoxide to immobilize Hb for fabrication of a H_2O_2 sensor [13]. This sensor shows a high thermal stability up to 74 °C. Another example is the treatment of the surface of a platinum electrode with nanoporous ZrO_2/chitosan composite matrix to fabricate a glucose biosensor [103]. Other studies reveal that nanoporous ZrO_2 has general affinity for the binding of proteins because the amine and carboxyl groups on the surface of enzyme can act as ligands to ZrO_2. Thus, the usage of glutaraldehyde in crosslinking, which always denatures an enzyme, can be avoided, and a HRP-based biosensor for H_2O_2 is correspondingly constructed [104]. Moreover, H_2O_2 sensors are constructed based on the self-assembly of ZrO_2 nanoparticles with heme proteins (Hb and Mb) on a functional glassy carbon electrode [105].

Zinc oxide (ZnO) is another attractive semiconductor material. It has been demonstrated that ZnO with a high isoelectric point (~9.5) is suitable for the adsorption of low-isoelectric point proteins [106]. For example, low-isoelectric point tyrosinase is reported to be adsorbed on the surface of such ZnO nanoparticles, facilitated by the electrostatic interactions, and is immobilized on a glassy carbon electrode via film formation to develop a mediator-free phenol biosensor [107]. Therefore, ZnO nanoparticles deserve further investigation as an important promising candidate as support material in the fabrication of biosensors.

10.4 Carbon Nanotubes

Carbon nanotubes (CNTs), discovered in 1991, are a new type of carbon material obtained by folding grapheme layers into carbon cylinders. They present a closed topology and tubular structure with diameters of several nanometers and lengths to several microns. Basically, there are two types of carbon nanotubes, multiwall carbon nanotubes (MWCNTs) and single-wall carbon nanotubes (SWCNTs). They can behave as conductors or semiconductors depending on their structures, mainly on their diameter and helicity.

Carbon nanotubes have received enormous attention for the preparation of electrochemical sensors due to their unique properties, as reviewed in some publications [108–114]. Similar to the conventional electrochemical biosensors, proteins and CNTs based biosensors can also be, generally, divided into three categories: (a) first-generation biosensors via the detection of H_2O_2 or O_2 involved in the enzymatic reaction, (b) second-generation biosensors through the utilization of electron-transfer mediators, and (c) third-generation biosensors based on the direct electron transfer of enzymes or proteins. Especially, CNT materials have been proved to possess electrocatalytic activity toward the oxidation of H_2O_2 and the reduction of O_2 and can facilitate the direct electron transfer of proteins or enzymes. These features make CNTs particularly attractive for the development of first- and third-generation electrochemical biosensors [111]. Many second-generation protein-based biosensors have also been fabricated by using CNTs. For example, organic electrocatalysts, such as polynuclear aromatic dyes, have been employed, which can be stably immobilized onto CNTs and can be further used to accelerate the oxidation of H_2O_2 or shuttle electron transfer between the CNTs and enzymes or proteins. Ye et al. have functionalized the MWCNTs with iron-phthalocyanines to redox-catalyze H_2O_2 oxidation and have constructed a highly sensitive and selective glucose biosensor [115]. Chen et al. have co-immobilized methylene blue with HRP onto CNTs to shuttle the electron transfer between HRP and the CNTs. The modified electrode exhibited a good bioelectrocatalytic activity toward H_2O_2 reduction [116].

The nano-dimensions, graphitic surface chemistry and electronic properties of carbon nanotubes make them an ideal material for use in chemical and biochemical sensing [108–114, 117, 118]. In addition, the fabricated sensors using CNTs usually show high sensitivity and stability of electrocatalysis. Although the mechanism is still not fully understood, some researchers suggest that these advantages may be attributed to the special structure of nanotubes, such as their open ends [119, 120]. The electroactive ends of the nanotubes are readily accessible to species in solution. The opened nanotube ends allow the enzyme to enter the hollow CNTs. In addition, the rigidity of the tubes allows them to be plugged into biomolecules, so enabling electrical connection to the redox centers of the biomolecules. Furthermore, the large surface area per unit volume of CNTs allows a large amount of enzyme to be immobilized within nanotubes.

However, the potential application of CNTs is still limited by (a) the CNTs tend to aggregate in most solvents; (b) electronic communication between proteins and CNTs is rather slow at pristine CNTs; and (c) strong interactions between proteins and CNTs may distort the proteins. Therefore, the biosensing applications of CNTs require rational functionalization of the CNTs to improve their solubility and biocompatibility.

The combination of CNTs with redox active protein (enzymes) appears to offer a convenient platform for a fundamental understanding of biological redox reactions and for the development of third-generation biosensors. Direct electron transfer and bioelectrocatalysis of many proteins, such as GOD, cyt c, Mb, Hb, HRP, etc., have been investigated with CNTs-modified electrodes and their performance has been found to be much superior to those of other carbon electrodes in terms of reaction rate, reversibility, and detection limit [121–132].

GOD is the major protein employed for sensor fabrication by using CNTs. GOD has been immobilized by coating onto the surface of SWCNTs without a gross loss of enzyme activity [133]. Treatment of this bio-SWCNT sensor with both a diffusive mediator and equilibrated glucose substrate enhanced the catalytic signal by more than one order of magnitude compared with that observed at an activated macrocarbon electrode. This enhanced performance was partly due to the high enzyme loading and partly because of better electrical communication ability of the nanotubes. Other research on MWCNTs has revealed that a MWCNTs-modified glassy carbon electrode can be employed as an amperometric oxygen sensitive electrode to fabricate a glucose biosensor [134]. CNT-based glucose biosensor was also fabricated via polypyrrole [135, 136]. Recently, a CNTs-doped polypyrrole (PPy) glucose biosensor was reported [137]. Unlike other work on glucose biosensors based on CNT/PPy electrodes, where CNT was physically entrapped within the growing film, the PPy/GOD films were formed by using "oxidized" CNT as the sole charge-balancing anionic dopant. This is the first example of anionic CNT acting as dopant in the preparation of conducting-polymer enzyme electrodes. While the concept has been presented within the context of glucose sensing, it could readily be extended to other biocatalytic electrodes based on judicious selection of the enzyme.

Heme proteins have also been extensively employed to fabricate biosensors by using CNTs. The direct electrochemical response of Mb on MWCNT-modified glassy carbon electrode has been reported [16]. Mb exhibited elegant catalytic activity for electrochemical reduction of oxygen, based on which an unmediated biosensor for O_2 was developed. An unmediated H_2O_2 biosensor, based on the peroxidase-like activity of Mb on MWCNTs, has also been constructed [138]. Recently, the same research group reported that Mb can be strongly adsorbed onto the surface of MWCNTs with an approximate monolayer to develop an unmediated NO biosensor, with a low detection limit [139]. HRP is able to adsorb on a carbon nanotube microelectrode to transfer electrons directly with the electrode and retain its catalytic activity toward H_2O_2 [124]. A H_2O_2 biosensor based on Cyt c and MWCNTs has also been fabricated [140]. The enhanced electron-transfer rate was also observed in systems in which microperoxidase (MP-11) was attached to the end of aligned SWCNTs array and MWCNTs [141, 142].

Many other proteins (enzymes) have been employed to prepare various kinds of biosensors by using CNTs. For example, the direct electrochemistry of xanthine oxidase has been achieved on a SWCNTs-modified gold electrode, which may provide an attractive route for the development of biosensors [143]. Table 10.1 gives information on sensors that use other proteins (enzymes).

Understanding the interaction between CNTs and proteins should be helpful in developing protein-based biosensors using CNTs. Usually, proteins (enzymes) are immobilized onto CNTs in three ways, taking GOD as example.

Tab. 10.1. Protein-based biosensors using carbon nanotubes.

Protein(enzyme)	CNT	CNT immobilization matrix/electrode	Analyte	Ref.
Acetylcholinesterase (AChE)	MWCNT	Thick film strip electrode	Organophosphorus (OP) insecticides	219
L-Amino acid oxidase	CNT	Alkoxy silane sol–gel	L-Amino acid	220
Urease	CNT	Sol–gel	Urea	221
Acetylcholinesterase	CNT	Sol–gel	Acetylthiocholine	221
L-lactate oxidase	MWCNT	Sol–gel/GCE	L-Lactate	222
Putrescine oxidase	MWCNT	PDDA/GCE	Putrescine	223
Aflatoxin–detoxifizyme (ADTZ)	MWCNT	Au electrode	Sterigmatocystin	224
Choline oxidase (ChOx)	MWCNT	Sol–gel/platinum electrode	Choline	225
Cholesterol oxidase	MWCNT	Sol–gel chitosan hybrid film	Cholesterol	226
Dehydrogenase	CNT	CNT paste electrode	Glucose, NADH	227
		GCE	Glucose, NADH	228
		GCE	NADH	229
Catalase	MWCNT	GCE	H_2O_2	230
	SWCNT	Au electrode		231
Microperoxidase (MP-11)	MWCNT	MWCNT-Au NPs nanohybrid film	H_2O_2	232
Laccase	CNT	CNT-chitosan composite film	2,2-Azino-bis-(3-ethylbenzthiazoline-6-sulfonic acid) diammonium salt (ABTS), catechol, O_2	233
lactate oxidase (LOX)	SWCNT	Functionalized CNT–COOH electrode	Lactate	234
Cholesterol esterase, cholesterol oxidase, and peroxidase	MWCNT	Screen-printed carbon electrode	Cholesterol	235

(a) Physical adsorption [144]. GOD immobilized in this way usually maintains its substrate-specific enzyme activity in the presence of glucose. The tubular fibrils become positioned within tunneling distance of the cofactors without too much denaturation of the enzyme.

(b) GOD is covalently attached to the ends of the aligned tubes, which allows close approach to FAD [145]. In addition, some researchers have reported the synthesis of a fully integrated CNT protein composite system with high function density and with direct chemical bonding GOD to functionalized MWCNT under a high degree of control and specificity [146]. Recently, a highly ordered array of CNTs has been reported to serve as a universally direct nanoelectrode interface for redox proteins [147]. The site-selective, covalent docking of GOD on the CNTs tips has a marked effect on enhancing electron transfer properties.

(c) FAD is directly attached to the ends of the tubes [148]. A good example is given by Willner and his coworkers, who covalently immobilized the FAD redox center of GOD on one end of SWCNT, and the other end of the SWCNT was attached to a gold surface or another FAD to accelerate electrical communication between the electrode and redox proteins [149].

CNTs have been extensively employed in preparing different kinds of electrodes for biosensors on a larger scale. Wang et al. have reported a remarkable decrease in overvoltage for the oxidation of NADH using composites obtained by dispersing CNTs in a Teflon binder [150]. Compton et al. have proposed the use of MWCNTs abrasively attached to the basal plane pyrolytic graphite [151]. Unique electrochemistry of soluble molecules and adsorbed proteins on flat mat-like layers of SWCNTs or MWCNTs has been demonstrated. Rusling et al. have reported that Mb was covalently attached onto the ends of vertically oriented SWCNTs forest arrays used as electrodes [121]. The results suggest that the "trees" in the nanotube forest behaved electrically similar to a metal, conducting electrons from the external circuit to the redox sites of the proteins. Huang et al. [152] and Zhao el al. [153] have reported multilayers of MWCNTs and GOD through a layer-by-layer technique. Liu has reported the preparation of enzyme-polyion thin films consisting of GOD alternately assembled with poly(diallyldimethylammonium) chloride polymer on CNTs at a glassy carbon electrode [154].

CNTs have also been used to develop electrochemiluminescence (ECL) biosensors. For instance, Wohlstadter et al. have reported ECL biosensors for the assay of a-fetoprotein [113]. Studies revealed that CNTs were very useful for ECL-based assays. CNTs have also been used to fabricated sensors with screen-printed carbon electrode [155, 156].

CNTs array-based biosensors have also been reported. Aligned MWCNTs grown on a platinum substrate have been described for the development of an amperometric biosensor [114]. The two array systems in this work were either acid treated or air treated. The results showed that chemical etching was more effective in opening the carbon nanotubes and allowing the enzyme to enter the inner channel.

A so-called carbon nanotube paste electrode (CNTPE) has been reported in the fabrication of a glucose biosensor modified with some metallic particles [157]. The bioelectrodes were obtained by dispersing the metal particles, enzyme and

MWCNTs within a mineral oil binder. The resulting CNTPE combines the ability of CNTs to promote electron-transfer reactions with the attractive advantages of composite materials.

Microelectrodes and ultramicroelectrodes are important analytical tools because they exhibit fast response times, significantly improved Faradic-to-capacitive current ratios, and substantially reduced ohmic drops [158, 159]. Therefore, it is reasonable to develop so-called micro-biosensors. The electrocatalytic reduction of dissolved oxygen in acidic medium using microelectrodes constructed from MWCNTs has been reported [160]. Meanwhile, Martin and coworkers [161] have reported that the detection limits obtained at nanoscopic electrodes ensembles are much lower, three orders of magnitude, than the corresponding detection limits for a conventional macroscopic disk electrode. An important consideration in the future advancement of practical sensors is how to develop new immobilization strategies that are appropriate for the construction of miniature sensors, which requires precise control of film deposition on a small electrode. Electrochemical immobilization provides an elegant alternative for the deposition of enzymes on very small area electrodes of defined geometry.

Nanoelectrode ensembles based on low-site density, aligned CNTs have also been fabricated, and the electrochemical characteristics have been investigated [162]. Subsequently, the same research group developed a glucose biosensor based on CNT nanoelectrode ensembles [163]. The operation eliminates the need for permselective membrane barriers or artificial electron mediators, thus greatly simplifying the sensor design and fabrication. The same group have also fabricated CNT nanoelectrode ensembles by another method [164]. Platinum nanoparticles have been used to modify the CNT nanoelectrode ensembles and a new amperometric biosensor, based on adsorption of GOD at the platinum nanoparticle-modified CNT electrode [165]. This work combines the advantages of CNTs with Pt nanoparticles. The excellent electrocatalytic activity and special 3D structure of the enzyme electrode result in good characteristics such as a large determination range (0.1–13.5 mM), a short response time (within 5 s) and stability (73.5% remains after 22 days). Another reported glucose biosensor is also notable, due to its high sensitivity and selectivity coupled to a wide linear range, prolonged lifetime and oxygen independence [166].

CNTs are often used together with other materials (including nanomaterials) for biosensor fabrication. Early in 2002, a novel glassy carbon electrode modified by a gel containing MWCNTs and the ionic liquid 1-butyl-3-methylimidazolium hexafluorophosphate ($BMIPF_6$) was reported [167]. Wang et al. have reported on a simple avenue for preparing effective CNT-based electrochemical sensors and biosensors using CNT/Teflon composite materials [150]. Unlike early CNT-modified electrodes, the composite devices rely on the use of CNT as the sole conductive component rather than utilizing it as the modifier in connection with another electrode surface. Another novel glucose biosensor based on CNT epoxy resin biocomposite has also been reported [168]. Experimental results show that the CNT epoxy composite biosensor offers excellent sensitivity, a reliable calibration profile, and stable electrochemical properties together with a significantly lower detection potential (+0.55 V) than the sensor with no CNT. Recently, a nanobiocomposite material based on a MWCNT-Nafion thin film for bioanalytical application has been proposed [169]. By the combination of MWCNTs, Nafion, and GOD, a novel electroanalytical nanobiocomposite thin film can be produced by simple solvent casting. Compton et al. have proposed another glucose biosensor, fabricated by immobilizing GOD encapsulated in a sol–gel matrix on basal plane pyrolytic graphite modified with MWCNTs [151]. The present carbon nanotube sol–gel biocomposite glucose sensor showed excellent properties for the sensitive determination of glucose. More recently, Yang has used sol–gel solution with highly dispersed Pt nanoparticles as a binder for MWCNTs to fabricate glucose biosensors [170]. In another study by Yang, a CNT-cobalt hexacyanoferrate nanoparticle-chitosan biopolymer system was used in the fabrication of biosensors [171]. With the introduction of CNT, this system could show significant improvement of redox activity of cobalt nanoparticles due to the excellent electron-transfer ability of CNTs. Meanwhile, CNTs have also been doped in organically modified sol–gels (ormosils) to enhance the conductive properties of the sol–gel matrixes, since the conductivity of ormosils is not good, although the porosity and biocompatibility are very favorable. Accordingly, H_2O_2 and glucose sensors have been developed by entrapping enzymes in a new ormosil composite doped with ferrocene monocarboxylic acid-bovine serum albumin conjugate and MWCNTs [172, 173].

The insolubility of CNTs in most solvents is one of the limitations in the design of CNTs-based biosensing devices. A stable suspension of CNTs can be obtained by dispersing CNTs in a solution of surfactant, such as cetyltrimethylammonium bromide (CTAB, a cationic surfactant) [174]. CNTs have promotion effects on the direct electron-transfer of GOD immobilized onto the surface of CNTs. Solubilization of CNTs by Nafion, a widely used perfluorosulfonated polymer, has also been reported for both single- and multiwall CNTs [175]. Redox activity has also been dramatically enhanced at CNT/Nafion-coated electrodes. This gives a new way to fabricate amperometric biosensors. A recent interesting observation is that covalently linked composites of MWCNT and GOD composites are highly water soluble [146].

SWCNTs have been reported to combine platinum nanoparticles with a diameter of 2–3 nm for use in fabricating electrochemical sensors with remarkably improved sensitivity toward H_2O_2 and glucose [176]. This work demonstrates that it is possible to obtain very small Pt nanoparticles deposited onto Nafion-solubilized SWCNT. Platinum nanoparticles are in electrical contact, through the SWCNT, with the glassy carbon or carbon fiber backing, enabling the composite structure to be used as an electrode. This composite structure may offer an excellent platform for various biosensing applications. Finally, CNTs have been combined with Fe nanoparticles to develop a glucose biosensor [177].

10.5 Nanocomposite Materials

Collagen/poly(acrylic acid) bilayers were once reported to be added to nanoparticle CdTe/polycation layer-by-layer films to produce porous collagen bilayers. Such stratified multilayer systems showed successful cell attachment and survival while native nanoparticle CdTe/polycation films were strongly cytotoxic [178]. Hu et al. have made a detailed electrochemical investigation of Hb incorporated in collagen films [179]. The Hb-collagen film electrodes were also used to catalyze the reduction of nitrite, oxygen and H_2O_2, indicating potential applications of the films for the fabrication of a new type of biosensor. A novel amperometric glucose biosensor combined with an enzymatic assay of glucose level in rat brain has also been developed based on ferrocene-doped silica (FcDS) nanoparticles conjugated with a biopolymer chitosan (CHIT) membrane [180]. The formation through a water-in-oil (W/O) microemulsion method of such dye-doped silica particles clearly demonstrated their remarkable bioconjugation and photostability.

Gold nanoparticles have been widely used in the electrochemical study of proteins; however, the reversibility of Hb is not so good, because both gold nanoparticles and Hb are positive-charged, and the electrostatic repulsion may result in unfavorable orientation of Hb. It has been reported that α-zirconium phosphate is an anionic material with good biocompatibility, but direct electron transfer of the anchored protein is not easily realized perhaps for the weak conductivity and lower enzyme-loading ability of ZrP [181]. To solve the problem, Chen et al. have introduced the gold nanoparticles into ZrP, based on which an excellent H_2O_2 biosensor was constructed [182]. Liu et al. have also developed a H_2O_2 biosensor based on the immobilization of HRP to a nano-Au monolayer, which is supported by a PAMAM dendrimer/cystamine modified gold electrode [183]. Meanwhile, Yu et al. have reported a new electrode interface by using L-cysteine–gold particle nanocomposite immobilized in the network of a Nafion membrane on a glassy carbon electrode [184].

10.6 Nanoparticles with Special Functions

10.6.1 Semiconductor Nanoparticles

Some nanomaterials used for sensors fabrication have several special functions. For instance, semiconductor nanoparticles that have recently been used for sensing applications are a highly luminescent, photostable class of fluorophore [185–187]. Colloidal semiconductor nanocrystals, often referred to as "quantum dots" or "QDs", are single crystals a few nanometers in diameter whose size and shape can be precisely controlled by the duration, temperature, and ligand molecules used in their synthesis [185]. The synthesis process yields QDs that have composition- and size-dependent absorption and emission. Absorption of a photon with energy above the semiconductor band gap energy results in the creation of an electron–hole pair or exciton. Absorption has an increased probability at higher energies and results in a broadband absorption spectrum, in marked contrast to standard fluorophores. Luminescent QDs have unique spectroscopic properties that include a broad adsorption spectra: size-tunable, narrow, and symmetric photoluminescence emissions ranging from UV to the near-IR range; and exceptional resistance to chemical and photodegradation along with high photobleaching thresholds [188]. Nanocrystals composed of ZnSe, CdS, CdSe, and CdTe also have emission spectra that span the visible spectrum. Researchers have demonstrated that QDs can function as fluorescence resonance-energy transfer (FRET) donors in prototype hybrid QD–protein sensors [189]. Nanosized semiconductor crystals can be effectively coupled to biomolecular units such as enzyme, to generate novel photoelectrochemical systems [190].

CdSe nanocrystals have broad absorption spectra that progressively shift to longer wavelengths with increasing particle size. These nanocrystals are tunable fluorophores with absorption characteristics that allow simultaneous excitation of different particle sizes at a single wavelength, while they exhibit a luminescent emission that spans a wide range of wavelengths in the visible spectrum [188]. Researchers have fabricated a facile, reagentless method for generating protein-based semiconducting nanoparticle sensors for small molecules using CdSe nanoparticles [187]. They used maltose binding protein (MBP) for the sensor receptor. When MBP binds with maltose, the conformation of MBP is changed, and then the movement prohibits the electron transfer to CdSe, resulting in fluorescence emission. Researchers have also described a novel and direct method for conjugating protein molecules to luminescent CdSe–ZnS core–shell nanocrystals used as bioactive fluorescent probes in sensing, immunoassay, imaging, and other diagnostics applications [188].

CdS is also a promising nanomaterial that has also been used in biosensor fabrication. The direct electrochemistry of GOD adsorbed on CdS nanoparticles-modified pyrolytic graphite electrode has been investigated [191]. The results showed that the fabricated biosensor was sensitive and stable in detecting glucose, indicating that CdS nanoparticles are a good candidate material for the immobilization of enzyme in glucose biosensor construction. Hb has also been immobilized with CdS nanoparticles on pyrolytic graphite electrode to characterize the electrochemical reactivity and peroxidase activity of the protein, and the result demonstrates that good redox waves of Hb can be achieved after this protein is entrapped in CdS nanoparticles [192]. Combined with formaldehyde dehydrogenase enzyme, nanocrystalline CdS has also been immobilized by self-assembling on a gold electrode to prepare a biological–inorganic hybrid to perform catalytic oxidation of formaldehyde [190, 193]. Researchers have also prepared an acetylcholine esterase (AChE)/CdS nanoparticle hybrid system for the photoelectrochemical detection of AChE inhibitors [194]. The photoelectrochemical charging effect of a Au-CdS nanoparticle array has also been employed to develop a sensor for acetylcholine esterase inhibitors [195].

As mentioned above, TiO_2 has been a popular nanomaterial for use in sensor devices. Recently, Li et al. have successfully constructed a Hb-based H_2O_2 biosensor tuned by the photovoltaic effect of the nanoparticles [196]. The catalytic ability of the protein and the sensitivity of sensor could be greatly enhanced after UV light irradiation on the co-immobilized nanomaterial.

10.6.2 Magnetic Nanoparticles

Magnetic nanoparticles have been widely applied in various fields of biology and medicine, such as magnetic targeting, magnetic resonance imaging, diagnostics, immunoassays, RNA and DNA purification, gene cloning, and cell separation and purification etc. [197]. Thus, researchers have fabricated magneto-bioelectronics that rely on magnetic nanoparticles. Multilayer films have been prepared on the surface of a glassy carbon electrode by the deposition of chitosan/Fe_3O_4 nanoparticles and phytic acid via the layer-by-layer assembly technique [198]. The multilayer films exhibited good biocompatibility, and adsorbed Hb on the film could realize direct electron transfer reactions and maintain high catalytic activity. In contrast, the orientation of magnetic particles can be controlled with a magnet, which can in turn control the electron transfer of protein-based biosensors. Accordingly, a magneto-switchable bioelectrocatalysis glucose biosensor has been prepared with magnetic Fe_3O_4 nanoparticles as control factor [199]. Switchable systems that can be used as both H_2O_2 biosensor and glucose biosensor have also been prepared with magnetic nanoparticles by the same research group [197]. Other magnetic nanoparticles have also been used in fabricating protein-based biosensors with good characteristics. For example, a phenol biosensor has been developed based on the immobilization of tyrosinase on the surface of modified magnetic $MgFe_2O_4$ nanoparticles attached to the surface of a carbon paste electrode [200].

10.7 Other Nanomaterials

In addition to the above-mentioned nanomaterials, which have been largely used for sensor fabrication, there are also many other nano-materials, which either have been used for biosensing devices or may have the potential of biosensor application. One such nanostructure material is nanocrystalline silicon, often referred to as porous silicon, which, in fact, has been extensively used in sensors. Since the discovery of its strong visible luminescence at room temperature, porous silicon has attracted considerable interest in its possible use in the construction of biosensors. Its ability to emit light is due to its tiny pores that range from less than 2 nm to micrometer dimensions. In addition, porous silicon possesses a high surface to volume ratio (as much as 500 $m^2\ cm^{-3}$) and it can be fabricated easily using some of the established processes of silicon technology. Porous silicon has been used as an optical interferometric transducer for detecting small organic molecules (biotin and digoxigenin), 16-nucleotide DNA oligomers, and proteins (streptavidin and antibodies) at pico- and femtomolar analyte concentrations [201, 202]. Many potentiometric biosensors based on porous silicon have been reported [203–205]. In addition, the enzymes penicillinase and lipase have separately been immobilized on the surface of porous silicon to develop biosensors for penicillin and triglycerides [206].

SiO_2 nanoparticles are another often used nanomaterial. The electrochemical and electrocatalytic properties of heme proteins, including Mb, Hb, and HRP in layer-by-layer films assembled with SiO_2 nanoparticles and the proteins have been investigated. The proteins in the films can display good electrocatalytic activities toward various substrates such as oxygen, H_2O_2, trichloroacetic acid, and nitrite, which has showed a potential applicability in fabricating a new kind of biosensor without using mediators [14]. SiO_2 nanoparticles have also been used to immobilize GOD by the sol–gel method. Experimental results show that hydrophobic SiO_2 nanoparticles can immobilize enzyme well, providing a good and simple approach for preparing a high quality glucose biosensor [207, 208]. Moreover, SiO_2 nanoparticles have been introduced into the construction of enzyme field-effect transistors (ENFETs) to create a glucose-sensitive ENFET, and the SiO_2 nanoparticles can provide a biocompatible environment and improve the enzyme activity, and prevent the immobilized enzymes from leakage as well [209]. In addition, boron-doped silicon nanowires (SiNWs) have been reported by Cui et al. to create highly sensitive, real-time electrically based sensors for biological and chemical species [210]. Antigen-functionalized SiNWs showed reversible antibody binding and concentration-dependent detection in real time.

Recently, a disposable glucose biosensor was reported based on co-dispersion of a diffusion polymeric mediator, poly(vinylferroceneco-acrylamide) (PVFcAA), and GOD in an alumina nanoparticulate membrane on a screen-printed carbon electrode [211]. The nanoparticulate membrane served not only biosensing, but also analyte-regulating functions. Since the membrane is highly hydrophilic, the amount of dissolved oxygen, one of the main interferants, is greatly reduced, and, thus, little interference is observed.

Arrays of nanoscopic gold tubes have been prepared by electroless deposition of the metal within the pores of polycarbonate particle track-etched membranes [212]. GOD can be immobilized onto the self-assembled monolayers of gold tubes, via crosslinking with glutaraldehyde or covalent attachment by carbodiimide coupling. Based on a similar method of template synthesis, Miao et al. have immobilized GOD in the polypyrrole nanotubes and produced a biosensor [213]. Compared with conventional techniques, this immobilization strategy greatly enhances the amount of the enzyme immobilized, the retention of the immobilized activity and the sensitivity of the biosensor.

Polyaniline nanoparticles, synthesized with dodecylbenzylsulfonic acid, can be successfully electrodeposited on the surface of glassy carbon electrodes to form nanostructured films suitable as heterogeneous mediators [214], and nanoparticles have been applied to a glassy carbon electrode surface with HRP for H_2O_2 sensing. A sensitive and selective amperometric glucose biosensor based on platinum microparticles dispersed in nano-fibrous polyaniline has also been developed [215]. PANI has a large specific surface area, good conductivity, high reaction ability, and many microgaps between the fibers, which is very useful for sensor fabrication. This nanofibrous morphology is also beneficial to the dispersion of metal catalyst and the immobilization of enzyme.

Recently, there has been a tendency to extend the study of nanoparticles from metal and inorganic compounds to organic compounds because of the diversity offered by organic molecules [216]. It is expected that organic nanoparticles can be used in constructing sensors, bioprobes, devices, etc. [217, 218].

10.8 Conclusion

Nanobiotechnology has played an increasingly important role in the development of protein-based biosensors. Owing to the unique properties of nanomaterials, such as large surface-to-volume ratio, high surface reaction activity, high catalytic efficiency, and strong adsorption ability, numerous protein-based biosensors have been well constructed with higher selectivity, better stability and a lower detection limit for the detection of even more species. In addition, the special electronic, optical and magnetic characteristics of nanomaterials are providing a platform for fabricating more novel biosensors, which might also open up new topics in the field of protein-based biosensors. Greater achievements and advances will be made by combining nanotechnology with protein science and biosensing.

References

1 Li, G., *Protein-based Voltammetric Sensors*, Encyclopedia of Sensors, Eds. Grimes, C. A. et al., American Scientific Publishers Stevenson Ranch, CA, **2006**, Vol. 8, pp. 301–314.

2 Zhang, W. J., Li, G. X., Third-generation biosensors based on the direct electron transfer of proteins., *Anal. Sci.* **2004**, 20, 603–609.

3 Macus, R. A., Sutin, N., Electron transfers in chemistry and biology., *Biochim. Biophys. Acta*, **1985**, 811, 265–322.

4 Gorton, L., Carbon paste electrodes modified with enzymes, tissues, and cells., *Electroanalysis* **1995**, 7, 23–45.

5 Heller, A., Electrical connection of enzyme redox centers to electrodes., *J. Phys. Chem.* **1992**, 96, 3579–3587.

6 Bartlett, P. N., Tebbutt, P., Whitaker, R. P., Kinetic aspects of the use of modified electrodes and mediators in bioelectrochemistry., *Progr. React. Kinet.* **1991**, 16, 155.

7 Gorton, L., A. Lindgren, T. L., Munteanu, F. D., Ruzgas, T., Gazaryan, I., Direct electron transfer between heme-containing enzymes and electrodes as basis for third generation biosensors., *Anal. Chim. Acta* **1999**, 400, 91–108.

8 Hernandez-Santos, D., Gonzalez-Garcia, M. B., Garcia, A. C., Metal-nanoparticles based electroanalysis., *Electroanalysis* **2002**, 14, 1225–1235.

9 Kima, J., Gratea, J. W., Wang, P., Nanostructures for enzyme stabilization., *Chem. Eng. Sci.* **2006**, 61, 1017–1026.

10 Chen, J. R., Miao, Y. Q., He, N. Y., Wu, X. H., Li, S. J., Nanotechnology and biosensors., *Biotechnol. Adv.* **2004**, 22, 505–518.

11 Vaseashta, A., Dimova-Malinovska, D., Nanostructured and nanoscale devices, sensors and detectors., *Sci. Technol. Adv. Mater.* **2005**, 6, 312–318.

12 Li, Q. W., Luo, G., Feng, J., Direct electron transfer for heme proteins assembled on nanocrystalline TiO_2 film., *Electroanalysis* **2001**, 13, 359–363.

13 Liu, S., Dai, Z., Chen, H., Ju, H., Immobilization of hemoglobin on zirconium dioxide nanoparticles for preparation of a novel hydrogen peroxide biosensor., *Biosens. Bioelectron.* **2004**, 19, 963–969.

14 He, P., Hu, N., Electrocatalytic properties of heme proteins in layer-by-layer films assembled with SiO_2 nanoparticles., *Electroanalysis* **2004**, 16, 1122–1131.

15 Liu, S. Q., Ju, H. X., Renewable reagentless hydrogen peroxide sensor based on direct electron transfer of horseradish peroxidase immobilized on colloidal gold-modified electrode., *Anal. Biochem.* **2002**, 307, 110–116.

16 Zhang, L., Zhao, G. C., Wei, X. W., Yang, Z. S., Electroreduction of oxygen by myoglobin on multi-walled carbon nanotube-modified glassy carbon electrode., *Chem. Lett.* **2004**, 33, 86–87.

17 Cao, D. F., He, P. L., Hu, N. F., Electrochemical biosensors utilising electron transfer in heme proteins immobilised on Fe_3O_4 nanoparticles., *Analyst* **2003**, 128, 1268–1274.

18 Katz, E., Willner, I., Wang, J., Electroanalytical and bioelectroanalytical systems based on metal and semiconductor nanoparticles., *Electroanalysis* **2004**, 16, 19–44.

19 Lim, S. H., Wei, J., Lin, J., Li, Q., KuaYou, J., A glucose biosensor based on electrodeposition of palladium nanoparticles and glucose oxidase onto nafion-solubilized carbon nanotube electrode., *Biosens. Bioelectron.* **2005**, 20, 2341–2346.

20 Hrapovic, S., Liu, Y., Male, K. B., Luong, J. H. T., Electrochemical biosensing platforms using platinum nanoparticles and carbon nanotubes., *Anal. Chem.*, **2004**, 76, 1083–1088.

21 Liu, S., Leech, D., Ju, H., Application of colloidal gold in protein immobilization, electron transfer, and biosensing., *Anal. Lett.* **2003**, 36, 1–19.

22 Han, X., Cheng, W., Zhang, Z., Dong, S., Wang, E., Direct electron transfer between hemoglobin and a glassy carbon electrode facilitated by lipid-protected gold nanoparticles., *Biochim. Biophys. Acta Bioenerg.* **2002**, 1556, 273–277.

23 Yanez-Sedeno, P., Pingarron, J. M., Gold nanoparticle-based electrochemical biosensors., *Anal. Bioanal. Chem.* **2005**, 382, 884–886.

24 Xiao, Y., Patolsky, F., Katz, E., Hainfeld, J. F., Willner, I., Plugging into enzymes: Nanowiring of redox enzymes by a gold nanoparticle. *Science* **2003**, 299, 1877–1881.

25 Brown, K. R., Fox, A. P., Natan, M. J., Morphology-dependent electrochemistry of cytochrome c at Au colloid-modified SnO_2 electrodes., *J. Am. Chem. Soc.* **1996**, 118, 1154–1157.

26 Varfolomeev, S. D., Kurochkin, I. N., Yaropolov, A. I., Direct electron transfer effect biosensor. *Biosens. Bioelectron.*, **1996**, 11, 863–871.

27 Ghindilis, A. L., Atanasov, P., Wilkins, E., Enzyme-catalyzed direct electron transfer: Fundamentals and analytical applications., *Electroanalysis* **1997**, 9, 661–674.

28 Zhao, J., Henkens, R. W., Stonehuerner, J., O'Daly, J. P., Crumbliss, A. L., Direct electron transfer at horseradish peroxidase-colloidal gold modified electrodes., *J. Electroanal. Chem.* **1992**, 327, 109–119.

29 Luo, X. L., Xu, J. J., Zhang, Q., Yang, G. J., Chen, H. Y., Electrochemically deposited chitosan hydrogel for horseradish peroxidase immobilization through gold nanoparticles self-assembly., *Biosens. Bioelectron.* **2005**, 21, 190–196.

30 Xu, S., Han, X., A novel method to construct a third-generation biosensor: Self-assembling gold nanoparticles on thiol-functionalized poly(styrene-co-acrylic acid) nanospheres., *Biosens. Bioelectron.* **2004**, 19, 1117–1120.

31 Wang, L., Wang, E., A novel hydrogen peroxide sensor based on horseradish peroxidase immobilized on colloidal Au modified ITO electrode., *Electrochem. Commun.* **2004**, 6, 225–229.

32 Jia, J. B., Wang, B. Q., Wu, A. G., Cheng, G. J., Li, Z., Dong, S. J., A

method to construct a third-generation horseradish peroxidase biosensor: Self-assembling gold nanoparticles to three-dimensional sol-gel network., *Anal. Chem.* **2002**, 74, 2217–2223.

33 Xiao, Y., Ju, H. X., Chen, H. Y., Hydrogen peroxide sensor based on horseradish peroxidase-labeled Au colloids immobilized on gold electrode surface by cysteamine monolayer., *Anal. Chim. Acta* **1999**, 391, 73–82.

34 Crumbliss, A. L., Perine, S. C., Stonehuerner, J., Tubergen, K. R., Zhao, J. G., Henkens, R. W., O'Daly, J. P., Colloidal gold as a biocompatible immobilization matrix suitable for the fabrication of enzyme electrodes by electrodeposition., *Biotechnol. Bioeng.* **1992**, 40, 483–490.

35 Xu, X., Liu, S., Ju, H., A novel hydrogen peroxide sensor via the direct electrochemistry of horseradish peroxidase immobilized on colloidal gold modified screen-printed electrode., *Sensors* **2003**, 3, 350–360.

36 Lei, C. X., Hu, S. Q., Shen, G. L., Yu, R. Q., Immobilization of horseradish peroxidase to a nano-Au monolayer modified chitosan-entrapped carbon paste electrode for the detection of hydrogen peroxide., *Talanta* **2003**, 59, 981–988.

37 Yeh, J. I., Zimmt, M. B., Zimmerman, A. L., Nanowiring of a redox enzyme by metallized peptides., *Biosens. Bioelectron.* **2005**, 21, 973–978.

38 Xiao, Y., Ju, H. X., Chen, H. Y., Direct electrochemistry of horseradish peroxidase immobilized on a colloid/cysteamine-modified gold electrode., *Anal. Biochem.* **2000**, 278(1), 22–28.

39 Lei, C. X., Wang, H., Shen, G. L., Yu, R. Q., Immobilization of enzymes on the nano-Au film modified glassy carbon electrode for the determination of hydrogen peroxide and glucose., *Electroanalysis* **2004**, 16, 736–740.

40 Zhao, J., O'Daly, J. P., Henkens, R. W., Stonehuerner, J., Crumbliss, A. L., A xanthine oxidase/colloidal gold enzyme electrode for amperometric biosensor applications., *Biosens. Bioelectron.* **1996**, 11, 493–502.

41 Agui, L., Manso, J., Yanez-Sedeno, P., Pingarron, J. M., Amperometric biosensor for hypoxanthine based on immobilized xanthine oxidase on nanocrystal gold-carbon paste electrodes., *Sens. Actuators B* **2006**, 113, 272–280.

42 Liu, Y. J., Nie, L. H., Tao, W. Y., Yao, S. Z., Amperometric study of Au-colloid function on xanthine biosensor based on xanthine oxidase immobilized in polypyrrole layer., *Electroanalysis* **2004**, 16, 1271–1278.

43 Miscoria, S. A., Barrera, G. D., Rivas, G. A., Enzymatic biosensor based on carbon paste electrodes modified with gold nanoparticles and polyphenol oxidase., *Electroanalysis* **2005**, 17, 1578–1582.

44 Crumbliss, A. L., Stonehuerner, J. G., Henkens, R. W., Zhao, J., O'Daly, J. P., A carrageenan hydrogel stabilized colloidal gold multienzyme biosensor electrode utilizing immobilized horseradish peroxidase and cholesterol oxidase/cholesterol esterase to detect cholesterol in serum and whole blood., *Biosens. Bioelectron.* **1993**, 8, 331–337.

45 Shumyantseva, V. V., Carrara, S., Bavastrello, V., Riley, D. J., Bulko, T. V., Skryabin, K. G., Archakov, A. I., Nicolini, C., Direct electron transfer between cytochrome P450scc and gold nanoparticles on screen-printed rhodium–graphite electrodes., *Biosens. Bioelectron.* **2005**, 21, 217–222.

46 Yu, J., Ju, H., Amperometric biosensor for hydrogen peroxide based on hemoglobin entrapped in titania sol–gel film., *Anal. Chim. Acta* **2003**, 486, 209–216.

47 Wang, H. Y., Guan, R., Fan, C. H., Zhu, D. X., Li, G. X., A hydrogen peroxide biosensor based on the bioelectrocatalysis of hemoglobin incorporated in a kieselguhr film., *Sens. Actuators B* **2002**, 84, 214–218.

48 Dai, Z., Liu, S., Ju, H., Chen, H., Direct electron transfer and enzymatic activity of hemoglobin in a hexagonal mesoporous silica matrix., *Biosens. Bioelectron.* **2004**, 19, 861–867.

49 Gu, H. Y., Yu, A. M., Chen, H. Y., Direct electron transfer and characterization of hemoglobin immobilized on a Au colloid–cysteamine-modified gold electrode., *J. Electroanal. Chem.* **2001**, 516, 119–126.

50 Zhang, J. D., Oyama, M., A hydrogen peroxide sensor based on the peroxidase activity of hemoglobin immobilized on gold nanoparticles-modified ITO electrode., *Electrochim. Acta* **2004**, 50, 85–90.

51 Zhang, L., Jiang, X., Wang, E., Dong, S., Attachment of gold nanoparticles to glassy carbon electrode and its application for the direct electrochemistry and electrocatalytic behavior of hemoglobin., *Biosens. Bioelectron.* **2005**, 21, 337–345.

52 Yang, W. W., Li, Y. C., Bai, Y., Sun, C. Q., Hydrogen peroxide biosensor based on myoglobin/colloidal gold nanoparticles immobilized on glassy carbon electrode by a Nafion film., *Sens. Actuators B* **2006**, 115, 42–48.

53 Liu, S. Q., Ju, H. X., Electrocatalysis via direct electrochemistry of myoglobin immobilized on colloidal gold nanoparticles., *Electroanalysis* **2003**, 15, 1488–1493.

54 Ju, H. X., Liu, S. Q., Ge, B., Lisdat, F., Scheller, F. W., Electrochemistry of cytochrome c immobilized on colloidal gold modified carbon paste electrodes and its electrocatalytic activity., *Electroanalysis* **2002**, 14, 141–147.

55 Feng, J. J., Zhao, G., Xu, J. J., Chen, H. Y., Direct electrochemistry and electrocatalysis of heme proteins immobilized on gold nanoparticles stabilized by chitosan., *Anal. Biochem.* **2005**, 342, 280–286.

56 Luo, X. L., Xu, J. J., Du, Y., Chen, H. Y., A glucose biosensor based on chitosan–glucose oxidase–gold nanoparticles biocomposite formed by one-step electrodeposition., *Anal. Biochem.* **2004**, 334, 284–289.

57 Zhang, S. X., Wang, N., Niu, Y. M., Sun, C. Q., Immobilization of glucose oxidase on gold nanoparticles modified Au electrode for the construction of biosensor., *Sens. Actuators B* **2005**, 109, 367–374.

58 Zhang, S. X., Wang, N., Yu, H. J., Niu, Y., Sun, C. Q., Covalent attachment of glucose oxidase to an Au electrode modified with gold nanoparticles for use as glucose biosensor. *Bioelectrochemistry* **2005**, 67, 15–22.

59 Pan, M., Guo, X. S., Cai, Q., Li, G., Chen, Y. Q., A novel glucose sensor system with Au nanoparticles based on microdialysis and coenzymes for continuous glucose monitoring., *Sens. Actuators A* **2003**, 108, 258–262.

60 Mena, M. L., Yanez-Sedeno, P., Pingarron, J. M., A comparison of different strategies for the construction of amperometric enzyme biosensors using gold nanoparticle-modified electrodes., *Anal. Biochem.* **2005**, 336, 20–27.

61 Yanez-Sedeno, P., Pingarron, J. M., Gold nanoparticle-based electrochemical biosensors., *Anal. Bioanal. Chem.* **2005**, 382, 884–886.

62 Gooding, J. J., Hibbert, D. B., The application of alkanethiol self-assembled monolayers to enzyme electrodes., *Trends Anal. Chem.* **1999**, 18, 525–533.

63 Grabar, K. C., Freeman, R. G., Hommer, M. B., Natan, M. J., Preparation and characterization of Au colloid monolayers., *Anal. Chem.* **1995**, 67, 735–743.

64 Grabar, K. C., Smith, P. C., Musick, M. D., Davis, J. A., Walter, D. G., Jachson, M. A., Guthrie, A. P., Natan, M. J., Kinetic control of interparticle spacing in Au colloid-based surfaces: Rational nanometer-scale architecture., *J. Am. Chem. Soc.* **1996**, 118, 1148–1153.

65 Blonder, R., Sheeney, L., Willner, I., Three-dimensional redox-active layered composites of Au–Au, Ag–Ag and Au–Ag colloids., *Chem. Commun.* **1998**, 13, 1393–1394.

66 Fisheson, N., Shkrob, I., Lev, O., Gun, J., Modestov, A. D., Studies on charge transport in self-assembled

gold-dithiol films: Conductivity, photoconductivity, and photoelectrochemical measurements., *Langmuir* **2001**, 17, 403–412.

67 Doron, A., Katz, E., Willner, I., Organization of Au colloids as monolayer films onto ITO glass surfaces: Application of the metal colloid films as base interfaces to construct redox-active monolayers., *Langmuir* **1995**, 11, 1313–1317.

68 Lei, C. X., Hu, S. Q., Gao, N., Shen, G. L., Yu, R. Q., An amperometric hydrogen peroxide biosensor based on immobilizing horseradish peroxidase to a nano-Au monolayer supported by sol–gel derived carbon ceramic electrode., *Bioelectrochemistry* **2004**, 65, 33–39.

69 Shipway, A. N., Lahav, M., Willner, I., Nanostructured gold colloid electrodes., *Adv. Meterol.* **2000**, 12, 993–998.

70 Zhong, X., Yuan, R., Chai, Y., Liu, Y., Dai, J., Tang, D., Glucose biosensor based on self-assembled gold nanoparticles and double-layer 2d-network (3-mercaptopropyl)-trimethoxysilane polymer onto gold substrate., *Sens. Actuators B* **2005**, 104, 191–198.

71 Patolsky, F., Gabriel, T., Willner, I., Controlled electrocatalysis by microperoxidase-11 and Au-nanoparticle superstructures on conductive supports., *J. Electroanal. Chem.* **1999**, 479, 69–73.

72 Bharathi, S., Nogami, M., Ikeda, S., Novel electrochemical interfaces with a tunable kinetic barrier by self-assembling organically modified silica gel and gold nanoparticles., *Langmuir* **2001**, 17, 1–4.

73 Di, J. W., Shen, C. P., Peng, S. H., Tu, Y. F., Li, S. J., A one-step method to construct a third-generation biosensor based on horseradish peroxidase and gold nanoparticles embedded in silica sol–gel network on gold modified electrode., *Anal. Chim. Acta* **2005**, 553, 196–200.

74 Cai, W. Y., Xu, Q., Zhao, X. N., Zhu, J. J., Chen, H. Y., Porous gold-nanoparticle-$CaCO_3$ hybrid material: Preparation, characterization, and application for horseradish peroxidase assembly and direct electrochemistry., *Chem. Mater.* **2006**, 18, 279–284.

75 Liu, S. Q., Ju, H. X., Reagentless glucose biosensor based on direct electron transfer of glucose oxidase immobilized on colloidal gold modified carbon paste electrode., *Biosens. Bioelectron.* **2003**, 19, 177–183.

76 Liu, S. Q., Yu, J. H., Ju, H. X., Renewable phenol biosensor based on a tyrosinase-colloidal gold modified carbon paste electrode., *J. Electroanal. Chem.* **2003**, 540, 61–67.

77 Xu, X. X., Liu, S. Q., Li, B., Ju, H. X., Disposable nitrite sensor based on hemoglobin-colloidal gold nanoparticle modified screen-printed electrode., *Anal. Lett.* **2003**, 36, 2427–2442.

78 Liu, S. Q., Ju, H. X., Nitrite reduction and detection at a carbon paste electrode containing hemoglobin and colloidal gold., *Analyst* **2003**, 128, 1420–1424.

79 Carralero, V., Mena, M. L., Gonzalez-Cortes, A., Yanez-Sedeno, P., Pingarron, J. M., Development of a tyrosinase biosensor based on gold nanoparticles-modified glassy carbon electrodes: Application to the measurement of a bioelectrochemical polyphenols index in wines., *Anal. Chim. Acta* **2005**, 528, 1–8.

80 Gonzalez-Garcia, M. B., Fernandez-Sanchez, C., Costa-Garcia, A., Colloidal gold as an electrochemical label of streptavidin-biotin interaction., *Biosens. Bioelectron.* **2000**, 15, 315–321.

81 Velev, O. D., Kaler, E. W., In situ assembly of colloidal particles into miniaturized biosensors., *Langmuir* **1999**, 25, 3694–3698.

82 Niemeyer, C. M., Ceyhan, B., DNA-directed functionalization of colloidal gold with proteins., *Angew. Chem. Int. Ed.* **2001**, 40, 3685–3688.

83 Thanh, N. T. K., Rosenzweig, Z., Development of an aggregation-based immunoassay for anti-protein A using gold nanoparticles., *Anal. Chem.* **2002**, 74, 1624–1628.

84 Kerman, K., Morita, Y., Takamura, Y., Ozsoz, M., Tamiya, E., Modification of Escherichia coli single-stranded DNA binding protein with gold nanoparticles for electrochemical detection of DNA hybridization., *Anal. Chim. Acta* **2004**, 510, 169–174.
85 Kvitek, L., Prucek, R., The preparation and application of silver nanoparticles., *J. Mater. Sci.*, **2005**, in the press, available online, http://dx.doi.org/10.2007/s10853-005-0789-2.
86 Liu, T., Zhong, J., Gan, X., Fan, C., Li, G., Matsuda, N., Wiring electrons of cytochrome c with silver nanoparticles in layered films., *ChemPhysChem* **2003**, 4, 1364–1366.
87 Ren, C. B., Song, Y. H., Li, Z., Zhu, G. Y., Hydrogen peroxide sensor based on horseradish peroxidase immobilized on a silver nanoparticles/cysteamine/gold electrode., *Anal. Bioanal. Chem.* **2005**, 381, 1179–1185.
88 Gan, X., Liu, T., Zhong, J., Liu, X. J., Li, G. X., Effect of silver nanoparticles on the electron transfer reactivity and the catalytic activity of myoglobin., *ChemPhysChem* **2004**, 5, 1686–1691.
89 Gan, X., Liu, T., Zhu, X. L., Li, G. X., An electrochemical biosensor for nitric oxide based on silver nanoparticles and hemoglobin., *Anal. Sci.* **2004**, 20, 1271–1275.
90 Ren, X. L., Meng, X. W., Chen, D., Tang, F., Jiao, J., Using silver nanoparticle to enhance current response of biosensor., *Biosens. Bioelectron.* **2005**, 21, 433–437.
91 Ren, X. L., Meng, X. W., Tang, F. Q., Preparation of Ag–Au nanoparticle and its application to glucose biosensor., *Sens. Actuators B* **2005**, 110, 358–363.
92 Zhao, S., Zhang, K., Sun, Y. Y., Sun, C. Q., Hemoglobin/colloidal silver nanoparticles immobilized in titania sol–gel film on glassy carbon electrode: Direct electrochemistry and electrocatalysis., *Bioelectrochemistry* **2006**, 1, 10–15.
93 Reynolds, E. R., Yacynych, A. M., Platinized carbon ultramicroelectrodes as glucose biosensors., *Electroanalysis* **1993**, 5, 405–411.
94 Hartl, A., Schmich, E., Garrido, J., Hernando, J., Catharino, S. C. R., Walter, S., Feulner, P., Kromka, A., Steinmuller, D., Stutzmann, M., Protein-modified nanocrystalline diamond thin films for biosensor applications., *Nat. Mater.* **2004**, 3, 736–742.
95 Olivia, H., Sarada, B. V., Honda, K., Fujishim, A., Continuous glucose monitoring using enzyme-immobilized platinized diamond microfiber electrodes., *Electrochim. Acta* **2004**, 49, 2069–2076.
96 Zhang, F. F., Wan, Q., Li, C. X., Zhu, Z. Q., Xian, Y. Z., Jin, L. T., Yamamoto, K., A novel glucose biosensor based on palladium nanoparticles and its application in detection of glucose level in urine., *Chin. J. Chem.* **2003**, 21, 1619–1623.
97 Pan, D., Chen, J., Yao, S., Nie, L., Xia, J., Tao, W., Amperometric glucose biosensor based on immobilization of glucose oxidase in electropolymerized o-aminophenol film at copper-modified gold electrode., *Sens. Actuators B* **2005**, 104, 68–74.
98 Xu, J. J., Luo, X. L., Du, Y., Chen, H. Y., Application of MnO_2 nanoparticles as an eliminator of ascorbate interference to amperometric glucose biosensors., *Electrochem. Commun.* **2004**, 6, 1169–1173.
99 Wang, K., Xu, J. J., Chen, H. Y., Biocomposite of cobalt phthalocyanine and lactate oxidase for lactate biosensing with MnO_2 nanoparticles as an eliminator of ascorbic acid interference., *Sens. Actuators B* **2006**, 114, 1052–1058.
100 Yin, L. T., Chou, J. C., Chung, W. Y., Sun, T. P., Hsiung, K. P., Hsiung, S. K., Glucose ENFET doped with MnO_2 powder., *Sens. Actuators B* **2001**, 76, 187–192.
101 Luo, X. L., Xu, J. J., Zhao, W., Chen, H. Y., A novel glucose ENFET based on the special reactivity of MnO_2 nanoparticles., *Biosens. Bioelectron.*, **2004**, 19, 1295–1300.
102 Zhang, Y., He, P., Hu, N., Horseradish peroxidase immobilized

in TiO_2 nanoparticle films on pyrolytic graphite electrodes: Direct electrochemistry and bioelectrocatalysis, *Electrochim. Acta* **2004**, 49, 1981–1988.

103 Yang, Y., Yang, H., Yang, M., Liu, Y., Shen, G., Yu, R., Amperometric glucose biosensor based on a surface treated nanoporous ZrO_2/chitosan composite film as immobilization matrix., *Anal. Chim. Acta* **2004**, 525, 213–220.

104 Liu, B., Cao, Y., Chen, D., Kong, J., Deng, J., Amperometric biosensor based on a nanoporous ZrO_2 matrix. *Anal. Chim. Acta* **2003**, 478, 59–66.

105 Zhao, G., Feng, J. J., Xu, J. J., Chen, H. Y., Direct electrochemistry and electrocatalysis of heme proteins immobilized on self-assembled ZrO_2 film., *Electrochem. Commun.* **2005**, 7, 724–729.

106 Topoglidis, E., Cass, A. E. G., O'Regan, B., Durrant, J. R., Immobilisation and bioelectrochemistry of proteins on nanoporous TiO_2 and ZnO films., *J. Electroanal. Chem.* **2001**, 517, 20–27.

107 Li, Y. F., Liu, Z. M., Liu, Y. L., Yang, Y. H., Shen, G. L., Yu, R. Q., A mediator-free phenol biosensor based on immobilizing tyrosinase to ZnO nanoparticles., *Anal. Biochem.* **2006**, 349, 33–40.

108 Katz, E., Willner, I., Biomolecule-functionalized carbon nanotubes: Applications in nanobioelectronics., *ChemPhysChem* **2004**, 5, 1084–1104.

109 Merkoci, A., Pumera, M., Llopis, X., Perez, B., Valle, M. D., Alegret, S., New materials for electrochemical sensing VI: Carbon nanotubes., *Trends Anal. Chem.* **2005**, 24, 826–838.

110 Rubianes, M. D., Rivas, G. A., Enzymatic biosensors based on carbon nanotubes paste electrodes., *Electroanalysis* **2005**, 17, 73–78.

111 Gong, K. P., Yan, Y. M., Zhang, M. N., Su, L., Xiong, S. X., Mao, L. Q., Enzymatic biosensors based on carbon nanotubes paste electrodes., *Anal. Sci.* **2005**, 21, 1383–1393.

112 Lin, Y. H., Yantasee, W., Wang, J., Carbon nanotubes (CNTs) for the development of electrochemical biosensors., *Front. Biosci.* **2005**, 10, 492–505.

113 Wohlstadter, J. N., Wilbur, J. L., Sigal, G. B., Biebuyck, H. A., Billadeau, M. A., Dong, L., Fischer, A. B., Gudibande, S. R., Jameison, S. H., Kenten, J. H., Leginus, J., Leland, J. K., Massey, R. J., Wohlstadter, S. J., Carbon nanotube-based biosensor., *Adv. Mater.* **2003**, 15, 1184–1187.

114 Sotiropoulou, S., Chaniotakis, N. A., Carbon nanotube array-based biosensor., *Anal. Bioanal. Chem.* **2003**, 375, 103–105.

115 Ye, J. S., Wen, Y., Zhang, W. D., Cui, H. F., Xu, G. Q., Sheu, F. S., Electrochemical biosensing platforms using phthalocyanine-functionalized carbon nanotube electrode., *J. Electroanal. Chem.* **2005**, 17, 89–96.

116 Xu, J. Z., Zhu, J. J., Wu, Q., Hu, Z., Chen, H. Y., An amperometric biosensor based on the coimmobilization of horseradish peroxidase and methylene blue on a carbon nanotubes modified electrode., *Electroanalysis* **2003**, 15, 219–224.

117 Dresselhaus, M. S., Dresselhaus, G., Avouris, P., Carbon nanotubes synthesis, structure, properties, and applications., *Fullerenes, Nanotubes, Carbon Nanostruct.* **2002**, 10(2), 181–182.

118 Davis, J. J., Coleman, K. S., Azamian, B. R., Bagshaw, C. B., Green, M. L. H., Chemical and biochemical sensing with modified single walled carbon nanotubes., *Chem. Eur. J.* **2003**, 9, 3732–3739.

119 Banks, C. E., Davies, T. J., Wildgoose, G. G., Compton, R. G., Electrocatalysis at graphite and carbon nanotube modified electrodes: Edge-plane sites and tube ends are the reactive sites., *Chem. Comm.* **2005**, 7, 829–841.

120 Wang, S. G., Zhang, Q., Wang, R., Yoon, S. F., A novel multi-walled carbon nanotube-based biosensor for glucose detection., *Biochem. Biophys. Res. Commun.* **2003**, 311, 572–576.

121 Yu, X., Chattopadhyay, D., Galeska, I., Papadimitrakopoulos, F.,

Rusling, J. F., Peroxidase activity of enzymes bound to the ends of single-wall carbon nanotube forest electrodes., *Electrochem. Commun.* **2003**, 5, 408–411.

122 Wang, J., Li, M., Shi, Z., Li, N., Gu, Z., Direct electrochemistry of cytochrome c at a glassy carbon electrode modified with single-wall carbon nanotubes., *Anal. Chem.* **2002**, 74, 1993–1997.

123 Cai, C., Chen, J., Direct electron transfer and bioelectrocatalysis of hemoglobin at a carbon nanotube electrode., *Anal. Biochem.* **2004**, 325, 285–292.

124 Zhao, Y., Zhang, W., Chen, H., Luo, Q., Li, S., Direct electrochemistry of horseradish peroxidase at carbon nanotube powder microelectrode., *Sens. Actuators B* **2002**, 87, 168–172.

125 Sotiropoulou, S., Gavalas, V., Vamvakaki, V., Chaniotakis, N. A., Novel carbon materials in biosensor systems., *Biosens. Bioelectron.* **2003**, 18, 211–215.

126 Zhao, G. C., Yin, Z. Z., Wei, X. W., A reagentless biosensor of nitric oxide based on direct electron transfer process of cytochrome c on multi-walled carbon nanotube., *Front. Biosci.* **2005**, 10, 2005–2010.

127 Zhao, G. C., Zhang, L., Wei, X. W., Yang, Z. S., Myoglobin on multi-walled carbon nanotubes modified electrode: Direct electrochemistry and electrocatalysis., *Electrochem. Commun.* **2003**, 5, 825–829.

128 Zhao, L., Liu, H., Hu, N., Electroactive films of heme protein-coated multiwalled carbon nanotubes., *J. Colloid Interface Sci.* **2006**, 296, 204–211.

129 Qi, H., Zhang, C., Li, X., Amperometric third-generation hydrogen peroxide biosensor incorporating multiwall carbon nanotubes and hemoglobin., *Sens. Actuators B* **2006**, 114, 364–370.

130 Liu, Y., Wang, M., Zhao, F., Xu, Z., Dong, S., The direct electron transfer of glucose oxidase and glucose biosensor based on carbon nanotubes/chitosan matrix., *Biosens. Bioelectron.* **2005**, 21, 984–988.

131 Yamamoto, K., Shi, G., Zhou, T. S., Xu, F., Xu, J. M., Kato, T., Jin, J. Y., Jin, L., Study of carbon nanotubes-HRP modified electrode and its application for novel on-line biosensors., *Analyst* **2003**, 128, 249–254.

132 Zhao, Y. D., Bi, Y. H., Zhang, W. D., Luo, Q. M., The interface behavior of hemoglobin at carbon nanotube and the detection for H_2O_2., *Talanta* **2005**, 65, 489–494.

133 Azamian, B. R., Davis, J. J., Coleman, K. S., Bagshaw, C., Green, M. L. H., Bioelectrochemical single-walled carbon nanotubes., *J. Am. Chem. Soc.* **2002**, 64, 124–126.

134 Dai, Y. Q., Shiu, K. K., Glucose biosensor based on multi-walled carbon nanotube modified glassy carbon electrode., *Electroanalysis* **2004**, 16, 1697–1703.

135 Gao, M., Dai, L., Wallace, G., Glucose sensors based on glucose-oxidase-containing polypyrrole/aligned carbon nanotube coaxial nanowire electrodes., *Synth. Met.* **2003**, 137, 1393–1394.

136 Loh, K. P., Zhao, S. L., Zhang, W. D., Diamond and carbon nanotube glucose sensors based on electro-polymerization., *Diamond Relat. Mater.* **2004**, 13, 1075–1079.

137 Wang, J., Musameh, M., Carbon-nanotubes doped polypyrrole glucose biosensor., *Anal. Chim. Acta* **2005**, 539, 209–213.

138 Zhao, G. C., Zhang, L., Wei, X. W., An unmediated H_2O_2 biosensor based on the enzyme-like activity of myoglobin on multi-walled carbon nanotubes., *Anal. Biochem.* **2004**, 329, 160–161.

139 Zhang, L., Zhao, G. C., Wei, X. W., Yang, Z. S., A nitric oxide biosensor based on myoglobin adsorbed on multi-walled carbon nanotubes., *Electroanalysis* **2005**, 17, 630–634.

140 Zhao, G. C., Yin, Z. Z., Zhang, L., Wei, X. W., Direct electrochemistry of cytochrome c on a multi-walled carbon nanotubes modified electrode

and its electrocatalytic activity for the reduction of H_2O_2., *Electrochem. Commun.* **2005**, 7, 256–260.

141 Gooding, J. J., Wibowo, R., Liu, J., Yang, W., Losic, D., Orbons, S., Mearns, F. J., Shapter, J. G., Hibbert, D. B., Protein electrochemistry using aligned carbon nanotube arrays., *J. Am. Chem. Soc.* **2003**, 125, 9006–9007.

142 Wang, M. K., Shen, Y., Liu, Y., Wang, T., Zhao, F., Liu, B. F., Dong, S. J., Direct electrochemistry of microperoxidase-11 using carbon nanotube modified electrodes., *J. Electroanal. Chem.* **2005**, 578, 121–127.

143 Wang, L., Yuan, Z., Direct electrochemisty of xanthine oxidase at a gold electrode modified with single-wall carbon nanotubes., *Anal. Sci.* **2004**, 20, 635–638.

144 Guiseppi-Eli, A., Lei, C., Baughman, R. H., Direct electron transfer of glucose oxidase on carbon nanotubes., *Nanotechnology* **2002**, 13, 559–564.

145 Xue, H., Sun, W., He, B., Shen, Z., Single-wall carbon nanotubes as immobilization material for glucose biosensor., *Synth. Met.* **2003**, 135, 831–832.

146 Li, J., Wang, Y. B., Qiu, J. D., Sun, D. C., Xia, X. H., Biocomposites of covalently linked glucose oxidase on carbon nanotubes for glucose biosensor., *Anal. Bioanal. Chem.* **2005**, 383, 918–922.

147 Withey, G. D., Lazareck, A. D., Tzolov, M. B., Yina, A., Aich, P., Yeh, J. I., Xua, J. M., Ultra-high redox enzyme signal transduction using highly ordered carbon nanotube array electrodes., *Biosens. Bioelectron.* **2006**, 21, 1560–1565.

148 Liu, J., Chou, A., Rahmat, W., Paddon-Row, M. N., Gooding, J. J., Achieving direct electrical connection to glucose oxidase using aligned single walled carbon nanotube arrays., *Electroanalysis* **2005**, 17, 38–46.

149 Patolsky, F., Weizmann, Y., Willner, I., Long-range electrical contacting of redox enzymes by SWCNT connectors., *Angew. Chem. Int. Ed.* **2004**, 43, 2113–2117.

150 Wang, J., Musameh, M., Carbon nanotube/teflon composite electrochemical sensors and biosensors., *Anal. Chem.* **2003**, 75, 2075–2079.

151 Salimia, A., Compton, R. G., Hallaj, R., Glucose biosensor prepared by glucose oxidase encapsulated sol-gel and carbon-nanotube-modified basal plane pyrolytic graphite electrode., *Anal. Biochem.* **2004**, 333, 49–56.

152 Huang, J., Yang, Y., Shi, H., Song, Z., Zhao, Z., Anzai, J. I., Osa, T., Chen, Q., Multi-walled carbon nanotubes-based glucose biosensor prepared by a layer-by-layer technique., *Mater. Sci. Eng. C* **2006**, 26, 113–117.

153 Zhao, H., Ju, H. X., Multilayer membranes for glucose biosensing via layer-by-layer assembly of multiwall carbon nanotubes and glucose oxidase., *Anal. Biochem.* **2006**, 350, 138–144.

154 Liu, G., Lin, Y., Amperometric glucose biosensor based on self-assembling glucose oxidase on carbon nanotubes., *Electrochem. Commun.* **2006**, 8, 251–256.

155 Wang, J., Musameh, M., Carbon nanotube screen-printed electrochemical sensors., *Analyst* **2004**, 129, 1–2.

156 Guan, W. J., Li, Y., Chen, Y. Q., Zhang, X. B., Hu, G. Q., Glucose biosensor based on multi-wall carbon nanotubes and screen printed carbon electrodes., *Biosens. Bioelectron.* **2005**, 21, 508–512.

157 Luque, G. L., Ferreyra, N. F., Rivas, G. A., Glucose biosensor based on the use of a carbon nanotube paste electrode modified with metallic particles., *Microchim. Acta* **2006**, 152, 277–283.

158 Malinski, K., Taha, Z., Nitric oxide release from a single cell measured in situ by a porphyrinic-based microsensor., *Nature* **1992**, 352, 676–678.

159 Troyer, K. P., Wightman, R. M., Dopamine transport into a single cell in a picoliter., Vial. *Anal. Chem.* **2002**, 74, 5370–5375.

160 Britto, P. J., Santhanam, K. S. V., Rubio, A., Alonso, J. A., Ajayan, P. M., Improved charge transfer at carbon nanotube electrodes., *Adv. Mater.* **1999**, 11, 154–157.

161 Menon, V. P., Martin, C. R., Fabrication and evaluation of nanoelectrode ensembles., *Anal. Chem.* **1995**, 67, 1920–1928.

162 Tu, Y., Lin, Y. H., Ren, Z. F., Nanoelectrode arrays based on low site density aligned carbon nanotubes., *Nano Lett.* **2003**, 3, 107–109.

163 Lin, Y., Lu, F., Tu, Y., Ren, Z., Glucose biosensors based on carbon nanotube nanoelectrode ensembles., *Nano Lett.* **2004**, 4, 191–195.

164 Tu, Y., Huang, Z. P., Wang, D. Z., Wen, J. G., Ren, Z. F., Growth of aligned carbon nanotubes with controlled site density., *Appl. Phys. Lett.* **2002**, 80, 4018–4020.

165 Tang, H., Chen, J., Yao, S., Nie, L., Deng, G., Kuang, Y., Amperometric glucose biosensor based on adsorption of glucose oxidase at platinum nanoparticle-modified carbon nanotube electrode., *Anal. Biochem.* **2004**, 331, 89–97.

166 Wang, J., Musameh, M., Enzyme-dispersed carbon-nanotube electrodes: A needle microsensor for monitoring glucose., *Analyst* **2003**, 128, 1382–1385.

167 Zhao, Q., Zhan, D. P., Ma, H. Y., Zhang, M. Q., Zhao, Y. F., Jing, P., Zhu, Z. W., Wan, X. H., Shao, Y. H., Zhuang, Q. K., Direct proteins electrochemistry based on ionic liquid mediated carbon nanotube modified glassy carbon electrode., *Front. Biosci.* **2002**, 7, 752–764.

168 Perez, B., Pumera, M., del Valle, M., Merkoci, A., Alegret, S., Glucose biosensor based on carbon nanotube epoxy composites., *J. Nanosci. Nanotechnol.* **2005**, 5, 1694–1698.

169 Tsai, Y. C., Li, S. C., Chen, J. M., Cast thin film biosensor design based on a nafion backbone, a multiwalled carbon nanotube conduit, and a glucose oxidase function., *Langmuir* **2005**, 21, 3653–3658.

170 Yang, M., Yang, Y., Liu, Y., Shen, G., Yu, R., Platinum nanoparticles-doped sol–gel carbon nanotubes composite electrochemical sensors and biosensors., *Biosens. Bioelectron.* **2006**, 21, 1125–1131.

171 Yang, M., Jiang, J., Yang, Y., Chen, X., Shen, G., Yu, R., Carbon nanotube cobalt hexacyanoferrate nanoparticle-biopolymer system for the fabrication of biosensors., *Biosens. Bioelectron.* **2006**, 21, 1791–1797.

172 Kandimalla, V. B., Tripathi, V. S., Ju, H. X., A conductive ormosil encapsulated with ferrocene conjugate and multiwall carbon nanotubes for biosensing application., *Biomaterials* **2006**, 27, 1167–1174.

173 Tripathi, V. S., Kandimalla, V. B., Ju, H. X., Amperometric biosensor for hydrogen peroxide based on ferrocene-bovine serum albumin and multiwall carbon nanotube modified ormosil composite., *Biosens. Bioelectron.* **2006**, 21, 1529–1535.

174 Cai, C., Chen, J., Direct electron transfer of glucose oxidase promoted by carbon nanotubes., *Anal. Biochem.* **2004**, 332, 75–83.

175 Wang, J., Musameh, M., Lin, Y., Solubilization of carbon nanotubes by nafion toward the preparation of amperometric biosensors., *J. Am. Chem. Soc.* **2003**, 125, 2408–2409.

176 Hrapovic, S., Liu, Y. L., Male, K. B., Luong, J. H. T., Electrochemical biosensing platforms using platinum nanoparticles and carbon nanotubes., *Anal. Chem.* **2004**, 76, 1083–1088.

177 Wu, J., Zou, Y., Gao, N., Jiang, J., Shen, G., Yu, R., Electrochemical performances of C/Fe nanocomposite and its use for mediator-free glucose biosensor preparation., *Talanta* **2005**, 68, 12–18.

178 Sinani, V. A., Koktysh, D. S., Yun, B. G., Matts, R. L., Pappas, T. C., Motamedi, M., Thomas, S. N., Kotov, N. A., Collagen coating promotes biocompatibility of semiconductor nanoparticles in stratified LBL films., *Nano Lett.* **2003**, 3, 1177–1182.

179 Li, M., He, P. L., Zhang, Y., Hu, N. F., An electrochemical investigation of hemoglobin and catalase incorporated in collagen films., *Biochim. Biophys. Acta* **2005**, 1749, 43–51.

180 Zhang, F. F., Wan, Q., Wang, X. L., Sun, Z. D., Zhu, Z. Q., Xian, Y. Z., Jin, L. T., Yamamoto, K., Amperometric sensor based on ferrocene-doped silica nanoparticles as an electron transfer mediator for the determination of glucose in rat brain coupled to in vivo microdialysis., *J. Electroanal. Chem.* **2004**, 571, 133–138.

181 Bellezza, F., Cipiciani, A., Costantino, U., Nicolis, S., Catalytic activity of myoglobin immobilized on zirconium phosphonates., *Langmuir* **2004**, 20, 5019–5025.

182 Feng, J. J., Xu, J. J., Chen, H. Y., Synergistic effect of zirconium phosphate and Au nanoparticles on direct electron transfer of hemoglobin on glassy carbon electrode., *J. Electroanal. Chem.* **2005**, 585, 44–50.

183 Liu, Z. M., Yang, Y., Wang, H., Liu, Y. L., Shen, G. L., Yu, R. Q., A hydrogen peroxide biosensor based on nano-Au/PAMAM dendrimer/ cystamine modified gold electrode., *Sens. Actuators B* **2005**, 106, 394–400.

184 Li, X. L., Wu, J., Gao, N., Shen, G. L., Yu, R. Q., Electrochemical performance of L-cysteine–goldparticle nanocomposite electrode interface as applied to preparation of mediator-free enzymatic biosensors., *Sens. Actuators B* **2006**, 177, 35–42.

185 Michalet, X., Pinaud, F. F., Bentolila, L. A., Tsay, J. M., Doose, S., Li, J. J., Sundaresan, G., Wu, A. M., Gambhir, S. S., Weiss, S., Quantum dots for livecells, in vivo imaging, and diagnostics., *Science* **2005**, 307, 538–544.

186 Wang, Y., Tang, Z., Kotov, N. A., Bioapplication of nanosemiconductors., *Mater. Today* **2005**, 8, 20–31.

187 Marinella, G., Sandros, D. G., Benson, D. E., A modular nanoparticle-based system for reagentless small molecule biosensing., *J. Am. Chem. Soc.* **2005**, 127, 12 198–12 199.

188 Mattoussi, H., Mauro, J. M., Goldman, E. R., Anderson, G. P., Sundar, V. C., Mikulec, F. V., Bawendi, M. G., Self-assembly of CdSe-ZnS quantum dot bioconjugates using an engineered recombinant protein., *J. Am. Chem. Soc.* **2000**, 122, 12 142–12 150.

189 Goldman, E. R., Medintz, I. L., Whitley, J. L., Hayhurst, A., Uyeda, H. T., Deschamps, J. R., Lassman, M. E., Mattoussi, H., A hybrid quantum dot-antibody fragment fluorescence resonance energy transfer-based TnT sensor., *J. Am. Chem. Soc.* **2005**, 127, 6744–6751.

190 Curri, M. L., Agostian, A., Leo, G., Mallardi, A., Cosma, P., Monica, M. D., Development of a novel enzyme/semiconductor nanoparticles system for biosensor application., *Mater. Sci. Eng. C* **2002**, 22, 449–452.

191 Huang, Y. X., Zhang, W. J., Xiao, H., Li, G. X., An electrochemical investigation of glucose oxidase at a CdS nanoparticles modified electrode., *Biosens. Bioelectron.* **2005**, 21, 817–821.

192 Zhou, H., Gan, X., Liu, T., Yang, Q. L., Li, G. X., Effect of nano cadmium sulfide on the electron transfer reactivity and peroxidase activity of hemoglobin., *J. Biochem. Biophys. Methods* **2005**, 64, 38–45.

193 Vastarella, W., Nicastri, R., Enzyme/semiconductor nanoclusters combined systems for novel amperometric biosensors., *Talanta* **2005**, 66, 627–633.

194 Pardo-Yissar, V., Katz, E., Wasserman, J., Willner, I., Acetylcholine esterase-labeled CdS nanoparticles on electrodes: Photoelectrochemical sensing of the enzyme inhibitors., *J. Am. Chem. Soc.* **2003**, 125, 622–623.

195 Zayats, M., Kharitonov, A. B., Pogorelova, S. P., Lioubashevski, O., Katz, E., Willner, I., Probing photoelectrochemical processes in Au-CdS nanoparticle arrays by surface

plasmon resonance: Application for the detection of acetylcholine esterase inhibitors., *J. Am. Chem. Soc.* **2003**, 125, 16 006–16 014.

196 Zhou, H., Gan, X., Wang, J., Zhu, X., Li, G., Hemoglobin-based hydrogen peroxide biosensor tuned by the photovoltaic effect of nano titanium dioxide., *Anal. Chem.* **2005**, 77, 6102–6104.

197 Katz, E., Willner, I., Switching of directions of bioelectrocatalytic currents and photocurrents at electrode surfaces by using hydrophobic magnetic nanoparticles., *Angew. Chem. Int. Ed.* **2005**, 44, 4791–4794.

198 Zhao, G., Xu, J. J., Chen, H. Y., Fabrication, characterization of Fe_3O_4 multilayer film and its application in promoting direct electron transfer of hemoglobin., *Electrochem. Commun.* **2006**, 8, 148–154.

199 Hirsch, R., Katz, E., Willner, I., Magneto-switchable bioelectrocatalysis., *J. Am. Chem. Soc.* **2000**, 122, 12 053–12 054.

200 Liu, Z., Liu, Y., Yang, H., Yang, Y., Shen, G., Yu, R., A phenol biosensor based on immobilizing tyrosinase to modified core-shell magnetic nanoparticles supported at a carbon paste electrode., *Anal. Chim. Acta* **2005**, 533, 3–9.

201 Lin, V. S., Motesharei, K., Dancil, K. P., Sailor, M. J., Ghadiri, M. R., A porous silicon-based optical interferometric biosensor., *Science* **1997**, 278, 840–843.

202 Di Francia, G., Quercia, L., La Ferrara, S., Manzo, S., Chiavarini, S., Cerullo, F., De Filippo, F., La Ferrara, V., Maddalena, P., Vitiello, R., *Proceedings of the 2nd Workshop on Chemical Sensors and Biosensors*, Amzzei, F. et al. (eds.), Evea Casaccia, Rome, **1999**, pp. 18–19.

203 Thust, M., Schoening, M. J., Frohnhoff, S., Arens-Fischer, R., Kordos, P., Lueth, H., Porous silicon as a substrate material for potentiometric biosensors., *Meas. Sci. Technol.* **1996**, 7, 26–29.

204 Reddy, R. R., Chadha, A., Bhattacharya, E., Porous silicon based potentiometric triglyceride biosensor., *Biosens. Bioelectron.* **2001**, 16, 313–317.

205 Reddy, R. R. K., Basu, I., Bhattacharya, E., Chadha, A., Estimation of triglycerides by a porous silicon based potentiometric biosensor., *Curr. Appl. Phys.* **2003**, 3, 155–161.

206 Schoning, M. J., Kurowski, A., Thust, M., Kordos, P., Schultze, J. W., Luth, H., Capacitive microsensors for biochemical sensing based on porous silicon technology., *Sens. Actuators B* **2000**, 64, 59–64.

207 Lin, Z., Fangqiong, T., Jinsuo, Y., Long, J., Preparation of glucose oxidase electrode containing hydrophobic silica nanoparticles by the sol-gel method., *Sci. Chin.* **1995**, 38, 1434–1438.

208 Yang, H., Zhu, Y., A high performance glucose biosensor enhanced via nanosized SiO2., *Anal. Chim. Acta* **2005**, 554, 92–97.

209 Luo, X. L., Xu, J. J., Zhao, W., Chen, H. Y., A glucose biosensor based on ENFET doped with SiO_2 nanoparticles., *Sens. Actuators B* **2004**, 97, 249–255.

210 Cui, Y., Wei, Q., Park, H., Lieber, C. M., Nanowire nanosensors for highly sensitive and selective detection of biological and chemical species., *Science* **2001**, 293, 1289–1292.

211 Gao, Z. Q., Xie, F., Shariff, M., Arshad, M., Ying, J. Y., A disposable glucose biosensor based on diffusional mediator dispersed in nanoparticulate membrane on screen-printed carbon electrode., *Sens. Actuators B.* **2005**, 111–112, 339–346.

212 Marc, D., Sophie, D. C., Immobilisation of glucose oxidase within metallic nanotubes arrays for application to enzyme biosensors., *Biosens. Bioelectron.* **2003**, 18, 943–951.

213 Miao, Y. Q., Qi, M., Zhan, S. Z., He, N. Y., Wang, J., Yuan, C. W., Construction of a glucose biosensor immobilized with glucose oxidase in the film of polypyrrole nanotubules., *Anal. Lett.* **1999**, 32, 1287–1299.

214 Morrin, A., Ngamna, O., Killard, A. J., Moulton, S. E., Smyth, M. R.,

Wallace, G. G., An amperometric enzyme biosensor fabricated from polyaniline nanoparticles., *Electroanalysis* **2005**, 17, 423–429.

215 Zhou, H., Chen, H., Luo, S., Chen, J., Wei, W., Kuang, Y., Glucose biosensor based on platinum microparticles dispersed in nano-fibrous polyaniline., *Biosens. Bioelectron.* **2005**, 20, 1305–1311.

216 Gong, X., Milic, T., Xu, C., Batteas, J. D., Drain, C. M., Preparation and characterization of porphyrin nanoparticles., *J. Am. Chem. Soc.* **2002**, 124, 14 290–14 291.

217 Chen, H. Z., Pan, C., Wang, M., Characterization and photoconductivity study of tiopc nanoscale particles prepared by liquid phase direct reprecipitation., *NanoStruct. Mater.* **1999**, 11(4), 523–530.

218 Wang, Y., Deng, K., Gui, L., Tang, Y., Zhou, J., Cai, L., Qiu, J., Ren, D., Wang, Y., Preparation and characterization of nanoscopic organic semiconductor of oxovanadium phthalocyanine., *J. Colloid Interface Sci.* **1999**, 213, 270–272.

219 Joshi, K. A., Tang, J., Haddon, R., Wang, J., Chen, W., Mulchandani, A., A disposable biosensor for organophosphorus nerve agents based on carbon nanotubes modified thick film strip electrode., *Electroanalysis* **2005**, 17, 54–58.

220 Gavalas, V. G., Law, S. A., Ball, J. C., Andrews, R., Bachas, L. G., Carbon nanotube aqueous sol-gel composites: Enzyme-friendly platforms for the development of stable biosensors., *Anal. Biochem. Biophy. Res. Commun.* **2004**, 329, 247–252.

221 Xu, Z., Chen, X., Qu, X., Jia, J., Dong, S., Single-wall carbon nanotube-based voltammetric sensor and biosensor., *Biosens. Bioelectron.* **2004**, 20, 579–584.

222 Huang, J., Song, Z., Li, J., Yang, Y., Shi, H. B., Wu, B. Y., Anzai, J. I., Osa, T., Chen, Q., A highly-sensitive L-lactate biosensor based on sol-gel film combined with multi-walled carbon nanotubes (MWCNTs) modified electrode., *Mater. Sci. Eng. C* **2006**, in the press, available online, http://dx.doi.org/10.1016/j.msec.2006.01.001.

223 Rochette, J. F., Sacher, E., Meunier, M., Luong, J. H. T., A mediatorless biosensor for putrescine using multiwalled carbon nanotubes. *Anal. Biochem.* **2005**, 336, 305–311.

224 Yao, D. S., Cao, H., Wen, S., Liu, D. L., Bai, Y., Zheng, W. J., A novel biosensor for sterigmatocystin constructed by multi-walled carbon nanotubes (MWNT) modified with aflatoxin–detoxifizyme (ADTZ)., *Bioelectrochemistry* **2005**, 68, 131–138.

225 Song, Z., Huang, J. D., Wu, B. Y., Shi, H. B., Anzai, J. I., Chen, Q., Amperometric aqueous sol–gel biosensor for low-potential stable choline detection at multi-wall carbon nanotube modified platinum electrode., *Sens. Actuators B* **2006**, 115, 626–633.

226 Tan, X. C., Li, M. J., Cai, P. X., Luo, L. J., Zou, X. Y., An amperometric cholesterol biosensor based on multiwalled carbon nanotubes and organically modified sol-gel/chitosan hybrid composite film., *Anal. Biochem.* **2005**, 337, 111–120.

227 Antiochia, R., Lavagnini, I., Pastore, P., Magno, F., A comparison between the use of a redox mediator in solution and of surface modified electrodes in the electrocatalytic oxidation of nicotinamide adenine dinucleotide., *Bioelectrochemistry* **2004**, 64, 157–163.

228 Zhang, M., Smith, A., Gorski, W., Carbon nanotube-chitosan system for electrochemical sensing based on dehydrogenase enzymes., *Anal. Chem.* **2004**, 76, 5045–5050.

229 Zhang, M., Gorski, W., Electrochemical sensing platform based on the carbon nanotubes/redox mediators-biopolymer system., *J. Am. Chem. Soc.* **2005**, 127, 2058–2059.

230 Salimi, A., Noorbakhsh, A., Ghadermarz, M., Direct electrochemistry and electrocatalytic activity of catalase incorporated onto multiwall carbon nanotubes-modified glassy carbon electrode., *Anal. Biochem.* **2005**, 344, 16–24.

231 Wang, L., Wang, J. X., Zhou, F. M., Direct electrochemistry of catalase at a gold electrode modified with single-wall carbon nanotubes. *Electroanalysis* **2004**, 16, 627–632.

232 Liu, Y., Wang, M., Zhao, F., Guo, Z., Chen, H., Dong, S., Direct electron transfer and electrocatalysis of microperoxidase immobilized on nanohybrid film., *J. Electroanal. Chem.* **2005**, 581, 1–10.

233 Liu, Y., Qu, X. H., Guo, H. W., Chen, H. J., Liu, B. F., Dong, S., Facile preparation of amperometric laccase biosensor with multifunction based on the matrix of carbon nanotubes–chitosan composite., *Biosens. Bioelectron.* **2006**, 21, 2195–2201.

234 Weber, J., Kumarc, A., Kumar, A., Bhansali, S., Novel lactate and pH biosensor for skin and sweat analysis based on single walled carbon nanotubes., *Sens. Actuators B* **2006**, 117, 308–313.

235 Li, G., Liao, J. M., Hu, G. Q., Ma, N. Z., Wu, P. J., Study of carbon nanotube modified biosensor for monitoring total cholesterol in blood., *Biosens. Bioelectron.* **2005**, 20, 2140–2144.

11
Biomimetic Nanosensors

Raz Jelinek and Sofiya Kolusheva

11.1
Introduction

Within the burgeoning field of nanotechnology, *bio*nanotechnology is an important area. Bionanotechnology is generally defined as the ability to produce biological devices and materials through molecular manipulation and specific control of the physical, chemical and biological properties of the materials employed. The goals of this increasingly visible discipline focus on the creation of functional systems that could be employed in electronic circuitry and devices, advanced materials, biomedical devices, sensors, and others. Many excellent reviews have been published in recent years on subjects at the interface between biology and nanotechnology, reflecting the increasing interest in its potential for addressing diverse scientific and technological challenges [1].

In the field of *sensor* research and development, bionanotechnology is poised to make significant contributions, and has the potential to radically alter the way sensors are designed, constructed, and implemented. As a rule of thumb, the requirements for miniaturization are ever present and are often critical in sensor design. Accordingly, utilizing bio-inspired concepts and technical approaches and introducing biologically-based detection technologies have already had considerable impact in biosensor development. In particular, the ability to harness molecular recognition in the *literal* sense – employing fewer and fewer recognition units in sensor devices – could revolutionize the way biosensors operate. Several reviews on varied aspects and exciting developments in this field have been published recently [2–6].

The present chapter discusses the nascent but rapidly expanding field of nanobiosensors, putting particular emphasis on the way concepts and methodologies borrowed from the biological world – termed "biomimetics" – contribute and shape the chemical/physical characteristics of sensor devices. Put the other way around – we have also summarized here reports depicting nanosensors designed to analyze biological and cellular systems without interference in their functions or structures. Such goals are sometimes satisfied by nanosensor assemblies that mimic certain components within the biological or cellular system investigated.

Nanotechnologies for the Life Sciences Vol. 8
Nanomaterials for Biosensors. Edited by Challa S. S. R. Kumar

ISBN: 978-3-527-31388-4

Biomimetic nanosensors are referred to in a rather broad sense, including various systems and applications that can be depicted as "nanotechnology", and focusing on the interface between biology and chemistry, physics, or engineering. Similarly, "biomimetic sensors" depicted here represent important concepts, design criteria, or molecular targets that are "biological", but include innovative components from other disciplines. The explosive growth ofpublications that touch in different ways on the subject of this chapter (and due to space limitations) necessitated summarizing only selected studies.

Thematically, this chapter is organized according to the roles played by biology and biomimetic assemblies in sensing devices, and mirroring this – the development of nanotechnology-based (or nanotechnology-inspired) sensing of biological and cellular systems. Specifically, Section 11.2 focuses on the description of biosensing applications relying on nanostructures, such as nanoparticles and nanotubes, for studying and detection of biological systems. Emphasis in this section is on the *nanostructures* and how their unique physicochemical properties contribute to bio-sensor designs. The spotlight in Section 11.3 is rather on biological (and cellular) systems, and the solutions nanotechnology offers to detect and elucidate molecular events in those assemblies. Section 11.4 summarizes the contribution of biological molecules as building blocks and structural components in biomimetic nano-scale sensors. Finally, Section 11.5 discusses the increasing prominence of nano-biotechnology in biomedical diagnosis. Certain overlap exists among the topics, primarily due to the interchanging roles of the sensors and biological molecules in many bio-inspired devices and applications.

11.2 Nanostructures in Biosensor Design

For such a young scientific endeavor, nanotechnology has already demonstrated a noticeable role in biosensor research and development. The contributions of nanotechnology can be roughly divided between the more "conceptual" impact – pointing and demonstrating new approaches and models for biosensor design – and the practical utilization of novel or superior physicochemical properties of nanostructures in certain applications. This subsection will discuss examples for both aspects of biomimetic nanotechnology in sensors.

One of the most important recent "meeting points" between nanotechnology and the biological world has been the demonstration that nanoparticles (NPs) can be successfully incorporated and used for detection and imaging biological assemblies and cell systems. Semiconductor nanocrystals or "quantum dots", in particular, have become an important tool in this field. The concept underlying the use of quantum dots, particularly for biological and cellular imaging, is simple and elegant (Fig. 11.1). Essentially, NPs with different sizes and compositions exhibit distinct optical properties (mainly fluorescence or luminescence) [7, 8]. NPs can be conjugated to varied biological targets, including proteins, enzymes, DNA and others, thus facilitating detection, imaging and functional analysis of biolog-

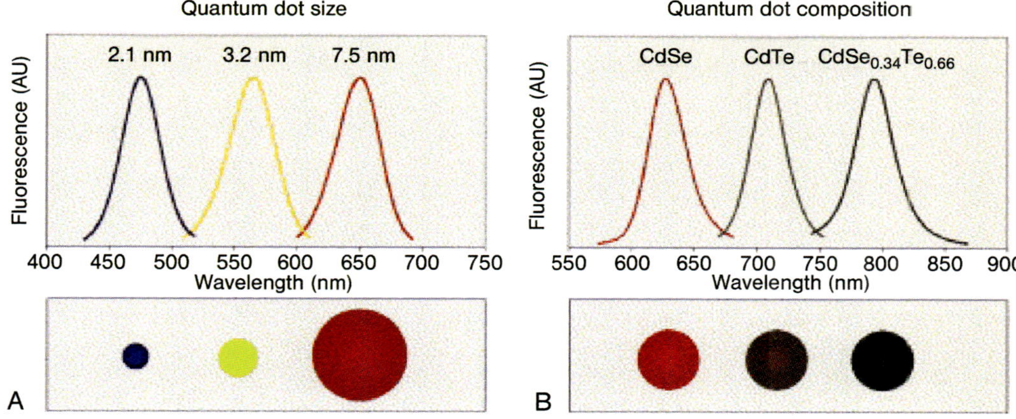

Fig. 11.1. Variation of fluorescence emitted from semiconductor nanoparticles (quantum dots), in relation to particle size and composition. (Reprinted from Ref. [7]. Copyright (2004), with permission from Elsevier.)

ical processes via the differently-labeled NPs [8–11]. Coupling of the NPs to biological molecules provides these nanosensing particulates with means of exploiting the diversity of the biological universe for varied applications. In particular, bioconjugation allows probing and pinpointing specific cellular pathways and physical environments within the cell. Semiconductor NPs were used for imaging of single cells [11], cell components [12], tissues [13], plant cells [14], and living animals [15].

An important advantage of the use of NPs for biological imaging is their often superior spectral and optical properties compared with conventional dyes [16]. In that regard, the ability to tune their optical properties through varying the particle size and molecular composition, the coupling of the recognition element to NPs with different emission wavelengths, and the relative biological inertness of the NP core all make conjugated NPs an attractive labeling tool. Semiconductor NPs are currently widely used in scientific and applied biotechnology research, e.g., for intra- and extra-cellular imaging of cancer markers, and for orthogonal detection of different molecular targets within a cell [16].

Diverse techniques have been developed for making NPs generally biocompatible, and to overcome problems of nonspecific adsorption, aggregation, and intracellular targeting in cell imaging. A simple and innovative approach has been the inclusion of semiconductor NPs within phospholipids micelles, making such assemblies easy to manipulate and introduce into cellular systems [17]. The creation of lipid/NP particles has certain advantages over covalent coupling of NPs to biological molecules. First, the assembly process is simple and does not require formation of chemical bonds between the lipids and the NPs. Furthermore, the small lipid-surrounded particulates can transport into cellular systems and microorganisms without adversely affecting their viability [17].

A critical test for the use of NPs for cellular imaging and analysis is whether these particles can be depicted as "inert" markers of cells (or specific organelles within the cell). Many studies have shown that NP incorporation does not adversely affect cell processes or cell viability [8, 11]. In that sense, NPs can be referred to as "biomimetic" in that they can become part of a biological system (the cell) without affecting its viability and biological functionalities, and augment it through providing the imaging capabilities.

Semiconductor NPs are not the only nano-scale probes used in biological imaging; however, their remarkable biological and optical versatility make them promising components for imaging and bioanalysis applications. Other types of NPs have been reported in biosensing applications. NPs composed of silica or metal atoms, for example, also allow easy surface modification aimed to incorporate labile biological functionalities [18, 19]. Zhao et al. have developed an elegant sensing technique based on bioconjugated silica NPs containing fluorescent dyes [20] (Fig. 11.2). The silica NPs were coupled to recognition elements such as antibodies recognizing specific antigens on bacterial surfaces, and additionally encapsulated high concentration of fluorescence dyes within the NP matrix. Consequently, binding of even a *single* bacterial cell yielded a detectable fluorescence signal from the intensely fluorescent NP. This approach could easily be extended to rapid pathogen detection and high-throughput screening using simple chip devices.

Gold NPs, in particular, exhibit varied uses in biological detection, in some instances comprising the basic building blocks for the biosensors, in other cases as vehicles of transmission of the induced signals. An example of the former design is the use of gold NPs in monolayers for the construction of amperometric immunosensors [21]. Here, the NPs constituted both the sensing interface as well as the substrate for biological immobilization. Another study has described the use of

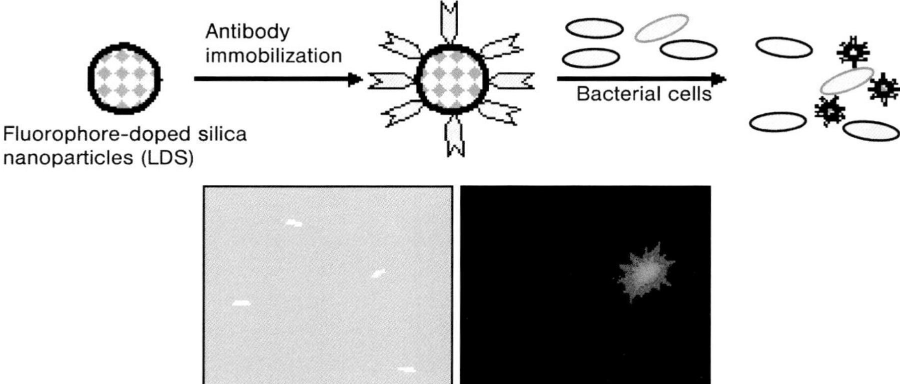

Fig. 11.2. Schematic depiction of bacterial detection using fluorophore-containing silica NPs [20]. Binding of bacterial cells to antibody-labeled NPs causes the release of the fluorescence dyes.

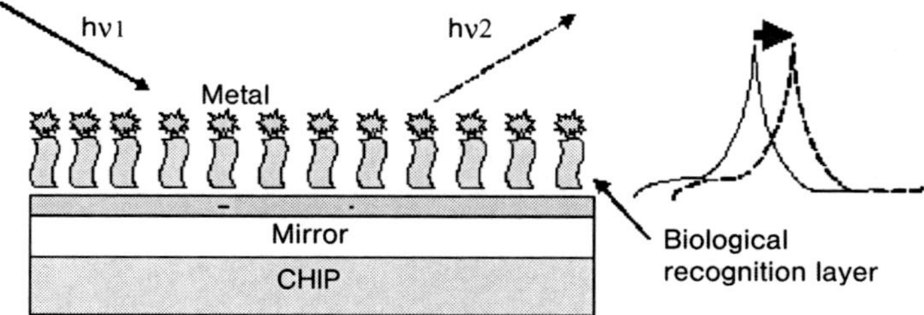

Fig. 11.3. A biosensor design based on a metal cluster resonance bio-chip; $h\nu 1$ and $h\nu 2$ are the incoming and reflected light, respectively, and the shift in color is caused by absorption due to mirror–NP layer resonance. Light absorption from the metal layer is modified by the thickness of the biological recognition layer [24].

gold NPs for assembling a surface sensor for DNA hybridization [22]. Silica NPs encapsulating luminescent markers have been prepared and used for biosensing through conjugation with biological molecules and bio-surfaces [23]. The latter work exploited advanced synthesis methods to assemble NP in distinct sizes, thus achieving controlled luminescence. This, in turn, has endowed the system with "molecular encoding" capabilities through multiplexed signaling.

Films composed of NPs forming parts of *biosensor chip* assemblies have been reported [24]. Metal nanoclusters were positioned on top of a biological recognition layer (receptor/ligands) at a nanometer-scale distance from a reflecting layer (Fig. 11.3). The surface-enhanced light absorption by the NP layer was shown to be highly sensitive to the thickness of the biological recognition assembly, thus providing means for sensing biomolecular binding [24]. The simplicity of this application stems from the enhancement of visible signals – essentially color changes detected by the naked eye – through deposition of the NPs on surfaces functionalized with macromolecules. Such bio-chip devices could find varied uses in proteomics applications, and their extension to single molecule detection is possible through surface micro-patterning and nanolithography.

Confining NPs on solid surfaces could facilitate the application of sensitive surface spectroscopic methods in a highly localized manner. Metallic NPs, particularly gold or silver, placed on solid substrates yielded a dramatic increase in the intensity of surface-enhanced Raman scattering (SERS) signals, enabling trace-level detection of varied biological compounds, including, for example, viral DNA [25]. Silver NPs coupled to biological ligands yielded improved localized surface plasmon resonance (LSPR) signals following ligand/receptor binding, making such systems potentially useful in immunoassays [26].

Biologically functionalized magnetic nanoparticles have been also contributed to bio-sensing applications. The sensing schemes in such designs do not involve *optical detection* but rather rely on the modification of the *magnetic properties* of the particles, induced through biological interactions. One such system for sensing biomolecular interactions has utilized biocompatible magnetic NPs onto which biological molecules could be chemically attached [27]. The researchers demonstrated that differences in the aggregation state of the magnetic NPs, affected through specific biomolecular interactions involving the functionalized NPs, could dramatically alter the magnetic relaxation properties of the particles – making the system a "magnetic relaxation switch" for identification of biological analytes. A notably practical advantage of the technique is that such "relaxation sensing" could be implemented by using conventional magnetic resonance instruments.

One of nanotechnology greatest molecular players – the carbon nanotube – could also form a basis for a wide range of biosensor applications. The remarkable chemical properties of carbon nanotubes [particularly single-wall (SW) nanotubes] have already been exploited in biosensor applications. Particularly important factors when considering carbon nanotubes as nanobiosensors are their nano-scale dimensions, versatile chemical functionalization possibilities, and the ready adsorption of proteins and other biomolecules onto their surface. All these factors make carbon nanotubes ideal candidates for sensing platforms, capable of detecting analytes at extremely low molecular concentrations. Indeed, the field of carbon nanotube sensors has developed rapidly in recent years due in large part to the spectacular progress in nanotube surface chemistries [28]. Biological molecules have been attached to nanotube surfaces through covalent and noncovalent bonding, with and without pre-activation, and through the use of diverse synthetic procedures [28]. Carbon nanotubes derivatized with oligonucleotides and embedded in silica have been used, for example, for ultrasensitive DNA detection [29]. That study and other reports point to an important advantage of carbon nanotubes: their simultaneous role both as *immobilization matrixes* for the detected analytes, as well as mediators of the induced signals (generally amperometric or electrical) within the bio-sensor assemblies [30, 31].

Other recent developments straddling the chemistry/nanotechnology interface point to technologies that could play a role in biomimetic nanosensors. Molecular imprinting, for example, could become a useful tool for biosensor architectures. Molecular imprinting creates recognition sites in polymers by using template molecules; the templates are prepared by initiation of the polymerization processes while specific molecules of a particular analyte (or resembling the analyte) are incorporated within the solidifying polymeric material [32, 33]. Following the removal of the embedded analyte molecules, the polymer essentially becomes a porous framework that selectively adsorbs only the analyte molecules within the pre-shaped binding sites [34]. This kind of "template biosensing" could be compatible with nanosensor design through accurately controlling the pore dimensions, or through manipulation of the recognition surface, for example by using microgel spheres that determine the overall size of the porous framework [35]. Recent studies have employed molecularly imprinted polymers as bio-sensing templates, achieving extremely sensitive (subpicomolar) detection of biological ligands [36]. A potentially useful implementation of molecular imprinting focuses on separation applications, particularly capillary chromatography and electrophoresis [37]. Such techniques, however, often face considerable hurdles when applied to minute sample quantities. Indeed, technical challenges still exist for transforming molecular imprinting into a viable sensing technology.

11.3 Nanosensors for Probing Biological and Cellular Systems

Nanosensors applied in the context of biological systems can be roughly divided into two broad applications. One route focuses on sensor technologies designed to investigate biological *molecules* and physiological/biological *processes* (rather than the complete "living systems"), while another active research track employs nanosensing for investigating actual *living* cells or tissues. This section summarizes the contribution of nanotechnology to both applications, beginning with sensors aiming to elucidate various aspects in the primary constituents of biology: proteins, nucleic acids, carbohydrates, and others.

The protein world has probably attracted the most intense activity in nanosensor development, undoubtedly due to the broad knowledge-base and the significance of proteins in numerous biochemical and therapeutic processes. The main difference between the established and highly versatile protein bio-assays available and newly introduced nanosensor methodologies is the focus on a very limited number of molecules (preferably, sometimes, single protein molecules) in nanosensor design, rather than looking at greater populations. Similar to fluorescent markers, luminescent dyes, or radioactive tracers, a primary challenge in protein nanosensor research and development is not to affect or interfere with the structural and functional properties of the inspected molecule. Indeed, the unprecedented technical advances in biomolecular manipulation in recent years have greatly contributed to progress in this field. An experiment exemplifying these concepts was the analysis of the molecular interactions in metal transfer by the protein metallothionein. The protein was labeled at two specific locations with a fluorescent-energy donor and an acceptor, respectively [38]. Such labeling was achieved through introducing a mutation in the protein that did not affect the protein properties. Monitoring the fluorescence resonance energy transfer (FRET) between the donor and acceptor bound to the protein allowed probing fine details pertaining to varied molecular processes of the protein.

An innovative method for detection of single antigen–antibody binding events at a bio-chip surfaces has been described recently [39] (Fig. 11.4). The sensing device relied on the fabrication of an artificial pore between two electrode surfaces; the electrical current measured through the pore (or more accurately its resistivity) was sensitive to the passage of streptavidin-derivatized colloids. When antigens, for example, were specifically bound to antibodies crosslinked to the colloid surface, the size change of the colloid was reflected in the recorded current. This microfluidic bio-sensor can be constructed through conventional lithography techniques – and points to the substantial potential of "nanolithography" to produce truly "single molecule" nanosensors.

Oligonucleotide detection has also been at the forefront of nanotechnology-based sensing techniques. This is mostly due to the explosive growth of genomics as a tool for biological research and pharmaceutical R&D. The introduction of the "gene-chip" and the quest for ever higher molecular densities and sophistication of chip surfaces have been major driving forces for new scientific and technical

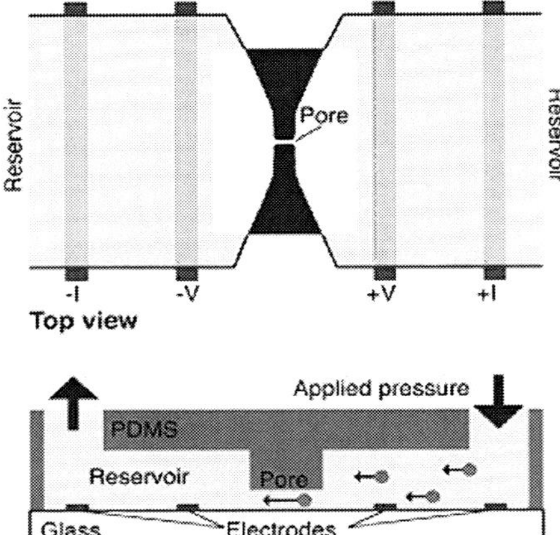

Fig. 11.4. Schematic depiction of apparatus designed to detect antigen–antibody interactions through measuring the electrode voltage across a small pore. The measured current was sensitive to the size of antibody-functionalized colloids passing through the pores; antigen binding to the colloids modified the measured current [39]. (From Ref. [39]. Copyright (**2003**), National Academy of Science USA.)

breakthroughs involving nanobiosensors. Pioneering work by Letsinger and Mirkin has demonstrated that gold NPs can be used as sensitive probes for single-strand oligonucleotides [40]. The novel detection scheme was based on the colorimetric transitions of the gold NP suspension following the formation of a polymeric network, owing to hybridization between the target solubilized oligonucleotides and complementing oligonucleotide strands attached to the gold NPs. Furthermore, the red–blue transformations of the NPs were highly dependent upon hybridization mismatches – providing a unique tool for detection of DNA damage. Extending the gold NP technology, the same researchers demonstrated a simple scanner-based detection of target DNA segments using derivatized gold NPs [10]. This work also showed that the sensor signal can be further enhanced through addition of *silver* NPs to the surface-immobilized gold NPs. Of particular importance, single nucleotide mismatch could be resolved by achieving heat-induced dissociation of the hybridized gold NP network. These seminal studies have demonstrated the power of nanotechnology to introduce a new tool-kit for DNA sensing in particular, and biological sensing in general.

Other, diverse techniques have been developed to identify DNA strands with high fidelity and sensitivity. A critical requirement in many methods has been amplification of signals arising from recognition events between very dilute DNA analytes and their complementary oligonucleotides. "Nanopores" fabricated in polymer surfaces succeeded in detecting single DNA molecules, and achieved size discrimination based on strand length [41]. Another approach for high sensitivity DNA sensing involves the construction of optical fiber bundles, in which each fiber within such "nano-arrays" can transmit precise optical signals induced by various physical changes, e.g., complementary strand binding of functionalized nanobeads [42, 43]. Detection of DNA binding through force deflection measurements on a nanomechanical cantilever array has been reported [44].

Carbohydrates appear increasingly important in different biological processes and therapeutic applications. Similar to nucleic acids and proteins, some studies have focused on the integration between carbohydrates and nanometer-size systems and devices, while other efforts were directed to integrate advanced nanotechnology-oriented concepts and instrumentation for specific detection of carbohydrates. In a similar way to the approach described above for protein engineering with fluorescent probes, ribose uptake and metabolism was monitored by flaking a bacterial ribose-binding protein with two variants of the green fluorescent protein (GFP), a popular macromolecular marker [45]. The FRET rate in such a system was inversely correlated to ribose concentration, allowing evaluation of free ribose concentrations within cells.

An intriguing study has described an amperometric biosensor facilitating high-sensitivity detection of carbohydrates through embedding nickel nanoparticles within a graphite-film electrode [46]. Essentially, the dispersion of the Ni nanoparticles within the carbon film yielded an order-of-magnitude lower detection threshold compared with conventional electrode arrangements. Similar to other NP-assisted biosensors described above, the nanoparticles in the device did not participate in the actual detection of the carbohydrate molecules, but rather provided the means for improving technical performance of the electrode. Another report demonstrated that nanometer-size amphiphilic C_{60}-dendrimers could be employed in a biosensor construct for achieving improved association between the sensor surface and the carbohydrate analytes [47]. Binding was achieved through creating ordered Langmuir monolayers of the C_{60} conjugates. The films could be further transferred to solid quartz surfaces, pointing to their potential applicability in biosensor design.

Atomic force microscopy (AFM), a major driving force in nanotechnology research, has also contributed to development of nanobiosensor for carbohydrates and other biomolecules. Among the most abundant uses of AFM in nanobiosensor research has been imaging of single biomolecules [48–50]. A potential high impact application of AFM has been the introduction of "dip-pen nanolithography" – placing molecules at desired locations on surfaces, and chemically manipulating molecular entities at specific locations using chemically-coated AFM tips [49]. A recent demonstration of dip-pen nanolithography for biosensing application has been the construction of bioactive protein nano-arrays on a conductive surface for detection of protein–protein interactions and proteomics applications [51]. Other biological applications of AFM have been reported. The technique was used as a tool for determination of carbohydrate heterogeneity on bacterial surfaces [52], or the observation of a non-homogeneous distribution of specific oligosaccharide units on the surface of yeast cells through derivatization of the AFM tip with lectins [53]. A novel saccharide "force fingerprinting" technique, based on the single-molecule imaging capabilities of AFM, has been reported [54].

The numerous AFM studies of biomolecules both illustrate the potential of this technique as well as the significant challenges for using single-molecule imaging and force measurements for routine biosensor applications. On the one hand, the atomic-level resolution of AFM could provide "imaging fingerprinting" for varied biological surfaces and surface-bound molecules. It might be possible, for example, to assemble an AFM image database for bacterial surfaces that might be used for rapid pathogen identification. Further contributions could be envisaged from integration of computer-aided image analysis into AFM-biosensor applications. On the other hand, formidable technical difficulties could arise from using AFM for sufficiently fast and reliable detection in biosensors. In particular, the wealth of atomic details and sensitivity of the method might lead to impracticality as a sensing method.

Nanotechnology approaches could yield information on processes occurring within lipid bilayers comprising the cell membrane. Jang et al. have inserted silicon NPs capped with a hydrophobic organic layer into the interior of lipid bilayer vesicles [55]. Photoluminescence from the highly optically-sensitive silicon NPs could be quenched by external quenchers – opening up possibilities to use the technique for detection of membrane properties such as the communication between membrane-embedded species and the external environment of a cell.

The use of biomimetic vesicles themselves as bio-sensing platforms has gained acceptance. Vesicles (also referred to as liposomes), having a simple lipid bilayer structure, can be particularly attractive for bio-sensing applications because they can be created in a range of sizes, and could enclose different dyes and optically-active compounds. Liposomes can also be made reactive/non-reactive to biological molecules, other vesicles, or cells, depending upon their lipid components and the membrane compositions of their biological target [56]. Thus, such aggregates can serve as specific detectors for pre-selected molecules, cell locations, or tissues. Somewhat related to liposomes, newly-developed "nanogels" (Fig. 11.5) consisting of nano-scale polymeric hydrogels coated with phospholipids [57] could be attractive for bio-sensing applications since they both mimic the bilayer membrane as

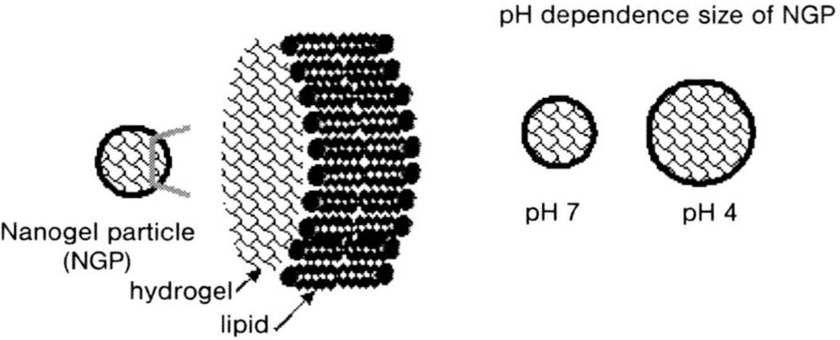

Fig. 11.5. "Nanogel" particles employed for bio-sensing applications [57]. The particle size is dependent upon the pH, making it a sensitive probe of different cellular compartments.

well as providing a rigid scaffolding for addition of optical transduction components. Nanogel particles (NGP) were shown to exhibit size sensitivity to the solution pH, making these particles potential sensor for tissue or intracellular pH changes.

Several studies have exploited the properties of *whole microorganisms* for constructing nano-scale sensor assemblies with unique features. Genetic engineering of filamentous bacteriophage allowed expression of a recognition element for ZnS nanoparticles on the phage surface (through peptide selection by affinity screening), and the NPs could then be ordered via a liquid crystal alignment of the phage [58]. This clever design resulted in NP "patterning" that could be further employed in varied nanosensor constructs.

The primary aim of biosensor design for *cellular* sensing technology is the desire to expose biochemical phenomena at the *single cell* level, and, preferably, maintain cell viability throughout the measurement. Indeed, miniaturization constitutes the core of different cellular sensing technologies, and nanotechnology could open new avenues for such applications. An interesting device addressing that goal was a nano-calorimetric sensor for measuring minute temperature changes in isolated cell suspensions, generated by biological or pharmaceutical stimuli [59]. The cellular sensor consisted of a gold and nickel thermoelectric transducer, and could produce signals upon extremely small temperature and heat changes. That report both exemplifies sensor fabrication achievements, as well as the technical and technological challenges nanobiosensor research is still facing.

Because *aqueous* environments are essential for cell viability, development of nanobiosensors for cell activity and cellular processes have greatly benefited from technical advances in micro- and nano-fluidics. Indeed, the increasing sophistication of fluidic biochips and cells has already contributed to demonstrating intriguing biosensor designs up to a single cell level. Detection of metabolic processes in few dozens bacterial cells was achieved using a microfluidic biochip prototype that could hold as little as 5 nL of a bacterial suspension [60]. Such an impedance-based biosensor of cellular metabolism could, in principle, accomplish detection of a single bacterial cell through lithographic design that would decrease the electrode size and distance between the sensor electrodes.

Being a "silent observer" to processes within a cell is a central goal in nanobiosensor research and development. An innovative method for literally "illuminating" the cell interior with biologically-inert nano-capsules has been developed by R. Kopelman [61–63]. The cell-inserted nanoprobes, denoted PEBBLEs (photonic explorers for bioanalysis with biologically localized embedding), have been fabricated from different polymeric matrixes encapsulating both a reference dye and a sensing dye [63, 64]. Owing to their small size (20–100 nm) and biological inertness, PEBBLE nanosensors can be delivered into the cell interior without compromising cell viability. Inside the cell, PEBBLEs essentially mimic cellular compartments, as they interact with and "sense" the intracellular environment. The sensitivity of the encapsulated dyes to ions and molecules have made PEBBLEs a useful platform for measuring intracellular concentrations of protons (pH), K^+, Mg^{2+}, Ca^{2+}, Cl^-, oxygen, and glucose [65, 66].

Antibody-based nanoprobes have been a recent intriguing nanosensor approach for probing the interior of living cells. These devices utilize extremely thin optical fibers – less than 50 nm in diameter – derivatized with antibodies and inserted into single cells [67]. Similar to PEBBLEs, the unique advantages of these nanoprobes stem from their exceedingly small diameters, making them almost inert and non-perturbing to the inspected cells.

While different methods have been developed for probing cell interior, relatively fewer nanotechnology applications have been designed to explore processes at the cell-surface or at the external membrane of living cells. A recent study has depicted the construction of new membrane-fused fluorescent/colorimetric nano-patches for probing membrane events in living cells [68]. The nano-scale probes in that work were based on a conjugated polymer – polydiacetylene – that changes its visible color and its fluorescence emission following structural perturbations in its vicinity [68, 69]. Polydiacetylene nano-patches were attached to the cell surface through coupling to phospholipid moieties. Such membrane-incorporated lipid/polymer patches do not adversely affect cell viability, and respond to local structural perturbations within the membranes both through induction of fluorescence as well as by undergoing blue–red color changes [67] (Fig. 11.6). These chemo-engineered nanopatch-labeled cells could be used for microscopic imaging and fluorescence or visible spectroscopic analyses of physiological processes affecting the cell membrane, its structure or morphology.

One of the challenges in coupling sensor devices to biological or cellular systems is the efficient transduction of the signals (optical, electronic) from the biological assembly into the recording/analysis units. Varied methods have been developed

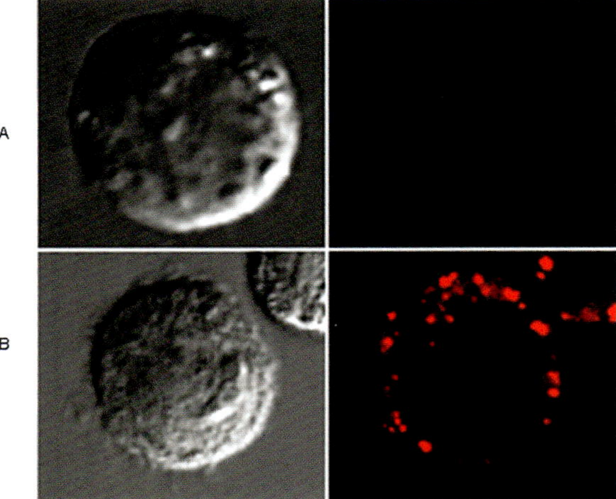

Fig. 11.6. Phase contrast microscopic images (left) and fluorescence confocal images (right) of nanopatch-labeled cells [67]. (A) Control "blue" cells; (B) cells treated with a membrane-active substance (polymyxin-B). Fluorescence spots corresponding to the "red" nanopatches indicate membrane disruption by polymyxin-B.

to interface the biological constructs investigated with opto-electronic detectors, and this is an active field of research. The fabrication of tiny micro-electrodes has been reported, and these were successfully used for monitoring adhesion and movement of different cells [70]. The use of organic dyes as reporters for the biological/inorganic interface has also been described [71]. Surface patterning of biological chips has further allowed fabrication of multi-analyte biosensors, as well as monitoring pathological conditions *in vitro* [70].

The considerable advancement in construction of optical nano-fibers has been central to developing nanosensors for investigating and imaging cell interiors [3]. Coupling of biological recognition elements to optical fibers (generally optical fiber tips) has been another active field of research aiming to efficiently transmit biological signals to opto-electronic detection systems. Varied types of biological molecules were attached to optical fiber surfaces, including enzymes, antibodies, nucleotides, and even whole cells, making this approach highly versatile as a tool for detection of diverse analytes [42]. The increasingly sophisticated technologies for manufacturing thin, resilient optical fibers with highly uniform consistencies should significantly contribute to this nanosensing track.

Several laboratories reported the incorporation of nano-electronic devices *inside* biological systems, thus creating new biosensor designs. Ritter et al. have described the inclusion of nano-electronic sensors in a cellular neural/nonlinear network for investigating fundamental physiological aspects, including image recognition, target tracking, and others [72]. Miniaturization of the devices through nanotechnology makes possible their use as tiny probes within cells and tissues.

Technological progress in cellular nanosensor fabrication also promises to yield important insight into cell metabolite levels, also known as "metabolomics". The capability for detecting changes in metabolite patterns at a *single cell* level (rather than in *ensembles* of cells as mostly done now) is extremely important for elucidation of specific factors and substances affecting cell functions [73]. In particular, the recent introduction of protein-based nanosensors, expressed via conventional biotechnology methods, could play prominent roles in intracellular metabolite sensing. One successful design was based on the attachment of two fluorescent proteins through a fusion protein, in which one of the proteins had calcium-binding properties [74, 75]. This nanosensor made possible analysis of Ca^{2+} levels in isolated cells as well as in living microorganisms. Another reported protein-based nanosensor employed a fluorescent marker (green fluorescent protein) fused to a bacterial periplasmatic binding protein, facilitating real-time sensing of a range of cellular metabolites [75].

11.4 Biological Components in Nanosensors

The extraordinary diversity of the biological universe has inspired numerous devices and structures in the materials science realm. This is particularly the case in biosensor design, in which varied biological concepts, or even actual biological molecules, can be utilized for construction of novel sensing assemblies. In that sense, miniaturization, which constitutes the core of nano-biotechnology and nanobiosensor development, increasingly relies on the incorporation of biomolecules from different sources into the sensing devices. In most instances, biological molecules and entire microorganisms have been used as templates for construction of organized nanostructures.

"Classical" biosensor designs have generally exploited enzymatic reactions as the biological transduction component. Accordingly, enzyme immobilization and maintenance of enzyme activity have been primary tasks in biosensor development. Nanotechnology offers significant advantages and new approaches for enzyme-based biosensors. In a complementary sense, enzymatic redox systems by themselves are ideal candidates for nanobiosensor applications: electron transfer is generally highly localized and can be detected in many instances both electronically and optically. Furthermore, enzymes can be readily chemically and biologically manipulated through self-assembly and protein engineering approaches and can be assembled in modular building blocks [76].

Recent reports have described the incorporation of gold nanoparticles as "mediators" of the electrochemical signals produced by immobilized enzymes in biosensors [77]. The NPs in such studies have the potential not only to improve the electrode performance but also to participate in immobilization of the enzyme and retaining of biological activities. Another study has used the aggregation properties of gold NPs for *indirect detection* of enzyme inhibitors, some of which are known nerve gases [78]. The sensing system relied on prevention of thiocholine-mediated transformation of $AuCl_4^-$ into growing nanoparticles through inhibiting the action of the enzyme acetylcholine-esterase needed to produce thiocholine.

Peptides and proteins have also constituted important nano-structural components in sensor systems. Diverse nano-assemblies, some with potential bio-sensor applications, have relied on controlling and manipulating the *self-assembly* properties of short peptide sequences [79]. The extensive and advanced technical capabilities of peptide and protein engineering have facilitated a remarkable plethora of peptide- and protein-based nanostructures, which could furthermore respond to external stimuli by emission of detectable signals – essentially performing a biosensor task [80].

The use of biomimetic receptors in biosensor design has led to development of fascinating hybrid systems. Artificial receptors that mimic ion channel proteins have been used for detection of diverse ionic species [81]. Interestingly, that work has shown that molecules *much larger* than the ions were actually recognized and discriminated by the receptors. A chromatic detection platform for catecholamines (such as adrenaline, noradrenaline, and dopamine) based upon nanoscale vesicles displaying synthetic catecholamine receptors at their surface has been reported recently [82]. The synthetic hosts in the vesicle assembly were incorporated within a matrix that contained both biomimetic membrane bilayers, as well as domains of a chromatic polymer reporter. This new assembly featured remarkable sensitivity and selectivity among similar ligands compared with existing receptor-based catecholamine biosensors.

DNA has been a natural component in diverse assays and sensors because molecular recognition is an intrinsic property of the molecule. The introduction of aptamer technology has been a particularly significant driving force for oligonucleotide-based molecular sensing [83]. Particularly important in this context is the possibility for carrying out various chemical modifications of aptamers, thereby attaching specific probes to the molecules [84]. The DNA ligands could then emit the desired signal (optical, fluorescent, etc.) upon binding their protein target. An example is the construction of "molecular beacon aptamers" for fluorescent detection of even minute amounts of a platelet-derived growth factor through their high affinity binding to the protein [85]. The notable strength of aptamer technology for nano-biosensing applications stems from the versatility of ligands that can be constructed and their high specific affinity, which allows precise localization onto target molecules or even cells and tissues.

An elegant nanosensor design for DNA detection using only biological components has been reported recently [86]. That construct contained an aptamer unit bound tightly to thrombin (through the aptamer peptide recognition element), thereby disrupting the catalytic action of the enzyme towards a fluorescently-labeled peptide. However, in the presence of an oligonucleotide strand complementary to a nucleic acid tethered to the aptamer, the aptamer was released from the binding pocket, facilitating peptide cleavage by thrombine and generation of a fluorescence signal (Fig. 11.7). The important aspect of this nano-scale system is

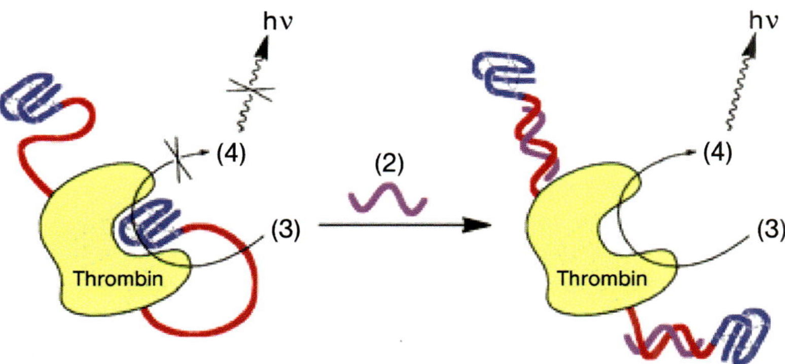

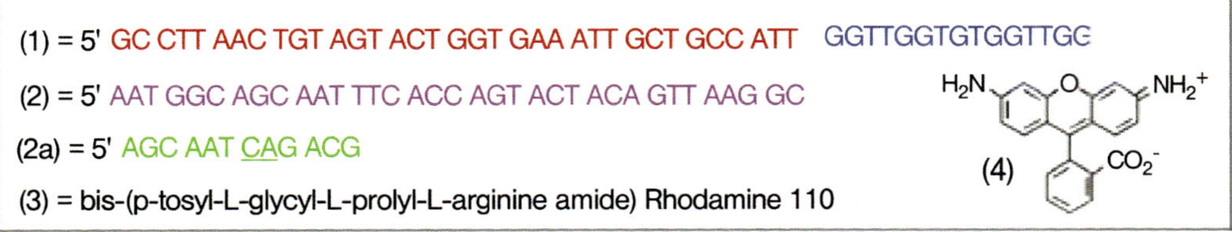

Fig. 11.7. Oligonucleotide detection by an all-biological sensor. Double strand recognition frees the thrombin binding pocket, facilitating enzymatic activity and fluorescence signal. (From Ref. [86]. Copyright (2005), The American Chemical Society.)

the use of only biological building blocks to produce a highly sensitive and specific signal. Such an assembly could be easily coupled to opto-electronic components, making a complete biological DNA nanosensor unit. The same research group has reported the application of magnetic NPs functionalized with sequence-specific DNA as substrates for endonuclease activity both as detector for complementary DNA strands, but also as novel biological "logic" circuitry [87].

DNA and oligonucleotide derivatives have been used for varied other sensing applications. DNA enzymes (referred to also as "DNAzymes"), for example, were the basis for a highly sensitive lead biosensor [88]. This clever application relied on the high specificity of DNA enzymes to metal ions such as Pb^{2+}, in which metal binding consequently resulted in a catalytic reaction – generally cleavage of the substrate strand of the DNA. In the reported application, the DNA enzyme formed a cluster of gold NPs; the presence of lead ions resulted in DNA cleavage and prevention of NP aggregation – leading to different visible colors.

DNA was one of the early examples of biological scaffolding for nano-fabrication. The molecular recognition and self-assembly properties inherent in DNA have been powerful tools for assembling complex structures that could have sensing properties. A research group at the Israel Institute of Technology has demonstrated that DNA could form the basis for varied organized nanostructures [89]. These studies demonstrated that creation of highly defined carbon nanotube constructs using DNA as a self-assembled framework had potential for bio-sensing applications [90]. The researchers exploited the "coding" inherent in DNA by directing a DNA-binding protein (a bacterial RecA protein) towards the DNA strand. This groundbreaking experiment demonstrated that DNA can constitute a molecular "toolbox" for assembling organized nonbiological nanostructures.

Nano-scale patterns constructed through "bio-inspired" approaches, such as DNA hybridization, could, indeed, become an important tool in sensor technologies. Such molecular patterning could be crucial for improving sensitivity and selectivity of the sensors, for integrating the biosensor within non-biological assemblies such as silicon chips, and for coupling of the nano-scale sensors to electro-optic devices and circuits. Different techniques have been developed for producing biomolecular patterning, e.g., microelectronic fabrication combined with DNA hybridization and chemical immobilization on chip surfaces [91, 92]. DNA and antibodies have been used for extremely sensitive detection of molecular interactions through attachment to tiny cantilevers [93]. Such sensing devices, conceptually and practically, resemble force microscopes in that the cantilever moves (or changes its shape) in response to molecular attractions between the immobilized biomolecules and their targets. Such constructs could, in principle, facilitate detection at a single molecule level; this property would depend upon the cantilever size and the immobilization methods. The "functionalized cantilever" technology could form the basis of extremely sensitive sensing devices capable of detecting minute analyte concentrations.

Several reports have depicted the coupling between nano-scale assemblies and biological molecules in sensor applications. An interesting, albeit complex, nanosensor platform has been described in which the luminescent protein firefly luciferase was used to assemble colloids deposited on an optical sensor surface [94]. The luciferase was utilized in this system for monitoring binding events occurring at the sensor surface, but also as a participant in the sensing mechanisms through its catalytic activity. The integration of whole cells or cell elements into inorganic biosensor devices is another promising avenue for new sensing applications. Recent progress in engineering of molecular circuits has opened the way for construction of microelectronic devices containing whole cells [95]. Indeed, the coupling of nanofabrication and bio-microelectronics could revolutionize microbiological and environmental monitoring [96].

Developments in surface manipulation through microprinting, lithography, and microfluidics continue to contribute significantly to nanobiosensor research. A recent report demonstrated, for example, the construction of a "lab-on-a-chip" using a modular architecture [97]. One of the most remarkable aspects of that work was the creation of an integrated chip for detection of biological markers through the use of highly spatially-defined microfluidic channels and reaction chambers. This device is a fine representative of the large variety of microfluidic labs-on-chips demonstrated or already commercialized for detection of minute amounts of analytes in liquid samples.

11.5 Nano-biotechnology and Biomedical Diagnosis

The emphasis in nanotechnology research on manipulation of ever smaller structures has opened revolutionary new avenues for biomedical applications and strategies, an approach referred to as "nano-medicine" [98]. An often touted futuristic application of nanotechnology in the field of medical diagnosis has been the introduction of tiny "nano-robots" traveling within blood vessels and tissues, scanning for malfunction and disease. Even though such applications are still far off, initial steps towards "molecular diagnostics" have been reported. A fascinating example of a nanosensor for *in vivo* analysis has been the implanting of nanoprobes for monitoring neural tissues and therapeutic treatments of neural diseases [99, 100]. Carbon nanotubes and nanofibers, in particular, are promising candidates for diverse biomedical uses because of their unique chemical and biological reactivity profiles. Several breakthrough studies have demonstrated effects of composites containing carbon nanotubes on functionalities and behavior of nerve cells and bone-forming cells [100].

Cancer diagnosis using nano-scale devices has gained considerable interest. Two main directions are pursued in this field: (a) *in vivo* imaging – using biomimetic nano-constructs for delivering imaging agents to malignant tissues and suspected areas within the body, and (b) *ex vivo* analysis – early detection of precancerous and malignancies in fluids extracted from the body using nanostructures as platforms, generally in high-throughput screening arrays [101]. NPs are particularly attractive candidates for *in vivo* imaging. In principle, they can undergo "orthogonal" couplings – both to molecular markers, such as fluorophores or radio-isotopes, as well as conjugated with molecular entities necessary for targeting the particles to their cellular or tissue destinations [102]. Furthermore, in addition to *imaging*, NPs can be simultaneously used as vehicles for *delivery* of therapeutic substances to desired targets – making them a highly versatile platform for cancer diagnostics and treatment.

Nanoparticles have been employed for other cancer detection applications. A recent study has described the attachment of biocompatible superparamagnetic iron oxide NPs to nucleotide repeats produced by telomerase – an enzyme for which elevated levels are associated with many malignancies [103]. Intriguingly, such magnetic nanoparticles switched their magnetic state upon binding – facilitating detection of the biologically originated signals using conventional magnetic readers [103]. Such "magnetic illumination" from the bound nanoparticles could be also captured easily by modified magnetic resonance imaging apparatus and forms the basis for high-throughput sample screening.

A different type of NPs – the "nanoshells" – has been implemented in cancer diagnostics and imaging. Nanoshells, spherical metallic shells encapsulating a dielectric core, have been successfully used for tissue imaging because of the remarkable tunability of their optical properties [104]. Specifically, the optical resonance of the particles can be tuned to the near-IR (NIR) range – in which tissues are transparent. Accordingly, bioconjugation of the nanoshells with molecular markers and contrast agents allows illumination of inner body parts and disease diagnostics.

Detection of disease biomarkers, combined with micro-array technologies and biochips, has been an increasingly active field in nanobiosensor research, with significant scientific and commercial potential at stake. Several innovative nanotechnology-based techniques have been reported in recent years for detection of biomarkers for Alzheimer's disease. Haes et al. have employed LSPR in a nanoscale optical sensor for immunological detection of amyloid-derived ligands [105]. In this application, which is amenable for high-throughput screening, the nanosensor surface successfully mimicked a physiological environment, facilitating antibody–antigen recognition. Nanoparticles have been employed in a recent breakthrough application for detection of a soluble biomarker for Alzheimer's disease in cerebral spinal fluid [106]. The detection scheme was based on the onset of aggregation by oligonucleotide-modified gold NPs following binding of an Alzheimer's disease pathogenic peptide to its specific antibody, where the antibody was also attached to the gold NPs (Fig. 11.8). This elegant approach not only achieved very high sensitivity due to signal amplification from the gold NPs, but is also notable for being carried out in an actual physiological solution. Such biomimetic diagnostic systems point to future development of molecular diagnostic kits in which pathogenic markers could be identified at very early stages of disease progression.

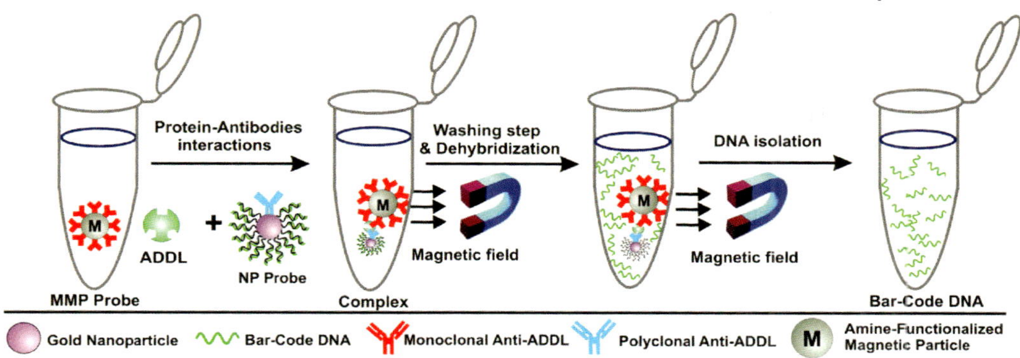

Fig. 11.8. Bio-barcode amplification assay [106]. The assay uses gold nanoparticles functionalized with monoclonal antibodies that recognize and bind a molecular marker for Alzheimer's disease (ADDLs). Following separation and dehybridization, the oligonucleotides can be identified. (From Ref. [106]. Copyright (2005), the National Academy of Science USA.)

11.6 Conclusions and Future Directions

Biosensors (or "biological detection devices" in the broad sense of the word) are becoming ubiquitous and increasingly complex. The proliferation of biological and biomedical applications and technologies, such as high-throughput screening, biomedical monitoring and diagnosis, and molecular arrays for functional genomics and proteomics analyses, all point to the pressing need for developing new biosensor platforms, technologies, and concepts. The biological and physiological spheres would most likely be increasingly exploited for ideas and building blocks in biosensor design, a process that is already taking place.

Nature provides abundant examples for innovative nanobiosensors. Snakes' "thermal vision" organs [i.e., sensitive to infra-red (IR) electromagnetic radiation] were found to contain an intricate array of nano-pits containing specific IR receptors [107]. The highly-defined spacings between these nano-pits are assumed to play a critical role in selective light absorption at particular wavelengths within the IR spectrum – endowing the snake with environmental sensing and navigation. Elaborate multilayered spheres of IR receptors have been similarly identified in *Melanophila acuminate* beetles [108]. Such dazzling arrays could inspire new sensors and IR observation devices.

In several fields, the advantages offered by nanobiosensors are obvious. The use of nanostructures as platforms for screening of large pool of compounds, i.e., genomics, proteomics, metabolomics, etc., would have a significant impact since this would increase array densities and assay efficiency. In biomedicine and medical diagnosis, nanosensors could have a profound effect on the way diagnosis is carried out, both inside and outside the human body. "Molecular imaging" and "molecular diagnostics" point to a future in which "smart agents" consisting of tiny molecular entities will circulate in the blood stream, aiming to identify and report on various diseases and pathological conditions. *In vitro* diagnostics will rely on nanostructures to form devices, making possible high-throughput screening and rapid analysis of millions of genetic and other biological markers for identification of disease or health risks in individuals. Such applications could realize the much touted "personalized medicine" revolution.

Close collaboration among biologists, chemists, physicists, and engineers have become a defining feature in nanobiosensor research and development, and promise to open new horizons in the field. Such efforts have led to extraordinary advances in molecular manipulation, using physical, chemical, and biological tools. The general interest in nanotechnology commercialization and increasing activity in this field by industrial entities will undoubtedly shorten the routes from bench-top to commercial applications, a process that is already happening, as witnessed by the proliferation of start-up companies dealing mainly with nanobiosensor development.

Abbreviations

AFM	Atomic-force microscopy
FRET	Fluorescence resonance energy transfer
GFP	Green fluorescent protein
LSPR	Localized surface plasmon resonance
NP	Nanoparticle
PEBBLE	Photonic explorer for bioanalysis with biologically localized embedding
SERS	Surface-enhanced Raman scattering

References

1 Sarikaya, M., Tamerler, C., Jen, A. K.-Y., Schulten, K., Baneyx, F. (**2003**) Molecular biomimetics: Nanotechnology through biology., *Nat. Mater.* 2, 577–585.

2 Katz, E., Willner, I. (**2004**) Nanobiotechnology: Integrated nanoparticle-biomolecule hybrid systems: Synthesis, properties, and applications., *Angew. Chem. Int. Ed.* 43, 6042–6108.

3 Cullum, B. M., Vo-Dinh, T. (**2000**) The development of optical nanosensors for biological measurements., *Trends Biotechnol.* 18, 388–393.

4 Fortina, P., Kricka, L. J., Surrey, S., Grodzinski, P. (**2005**) Nanobiotechnology: The promise and reality of new approaches to molecular recognition., *Trends Biotechnol.* 23, 168–173.

5 Rosi, N. L., Mirkin, C. A. (**2005**) Nanostructures in biodiagnostics., *Chem. Rev.* 105, 1547–1562.

6 West, J., Halas, N. (**2003**) Engineered nanomaterials for biophotonics applications: Improving sensing, imaging, and therapeutics., *Annu. Rev. Biomed. Eng.* 5, 285–292.

7 Bailey, R. E., Smith, A. M., Nie, S. (**2004**) Quantum dots in biology and medicine, *Physica E.* 25, 1–12.

8 Chan, W. C., Nie, S. (**1998**) Quantum dot bioconjugates for ultrasensitive

nonisotopic detection., *Science*. 281, 2016–2018.

9 Cao, Y.-W. C., Jin, R., Mirkin, C. A. (**2002**) Nanoparticles with Raman spectroscopic fingerprints for DNA and RNA detection., *Science*. 297, 1536–1540.

10 Taton, T. A., Mirkin, C. A., Letsinger, R. L. (**2000**) Scanometric DNA array detection with nano-particle probes., *Science*. 289, 1751–1760.

11 Bruchez, M. P., Moronne, M., Gin, P., Weiss, S., Alivisatos, A. P. (**1998**) Semiconductor nanocrystals as fluorescent biological labels., *Science*. 281, 2013–2016.

12 Gao, X., Yang, L., Petros, J. A., Marshall, F. F., Simons, J. W., Nie, S. (**2005**) In vivo molecular and cellular imaging with quantum dots., *Curr. Opin. Biotechnol.* 16, 63–72.

13 Gao, X., Nie, S. (**2003**) Molecular profiling of single cells and tissue specimens with quantum dots., *Trends Biotechnol.* 21, 371–373.

14 Ravindran, S., Kim, S., Martin, R., Lord, E., Ozkan, C. S. (**2005**) Quantum dots as bio-labels for the localization of a small plant adhesion protein., *Nanotechnology* 16, 1–4.

15 Larson, D. R., Zipfel, W. R., Williams, R. M., Clark, S. W., Bruchez, M. P., Wise, F. W., Webb, W. W. (**2003**) Water-soluble quantum dots for multiphoton fluorescence imaging in vivo., *Science* 300, 1434–1437.

16 Wu, X., Liu, H., Liu, J., Haley, K. N., Treadway, J. A., Larson, J. P., Ge, N., Peale, F., Bruchez, M. P. (**2003**) Immunofluorescent labeling of cancer marker Her2 and other cellular targets with semiconductor quantum dots., *Nat. Biotechnol.* 21, 41–46.

17 Dubertret, B., Skourides, P., Norris, D. J., Noireaux, V., Brivanlou, A. H., Libchaber, A. (**2002**) In vivo imaging of quantum dots encapsulated in phospholipid micelles., *Science* 298, 1759–1762.

18 Tan, W. B., Zhang, Y. (**2005**) Surface modification of gold and quantum dot nanoparticles with chitosan for bioapplications., *J. Biomed. Mater. Res., Part A* 75, 56–62.

19 Kim, D. K., Toprak, M., Mikhailova, M., Zhang, Y., Bjelke, B., Kehr, J., Muhammed, M. (**2002**) Surface modification of superparamagnetic nanoparticles for in-vivo bio-medical applications., *Mater. Res. Soc. Symp. Proc.* 704, 369–374.

20 Zhao, X., Hilliard, L. R., Mechery, S. J., Wang, Y., Bagwe, R. P., Jin, S., Tan, W. (**2004**) A rapid bioassay for single bacterial cell quantitation using bioconjugated nanoparticles., *Proc. Natl. Acad. Sci. U.S.A.* 101, 15 027–15 032.

21 Lei, C.-X., Yang, Y., Wang, H., Shen, G.-L., Yu, R.-Q. (**2004**) Amperometric immunosensor for probing complement III (C3) based on immobilizing C3 antibody to a nano-Au monolayer supported by sol-gel-derived carbon ceramic electrode., *Anal. Chim. Acta.* 513, 379–384.

22 Olofsson, L., Rindzevicius, T., Pfeiffer, I., Kaell, M., Hoeoek, F. (**2003**) Surface-based gold-nanoparticle sensor for specific and quantitative DNA hybridization detection., *Langmuir.* 19, 10 414–10 419.

23 Wang, L., Yang, C., Tan, W. (**2005**) Dual-luminophore-doped silica nanoparticles for multiplexed signaling., *Nano Lett.*. 5, 37–43.

24 Mayer, C., Palkovits, R., Bauer, G., Schalkhammer, T. (**2001**) Surface enhanced resonance of metal nano clusters: A novel tool for proteomics., *J. Nanoparticle Res.* 3, 361–371.

25 Isola, N. R., Stokes, D. L., Vo-Dinh, T. (**1998**) Surface-enhanced Raman gene probe for HIV detection., *Anal. Chem.* 70, 1352–1356.

26 Riboh, J. C., Haes, A. J., McFarland, A. D., Yonzon, C. R., Van Duyne, R. P. (**2003**) A nanoscale optical biosensor: Real-time immunoassay in physiological buffer enabled by improved nanoparticle adhesion., *J. Phys. Chem. B* 107, 1772–1780.

27 Perez, J. M., Josephson, L., O'Loughlin, T., Hogemann, D., Weissleder, R. (**2002**) Magnetic relaxation switches capable of sensing

molecular interactions., *Nat. Biotechnol.* 20, 816–820.

28 Davis, J. J., Coleman, K. S., Azamian, B. R., Bagshaw, C. B., Green, M. L. H. (**2003**) Chemical and biochemical sensing with modified single walled carbon nanotubes., *Chem.-A Eur. J.* 9, 3732–3739.

29 Li, J., Ng, H. T., Cassell, A., Fan, W., Chen, H., Ye, Q., Koehne, J., Han, J., Meyyappan, M. (**2003**) Carbon nanotube nanoelectrode array for ultrasensitive DNA detection., *Nano Lett.* 3, 597–602.

30 Sotiropoulou, S., Chaniotakis, N. A. (**2003**) Carbon nanotube array-based biosensor., *Anal. Bioanal. Chem.* 375, 103–105.

31 Valentini, F., Orlanducci, S., Terranova, M. L., Amine, A., Palleschi, G. (**2004**) Carbon nanotubes as electrode materials for the assembling of new electrochemical biosensors., *Sens. Actuators, B* 100, 117–125.

32 Roydhouse, R. H. (**1968**) Implant testing of polymerizing materials., *J. Biomed. Mater. Res.* 2, 265–277.

33 Laun, H. M. (**1992**) Rheometry towards complex flows: Squeeze flow technique., *Makromol. Chem., Macromol. Symp.* 56, 55–66.

34 Lee, B., Kim, Y., Lee, H., Yi, J. (**2001**) Synthesis of functionalized porous silicas via templating method as heavy metal ion adsorbents: The introduction of surface hydrophilicity onto the surface of adsorbents., *Microporous Mesoporous Mater.* 50, 77–90.

35 Ye, L., Cormack, P. A. G., Mosbach, K. (**2001**) Molecular imprinting on microgel spheres., *Anal. Chem. Data* 435, 187–196.

36 Devanathan, S., Salamon, Z., Nagar, A., Narang, S., Schleich, D., Darman, P., Hruby, V., Tollin, G. (**2005**) Subpicomolar sensing of delta-opioid receptor ligands by molecular-imprinted polymers using plasmon-waveguide resonance spectroscopy., *Anal. Chem.* 77, 2569–2574.

37 Kist, T. B. L., Mandaji, M. (**2004**) Separation of biomolecules using electrophoresis and nanostructures., *Electrophoresis* 25, 3492–3497.

38 Maret, W. (**2003**) Cellular zinc and redox states converge in the metallothionein/thionein pair., *J. Nutrition.* 133, 1460S–1462S.

39 Saleh, O. A., Sohn, L. L. (**2003**) Direct detection of antibody-antigen binding using an on-chip artificial pore., *Proc. Natl. Acad. Sci. U.S.A.* 100, 820–824.

40 Elghanian, R., Storhoff, J. J., Mucic, R. C., Letsinger, R. L., Mirkin, C. A. (**1997**) Selective colorimetric detection of polynucleotides based on the distance-dependent optical properties of gold nanoparticles., *Science.* 277, 1078–1080.

41 Mara, A., Siwy, Z., Trautmann, C., Wan, J., Kamme, F. (**2004**) An asymmetric polymer nanopore for single molecule detection., *Nano Lett.* 4, 497–501.

42 Monk, D. J., Walt, D. R. (**2004**) Optical fiber-based biosensors., *Anal. Bioanal. Chem.* 379, 931–945.

43 Tam, J. M., Song, L., Walt, D. R. (**2005**) Fabrication and optical characterization of imaging fiber-based nanoarrays., *Talanta* 67, 498–502.

44 McKendry, R., Zhang, J., Arntz, Y., Strunz, T., Hegner, M., Lang, H. P., Baller, M. K., Certa, U., Meyer, E., Guntherodt, H.-J., Gerber, C. (**2002**) Multiple label-free biodetection and quantitative DNA-binding assays on a nanomechanical cantilever array., *Proc. Natl. Acad. Sci. U.S.A.* 99, 9783–9788.

45 Lager, I., Fehr, M., Frommer, W. B., Lalonde, S. (**2003**) Development of a fluorescent nanosensor for ribose., *FEBS Lett.* 553, 85–89.

46 You, T., Niwa, O., Chen, Z., Hayashi, K., Tomita, M., Hirono, S. (**2003**) *Anal. Chem.* 75, 5191–5196.

47 Cardullo, F., Diederich, F., Echegoyen, L., Habicher, T., Jayaraman, N., Leblanc, R. M., Stoddart, J. F., Wang, S. (**1998**) *Langmuir* 14, 1955–1959.

48 Cardullo, F., Diederich, F., Echegoyen, L., Habicher, T., Jayaraman, N., Leblanc, R. M., Stoddart, J. F., Wang, S. (**1998**) Stable Langmuir and Langmuir-

Blodgett films of fullerene-glycodendron conjugates., *Langmuir* 14, 1955–1959.

49 Piner, R. D., Zhu, J., Xu, F., Hong, S., Mirkin, C. A. (**1999**) "Dip-pen" nanolithography., *Science* 283, 661–663.

50 You, T., Niwa, O., Chen, Z., Hayashi, K., Tomita, M., Hirono, S. (**2003**) An amperometric detector formed of highly dispersed Ni nanoparticles embedded in a graphite-like carbon film electrode for sugar determination., *Anal. Chem.* 75, 5191–5196.

51 Nam, J.-M., Han, S. W., Lee, K.-B., Liu, X., Ratner, M. A., Mirkin, C. A. (**2004**) Bioactive protein nanoarrays on nickel oxide surfaces formed by dip-pen nanolithography., *Angew. Chem. Int. Ed.* 43, 1246–1249.

52 Camesano, T. A., Abu-Lail, N. I. (**2002**) Heterogeneity in bacterial surface polysaccharides, probed on a single-molecule basis., *Biomacromolecules* 3, 661–667.

53 Gad, M., Itoh, A., Ikai, A. (**1997**) Mapping cell wall polysaccharides of living microbial cells using atomic force microscopy., *Cell Biol. Int.* 21, 697–706.

54 Wong, N. K. C., Kanu, N., Thandrayen, N., Rademaker, G. J., Baldwin, C. I., Renouf, D. V., Hounsell, E. F. (**2002**) Microassay analyses of protein glycosylation, *Protein Protocols Handbook*, 841–850.

55 Jang, H., Pell, L. E., Korgel, B. A., English, D. S. (**2003**) Photo-luminescence quenching of silicon nanoparticles in phospholipid vesicle bilayers., *J. Photochem. Photobiol., A* 158, 111–117.

56 Jelinek, R., Kolusheva, S. (**2005**) Membrane interactions of host-defense peptides studied in model systems., *Curr. Protein Peptide Sci.* 6, 103–114.

57 Kazakov, S., Kaholek, M., Teraoka, I., Levon, K. (**2002**) UV-induced gelation on nanometer scale using liposome reactor., *Macromolecules* 35, 1911–1920.

58 Lee, S.-W., Mao, C., Flynn, C. E., Belcher, A. M. (**2002**) Ordering of quantum dots using genetically engineered viruses., *Science* 296, 892–895.

59 Johannessen, E. A., Weaver, J. M. R., Bourova, L., Svooda, P., Cobbold, P. H., Cooper, J. M. (**2002**) Micromachined nanocalorimetric sensor for ultra-low-volume cell-based assays., *Anal. Chem.* 74, 2190–2197.

60 Gomez, R., Bashir, R., Bhunia, A. K. (**2002**) Microscale electronic detection of bacterial metabolism., *Sens. Actuators, B* 86, 198–208.

61 Clark, H. A., Hoyer, M., Philbert, M. A., Kopelman, R. (**1999**) Optical nanosensors for chemical analysis inside single living cells. 1. Fabrication, characterization, and methods for intracellular delivery of PEBBLE sensors., *Anal. Chem.* 71, 4831–4836.

62 Clark, H. A., Kopelman, R., Tjalkens, R., Philbert, M. A. (**1999**) Optical nanosensors for chemical analysis inside single living cells. 2. Sensors for pH and calcium and the intracellular application of PEBBLE sensors., *Anal. Chem.* 71, 4837–4843.

63 Buck, S. M., Koo, Y.-E. L., Park, E., Xu, H., Philbert, M. A., Brasuel, M. A., Kopelman, R. (**2004**) Optochemical nanosensor PEBBLEs: Photonic explorers for bioanalysis with biologically localized embedding., *Curr. Opin. Chem. Biol.* 8, 540–546.

64 Buck, S. M., Xu, H., Brasuel, M., Philbert, M. A., Kopelman, R. (**2004**) Nanoscale probes encapsulated by biologically localized embedding (PEBBLEs) for ion sensing and imaging in live cells., *Talanta* 63, 45–59.

65 Brasuel, M., Kopelman, R., Philbert, M., Aylott, J. W., Clark, H., Kasman, I., King, M., Monson, E., Sumner, J., Hoyer, M., Miller, T. J., Tjalkens, R. (**2002**) PEBBLE nanosensors for real time intracellular chemical imaging., *Opt. Biosens.*, 497–536.

66 Monson, E., Brasuel, M., Philbert, M. A., Kopelman, R. (**2003**) PEBBLE nanosensors for in vitro bioanalysis., *Biomed. Photonics Handbook*, 59/1–59/14.

67 Vo-Dinh, T., Alarie, J.-P., Cullum, B. M., Griffin, G. D. (**2000**) Antibody-based nanoprobe for measurement of a fluorescent analyte in a single cell., *Nat. Biotechnol.* 18, 764–767.

68 Orynbayeva, Z., Kolusheva, S., Livneh, E., Lichtenshtein, A., Nathan, I., Jelinek, R. (**2005**) Visualization of membrane processes in living cells by surface-attached chromatic polymer patches., *Angew. Chem., Int. Edn.* 44, 1092–1096.

69 Kolusheva, S., Wachtel, E., Jelinek, R. (**2003**) Biomimetic lipid/polymer colorimetric membranes: Molecular and cooperative properties., *J. Lipid Res.* 44, 65–71.

70 Connolly, P., Moores, G. R., Monaghan, W., Shen, J., Britland, S., Clark, P. (**1992**) Microelectronic and nanoelectronic interfacing techniques for biological systems., *Sens. Actuators, B* 6, 113–121.

71 Goepel, W., Schierbaum, K. D. (**1991**) Specific molecular interactions and detection principles [in chemical and biochemical sensing]. *Sensors* 2, 119–157.

72 Ritter, C., Heike, F., Herbert, K., Josef, K. F., Susanne, L., Christian, N., Helmut, O., Gabriele, P., Bernhard, S., Marieluise, S., Wolfgang, S., Gregor, S. (**2001**) Multiparameter miniaturized sensor arrays for multiple use., *Sens. Actuators, B* 76, 220–225.

73 Fehr, M., Ehrhardt, D. W., Lalonde, S., Frommer, W. B. (**2004**) Minimally invasive dynamic imaging of ions and metabolites in living cells., *Curr. Opini. Plant Biol.* 7, 345–351.

74 Schwaller, B., Durussel, I., Jermann, D., Herrmann, B., Cox, J. A. (**1997**) Comparison of the Ca2+-binding properties of human recombinant calretinin-22k and calretinin., *J. Biol. Chem.* 272, 29663–29671.

75 Fehr, M., Okumoto, S., Deuschle, K., Lager, I., Looger, L. L., Persson, J., Kozhukh, L., Lalonde, S., Frommer, W. B. (**2005**) Development and use of fluorescent nanosensors for metabolite imaging in living cells., *Biochem. Soc. Trans.* 33, 287–290.

76 Gilardi, G., Fantuzzi, A. (**2001**) Manipulating redox systems: Application to nanotechnology., *Trends Biotechnol.* 19, 468–476.

77 Lei, C.-X., Wang, H., Shen, G.-L., Yu, R.-Q. (**2004**) Immobilization of enzymes on the nano-Au film modified glassy carbon electrode for the determination of hydrogen peroxide and glucose., *Electroanalysis* 16, 736–740.

78 Pavlov, V., Xiao, Y., Willner, I. (**2005**) Inhibition of the acetylcholine esterase-stimulated growth of Au nanoparticles: Nanotechnology-based sensing of nerve gases., *Nano Lett.* 5, 649–653.

79 Zhang, S., Marini, D. M., Hwang, W., Santoso, S. (**2002**) Design of nanostructured biological materials through self-assembly of peptides and proteins., *Curr. Opin. Chem. Biol.* 6, 865–871.

80 Petka, W. A., Harden, J. L., McGrath, K. P., Wirtz, D., Tirrell, D. A. (**1998**) Reversible hydrogels from self-assembling artificial proteins., *Science* 281, 389–392.

81 Umezawa, Y., Roki, H. (**2004**) Ion channel sensors based on artificial receptors., *Anal. Chem.* 76, 320A–326A.

82 Kolusheva, S., Molt, O., Herm, M., Schrader, T., Jelinek, R. (**2005**) Selective detection of catecholamines by synthetic receptors embedded in chromatic polydiacetylene vesicles., *J. Am. Chem. Soc.* 127, 10 000–10 001.

83 Sampson, T. (**2003**) Aptamers and SELEX: The technology., *World Pat. Information* 25, 123–129.

84 Kandimalla, V. B., Ju, H. (**2004**) New horizons with a multi dimensional tool for applications in analytical chemistry – Aptamer., *Anal. Lett.* 37, 2215–2233.

85 Vicens, M. C., Sen, A., Vanderlaan, A., Drake, T. J., Tan, W. (**2005**) Investigation of molecular beacon

aptamer-based bioassay for platelet-derived growth factor detection., *Chembiochem* 6, 900–907.

86 Pavlov, V., Shlyahovsky, B., Willner, I. (**2005**) Fluorescence detection of DNA by the catalytic activation of an aptamer/thrombin complex., *J. Am. Chem. Soc.* 127, 6522–6523.

87 Weizmann, Y., Elnathan, R., Lioubashevski, O., Willner, I. (**2005**) Endonuclease-based logic gates and sensors using magnetic force-amplified readout of DNA scission on cantilevers., *J. Am. Chem. Soc.* 127, 12 666–12 672.

88 Liu, J., Lu, Y. (**2005**) Stimuli-responsive disassembly of nanoparticle aggregates for light-up colorimetric sensing., *J. Am. Chem. Soc.* 127, 12 677–12 683.

89 Braun, E., Eichen, Y., Sivan, U., Ben-Yoseph, G. (**1998**) DNA-templated assembly and electrode attachment of a conducting silver wire., *Nature* 391, 775–778.

90 Keren, K., Berman, R. S., Buchstab, E., Sivan, U., Braun, E. (**2003**) DNA-templated carbon nanotube field-effect transistor., *Science* 302, 1380–1382.

91 Bashir, R. (**2001**) DNA nanobiostructures., *Mater. Today* 4, 30–39.

92 Iqbal, S. M., Balasundaram, G., Ghosh, S., Bergstrom, D. E., Bashir, R. (**2005**) Direct current electrical characterization of ds-DNA in nanogap junctions., *Appl. Phys. Lett.* 86, 153 901/1–153 901/3.

93 Morrow, K. J. J. (**2002**) Nanotechnology: Applications to biotechnology., *Genomic/Proteomic Technol.* 2, 26–29.

94 Pastorino, L., Disawal, S., Nicolini, C., Lvov, Y. M., Erokhin, V. V. (**2003**) Complex catalytic colloids on the basis of firefly luciferase as optical nanosensor platform., *Biotechnol. Bioeng.* 84, 286–291.

95 Simpson, M. L., Sayler, G. S., Applegate, B. M., Ripp, S., Nivens, D. E., Paulus, M. J., Jellison, G. E. J. (**1998**) Bioluminescent-bioreporter integrated circuits form novel whole-cell biosensors., *Trends Biotechnol.* 16, 332–338.

96 Sayler, G. S., Simpson, M. L., Cox, C. D. (**2004**) Emerging foundations: Nano-engineering and bio-microelectronics for environmental biotechnology., *Curr. Opin. Microbiol.* 7, 267–273.

97 Shaikh, K. A., Ryu, K. S., Goluch, E. D., Nam, J.-M., Liu, J., Thaxton, C. S., Chiesl, T. N., Barron, A. E., Lu, Y., Mirkin, C. A., Liu, C. (**2005**) A modular microfluidic architecture for integrated biochemical analysis., *Proc. Natl. Acad. Sci. U.S.A.* 102, 9745–9750.

98 Emerich, D. F. (**2005**) Nanomedicine – prospective therapeutic and diagnostic applications., *Expert Opin. Biol. Therap.* 5, 1–5.

99 McKenzie, J. L., Shi, R., Kalkhoran, N. M., Sambito, M. A., Webster, T. J. (**2004**) In vitro analysis of carbon nanofiber and mesoscale porous silicon materials with nanoscale roughness for neural applications., *Mater. Res. Soc. Symp. Proc.* 311–316.

100 Webster, T. J., Waid, M. C., McKenzie, J. L., Price, R. L., Ejiofor, J. U. (**2004**) Nano-biotechnology: Carbon nanofibers as improved neural and orthopedic implants., *Nanotechnology* 15, 48–54.

101 Ferrari, M. (**2005**) Cancer nanotechnology: Opportunities and challenges., *Nat. Rev. Cancer* 5, 161–171.

102 Sullivan, D. C., Ferrari, M. (**2004**) Nanotechnology and tumor imaging: Seizing an opportunity., *Mol. Imag.* 3, 364–369.

103 Grimm, J., Perez, J. M., Josephson, L., Weissleder, R. (**2004**) Novel nanosensors for rapid analysis of telomerase activity., *Cancer Res.* 64, 639–643.

104 Loo, C., Lin, A., Hirsch, L., Lee, M.-H., Barton, J., Halas, N., West, J., Drezek, R. (**2004**) Nanoshell-enabled photonics-based imaging and therapy of cancer., *Technol. Cancer Res. Treatment* 3, 33–40.

105 Haes, A. J., Chang, L., Klein, W. L., Van Duyne, R. P. (**2005**) Detection of a biomarker for Alzheimer's disease from synthetic and clinical samples using a nanoscale optical biosensor., *J. Am. Chem. Soc.* 127, 2264–2271.

106 Georganopoulou, D. G., Chang, L., Nam, J.-M., Thaxton, C. S., Mufson, E. J., Klein, W. L., Mirkin, C. A. (**2005**) Nanoparticle-based detection in cerebral spinal fluid of a soluble pathogenic biomarker for Alzheimer's disease., *Proc. Natl. Acad. Sci. U.S.A.* 102, 2273–2276.

107 Fuchigami, N., Hazel, J., Gorbunov, V. V., Stone, M., Grace, M., Tsukruk, V. V. (**2001**) Biological thermal detection in infrared imaging snakes. 1. Ultramicrostructure of pit receptor organs., *Biomacromolecules* 2, 757–764.

108 Hazel, J., Fuchigami, N., Gorbunov, V. V., Schmitz, H., Stone, M., Tsukruk, V. V. (**2001**) Ultramicrostructure and microthermomechanics of biological IR detectors: Materials properties from a biomimetic perspective., *Biomacromolecules* 2, 304–312.

12
Reagentless Biosensors Based on Nanoparticles

David E. Benson

12.1
Introduction

The reagentless criterion for biosensors is one of the more stringent criteria for biosensor development. However, biosensors based on nanoparticles have progressed to the point that the reagentless criterion can be applied. The advantages of reagentless biosensors are a decreased need for recalibration, increased reproducibility, and minimized leaching of molecules used for analyte detection. From a development standpoint, concepts used for developing reagentless biosensors can be applied to various optical, electrochemical and magnetic detection methods. For a biosensor to be considered reagentless, the entire sensing modality needs to be unimolecular and connected through either covalent or kinetically-stable coordinate bonds. Multimolecular sensor ("non-reagentless") systems are well known and provide an avenue for rapid concept development. However, artifacts arise from multimolecular systems as one component is diluted more than another. A classic example of this is the constant recalibration required of glucose oxidase-based electrochemical biosensors due to mediator dilution. The advantage of a reagentless biosensor is that differential dilution artifacts have been removed. The solution to the reagentless criterion for biomolecule-modified nanoparticles is addressed through biomolecule adhesion, specifically or non-specifically, to the nanoparticle surface. Certain aspects of attachment chemistry will be discussed in this chapter with regard to the effect on biosensor performance; however, the bulk of this literature is left to other chapters in this book series and many excellent review articles [1–5]. Here we discuss the growing literature of reagentless biosensors that are based on nanoparticles.

Notably, many nanoparticle-based biosensors function well but do not meet the reagentless criteria. One methodology is to encapsulate various chemosensors and biosensors within a nanometer scale liposome or polymer (PEEBLEs) [6–9]. While this methodology is quite successful, it is beyond the scope of this chapter. This chapter covers methods to perturb nanoparticle properties to produce unimolecular biosensors, since this methodology will allow the unique properties of nanoparticulate materials to be harnessed for analytical applications. Additionally,

Nanotechnologies for the Life Sciences Vol. 8
Nanomaterials for Biosensors. Edited by Challa S. S. R. Kumar

ISBN: 978-3-527-31388-4

biotin–streptavidin, DNA hybridization probes and antibody sandwich assays are superb systems for concept demonstration, but do not meet the reagentless criterion for this chapter. Coverage of these assays is left for other chapters in this volume. Finally, several cellular biosensors meet the reagentless criterion [5]. Discussion of these biosensors is deferred to other chapters in that the sensing modality still requires the multimerization of these probes to produce changes in optical absorbance, scattering, and fluorescence. Despite these biosensors being a single molecular species, the multimerization of these species is required for biosensing, which does not meet the reagentless criterion.

Nanoparticle-based reagentless biosensor development tests the current scaling law description of nanoparticulate systems. Creation of a nanoparticle-based reagentless biosensor must center on the nanoparticle as the readout element. The biomolecule–nanoparticle interface must be altered to translate the (bio)molecular–biomolecular binding event into a change in nanoparticle properties. The scaling rules of both nanoparticles and (bio)molecular phenomena must be interlaced for reagentless biosensing with nanoparticles. Thus, the reagentless criterion provides a method to compare nanosystem scaling rules to the scaling rules that are well known for material and molecular systems.

There are three general methods for interlacing nanoparticle and (bio)molecular scaling laws to provide nanoparticle-based reagentless biosensors (Fig. 12.1): surface dielectric enhancement, catalytic activation, and biomolecular conformational modulated effects. Optical, electrochemical, or magnetic detection methods have been reported for at least one of these classes. Additionally, surface immobilized and solution-based biosensors have differential representation in each class. Dielectric enhancement, detected by surface plasmon resonance or electrical conductivity, is induced by the analyte adsorbing to a nanoparticle surface. The nanoparticles in this class are typically surface immobilized or embedded; however, local surface plasmon resonance scaling rules suggest a solution-based approached could work as well. Catalytic activation of a biomolecule–nanoparticle scaffold by a small molecule analyte provides excellent signal responses using optical, electrochemical, and magnetic detection methods. The optical and magnetic detected biosensors from this class are solution-based, while the electrochemical is inherently surface-immobilized. The use of analyte-induced biomolecule conformation changes provides a flexible method for reagentless biosensor design. While, optical solution-state biosensors in this class are typical, nanoparticles are an excellent coupling element for developing electrochemical surface-immobilized biosensors of this class. In each of these reagentless biosensor classes, the analyte–biomolecule interaction must change either the nanoparticle electronic structure or influence the environment surrounding the nanoparticle.

Here we discuss reagentless nanoparticle-based biosensors in the context of three general methods for interlacing nanoparticle and (bio)molecular scaling laws. Classification by these methods will be used to discuss relationships between detection methods, types of analyte–biomolecule interaction, and the nanoparticle scaling rules. A caveats and advantages subsection points out this relationship and finishes the discussion of each general method. Each method relies on one of

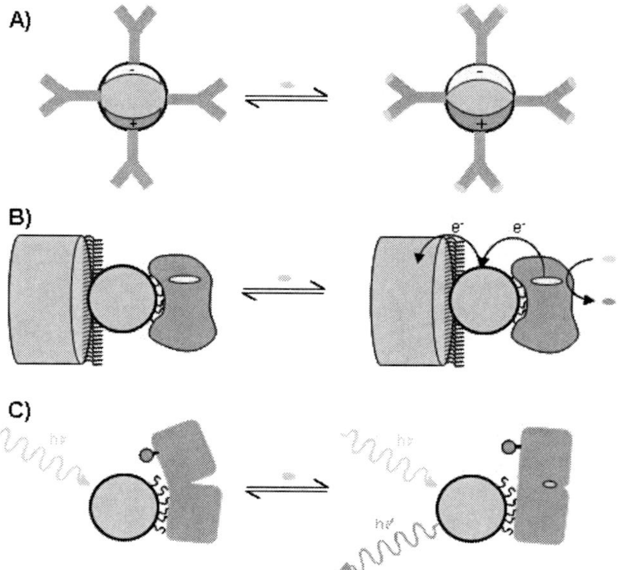

Fig. 12.1. Three classes of reagentless nanoparticle-based biosensors. (A) Analyte-induced changes in surface dielectric. The surface plasmon (+/−) is enhanced upon analyte binding to a biomolecule. (B) Catalytic enhancement of nanoparticle-mediated signals. Nanoparticle-facilitated electron (e^-) transfer from an enzyme to a working electrode is shown. (C) Biomolecular conformation changes that alter nanoparticle properties. Analyte-induced changes in protein conformation alters the distance between a nanoparticle and a reporter group, where the reporter group–nanoparticle distance changes the properties of the nanoparticle.

these related aspects more than another, which will be discussed in the conclusion section.

12.2
Surface Dielectric Enhancement

Gold nanoparticles, silicon nanowires, and single-wall carbon nanotubes enhance the sensitivity in detecting analytes at material–liquid interfaces by changes in surface dielectric (Fig. 12.2). Most biosensors in this category are reagentless, in that biomolecule surface immobilization is necessary for providing selectivity. Surface dielectric changes can be detected either by changes in surface plasmon resonance (SPR) of gold coated materials or by changes in electrical conductivity of nanoscale circuits. In each of these approaches, an analyte is concentrated by a biomolecule and the build up of this analyte changes the solvent dielectric at the material–liquid interface. Sensitivity enhancements observed for both techniques come from increasing the number/strength of analyte-material interactions. Thus, intro-

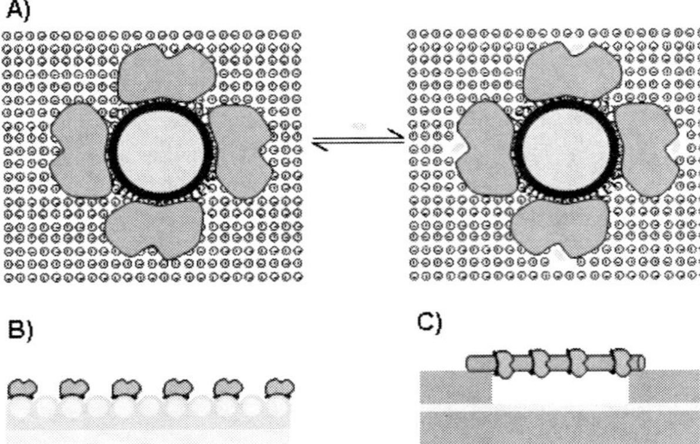

Fig. 12.2. Analyte-induced changes in surface solvent dielectric. (A) Upon analyte binding aqueous ions and water molecules are displaced, which alters the solvent dielectric at the nanoparticle surface. Two possible configurations are shown for detection of these effects: (B) Au bilayer films with proteins or nucleic acids attached to the solvent exposed surface and (C) Si nanowire FET devices, where the bottom is the drain and the *n*- and *p*-gate are on either side of the Si nanowire (top). Proteins or nucleic acids are attached to the Si nanowire or the exposed surface of the Au film to increase solvent and ion displacement upon analyte introduction.

duction of nanoparticles, nanotubes, or nanowires enhance preexisting material properties.

12.2.1
Gold Nanoparticle Enhanced Surface Plasmon Resonance

Various Au/Ag nanoparticle modified surfaces enhance the SPR effect, but only two primary configurations have been used for reagentless biosensing. Additional localized SPR (LSPR) detection methods have been shown in multimolecular biosensor formats; however, discussion of these methods is left for another chapter in this series [10]. For planar Au/Ag surfaces the surface plasmon propagates along the metal surface, where changes in the refractive index or surface dielectric at the metal–liquid interface will alter plasmon propagation. For Au/Ag nanoparticle templated or fabricated surfaces, the surface plasmon scattering is redirected from within the material through the nanoparticles and into solution. These corrugated Au/Ag surfaces provide enhanced detection of changes in the solvent dielectric above these nanometer scale Au/Ag surface features [11, 12]. Two experimental configurations have been used to measure the LSPR response. The classic measurement determined the minimum absorbance by a Au/Ag coated slide as a function of the angle of incident light. A change in the angle of incidence for minimum absorbance indicates a change in solvent dielectric at the Au/Ag surface. Alternatively, the amount of light reflected at the surface plasmon resonance wavelength of Au/Ag nanoparticles has been determined using a backscattering fiber optic setup. In this approach, changes in the solvent dielectric around the nanoparticle surface will change the amount of light back-scattered to the fiber optic cable. Backscattered reflection spectra are recorded and the intensity at the surface plasmon resonance peak is monitored. Both detection systems rely on the Au/Ag nanoparticle modified surface changing the SPR response as the analyte binds to surface-immobilized biomolecules.

One reagentless LSPR-based biosensor uses an Au nanoparticle coated quartz slide for optical detection. This work is ultimately based on earlier observations that Au nanoparticles adsorbing to Au surfaces shift the minimum angle of reflection from 54.2° to 55.8° [13]. While the binding of these nucleic acid probes is non-reagentless, reversibility measurements did provide reagentless detection of the HinF I restriction enzyme. Discussion of the restriction enzyme reagentless biosensing is left until Section 12.3. However, changes in the intensity of the Au/Ag nanoparticle surface plasmon absorbance can also be used as a detection strategy. Frederix and coworkers [14] have demonstrated LSPR-based human serum albumin (HSA) biosensing using a quartz substrate mounted in a cuvette using a standard UV/Vis spectrophotometer. The quartz substrate had Au nanoparticles adsorbed to the surface through 3-mercaptopropyl triethoxysilane and enlarged the Au nanoparticles under electroless reduction of $HAuCl_4$. Anti-HSA was covalently attached to the Au modified surface [14, 15] causing an increase in surface plasmon resonance (~634 nm, Au; ~445 nm, Ag) and, after subtraction of the protein contribution, the interband absorbance (250–400 nm, Au and Ag). Addition of 2.5 mg mL^{-1} HSA (~40 μM) to anti-HSA modified Au nanoparticle surfaces caused a 0.002 and a 0.004 absorbance unit increase after 30 min and overnight incubation in the absence of stirring. These small, but detectable, absorbance signals demonstrated reagentless biosensing could be obtained using common instrumentation. As noted by the authors, addition of stirring to this system will increase mass transport and facilitate a more rapid response. The small absorbance response could be enhanced by using the biotin-immobilized surface shown by Nath and Chilkoti [12] with a tightly packed array of monodisperse Au nanoparticles. This work used a flow cell variant of the Frederix work to increase mass transport, but maintained detection by a UV/Vis spectrophotometer. In this study, a biotin-immobilized array of 39-nm diameter Au nanoparticles on a glass slide showed a 0.050 absorbance increase upon the binding of 1 μg mL^{-1} (~20 nM) streptavidin [12]. Thus, reagentless biosensing by an array of glass-immobilized nanoparticles provides a highly accessible platform for Au nanoparticle-based reagentless biosensor development.

Au bilayer films with 100 nm spherical features allowed for wavelength-resolved backscattered reflectance detection of single-stranded DNA and fibrinogen (Fig. 12.2B) [16, 17]. In these reports, thiobutyric acid coated Au films on glass slides were EDC coupled with 100 nm diameter aminopropyltriethoxysilane coated silica nanoparticles and the second Au layer was vapor deposited over these 100 nm silica particles. Near monolayer coverage of these 100 nm features yielded an optimal LSPR response [17]. The 100 nm features of these surfaces create an array of Au nanoparticle shells that have been shown previously to provide large LSPR enhancements [18]. These Au bilayers are similar, but with larger features, to the nanoprism arrays on Au surfaces reported by Van Duyne and coworkers [11]. Antifibrinogen was immobilized on these structures through biotin–streptavidin chemistry, which increased the reflected absorbance at 560 nm by 0.2 absorbance units. Addition of 100 μg mL^{-1} (~2 μM of monomers) fibrinogen yielded an additional 0.2 absorbance unit increase, while bovine serum albumin addition (nonspecific control) yielded a 0.002 absorbance unit increase. When the neutral backbone peptide nucleic acid (PNA) or DNA is substituted for the antibody, single-stranded nucleic acids were detected in a reagentless format [16]. Biotinylated DNA or PNA probes (10-mers) from the tumor necrosis factor-α gene were attached to the Au corrugated surface, causing a 0.10 to 0.15 increase in absorbance at 550 nm. Addition of a complementary DNA target (1 μM) with a single nucleotide mismatch yielded a less than a 0.003 absorbance unit increase, while addition of a completely complementary DNA target (1 μM) yielded a 0.15 to 0.2 absorbance unit increase. The limit of detection of both the PNA and DNA immobilized biosensors were 0.1 pM of the target DNA with a 0.03–0.04 absorbance unit increase. PCR-amplified DNA targets were also detected down to 5 μL of the 1:40 dilutions from the initial amplified mixture. Application of this reagentless biosensing strategy to the smaller Au/Ag nanoisland structures fabricated and examined by Van Duyne and coworkers [11, 19] should produce similar absorbance changes, along with over 100 nm redshifts in the surface plasmon absorbance. Therefore, these nanotemplated Au/Ag surfaces provide a surface-immobilized platform for reagentless biosensing with inexpensive instrumentation.

A hybrid photoelectrical SPR-detected device has been constructed for reagentless biosensing of acetylthiocholine [20]. Using an open circuit configured Au electrode, monolayers of Au and CdSe nanoparticles were successively chemisorbed to the Au surface (Au/Au np/CdS np surfaces, Fig. 12.3). A similar surface had been shown previously to display a photoinduced change in the minimum angle of incidence for SPR reflected light under closed circuit constant potential electro-

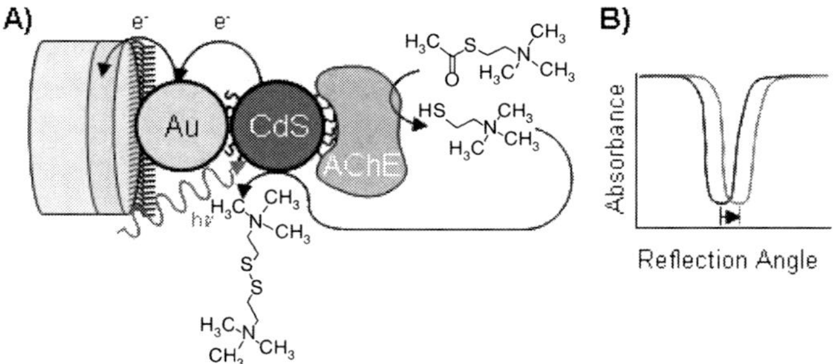

Fig. 12.3. Example of a hybrid biosensing device. Enzymatic hydrolysis of acetylthiocholine into thiocholine and thiocholine oxidation provides a pseudo-reagentless method for photoelectrochemical injections of electrons into thin Au films (A). Electron injection to the Au surface changes the incident angle of surface plasmon resonance (B).

chemical conditions [21]. With these Au/Au np/CdS np surfaces, applying mM concentrations of triethanolamine in the presence of 1.5 mW blue LED source caused up to a 0.4° shift in the angle of incidence minimum. This open circuit photostimulated SPR effect comes from a ~−150 mV bias across the Au/Au nanoparticle junction and a −25 mV photoinduced potential from the CdSe nanoparticle. The triethanolamine dependence of this process stems from oxidation of triethanolamine by the valence band holes of CdS, which are formed by photoexcitation. Back of the envelope calculations suggests four additional electrons are collected per Au nanoparticle during this photostimulated process. Pseudo-reagentless biosensing was then demonstrated, where acetylcholine esterase was attached to the CdS nanoparticle that in the presence of acetylthiocholine produced thiocholine. The acetylcholine esterase-produced thiocholine then mediated the photostimulated 0.4° shift in the minimal SPR absorbance angle of incidence at 25 mM acetylthiocholine and a 2 mM LOD. The acetylcholine LOD in this system is limited by the K_M of acetylcholine esterase, as observed by the μM LOD for an acetylcholine esterase inhibitor. This strategy has also been applied to an electrocatalytic biosensor (Section 12.3.1). This strategy not only demonstrates an exciting hybrid technique for nanoparticle-based reagentless biosensing, but also demonstrates the detection of a low molecular weight analyte and that electrode-immobilized metallic nanoparticles function as electron reservoirs.

12.2.2
Carbon Nanotube and Silicon Nanowire Enhanced Conductivity

Altering the conductivity of nanometer-scale electrode junctions is an alternative technique to optical surface plasmon resonance techniques for biosensor development (Fig. 12.2C). Single-wall carbon nanotubes (SWCNTs) [22] and Si nanowires (SiNWs) [23] have been used for biosensing [24, 25]. The conductivities of these devices are intimately linked to the electrostatics surrounding the SWCNTs and SiNWs. While there are different scaling effects for SWCNTs [26] and SiNWs [23], these nanotubes increase the sensitivity of conductivity measurements by increasing the interaction between the electrode carriers and the surrounding solution. Electrical conductivity detected biosensing with nanotubes is similar to SPR detected biosensing, in that the non-specific adsorption needs to be carefully excluded. While the sensing methodology is generally similar between SWCNTs and SiNWs, there are differences in device fabrication and biomolecule attachment chemistry.

Carbon nanotubes are attractive as biosensing transducers from a biocompatibility and synthetic standpoint. Covalent modification of SWCNTs for biomolecule attachment is an active area of research [24, 25]. Despite reported methods, biomolecular attachment to SWCNTs for reagentless biosensing has relied on surfactant physisorption with Tween [22]. Covalent coupling between Tween-20 and the biomolecule of interest then provides the specificity for conductivity measurements. Surfactant coating is an important issue in that the sensitivity of the conductivity response is minimized. An additional issue comes from the observation that conductivity detection occurs primarily at the SWCNT–metal electrode junction, or Schottky barrier modulation [26]. Adsorption of biomolecules with an isoelectric point similar to the solution pH to the SWCNT–metal junction will give the largest change in conductivity [26]. If SWCNTs are left untreated, non-specific protein adsorption abounds due to the large hydrophobic surface area of the SWCNTs, as observed by conductivity changes. Tween-20 and other surfactants also adsorb readily and irreversibly to SWCNT surfaces [22], forming the basis for biomolecule selective attachment strategies discussed here.

Selective protein detection has been displayed by SWCNT-based biosensors in a reagentless format. Initial studies from Dai and coworkers were performed with immunoglobulin G (IgG) and the 10E3 antibody that selectively binds the U1A RNA splicing factor [22]. Each of these selective sensors were generated through carbodiimide coupling between SWCNT adsorbed Tween-20 and either staphylococcal protein A (IgG biosensor) or U1A RNA splicing factor (10E3 biosensor). The staphylococcal protein A immobilized sensor showed a 1.02-fold decrease in conductance with the addition of 100 nM IgG after 5–10 min. The IgG LOD could be 20–50 nM for this system. The U1A-immobilized sensor gave a 1.03-fold decrease in conductivity with the addition of 1 nM 10E3 antibody after 10 min, placing the LOD for 10E3 around 200 pM. As a proof of principle after a mechanistic report, human chondrionic gonadotropin (hCG) was immobilized on a SWCNT device to detect a monoclonal antibody that specifically binds hCG, termed α-hCG. Addition of 10 nM α-hCG to the hCG immobilized sensor yielded a 1.03-fold decrease in conductivity after 15 min. The α-hCG LOD is approximately 5 nM from this report. With known solution state affinities for at least the SpA-IgG complex ($K_D \sim 1$ pM) the sensitivity of these biosensors seems to be limited by the SWCNT conductivity. Recently, SpA and hCG immobilized SWCNT biosensors were reported with 1 pM detection of IgG or α-hCG, respectively, within one minute [27]. This increased sensitivity was provided by a shadow mask lithography technique, which increased the area of SWCNT-Si/SiO_2 contact and increased the Schottky barrier where detection occurs. An increased conductivity response has been reported for an aptamer-immobilized SWCNT sensor [28]. The thrombin-specific DNA aptamer used in this report is one of the classic protein-specific aptamers [4]. The thrombin aptamer was carbodiimide coupled to Tween-coated SWCNTs, which yielded a sensor with 1.08-fold decrease in conductivity with the addition of 1 μM thrombin. The thrombin LOD for this biosensor was 10 nM and the response saturated at 300 nM. Since the thrombin–aptamer dissociation constant is 100 pM in solution the sensitivity of this biosensor can be improved. Finally, glucose/glucose oxidase production of H_2O_2 has provided a pseudo-reagentless biosensing strategy for glucose [29]. This report immobilized glucose oxidase to SWCNTs through a pyrene-modification so that, in the presence of glucose, O_2 would be converted into H_2O_2 and increase the conductivity around the Schottky junction. This biosensor showed a 1.12-fold increase in conductivity after the addition of 10 μM with a 30 s response time that lasted for at least 10 min. With the advent of SWCNT covalent modification chemistries [30] and the increased Schottky barrier area fabrication technique [27], the sensitivity and response of SWCNT conductivity sensors can be improved from these exciting reports.

Lieber and coworkers have demonstrated SiNWs as a conductivity detected biosensing platform. SiNW-based sensors rely on the solvent dielectric surrounding the nanowire surface and not the nanowire termini at the device interface. This chemical potential of SiNWs allows for an increased interaction between the solution dielectric and carriers in the nanowire. Thus, SiNWs are potentially more sensitive than SWCNT-based conductivity devices. Additionally, the passivated SiO_2 layer on the SiNW surface allows for facile biomolecule immobilization chemistry. 3-(Trimethoxysilyl)propyl aldehyde treatment is typically used to provide biotin or protein modified surfaces. The increased synthetic control of doping in SiNWs also allows analyte response validation by observing the opposite conductivity response in the oppositely doped SiNW. SiNW-based conductivity devices that detect a metal ion, DNA, and proteins have been reported.

SiNW-based biosensors displayed higher sensitivities but similar conductance changes compared with SWCNT-based biosensors. The initial report of SiNW-based biosensors described a calmodulin-terminated SiNW sensor [23]. This sensor showed a reversible 1.03-fold decrease in conductance within 15 s of 25 μM Ca^{2+} ions. Two systems were reported for SiNW-based biosensing of DNA. Li and coworkers used a top-down lithographic method to fabricate larger diameter SiNWs [31]. Chemical immobilization of the probe DNA oligonucleotide to SiNWs yielded a biosensor with a 1.13-fold increase (*p*-type SiNW) and a 1.8-fold decrease (*n*-type SiNW) in conductivity within 30 s of adding 25 pM complementary DNA. Lieber and coworkers used synthetic SiNWs with biotin-modified surfaces. Avidin attachment to these biotinylated SiNWs then facilitated biotinylated PNA immobilization [32]. Despite the larger distance of the probe PNA from the *p*-type SiNW surface, within 5 min a 1.25-fold increase in conductivity was observed with the addition of 100 fM complementary DNA to this biosensor. This biosensor was specific for the complementary DNA over a DNA containing a single base deletion, responsible for cystic fibrosis. A 10 fM detection limit of this PNA-modified SiNW device places this as one of the most sensitive detection methods for DNA. Direct protein adhesion was used for cancer marker biosensing, where monoclonal antibodies for prostate specific antigen (PSA) and carcinoembyronic antigen (CEA) were immobilized through aldehyde crosslinking to an array of *n*-type and p-type SiNWs [33]. This sensor showed a 1.10-fold increase (*p*-type) and a 1.12-fold decrease (*n*-type) in conductivity with the addition of 20 pM PSA. The LOD for PSA in buffer was 25 fM (1.01-fold change in conductivity) while a 0.8 pM LOD was found for PSA detection in donkey serum. Immobilization of a protein tyrosine kinase, Abl, to *p*-type SiNWs through aldehyde crosslinking chemistries yielded reagentless ATP biosensors with a 1 nM LOD [34]. The addition of 20 nM ATP caused, surprisingly, a 2.0-fold increase in conductivity. Addition of known Abl inhibitors decreased the ATP-dependent increase of the *p*-type SiNW conductivity to about 25% of the initial increase (1.25-fold increase). Given the large change in conductivity relative to the molecular weight of ATP, a protein conformation change could be responsible for this large relative increase in conductivity but the lack of site-specific attachment chemistry makes investigation of this point difficult. Taken together, SiNWs provide a versatile platform for biomolecule attachment, microfluidic interfacing, and high analyte sensitivity for reagentless biosensing.

12.2.3
Advantages and Caveats

The ability to directly couple biomolecules to nanoparticles for analytical readout is a clear advantage of the surface dielectric technique. The use of nanoparticles provides an exquisite sensitivity to solvent dielectric changes localized at the nanoparticle surface. Proper attachment of biomolecules to nanoparticle surfaces adds the molecular selectivity necessary for analyte detection. Most of these devices are surface immobilized, which minimize the amount of solution handling for biosensor activation. The surface immobilized nature of these devices allows for multiplexed device fabrication and application to microfluidic separations. Integration of these sensors with microfluidic separation is required due to the necessity of sample clean-up. Changes in surface dielectric effectively increase the local analyte concentration around the nanoparticle surface. Therefore, analyte detection in complex environments will be difficult based on the small signal changes and large changes in bulk solvent dielectric. By comparison of *n*- and *p*-type Si nanowires Lieber and coworkers have addressed this issue [33, 34]; however, performing this analysis in salt water, for example, would be difficult. Therefore, additional methods for reagentless biosensing have been explored to minimize the amount of sample clean-up necessary before analysis.

12.3
Catalytic Activation

An alternative method for developing reagentless biosensors is to couple nanoparticle–biopolymer composites with catalytic activity. The nanoparticle scaling rules used by biosensors in this class depends on whether the detection method is electrochemical or optical/magnetic. The distance dependent dissipation of this sensitivity should decay rapidly. Electrochemical biosensors that use nanoparticles take advantage of the improved electron flux at the enzyme–electrode interface that the nanoparticle provides. Scaling laws similar to the previously discussed LSPR-scaling laws are used in optical or magnetic detected biosensors in this class. The catalytic strategy can be readily employed to translate concepts from hybridization-based biosensors into reagentless biosensors for hydrolytic enzymes or reagents. While the surface dielectric methodology provides a method that integrates surface immobilization chemistry with analyte–biomolecule binding, the catalytic methodology translates multimolecular biosensing strategy into a reagentless biosensing strategy.

12.3.1
Electrocatalytic Detection

Nanoparticle-modified electrode surfaces allow multimolecular electrocatalytic biosensors to be translated into reagentless biosensors. Electrode immobilized glucose oxidase (GOx) is the classic electrocatalytic biosensor example. GOx-based biosensors provide a catalytic amplification of Faradaic current to overwhelm capacitive currents that typically dominate enzyme-immobilized electrochemical responses. Such catalytic enhancement occurs by the detection of H_2O_2 resulting from O_2 reduction during glucose oxidation by GOx. Catalytic amplification stores electron equivalents in small molecular products that are more rapidly transported to the electrode surface. Carbon nanotubes [35–38], Au nanoparticles [39–42], CdS nanoparticles [43], and Fe_3O_4 nanoparticles [44] that are immobilized on glassy carbon electrodes also store electron, or hole, equivalents to produce electrocatalytically amplified Faradaic currents for biosensing (Fig. 12.1B). The photovoltaic LSPR biosensing scaffold discussed in the previous section calculated under photostimulated, open circuit conditions that four electrons were injected per Au nanoparticle [20]. Both increased electronic communication and enzyme stability of colloidal Au adsorbed electrodes were demonstrated ten years ago [45–47]. Integration of the reagentless electrocatalytic biosensors discussed here shows a ~five-fold integrated current increase, by cyclic voltammetry, in non-mediated background electrocatalytic currents of enzyme-immobilized glass carbon electrodes. Most of these methods employ non-oriented physisorption or crosslinking chemistries for enzyme film formation on the electrode interface. The five-fold Faradaic current increase relative to the uncoated glassy carbon electrode was observed for carbon nanotube [35, 36, 38], Au nanoparticle [39, 42, 48], and Fe_3O_4 nanoparticle [44] adsorbed glassy carbon electrodes. The potential for increased enzyme stability and electrode interaction with active enzyme makes the correlation of this five-fold increase in integrated Faradaic current difficult to justify as an increased heterogeneous electron transfer rate. Since surface immobilized electrochemical behavior is not typically observed for bare glassy carbon electrodes, derivation of differences in the heterogeneous electron transfer rate is difficult. While no comparison of Faradaic current increases were provided by adsorbing CdS nanoparticles to glassy carbon electrodes, Willner and coworkers have shown photocatalytic electron transfer in enzymes adsorbed to CdS nanoparticle [21] and CdS nanoparticle–Au nanoparticle [20] modified Au electrodes. These nanoparticle-modified Au electrodes clearly point out that electron injection to Au nanoparticles occurs [20]. Also, photocatalytic electron injection is provided by CdS nanoparticle modified Au electrodes [21]. Finally, enzymatic immobilized versions of nanoparticle-modified electrodes can reversibly activate these photostimulated effects [20, 21]. Therefore, integration of nanoparticles into enzyme immobilized electrodes provides mediator-free electrocatalytic amplification.

For reagentless electrocatalytic biosensors the detectable analytes are predominantly low molecular weight. Glucose oxidase based electrocatalytic biosensors displayed different LODs for glucose. Single-wall carbon nanotube adsorbed electrodes with GOx have shown an increase in catalytic current with a LOD of 2 mM glucose and a response time of 30 s [35]. CdS nanoparticle modified glassy carbon electrodes were able to detect glucose by a decrease in the reductive current in cyclic voltammetry with a glucose LOD of 2 mM [43]. The decrease in reductive current upon glucose addition demonstrated slow electron injection from CdS for FAD reduction in the absence of photostimulation. The GOx adsorbed Au nanoparticle modified electrodes showed a LOD of 50 μM but a decrease in background current with a −600 mV potential (vs. SCE) upon glucose addition [41]. The Au nanoparticle based biosensor was able to detect accurately blood glucose levels (8.4 mM). Carbon nanotube modified electrodes with alcohol dehydrogenase immobilized showed an ethanol LOD of 1 mM [37], consistent with the K_M for the enzyme. Immobilization of hemeproteins provides bioelectrochemical sensors for various low molecular weight redox-active analytes. Hydrogen peroxide is a common analyte for these sensors, where the LOD for the multiwall carbon nanotube biosensors [36, 38] was ~1 mM. The Fe_3O_4 nanoparticle modified electrodes provided a 80 μM LOD for hemoglobin- or myoglobin-immobilized electrodes and a 2 μM LOD for catalase-immobilized electrodes. Fe_3O_4 nanoparticle modified electrodes demonstrated that the sensitivity was determined by the enzyme, not the electrode. Nitrite biosensing with these Fe_3O_4 nanoparticle modified electrodes [44] and with Au nanoparticle modified electrodes [42] both showed a 200 μM LOD for sodium nitrite in the electrocatalytic mode. However, the Au nanoparticle modified electrode with hemoglobin immobilized showed a 1 μM LOD for nitrite [42] and 50 nM LOD for NO [40] using reversible non-catalytic peak currents from cyclic voltammetry. The final example is the immobilization of the Cu-containing tyrosinase on Au nanoparticle modified electrodes, which demonstrated LODs for phenolic compounds from 150 nM (catechol) to 70 μM (gallic acid) [39]. Tyrosinase-modified electrodes were able to detect the concentration of caffeic acid (LOD 6 μM) from three red and three white wines, confirmed by a colorimetric test. Overall, LODs that reflect the Michaelis constants of the immobilized enzymes are observed for electrocatalytic biosensors that use nanoparticles as an electron reservoir.

A photoelectrochemical methodology demonstrates the potential for integrating semiconducting nanoparticles into the bioelectrocatalytic systems above. Willner and coworkers have demonstrated that illumination of CdS-modified Au electrodes with acetylcholine esterase attached to the CdS nanoparticle turns on electrocatalytic current [21]. Enzymatic production of thiocholine governs this process, in that thiocholine reduces CdS valance band holes, which are formed upon photoexciting the CdS nanoparticle. Monitoring the current as a function of wavelength (6–30 mM acetylthiocholine) produced an action spectra consistent with the band edge absorbance of the CdS nanoparticles. In the presence of 380 nm wavelength light, a LOD of 5 mM acetylthiocholine was detected, and the response saturated above 30 mM. In comparison, CdS nanoparticles adsorbed on glassy carbon electrodes were able to detect the slow reduction of glucose oxidase FAD in the absence of light and glucose [43]. Taken together these experiments demonstrate that electron flow between CdS and enzymes/substrates is reversible. Application of photostimulated electron transfer to the electrochemical-based biosensors discussed here provides a method to introduce additional detection selectivity. Semiconducting nanoparticles allow the development of such hybrid techniques, like the photostimulated LSPR effect in Section 12.2 [20], to be developed.

12.3.2 Catalytically Enabled Optical and Magnetic Detection

Enzymatic cleavage of biomolecule-tethered nanoparticles provides an alternative method for detecting enzymatic activity (Fig. 12.4). In this particular type of sensing, the biomolecule attached to the nanoparticle becomes the substrate to detect the enzyme or reagent catalyzing biomolecule hydrolysis. As biomolecule hydrolysis occurs the physical link between the nanoparticles or nanoparticle–organic fluorophore is broken. Once the nanoparticles or nanoparticle–organic fluorophore assembly dissociate, the properties of the non-modified nanoparticles and fluorophores are restored [1, 10, 49, 50]. Such a strategy has yielded pseudo-reagentless biosensors based on Au nanoparticle quenched CdSe@CdS nanoparticle emission [51], Fe_3O_4 nanoparticle associated enhancement of spin–spin relaxation time (T_2) of adjacent water protons [52], or magnetic field modulated cantilever deflection [53, 54]. This method might be extended to optically detected surface plasmon resonance from an Au nanoparticle assembly. While optical assays based on hydrolysis of DNA-modified Au nanoparticle assemblies exist [55, 56], the assembly and thermal denaturation steps seem to ensure specificity. It is assumed that the forces applied by magnetic field detection methods [52, 53] overcome the need for assembly and thermal denaturation. Since enzymatic cleavage uses an initial associated complex, the more effective scaling laws for nanoparticulate materials can be more readily employed for biosensor development.

To date, there is one example of a reagentless, optically-detected biosensor for collagenase [51]. In this system, a peptide that is specifically degraded by collagenase (GGLGPAGGCG) is covalently coupled between CdSe@CdS nanoparticles on the amino-terminus and Au nanoparticles on the carboxy-terminus. The covalently coupled CdSe@CdS-peptide-Au complex has the CdSe@CdS emission intensity partially quenched by energy transfer to the Au nanoparticles. Incubation with ~2 μM collagenase for 18, 39, and 47 h showed a 1.10, 1.20, and 1.35-fold increase, respectively, in CdSe@CdS emission intensity. While the response times for these

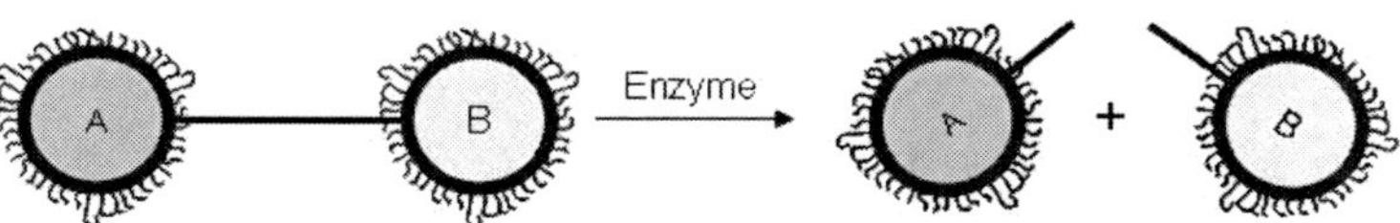

Fig. 12.4. Enzymatic cleavage of biomolecule tethered nanoparticle–nanoparticle or reporter group–nanoparticle pairs. Upon cleavage, the pair of chromophores no longer interact and restore the original signal of these chromophores.

collagenase biosensors are slow, the response time could be enhanced through increasing the accessibility of the peptide substrate.

DNA–magnetic nanoparticle assemblies that have been termed magnetic relaxation switches (MRS) provide magnetic resonance imaging (MRI) detected reagentless biosensing [52]. This method hybridizes two pools of covalently modified DNA–Fe_3O_4 nanoparticle aggregates with complementary nucleotide sequences. Once assembled, the 50–100 nm DNA–nanoparticle aggregates decreases the water spin–spin relaxation time (T_2) by two-fold. Duplex DNA formed in these hybridized assemblies contains a recognition site for the BamH I restriction endonuclease. Digestion with BamH I, but not EcoR I digestion, caused a disassembly of the aggregates and a restoration of the original T_2. Imaging microtiter plate samples within a MRI spectrometer, the Dpn I restriction endonuclease was specifically detected with methylated DNA–Fe_3O_4 nanoparticle assemblies, while unmethylated DNA–Fe_3O_4 nanoparticle assemblies showed no change in T_2. A similar strategy was reported where rennin and matrix metalloprotease-2 proteolytic activities were determined using peptide substrates that were biotinylated at both termini [57]. Unfortunately, the peptide substrate had to be hydrolyzed before nanoparticle assembly occurred, rendering this as a multimolecular assay. Despite the drawbacks of this second assay, catalytic hydrolysis has demonstrated reagentless biosensing can be performed using MRI.

Using larger magnetic particles, Willner and coworkers have demonstrated reagentless biosensors that detect restriction endonucleases with the application of a magnetic field [53, 54]. Atomic force microscope cantilevers were coated with Au and thiol-terminated DNA that was complementary to DNA modified magnetic particles. The duplex DNA formed on the cantilever surface contained recognition sites for restriction endonucleases. The cantilever deflection could be reversibly controlled by application of a magnet to each system. The force experienced by these cantilever systems was around 1–3 nN. In one system, ten units of either Apa I or Mse I restriction endonucleases were detected with a 5 minute response time [53]. In a different set of systems, two restriction enzyme sites were included on one DNA duplex providing an OR logic gate detection for either Apa I or EcoR I [54]. Alternatively, a mixed monolayer of duplex DNA was formed between the cantilever surface and the magnetic particle that contained recognition sites for EcoR I and Asc I on different duplex DNA strands. The mixed monolayer provides an AND logic gate detection of EcoR I and Asc I. The extension of logic gated detection strategies to microarray sensors is a particularly interesting application for biosensor development.

12.3.3
Advantages and Caveats

Catalytic activation schemes for biosensing with nanoparticles enhance the analyte-dependent signal. Enhancement occurs by one of two methods, either using nanoparticles to interface enzymatic electrochemical reactions or to adapt analyte–biomolecule recognition to the scaling laws of nanoparticle mediated effects. These methods harness the unique properties of nanoparticles to detect analytes in complex mixtures. Analyte specificity, in each case, is provided by the enzymatic amplification of a Faradaic current or enzyme (analyte) mediated cleavage. However, many additional factors contribute to enzyme inhibition and the degree of inhibition varies from enzyme to enzyme. The largest caveat with enzymatic activation schemes is that surface immobilization is necessary for analyte reversible detection. Thus, use of enzymatic activation schemes for reagentless biosensing for fluorescence contrast imaging agents will only provide single-use biosensors. An obvious alternative is to couple analyte-dependent biomolecular conformation changes to provide reversible analyte detection by fluorescence contrast imaging biosensors.

12.4 Biomolecule Conformational Modulated Effects

The final method for reagentless biosensing using nanoparticles is translating ligand-mediated biomolecule conformational motions into changes in nanoparticle properties (Fig. 12.5). While the previous methods have focused on directly coupling the ligand–biomolecule binding event to changes in the analytical signal, this method focuses on the detecting changes in the interaction between a reporter molecule and a nanoparticle. The reporter groups that have been described are organic compounds [58–62], a coordination complex [63, 64], or a nanoparticle [65]. With the reagentless criterion, these reporter groups have to be covalently attached to the biomolecule. Covalent bond formation is carried out through standard nucleophilic modification chemistries [66], where biomolecule-derived primary amines or thiols are modified with the reporter group. Reporter group co-

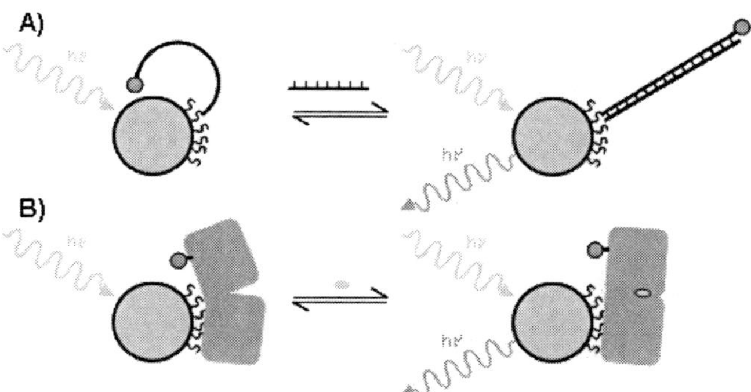

Fig. 12.5. Biomolecular conformation changes mediate reagentless nanoparticle-based biosensing. (A) Nucleic acid-based conformation changes will move a reporter group (the smaller ball) by 10 to 50 Å away from the nanoparticle upon complementary nucleic acid binding. (B) Protein-based conformation changes will move a reporter group by 1 to 6 Å away from the nanoparticle upon analyte binding.

valent attachment is necessary for ligand-mediated conformation changes to alter the reporter group–nanoparticle interaction. Since biomolecular conformation changes are the focal point of this method, the nanoparticle has to be site-specifically attached to the biomolecule as well. Thus, strategies for orthogonal site-specific attachment of reporter groups and nanoparticles are necessary for this method [55, 67–69]. These methods either use DNA synthesis and modification chemistry or recombinant DNA technology for protein production followed by modification chemistry. Once generated, the type of reporter compound–nanoparticle interaction that effectively reports analyte concentrations depends on the magnitude of biomolecule motion. DNA duplex formation provides the largest conformation change (10–50 Å, Fig. 12.5A), while analyte-induced protein conformation changes are much more subtle (1–10 Å, Fig. 12.5B). Therefore, this section will be organized based on the type of biomolecule (DNA or protein) used in analyte detection.

12.4.1 Biosensors Based on DNA Conformation Changes

With the larger analyte-induced conformation change of DNA, more nanoparticle-based detection methods have been reported (Fig. 12.5A). Energy transfer methods are a popular method for detecting changes in biomolecule conformation [70] and provide reagentless biosensors with DNA probes. Organic reporter molecules [58, 60–62] seem to be most effective in reagentless biosensor development. Au–Au nanoparticle interactions have also been used to provide colorimetric pseudo-reagentless biosensors [65]. Three types of analytes have been detected by these DNA–nanoparticle-based biosensors: oligonucleotides, proteins, and small molecules. The oligonucleotide biosensors [58, 60, 62] used base pair recognition for sequence specific recognition. With the development of selection strategies for small molecule and protein dependent conformation changes in nucleic acids (aptamers) [71, 72], small molecule [65] and protein [61] biosensors have been developed for reagentless DNA-nanoparticle based biosensors. Aptamers provide a complementary method for reagentless small molecule and protein biosensing relative to proteins [4]. The drawback of aptamer–nanoparticle based biosensors that are currently reported is that they are pseudo-reagentless where they begin as unimolecular species but after analyte addition part of the initial system is displaced. Irrespective of this critique, these aptamer-based biosensors have introduced an important concept in reagentless nanoparticle-based biosensor development.

Aptamer–nanoparticle based biosensors provide an avenue to rapidly expand the pool of analytes that are detected by reagentless nanoparticle-based biosensors. Using exponential selection strategies (SELEX) [71, 72], various RNA and DNA sequences have been identified that bind small molecules and proteins while inducing a change in nucleic acid tertiary structure [4]. The ligand-bound conformation is typically a stem-loop structure where the base of one of the nucleic acid termini forms a duplexed stem. Stem formation then mediates a reporter molecule that is attached to the stem forming nucleic acid terminus, moving either away from or towards the nanoparticle surface. As pointed out above, the change in the reporter group–nanoparticle distance can be between 10 to 50 Å in aptamers. Adenosine and cocaine aptamers were attached to Au nanoparticles through a thiolate-modified 3′-end of a 12-mer complementary DNA strand [65]. In the absence of adenosine or cocaine the oligonucleotide existed in an extended conformation allowing a second Au nanoparticle with a 5′ thiolate-modified DNA strand complementary to the aptamer DNA. The duplex DNA formed from this second Au nanoparticle DNA was composed of seven bases from the adenosine aptamer sequence and five additional bases used to separate the aptamer sequence from the hybridization sequence of the first Au nanoparticle. When these two pools of DNA-modified Au nanoparticles are combined, the classic redshifted absorbance (~580 nm) of aggregated Au nanoparticles was observed. Due to multiple DNA molecules attached to one Au nanoparticle, a simple two Au nanoparticle complex did not occur but a larger aggregate of Au nanoparticles formed. However, the addition of 0.5 to 5 mM adenosine changed the solution from blue to red within one minute. The equilibrium absorbance ratio between 522 nm (dissociated) and 700 nm (aggregated) increased up to a saturated value of 15 with the addition of adenosine or cocaine to the cognate aptamer assemblies. While the adenosine/cocaine-mediated dissociation kinetics look monoexponential, the absorbance ratios after one minute of reaction showed two binding affinities and saturated absorbance ratios. The low adenosine/cocaine effect could be explained by dissociation of Au nanoparticles on the surface of the Au nanoparticle aggregates. Despite these effects a colorimetric assay for adenosine and cocaine with LODs of 0.5 and 0.1 mM was developed by this strategy.

Thrombin-specific aptamer modified CdSe@ZnS nanoparticles were generated and demonstrated thrombin and DNA biosensing detected by fluorescence [61]. This strategy harnessed the high quantum yield and photostability of semiconducting nanoparticles for low micromolar analyte sensitivity. The system used commercially available streptavidin-coated CdSe@ZnS nanoparticles that emitted at 525 nm. A 5′-biotin modified DNA containing the thrombin aptamer sequence was hybridized with a complementary 12-mer DNA with a proprietary organic quencher ($\varepsilon_{525} \sim 66\,000$ M^{-1} cm^{-1}) attached to the 3′-terminus. Up to 40 quencher-containing DNA equivalents needed to be added per CdSe@ZnS nanoparticle to quench the fluorescence by 95%. Addition of 0.8 μM complementary DNA or 1 μM thrombin to this solution increased the emission intensity by 15- and 19-fold after 50 min. Based on the greater than 15-fold signal response the LODs should be significantly below the micromolar analyte concentrations reported. This report optimized this response by adopting a superquenched approach used for FRET based detection of maltose using maltose binding protein attached to CdSe@ZnS nanoparticles [73, 74]. Clearly, the slow response time of the thrombin aptamer system, relative to the Au nanoparticle aggregate disassembly, is due to the numerous organic quenchers that need to be removed per CdSe nanoparticle for a measurable response. Again, this work presents a pseudo-reagentless biosensor based on DNA modified CdSe nanoparticles. Nevertheless, the demonstration of a strategy for using aptamers to control the emission properties of semiconducting nanoparticles is highly significant.

The first demonstration of reagentless biosensing with nanoparticles through analyte-mediated biomolecule changes was reported by Nie and coworkers in 2002 [58]. This method was based on the observation of organic fluorophore adsorption to Au colloid surfaces as detected by fluorescence quenching and surface-enhanced Raman scattering [75, 76]. This design used a single DNA probe strand with a 3′-thiol group and a 5′-fluorescein or 5′-tetramethylrhodamine fluorophore. Upon addition of the probe DNA to the Au nanoparticle, the emission intensity from the organic fluorophore was completely quenched. After adding a four-fold excess (10 nM Au nanoparticle, 30 min incubation time, room temperature, 2–10 mM $MgCl_2$) of a complementary DNA strand, the emission intensity of tetramethylrhodamine or fluorescein was completely restored. Addition of single-base mismatch complementary DNA strand under the same conditions caused a 50% restoration of the fluorophore emission intensity. This study went on to carefully examine the complementary DNA strand binding kinetics and DNA–Au nanoparticle thermal stabilities. Up to 5 min (40% overall response), the binding kinetics were consistent with a single exponential binding event. After 5 min multiple phases were observed that slowed the response time. Notably, the response time of this biosensor [58] is similar, if not more rapid, than the thrombin biosensor based on aptamer attached CdSe nanoparticles [61]. Thermal stability studies of the probe DNA–Au nanoparticle conjugate showed minimal increase in emission intensity at higher temperatures as opposed to the analogous molecular beacon in solution. These results suggest substantial non-specific Au nanoparticle–DNA interactions, which were clearly defined by this report. Despite these caveats this report showed, for the first time, that biomolecule conformation could be used to couple ligand binding to nanoparticle-based signal enhancements.

This fluorophore modified DNA–Au nanoparticle methodology has been extended to a fluorescence microscopy detected microfluidic device [62]. Fluorescein and Cy3 were used as the fluorophores in this study and attached to the 5′ end of the DNA probe. The 3′-end of the DNA probe contained a thiol functional group that provided Au nanoparticle attachment. Au nanoparticle modified microchannels (50–90 μm) were generated by channel microfabrication followed by introducing 3-aminopropyltriethoxysilane, rinsing, and overnight incubation in Au nanoparticle solutions. The fluorophore-modified DNA probe was introduced to these microfluidic channels for 16 h and repeatedly rinsed. Minimal fluorescence was observed above background fluorescence in these rinsed channels. Fluorescence micrographs before and after complementary DNA addition (60 min incubation time) showed a significant increase in emission intensity for channels with 8 ppm of the complementary DNA. This detector was able to discriminate between DNA from danguevirus and enterovirus. This report demonstrates how a solution-based strategy with the DNA–Au nanoparticle based reagentless biosensor can be translated to a surface immobilized strategy.

Surface plasmon enhanced Raman spectroscopy has also been demonstrated for reagentless biosensing of DNA strands [60]. Plasmonic detection was provided by the use of a single strand of probe DNA with Rhodamine 6G attached to the 5′-end and a thiol on the 3′ end. The DNA recognition sequence was flanked with 6-mer complementary sequences on each end to induce stem-loop formation in the absence of analyte DNA. These DNA probes were adsorbed to 50 nm diameter Ag nanoparticles and investigated in by Raman spectroscopy with the 632.8 nm line from a HeNe laser excitation source. In the absence of any additional DNA the SERS enhanced spectrum for Rhodamine was observed. Addition of a four-fold excess of complementary analyte DNA caused an eight-fold decrease in the intensity of the prominent 1512 cm^{-1} vibration. All of the SERS vibrational bands decrease upon the addition; however, the higher energy and intensity bands provided the best spectroscopic handle for analysis. A sensor for the gene product of the HIV *gag* gene was then constructed. PCR amplified DNA (109 bases) was then added to this plasmonic detection biosensor. Addition of HIV *gag* amplified DNA cause a three- to six-fold decrease in SERS band intensities, while the addition of amplified DNA from human placenta (negative control) only caused a 1.5 to 2.5-fold decrease in SERS band intensities. This method removes the worry of reporter group photobleaching, since, even if photooxidation occurs, most of the SERS vibrational bands will be observed.

12.4.2
Biosensors Based on Protein Conformation Changes

Energy transfer methods are less prolific in reagentless fluorescence detected protein-based biosensors. The rationale for diminished reagentless energy transfer protein-based biosensors is that the ligand-induced atomic movements in proteins are smaller than in DNA. For an energy transfer methodology to function for a reagentless protein-based biosensor the distance for half-maximal energy transfer (R_o) must be similar to the ligand-dependent change in interatomic distance. Frommer and coworkers have shown reagentless energy transfer protein-based biosensors for maltose, ribose, and glutamate [77–79]. These systems use green fluorescent proteins (GFPs) that are engineered to absorb and emit at different visible wavelengths, genetically encoded on the amino- and carboxy-terminus of the sensor protein (Fig. 12.6A). The R_os for these GFP chromophore pairs are around 20–25 Å. Fluorophore pairs with R_os longer than 30 Å are difficult to use as reagentless protein-based biosensing. In this case, reagentless protein-based biosensing has been achieved by intercalating organic fluorophores into cracks in the protein structure that reversibly opens and closes as ligand-induced conformation changes occur [80]. The intercalation strategy has yielded various reagentless biosensors [3, 81]. Despite the success of these reagentless organic fluorophore modified protein-based biosensors, the susceptibility of organic molecules to photodegradation remains a significant difficulty in fluorescence contrast imaging. Semiconducting nanoparticles, of course, have minimal photodecomposition and have been an at-

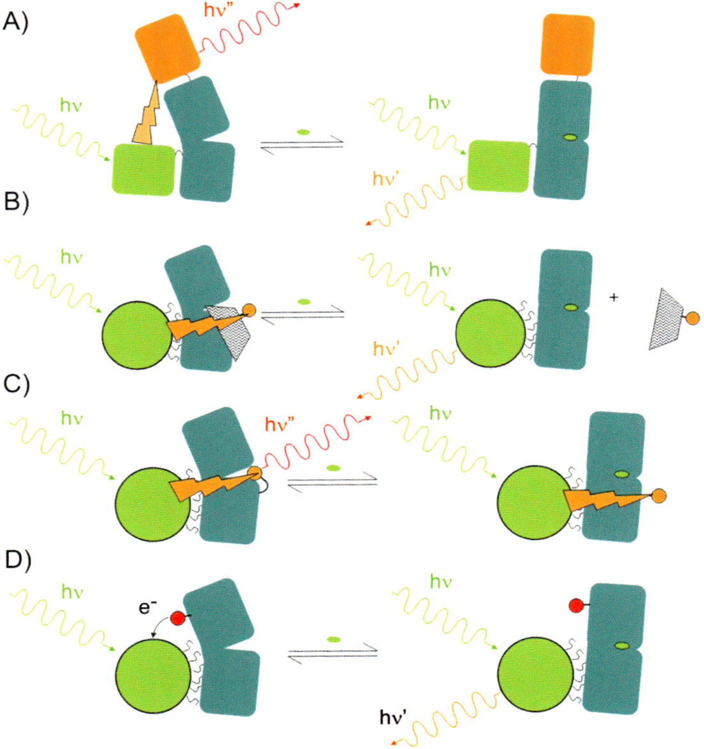

Fig. 12.6. Strategies for reagentless biosensing based on changes in protein conformation. (A) GFP-CFP protein chimeric strategy, where energy transfer efficiency from enhanced cyan fluorescent protein (green) to enhanced yellow fluorescent protein (orange) is altered as a function of analyte-induced change in protein conformation. (B) Displacement assay of a fluorescence quencher bound (shaded with orange ball) bound to a protein, where analyte binding displaces this quencher and restores emission intensity of the CdSe nanoparticle (green ball). (C) Method where the emission intensity of an organic fluorophore (orange ball) is altered as a function of analyte binding, where a CdSe nanoparticle (green ball) is used for photon collection via energy transfer. (D) An electron transfer quenching method where a reductant or oxidant (red ball) is placed close to the CdSe nanoparticle (green ball) surface in one conformation to provide electron transfer quenching (text and Fig. 12.7). Because the protein conformation changes upon analyte binding, the distance between the reductant/oxidant and the nanoparticle will be altered and the nanoparticle emission intensity will change.

tractive target to apply similar methodologies, energy transfer or intercalation, for reagentless protein-based biosensing.

To a first approximation, energy transfer can be used for reagentless protein-attached nanoparticle biosensing. Two energy transfer based biosensing methods using maltose binding protein (MBP) have been published [59, 74]. One system used the maltose selective displacement of β-cyclodextran from MBP as a pseudo-reagentless strategy (Fig. 12.6B) [74]. For this system, fluorophores were attached to a cyclodextran and energy transfer from the CdSe@ZnS nanoparticle to the cyclodextran-modified fluorophore reported the amount of maltose in solution. A 100 nM LOD for maltose was observed with a 0.1 to 10 μM dynamic range. The dynamic range is typical for this type of strategy; however, the LOD is higher than typical due the use of a competitive displacement strategy. This report did, though, demonstrate the effectiveness of energy transfer to proteins attached to semiconducting nanoparticles. Clearly, with this system, multiple organic fluorophores need to be adsorbed to the nanoparticle surface to substantially alter the nanoparticle emission intensity. A reagentless method based on nanoparticle–organic fluorophore energy transfer has been reported recently [59]. This report showed that CdSe@ZnS nanoparticles could function as high photon cross-section antenna. Because energy transfer directly couples the nanoparticle emission intensity to the organic fluorophore emission and absorbance properties, the interaction between the organic fluorophore (Cy3) and MBP actually mediated the changes in overall emission intensity (Fig. 12.6C). The linkage between Cy3-modified 41C MBP and emission intensity is similar to that demonstrated by previous studies [82–84]. In the end, maltose binding mediated a "turn off" signal and is still inherently linked to the relatively fast photodecomposition of organic fluorophores. Therefore, an alternative to energy transfer needs to be used for reagentless protein–nanoparticle based biosensing to take full advantage of the photonic properties of semiconducting nanoparticles.

Electron transfer quenching provides a route to reagentless protein–nanoparticle based biosensing that maintains the high photostability of the semiconducting nanoparticle. Electron transfer quenching involves an electron transfer to or from the excited state of a fluorophore followed by a back electron transfer to form the ground state to non-radiatively relax the fluorophore excited state [70]. Two proof-of-principle articles have been published for electron transfer quenching with a Ru^{II} complex modified MBP that was attached to either CdSe [63] or CdSe@ZnS nanoparticles [64]. In this system, the Ru^{II} complex reduces the valence band hole of the CdSe core, while the resulting Ru^{III} complex oxidizes the electron that remains in the conduction band of the *n*-type CdSe nanoparticle core (Fig. 12.7). When the cyclic nature of this reaction scheme is maintained, minimal photodegradation of this material occurs. This electron transfer quenching method has an exponential dependence [85] with respect to the distance between the Ru^{II} complex and the surface of the CdSe nanoparticle core. Energy transfer distance dependence dependences vary a function of $R_o^6/(R_o + r)^6$, where R_o is the distance at which half maximal energy transfer occurs [70]. However, the $r = 0$ point for energy transfer is at the center of the CdSe nanoparticle core [73, 74, 86] while the core–shell interface is the $r = 0$ point for electron transfer. Based on the R_os for energy transfer from core–shell CdSe nanoparticle donors to organic fluorophore acceptors (40–70 Å) [73, 86–88], energy transfer should have a shallow distance dependence relative to electron transfer quenching.

Benson and coworkers have shown electron transfer quenching provides a solution to reagentless small molecule biosensor design using protein–semiconducting

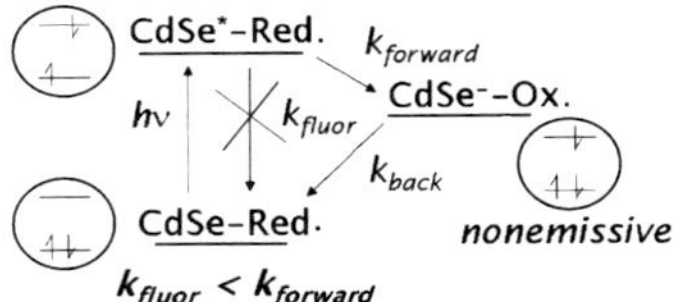

Fig. 12.7. Electron transfer quenching methodology. A nanoparticle ground state (CdSe) and a reductant (Red.) are placed close together. Upon photoexcitation ($h\nu$) the CdSe excited state is formed with an electron in the conduction band (upper molecular orbital) and a hole in the valence band (lower molecular orbital). For a close association of the reductant with the CdSe surface, electron (charge) transport can occur from the reductant to the valence band hole. The charge-separated state that forms ($CdSe^-$ and oxidized reductant, Ox.) is non-emissive. The electron in the conduction band of the CdSe core will then reduce the oxidized reductant to reform the ground state. As long as the first electron transfer rate ($k_{forward}$) is faster than the fluorescence emission rate ($k_{fluor.}$) the nanoparticle emission intensity will be diminished and the lifetime will be decreased.

nanoparticle assemblies (Fig. 12.6D) [63, 64, 67]. This work relies on the generation of a 1:1 complex between a Ru^{II} complex modified protein and a semiconducting nanoparticle. Such a complex has been obtained through the use of kinetically stable mercaptohexdecanoate capping groups [67], as opposed to the dihydrolipoic acid capping groups used by Medintz, Mattoussi, and coworkers to facilitate multiple proteins attached per semiconducting nanoparticle [74]. Four MBPs with different surface Cys attachment sites were surveyed. When the Ru^{II}-modified MBP was bound to either CdSe or CdSe@ZnS nanoparticles that emitted at 560–570 nm, the emission intensities decreased from 1.5 to 2.3-fold [63] and 1.2 to 1.8-fold [64]. Upon maltose addition to the Ru^{II} modified MBP attached to CdSe or CdSe@ZnS nanoparticles, the emission intensity increased by 1.39 to 1.45-fold [63] or 1.13 to 1.42-fold [64]. Recently, our laboratory has confirmed that maltose-dependent changes in CdSe emission intensities correlate with changes in CdSe emission lifetimes [89] and Faradaic electrochemical currents observed for these proteins adsorbed to Au working electrodes [90]. Maltose titrations yielded dissociation constants ranging from 0.20 to 0.75 μM (LOD 1 to 5 nM) and 0.25 to 1.00 μM (LOD 5 to 10 nM) for the CdSe and CdSe@ZnS nanoparticle systems. The similarity of these dissociation constants to the maltose dissociation constant for MBP in solution ($K_D \sim 0.70$ μM) suggests the function of MBP has not been altered by nanoparticle attachment. This is in direct contrast to the 200 to 800 μM dissociation constants reported for the Cy3-modified 41C MBP attached CdSe@ZnS nanoparticle biosensors [59] and the 0.8 to 8.0 μM dissociation constant determined for the fluorophore modified cyclodextran-MBP-CdSe@ZnS nanoparticle system [74]. The only nanoparticle-based perturbation seen for the Ru^{II} complex labeled systems is in the variation of the sample-to-sample dissociation constant reproducibility [63]. For MBPs that placed the Ru^{II} complex closer to the nanoparticle surface, predicted by quenching efficiency, a higher variation in the dissociation constants was observed. This suggested the Ru^{II} complex might be kinetically trapped in the hydrophobic layer surrounding the nanoparticle, which is formed by the fifteen methylenes of the capping ligand. However, electron transfer is still observed for MBPs with the Ru^{II} complex placed away from the nanoparticle surface [63, 64]. Using K46C and K25C MBP, where the Ru^{II} complex was away from the nanoparticle surface, these MBP-CdSe@ZnS nanoparticle biosensors were shown to be selective for maltose in a ten-sugar test panel and reversibly detect maltose in an increasing glucose background [64]. Taken together, the use of electron transfer quenching provides reagentless biosensing that modulates nanoparticle emission intensity using protein–semiconducting nanoparticle assemblies.

The distance dependent changes in electron transfer and energy transfer quenching efficiencies, based on reported data, are compared relatively in Fig. 12.8. As stated above, the delocalized hole–electron pair (exciton) density in core–

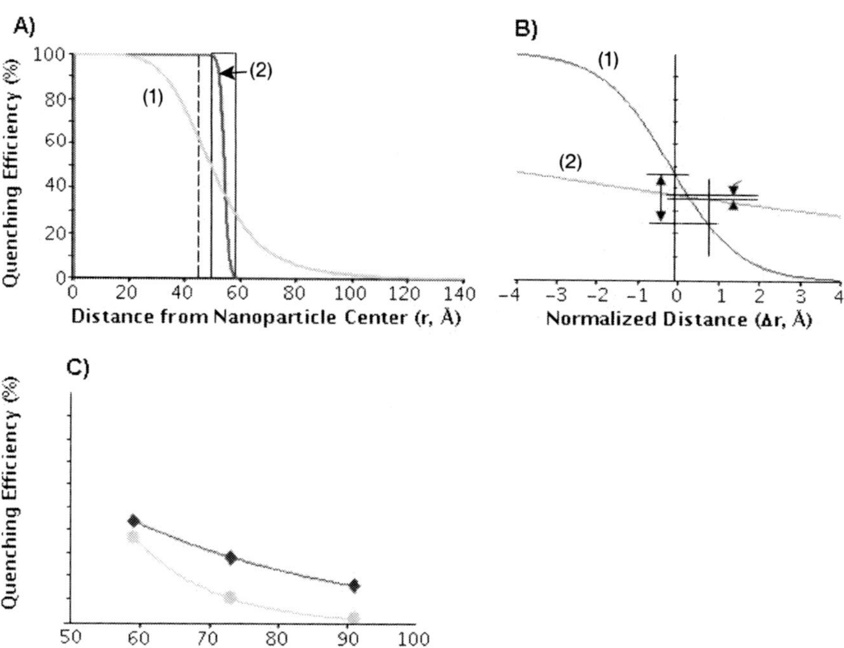

Fig. 12.8. Comparison of energy and electron transfer quenching efficiencies in maltose binding protein based biosensors. (A) Overall comparison of energy transfer distance dependence of quenching efficiency (line 1) and electron transfer quenching efficiency (line 2). The dotted line approximates the nanoparticle radius reported by Medintz and coworkers [86]. The range of the energy transfer based distance was calculated from assuming activation-less electron transfer. The area within the box (50–58 Å) is shown in panel B. (B) The region for nanoparticle distance determined for N282C (line 1, $r = 0$) and D80C (line 2, $r = 0$). Using the change in quenching efficiency for the electron transfer method [64], the expected FRET dependent change was calculated to be one tenth of the measured electron transfer change. (C) Quenching efficiency (electron transfer method [64], diamonds; energy transfer method [86], circles) for three structurally related attachment positions on MBP [64, 86] as a function of the FRET determined distances from the nanoparticle core [86].

shell nanoparticles creates different starting points for distance determinations by energy and electron transfer. Since the average emitted photon density comes from the center of the nanoparticle core the nanoparticle centroid is considered $r = 0$ R_o. Since exciton delocalization occurs significantly faster than nanoparticle emission, the closest approach for electron transfer is the nanoparticle core–shell interface and functions as the $r = 0$ for electron transfer. Energy transfer-determined distances have been reported for CdSe@ZnS nanoparticle–Cy3 fluorophore pairs using the derived R_o of 53 ± 3 Å [86]. Using this radius the Cy3 quenching efficiencies of 530-nm emitting CdSe@ZnS nanoparticles (2% to 37%) yielded RET-determined distances of 59 to 95 Å for six different surface Cys attachment points for Cy3 [86]. The orientation of MBP, in the absence of maltose, on the nanoparticle surface was derived to have both domains interacting with the nanoparticle surface and the maltose binding site pointing away from the nanoparticle surface. For the Ru^{II} complex modified MBP system, quenching efficiencies of 16% to 44% were obtained for CdSe@ZnS nanoparticle emission [64]. A rough comparison between these two published systems is presented in Fig. 12.8, where the RET determined distances of structurally related surface Cys attachment sites are compared.

Clearly, all three structurally related surface Cys attachment sites show the Ru^{II} complex with a higher quenching efficiency than Cy3. Furthermore, maltose addition decreases the Ru^{II} complex quenching efficiencies (−9% to −22%) [64], while no maltose dependent change in Cy3 mediated nanoparticle quenching efficiency has been reported. The differences in maltose-dependent quenching are no surprise based on the distance dependences of electron transfer and energy transfer discussed above. Figure 12.8(B) illustrates for the 282C/80C attachment site the comparison of these scaling rules, where energy transfer has a shallower distance dependence. Additionally, maltose dependent changes in quenching efficiencies for the electron transfer method are consistent with changes in the CdSe core fluorescence lifetimes [89]. However, the surprise in the Ru^{II} complex mediated CdSe@ZnS nanoparticle quenching is the shallow distance dependence with respect to RET-derived surface Cys distance to the nanoparticle centroid (Fig. 12.8C). From a technological standpoint the shallow distance dependence of Ru^{II} complex mediated quenching efficiencies abrogates [63] the need for Cys scanning mutagenesis methods as with organic fluorophore-MBP based biosensors [3, 81]. From a mechanistic standpoint, the Ru^{II} complex based quenching distance dependence might be more similar to protein–protein electron transfer as opposed to intraprotein electron transfer. Beratan and coworkers [91, 92] have recently found constructive interference from structural waters in electron transfer pathways provide very shallow distance dependences in electron transfer. Interestingly, the electron transfer distances calculated ruthenated MBP-CdSe@ZnS systems (8.0 to 9.5 Å) are similar to the distances for the cyt. b_5–cyt. c study [92]. In the end, the Ru^{II} complex [63, 64] provides reagentless biosensing that modulates the emission intensity of semiconducting nanoparticles in a ligand dependent fashion.

Clearly, from these examples, the scaling laws of the nanoparticle and the reporter group need to be carefully considered for reagentless biosensor construction. The biomolecular conformation change strategy clearly demonstrates this point. Nucleic acid–nanoparticle based biosensors can provide reagentless biosensors more readily using conformation change, in that the conformation change is typically greater than 10 Å. By moving a reporter group greater than 10 Å a substantial change in energy transfer and surface-enhanced Raman can be expected. Protein–nanoparticle based biosensors require more careful consideration of scaling rules since reporter groups typically are moved by 1–6 Å. Electron transfer quenching takes advantage of a less efficient methodology than energy transfer to provide a higher sensitivity to conformational motion. This is not to say that electron transfer quenching is the only solution to protein–nanoparticle based reagentless biosensing, but that it provides a scaling law that appropriately couples analyte-induced protein conformation motions into changes in the nanoparticle electronic structure.

12.5 Conclusion

The three general concepts outlined here provide reagentless biosensors from biomolecules attached to nanoparticles. As demonstrated by the literature cited in this chapter, these reagentless concepts provide sensors that can be adapted to various detection platforms. More importantly, the instrumentation used for biosensing is relatively inexpensive and modular. Extension of these systems to multiplexed or array assays has been reported and should be able to be extended for various other systems. The idea that the biomolecule provides the analyte sensitivity and selectivity is at the heart of the modular detection strategies. As outlined in this volume and this series, nanoparticles provide unique properties that can be detected by various analytical techniques. The translation between analyte–biomolecule binding and changes in nanoparticle properties requires the selection of a reporting methodology (e.g., FRET, SERRS, T_2 relaxation, electron transfer quenching) with an appropriate scaling law. The high surface area of nanoparticles provides a high sensitivity method to measure changes in local solvent dielectric. Biomolecule attachment to nanoparticle surfaces allows substrates for these biomolecules to alter the local solvent dielectric and properties such as localized surface plasmon absorption and electrical conductivity. Metallic and semiconducting nanoparticles can also be used as electron/hole buffers to increase electrochemical communication between enzymes and working electrodes. This strategy has allowed multimolecular bioelectrochemical sensors to be transformed into reagentless bioelectrochemical sensors. Energy transfer methods are the most popular method for coupling biomolecule and nanoparticle properties by energy transfer from organic fluorophores to metallic nanoparticles, semiconducting nanoparticles to organic fluorophores, or semiconducting nanoparticles to metallic nanoparticles. While energy transfer scaling laws facilitate enzymatic cleavage biosensors and nucleic acid–nanoparticle based biosensors, protein–nanoparticle biosensors require a coupling method with a more sensitive distance dependence. While reagentless protein–nanoparticle based biosensors that use energy transfer have been reported [59], this method monitors differences in the emission intensity of the organic fluorophore. Electron transfer quenching provides a method to effectively couple changes in protein conformation into changes in nanoparticle emission intensity [63, 64]. Magnetic force and T_2 magnetic relaxation methods, along with nanoparticle surface energy transfer [93, 94], have a much shallower distance dependence that could be used for reagentless biosensor development with enzymatic cleavage (pseudo-reagentless) or nucleic acid conformation changes. Another advantage of using nanoparticles is the advent of hybrid detection schemes, such as endonuclease detection by magnetic force measurements [53, 54], electrochemical surface plasmon resonance [20] and photoelectrochemical detection [21]. Such hybrid detection strategies provide signal selectivity on top of the analyte selectivity provided by the biomolecule. Such selectivity provides additional confidence necessary for device approval and commercialization. Finally, the development of nanoparticle based biosensors that are reagentless and reversible is an exciting frontier that requires a material, biomolecular, and molecular viewpoint that is producing sophisticated biosensors with inexpensive instrumentation.

Acknowledgments

The financial support of the National Science Foundation (DBI-0508134) is acknowledged. Marinella G. Sandros, Vivekanand Shete, and Kathleen Fleming are acknowledged for their help in bibliographic searches and reference acquisition.

References

1 Katz, E., Willner, I., Nanobiotechnology: Integrated nanoparticle-biomolecule hybrid systems: Synthesis, properties, and applications, *Angew. Chem. Int. Ed.* **2004**, 43, 6042–6108.

2 Dwyer, M. A., Hellinga, H. W., Periplasmic binding proteins: A versatile superfamily for protein engineering, *Curr. Opin. Struct. Biol.* **2004**, 14, 495–504.

3 Feltus, A., Daunert, S., Genetic engineering of signaling molecules. In *Optical Biosensors: Present and Future*, eds. Ligler, F. S., Lowe Taitt, C. A., Elsevier, Amsterdam, **2002**, pp. 307–329.

4 Rajendran, M., Ellington, A. D., Nucleic acids for reagentless biosensors. In *Optical Biosensors: Present and Future*, eds. Ligler, F. S., Lowe Taitt, C. A., Elsevier, Amsterdam, **2002**, pp. 369–396.

5 Medintz, I. L., Uyeda, H. T., Goldman, E. R., Mattoussi, H., Quantum dot bioconjugates for imaging, labelling and sensing, *Nat. Mater.* **2005**, 4, 435–446.

6 Brasuel, M., Kopelman, R., Aylott, J. W., Clark, H., Xu, H., Hoyer, M., Miller, T. J., Tjalkens, R., Philbert, M. A., Production, characteristics and applications of fluorescent PEBBLE nanosensors: Potassium, oxygen, calcium and pH imaging inside live cells, *Sens. Mater.* **2002**, 14, 309–338.

7 Xu, H., Aylott, J. W., Kopelman, R., Fluorescent nano-PEBBLE sensors designed for intracellular glucose imaging, *Analyst* **2002**, 127, 1471–1477.

8 Clark, H. A., Hoyer, M., Philbert, M. A., Kopelman, R., Optical nanosensors for chemical analysis inside single living cells. 1.

Fabrication, characterization, and methods for intracellular delivery of PEBBLE sensors, *Anal. Chem.* **1999**, 71, 4831–4836.

9 Rosenzweig, Z., Kopelman, R., Analytical properties and sensor size effects of a micrometer-sized optical fiber glucose biosensor, *Anal. Chem.* **1996**, 68, 1408–1413.

10 Comparelli, R., Curri, M. L., Cozzoli, P. D., Striccoli, M., Chapter 5 of this volume.

11 Haes, A. J., Van Duyne, R. P., A unified view of propagating and localized surface plasmon resonance biosensors, *Anal. Bioanal. Chem.* **2004**, 379, 920–930.

12 Nath, N., Chilkoti, A., Label-free biosensing by surface plasmon resonance of nanoparticles on glass: Optimization of nanoparticle size, *Anal. Chem.* **2004**, 76, 5370–5378.

13 He, L., Musick, M. D., Nicewarner, S. R., Salinas, F. G., Benkovic, S. J., Natan, M. J., Keating, C. D., Colloidal Au-enhanced surface plasmon resonance for ultrasensitive detection of DNA hybridization, *J. Am. Chem. Soc.* **2000**, 122, 9071–9077.

14 Frederix, F., Friedt, J.-M., Choi, K.-H., Laureyn, W., Campitelli, A., Mondelaers, D., Maes, G., Borghs, G., Biosensing based on light absorption of nanoscaled gold and silver particles, *Anal. Chem.* **2003**, 75, 6894–6900.

15 Frederix, F., Bonroy, K., Laureyn, W., Reekmans, G., Campitelli, A., Dehaen, W., Maes, G., Enhanced performance of an affinity biosensor interface based on mixed self-assembled monolayers of thiols on gold, *Langmuir* **2003**, 19, 4351–4357.

16 Endo, T., Kerman, K., Nagatani, N., Takamura, Y., Tamiya, E., Label-free detection of peptide nucleic acid-DNA hybridization using localized surface plasmon resonance based optical biosensor, *Anal. Chem.* **2005**, 77, 6976–6984.

17 Endo, T., Yamamura, S., Nagatani, N., Morita, Y., Takamura, Y., Tamiya, E., Localized surface plasmon resonance based optical biosensor using surface modified nanoparticle layer for label free monitoring of antigen-antibody reaction, *Sci. Technol. Adv. Mater.* **2005**, 6, 491–500.

18 Prodan, E., Nordlander, P., Halas, N. J., Electronic structure and optical properties of gold nanoshells, *Nano Lett.* **2003**, 3, 1411–1415.

19 Haes, A. J., Zou, S., Schatz, G. C., Van Duyne, R. P., A nanoscale optical biosensor: The long range distance dependence of the localized surface plasmon resonance of noble metal nanoparticles, *J. Phys. Chem. B* **2004**, 108, 109–116.

20 Zayats, M., Kharitonov, A. B., Pogorelova, S. P., Lioubashevski, O., Katz, E., Willner, I., Probing photoelectrochemical processes in Au-CdS nanoparticle arrays by surface plasmon resonance: Application for the detection of acetylcholine esterase inhibitors, *J. Am. Chem. Soc.* **2003**, 125, 16 006–16 014.

21 Pardo-Yissar, V., Katz, E., Wasserman, J., Willner, I., Acetylcholine esterase-labeled CdS nanoparticles on electrodes: Photoelectrochemical sensing of the enzyme inhibitors, *J. Am. Chem. Soc.* **2003**, 125, 622–623.

22 Chen, R. J., Bangsaruntip, S., Drouvalakis, K. A., Kam, N. W. S., Shim, M., Li, Y., Kim, W., Utz, P. J., Dai, H., Noncovalent functionalization of carbon nanotubes for highly specific electronic biosensors, *Proc. Natl. Acad. Sci.* **2003**, 100, 4984–4989.

23 Cui, Y., Wei, Q., Park, H., Lieber, C. M., Nanowire nanosensors for highly sensitive and selective detection of biological and chemical species, *Science* **2001**, 293, 1289–1292.

24 Kichambare, P. D., Star, A., Chapter 1 of this volume.

25 Wang, J., Lui, G., Lin, Y., Chapter 3 of this volume.

26 Chen, R. J., Choi, H. C., Bangsaruntip, S., Yenilmez, E., Tang, X., Wang, Q., Chang, Y.-L., Dai, H., An investigation of the mechanisms of electronic sensing of protein adsorption on carbon nanotube devices, *J. Am. Chem. Soc.* **2004**, 126, 1563–1568.

27 Byon, H. R., Choi, H. C., Network single-walled carbon nanotube-field effect transistors (SWNT-FETs) with increased Schottky contact area for highly sensitive biosensor applications, *J. Am. Chem. Soc.* **2006**, 128, 2188–2189.
28 So, H.-M., Won, K., Kim, Y. H., Kim, B.-K., Ryu, B. H., Na, P. S., Kim, H., Lee, J.-O., Single-walled carbon nanotube biosensors using aptamers as molecular recognition elements, *J. Am. Chem. Soc.* **2005**, 127, 11 906–11 907.
29 Besteman, K., Lee, J.-O., Wiertz, F. G. M., Heering, H. A., Dekker, C., Enzyme-coated carbon nanotubes as single-molecule biosensors, *Nano Lett.* **2003**, 3, 727–730.
30 Banerjee, S., Hemraj-Benny, T., Wong, S. S., Covalent surface chemistry of single-walled carbon nanotubes, *Adv. Mater.* **2005**, 17, 17–29.
31 Li, Z., Chen, Y., Li, X., Kamins, T. I., Nauka, K., Williams, R. S., Sequence-specific label-free DNA sensors based on silicon nanowires, *Nano Lett.* **2004**, 4, 245–247.
32 Hahm, J.-i., Lieber, C. M., Direct ultrasensitive electrical detection of DNA and DNA sequence variations using nanowire nanosensors, *Nano Lett.* **2004**, 4, 51–54.
33 Zheng, G., Patolsky, F., Cui, Y., Wang, W. U., Lieber, C. M., Multiplexed electrical detection of cancer markers with nanowire sensor arrays, *Nat. Biotechnol.* **2005**, 23, 1294–1301.
34 Wang, W. U., Chen, C., Lin, K.-h., Fang, Y., Lieber, C. M., Label-free detection of small-molecule-protein interactions by using nanowire nanosensors, *Proc. Natl. Acad. Sci.* **2005**, 102, 3208–3212.
35 Guiseppi-Elie, A., Lei, C., Baughman, R. H., Direct electron transfer of glucose oxidase on carbon nanotubes, *Nanotechnology* **2002**, 13, 559–564.
36 Salimi, A., Noorbakhsh, A., Ghadermarz, M., Direct electrochemistry and electrocatalytic activity of catalase incorporated onto multiwall carbon nanotubes-modified glassy carbon electrode, *Anal. Biochem.* **2005**, 344, 16–24.
37 Wang, J., Musameh, M., A reagentless amperometric alcohol biosensor based on carbon-nanotube/teflon composite electrodes, *Anal. Lett.* **2003**, 36, 2041–2048.
38 Zhao, G.-C., Yin, Z.-Z., Zhang, L., Wei, X.-W., Direct electrochemistry of cytochrome c on a multi-walled carbon nanotubes modified electrode and its electrocatalytic activity for the reduction of H_2O_2, *Electrochem. Commun.* **2005**, 7, 256–260.
39 Carralero Sanz, V., Mena, M. L., Gonzalez-Cortes, A., Yanez-Sedeno, P., Pingarron, J. M., Development of a tyrosinase biosensor based on gold nanoparticles-modified glassy carbon electrodes, *Anal. Chim. Acta* **2005**, 528, 1–8.
40 Gu, H.-Y., Yu, A.-M., Yuan, S.-S., Chen, H.-Y., Amperometric nitric oxide biosensor based on the immobilization of hemoglobin on a nanometer-sized gold colloid modified Au electrode, *Anal. Lett.* **2002**, 35, 647–661.
41 Liu, S., Ju, H., Reagentless glucose biosensor based on direct electron transfer of glucose oxidase immobilized on colloidal gold modified carbon paste electrode, *Biosens. Bioelectron.* **2003**, 19, 177–183.
42 Liu, S., Ju, H., Nitrite reduction and detection at a carbon paste electrode containing hemoglobin and colloidal gold, *Analyst* **2003**, 128, 1420–1424.
43 Huang, Y., Zhang, W., Xiao, H., Li, G., An electrochemical investigation of glucose oxidase at a CdS nanoparticles modified electrode, *Biosens. Bioelectron.* **2005**, 21, 817–821.
44 Cao, D., He, P., Hu, N., Electrochemical biosensors utilizing electron transfer in heme proteins immobilised on Fe3O4 nanoparticles, *Analyst* **2003**, 128, 1268–1274.
45 Crumbliss, A. L., Stonehuerner, J., Henkens, R. W., O'Daly, J. P., Zhao, J., The use of inorganic materials to control or maintain immobilized

enzyme activity, *New J. Chem.* **1994**, 18, 327–339.

46 Zhao, J., Henkens, R. W., Crumbliss, A. L., Mediator-free amperometric determination of toxic substances based on their inhibition of immobilized horseradish peroxidase, *Biotechnol. Prog.* **1996**, 12, 703–708.

47 Zhao, J., O'Daly, J. P., Henkens, R. W., Stonehuerner, J., Crumbliss, A. L., A xanthine oxidase/colloidal gold enzyme electrode for amperometric biosensor applications, *Biosens. Bioelectron.* **1996**, 11, 493–502.

48 Liu, J., Lu, Y., A colorimetric lead biosensor using DNAzyme-directed assembly of gold nanoparticles, *J. Am. Chem. Soc.* **2003**, 125, 6642–6643.

49 Gu, H., Xu, K., Xu, C., Xu, B., Biofunctional magnetic nanoparticles for protein separation and pathogen detection, *Chem. Commun.* **2006**, 941–949.

50 Zhelev, Z., Bakalova, Ohba, H., Baba, Y., Chapter 6 of this volume.

51 Chang, E., Miller, J. S., Sun, J., Yu, W. W., Colvin, V. L., Drezek, R., West, J. L., Protease-activated quantum dot probes, *Biochem. Biophys. Res. Commun.* **2005**, 334, 1317–1321.

52 Perez, J. M., O'Loughin, T., Simeone, F. J., Weissleder, R., Josephson, L., DNA-based magnetic nanoparticle assembly acts as a magnetic relaxation nanoswitch allowing screening of DNA-cleaving agents, *J. Am. Chem. Soc.* **2002**, 124, 2856–2857.

53 Weizmann, Y., Elnathan, R., Lioubashevski, O., Willner, I., Magnetomechanical detection of the specific activities of endonucleases by cantilevers, *Nano Lett.* **2005**, 5, 741–744.

54 Weizmann, Y., Elnathan, R., Lioubashevski, O., Willner, I., Endonuclease-based logic gates and sensors using magnetic force-amplified readout of DNA scission on cantilevers, *J. Am. Chem. Soc.* **2005**, 127, 12 666–12 672.

55 Liu, J., Lu, Y., A colorimetric lead biosensor using DNAzyme-directed assembly of gold nanoparticles, *J. Am. Chem. Soc.* **2003**, 125, 6642–6643.

56 Wang, Z., Kanaras, A. G., Bates, A. D., Cosstick, R., Brust, M., Enzymatic DNA processing on gold nanoparticles, *J. Mater. Chem.* **2004**, 14, 578–580.

57 Zhao, M., Josephson, L., Tang, Y., Weissleder, R., Magnetic sensors for protease assays, *Angew. Chem. Int. Ed.* **2003**, 42, 1375–1378.

58 Maxwell, D. J., Taylor, J. R., Nie, S., Self-assembled nanoparticle probes for recognition and detection of biomolecules, *J. Am. Chem. Soc.* **2002**, 124, 9606–9612.

59 Medintz, I. L., Clapp, A. R., Melinger, J. S., Deschamps, J. R., Mattoussi, H., A reagentless biosensing assembly based on quantum dot-donor Forster resonance energy transfer, *Adv. Mater.* **2005**, 17, 2450–2455.

60 Wabuyele, M. B., Vo-Dinh, T., Detection of human immuno-deficiency virus type 1 DNA sequence using plasmonics nanoprobes, *Anal. Chem.* **2005**, 77, 7810–7815.

61 Levy, M., Cater, S. F., Ellington, A. D., Quantum-dot aptamer beacons for the detection of proteins, *ChemBioChem* **2005**, 6, 2163–2166.

62 Li, Y.-T., Liu, H.-S., Lin, H.-P., Chen, S.-H., Gold nanoparticles for microfluidics-based biosensing of PCR products by hybridization-induced fluorescence quenching, *Electrophoresis* **2005**, 26, 4743–4750.

63 Sandros, M. G., Gao, D., Benson, D. E., A modular nanoparticle-based system for reagentless small molecule biosensing, *J. Am. Chem. Soc.* **2005**, 127, 12 198–12 199.

64 Sandros, M. G., Shete, V., Benson, D. E., Selective, reversible, reagentless maltose biosensing with core-shell semiconducting nanoparticles, *Analyst* **2006**, 131, 229–235.

65 Liu, J., Lu, Y., Fast colorimetric sensing of adenosine and cocaine based on a general sensor design involving aptamers and nanoparticles, *Angew. Chem. Int. Ed.* **2005**, 45, 90–94.

66 Hermanson, G. T., *Bioconjugate Techniques*, Academic Press, San Diego, **1996**.
67 Sandros, M. G., Gao, D., Gokdemir, C., Benson, D. E., General, high-affinity approach for the synthesis of fluorophore appended protein nanoparticle assemblies. *Chem. Commun.* **2005**, 2832–2834.
68 Goldman, E. R., Balighian, E. D., Mattoussi, H., Kuno, M. K., Mauro, J. M., Tran, P. T., Anderson, G. P., Avidin: A natural bridge for quantum dot-antibody conjugates, *J. Am. Chem. Soc.* **2002**, 124, 6378–6382.
69 Goldman, E. R., Medintz, I. L., Hayhurst, A., Anderson, G. P., Mauro, J. M., Iverson, B. L., Georgiou, G., Mattoussi, H., Self-assembled luminescent CdSe-ZnS quantum dot bioconjugates prepared using engineered poly-histidine terminated proteins, *Anal. Chim. Acta* **2005**, 534, 63–67.
70 Lakowicz, J. R., *Principles of Fluorescence Spectoscopy.* 2nd edn., Klewer Academic, New York, **1999**.
71 Ellington, A. D., Szostak, J. W., In vitro selection of RNA molecules that bind specific ligands, *Nature* **1990**, 346, 818–822.
72 Tuerk, C., Gold, L., Systematic evolution of ligands by exponential enrichment: RNA ligands to bacteriophage T4 DNA polymerase, *Science* **1990**, 249, 505–510.
73 Clapp, A. R., Medintz, I. L., Mauro, J. M., Fisher, B. R., Bawendi, M. G., Mattoussi, H., Fluorescence resonance energy transfer between quantum dot donors and dye-labeled protein acceptors, *J. Am. Chem. Soc.* **2004**, 126, 301–310.
74 Medintz, I. L., Clapp, A. R., Mattoussi, H., Goldman, E. R., Fisher, B., Mauro, J. M., Self-assembled nanoscale biosensors based on quantum dot FRET donors, *Nat. Mater.* **2003**, 2, 630–638.
75 Krug, J. T., II, Wang, G. D., Emory, S. R., Nie, S., Efficient Raman enhancement and intermittent light emission observed in single gold nanocrystals, *J. Am. Chem. Soc.* **1999**, 121, 9208–9214.
76 Nie, S., Emory, S. R., Probing single molecules and single nanoparticles by surface-enhanced Raman scattering, *Science* **1997**, 275, 1102–1106.
77 Deuschle, K., Okumoto, S., Fehr, M., Looger, L. L., Kozhukh, L., Frommer, W. B., Construction and optimization of a family of genetically encoded metabolite sensors by semirational protein engineering, *Protein Sci.* **2005**, 14, 2304–2314.
78 Fehr, M., Frommer, W. B., Lalonde, S., Visualization of maltose uptake in living yeast cells by fluorescent nanosensors, *Proc. Natl. Acad. Sci. U.S.A.* **2002**, 99, 9846–9851.
79 Lager, I., Fehr, M., Frommer, W. B., Lalonde, S., Development of a fluorescent nanosensor for ribose, *FEBS Lett.* **2003**, 553, 85–89.
80 Dattelbaum, J., Looger, L., Benson, D., Sali, K., Thompson, R., Hellinga, H., Analysis of allosteric signal transduction mechanisms in an engineered fluorescent maltose biosensor, *Protein Sci.* **2005**, 14, 284–291.
81 de Lorimier, R. M., Smith, J. J., Dwyer, M. A., Looger, L. L., Sali, K. M., Paavola, C. D., Rizk, S. S., Sadigov, S., Conrad, D. W., Loew, L., Hellinga, H. W., Construction of a fluorescent biosensor family, *Protein Sci.* **2002**, 11, 2655–2675.
82 Gilardi, G., Zhou, L. Q., Hibbert, L., Cass, A. E., Engineering the maltose binding protein for reagentless fluorescence sensing, *Anal. Chem.* **1994**, 66, 3840–3847.
83 Zhou, L. Q., Cass, A. E. G., Periplasmic binding protein based biosensors. 1. Preliminary study of maltose binding protein as sensing element for maltose biosensor. *Biosens. Bioelectron.* **1991**, 6, 445–450.
84 Brune, M., Hunter, J. L., Corrie, J. E., Webb, M. R., Direct, real-time measurement of rapid inorganic phosphate release using a novel fluorescent probe and its application to actomyosin subfragment 1 ATPase, *Biochemistry* **1994**, 33, 8262–8271.
85 Marcus, R. A., Sutin, N., Electron transfers in chemisty and biology,

Biochim. Biophys. Acta **1985**, 811, 265–322.

86 Medintz, I. L., Konnert, J. H., Clapp, A. R., Stanish, I., Twigg, M. E., Mattoussi, H., Mauro, J. M., Deschamps, J. R., A fluorescence resonance energy transfer-derived structure of a quantum dot-protein bioconjugate nanoassembly, *Proc. Natl. Acad. Sci. U.S.A.* **2004**, 101, 9612–9617.

87 Willard, D. M., Carillo, L. L., Jung, J., Van Orden, A., CdSe-ZnS quantum dots as resonance energy transfer donors in a model protein-protein binding assay, *Nano Lett.* **2001**, 1, 469–474.

88 Wang, S., Mamedova, N., Kotov, N. A., Chen, W., Studer, J., Antigen/antibody immunocomplex from CdTe nanoparticle bioconjugates. *Nano Lett.* **2002**, 2, 817–822.

89 Shete, V., Sandros, M. G., Benson, D. E., unpublsihed results.

90 Hernandez, F. J., Benson, D. E., unpublished results.

91 Kawatsu, T., Beratan, D. N., Kakitani, T., Conformationally averaged score functions for electronic propagation in proteins, *J. Phys. Chem. B* **2006**, 110, 5747–5757.

92 Lin, J., Balabin, I. A., Beratan, D. N., The nature of aqueous tunneling pathways between electron-transfer proteins, *Science* **2005**, 310, 1311–1313.

93 Lee, J., Govorov, A. O., Kotov, N. A., Nanoparticle assemblies with molecular springs: A nanoscale thermometer, *Angew. Chem. Int. Ed.* **2005**, 44, 7439–7442.

94 Yun, C. S., Javier, A., Jennings, T., Fisher, M., Hira, S., Peterson, S., Hopkins, B., Reich, N. O., Strouse, G. F., Nanometal surface energy transfer in optical rulers, breaking the FRET barrier, *J. Am. Chem. Soc.* **2005**, 127, 3115–3119.

13
Pico/Nanoliter Chamber Array Chips for Single-cell, DNA and Protein Analyses

Shohei Yamamura, Ramachandra Rao Sathuluri, and Eiichi Tamiya

13.1
Introduction

In recent years, miniaturized systems called Lab-on-a-chip or Micro-Total Analysis System (μ-TAS) have been examined as new systems for biochemical analyses. These systems are expected to perform DNA, protein and cell analysis for drug screening and development of novel therapies. Especially, microarray and microfluidic types of chip devices have been developed using micro- and nano-technological techniques. These lab-on-a-chip systems can be used for high-throughput identification of large numbers of potential drug targets, e.g., DNA, protein, chemicals [1, 2]. Recent advances in the human genome project have prompted the use of the miniaturized chip devices for high-throughput analysis of the vast amount of information potentially available. To obtain as much information as possible in a short time with a minimal use of reagents, researchers require highly integrated and sophisticated devices, such as microarrays. Thus, microarrays suitable for different biomaterial assay and detection technologies have been under intense investigation [1, 2]. Microarrays have been mostly applied to the assessment of the presence of a specific base sequence, or which genes are expressed and at what level [3, 4]. They have also been employed in identification of peptides and proteins as pharmaceutical drugs [5, 6]. Although the number of human genes was reported to be approximately 30 000 from the human genomic project, the functions and expression mechanism for most of the genes remain unknown, because the fate of the genes functionality is determined after protein expression, but proteins expression is controlled by cellular function. Therefore, it is necessary to develop the microarray chip devices that can perform high-throughput screening and analysis of proteins and cells at single-cell and single-molecule level.

To achieve single-cell or single-molecule analysis, highly integrated microarray systems that can perform assays at pico- and nano-liter volume level are greatly desirable to realize post-genomic research, such as proteomics and cellomics. Single-cell analysis contributes to elucidate the functional mechanism of genes, proteins and chemical responses, which lead to clinical diagnosis and drug discovery [7]. In

Nanotechnologies for the Life Sciences Vol. 8
Nanomaterials for Biosensors. Edited by Challa S. S. R. Kumar

ISBN: 978-3-527-31388-4

recent work, some researchers reported cellular microarrays using biomaterial spotting techniques for investigating gene expression or differentiation of some kinds of cells [8, 9]. These systems can screen and detect a group of cells but not single-cell based assay from bulk cell suspension, because they have a great possibility of cross-contamination with neighboring cells due to the absence of a physical boundary. Therefore, it is necessary to construct a microarray system that can perform high-throughout analysis of single-molecule or single-cell and quantitative detection.

However, there are no reports available, to the best of our knowledge, that describe high-throughput analysis of DNA, protein and cell at single-molecule and single-cell level together at one place. Therefore, we address the analysis of DNA, protein and cell using pico- or nano-liter chamber array system in this chapter. This chapter is divided into three topics: novel multiplexed PCR, cell-free protein synthesis, and high-throughput single-cell analysis system [10–13]. The chapter ends with a look at future prospects.

13.2
Multiplexed Polymerase Chain Reaction from A Single Copy DNA using Nanoliter-volume Microchamber Array

Polymerase chain reaction (PCR)-based techniques have become the most important part of DNA diagnostic laboratories since its first introduction in 1985 [14]. The discovery of this technology has earned its inventor, K. B. Mullis, a Nobel prize for his achievement [15], which has opened up new horizons for a limitless number of DNA-based research possibilities. Since then, qualitative PCR has been a well-established and straightforward technology, but the quantification of specific target DNA sequences in a complex sample has been a difficult task. Several variations, caused by the manipulation of nucleic acids that may occur during sample preparation, storage, or the course of the reaction hampered accurate quantification. The exponential nature of the PCR amplification can significantly magnify even minor variations in reaction conditions. Normalizing the amount of PCR products of the specific template with respect to an internal reference template has been partly successful against these variations. Since the challenge of accurate DNA quantification stimulated many researchers, a great variety of protocols already exist for the utility of quantitative PCR [16–18]. However, these methods are nearly exclusively restricted to be applied for research purposes only because of two factors they have in common: they are difficult tasks and are costly to run.

To supply the demand for faster, more accurate, and more cost-effective PCR devices with a high-throughput capacity, three important properties have directed the development of the next generation of PCR systems: automation, standardization, and miniaturization. Recently, Yang et al. have reported a high-sensitivity PCR assay in polycarbonate plastic, disposable PCR microreactors [19]. At a template concentration as low as 10 *Escherichia coli* cells (equivalent to 50 fg of genomic DNA),

221-bp product was successfully amplified within 30 min. Lee et al. [20] have described a microfabricated PCR device for simultaneous DNA amplification and electrochemical detection on gold or indium tin oxide (ITO) electrodes patterned on a glass substrate. A miniaturized flow-through PCR with different template types in a silicon chip thermocycler has also been reported to have a minimum power consumption [21]. With the flow-through PCR device of Fukuba et al. [22], 580 and 1450 bp of DNA fragments were successfully amplified from *E. coli* genomic DNA and directly from untreated cells. For temperature control of their chip, six heaters made of ITO were placed on a glass substrate to act as three uniform temperature zones. Lee et al. [23] have recently reported a bulk-micromachined PCR chip. They validated that the proposed chip amplified the DNA related to the tumor suppressor gene BRCA 1 (127 bp at 11th exon) after 30 thermal cycles in a 200-nL-volume chamber. Although most of the recent assays are accurate and sensitive, they involve the definition of very stringent limits. The PCR products are usually separated by gel electrophoresis, and the band intensities are quantified by video imaging and densitometry. Additionally, Lagally et al. [24, 25] have shown that microfluidic systems are capable of multiplexed PCR reactions and robust on-chip detection.

Microchamber arrays etched on silicon or glass have also been one of the most reported miniaturized devices for multiple simultaneous DNA amplification [26–28]. The minimum reported size for a microchamber for PCR was demonstrated by Leamon et al. in connection with PCR [29]. They reported a novel platform, namely PicoTiterPlate™, which enabled simultaneous amplification of 300 000 discrete PCR reactions in volumes as small as 39.5 pL. Following the PCR on the PicoTiterPlate™, the solution from each well was recovered, and then quantified by TaqMan assay. As for the easy integration with different applications, solid-phase amplification was also performed on PicoTiterPlate™ by immobilizing the PCR product to a DNA capture bead in each well. Thus, 370 000 beads bound with PCR product were obtained for parallel processing in numerous solid-phase applications. The volume of a microchamber for a successful PCR amplification was reduced to 86 pL by Nagai et al. [26]. However, as the sample volume was decreased, evaporation of sample solution and the introduction method of quite a small amount of solution into the reaction microchamber appeared as the major drawbacks.

In our research, for achieving simultaneous detection of several numbers of target DNA, the feasibility of our microchamber array was further improved by using TaqMan PCR [11, 12]. To the best of our knowledge, three different DNA sequences were amplified from three different DNA templates and detected in the same microchamber array simultaneously for the first time. In addition, the quantification of initial DNA concentration present in a microchamber was achieved from 0 to 12 copies per chamber, not only by monitoring the real-time fluorescence intensity but also by observing the end point fluorescence signal. Therefore, this system proves to be a promising device for the low-cost, high-throughput DNA amplification and detection for point-of-care clinical diagnosis, which can also be handled by non-specialist users.

13.2.1
PCR Microchamber Array Chip System

13.2.1.1 Microchamber Array Chip Fabrication

The microchamber array chip for DNA amplification was fabricated using micromachining techniques, including photolithography, and anisotropic wet etching on the optically polished side of a silicon (100) wafer. The chip substrate was designed to be 2.54 × 7.62 cm for compatibility with the dispensing system employed. Figure 13.1 shows the fabrication procedure, giving detailed schematic steps of the silicon chamber array chip.

Before proceeding to microfabrication, the silicon wafer was washed thoroughly with acetone and then immersed in a 60% (v/v) hydrogen fluoride solution and, finally, allowed to dry at room temperature. Subsequently, the silicon wafer surface was oxidized to silicon oxide (SiO_2) by wet thermal oxidation at 1000 °C for 8 h. A photo mask with 1248 chambers having a 24 × 52 pattern was then printed on the surface, after being coated with a positive photoresist (OFPR-800) layer, by a photolithographic process. Soaking the chip substrate in a NMD-3 solution for 5 min removes the light exposed OFPR portions and, followed by treatment with HF/NH_4F (v/v) 10:60 volumes solution, removed the exposed oxide layer and then remaining portions of photoresist by treating with acetone. The chip substrate was then anisotropically etched with 25% (w/v) tetramethylammonium hydroxide (TMAH) in an aqueous solution to a depth of 250 μm for 8 h at 80 °C. A layer of SiO_2 was grown on the surfaces, including the etched chamber walls and then OFPR-800 was spin coated and further photolithography was performed to leave the SiO_2 layer inside the microchamber walls. The oxidized layer and photoresist were removed by using HF/NH_4F (v/v) 10:60 volumes of solution and acetone, respectively. The microchamber feature was observed by a color laser 3D profile microscope VK-8500

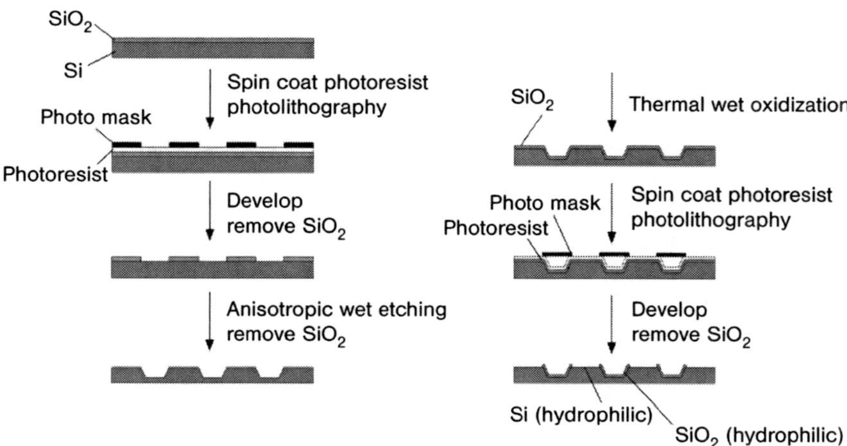

Fig. 13.1. Schematic illustration of microchamber array fabrication procedures.

(A)

(B)

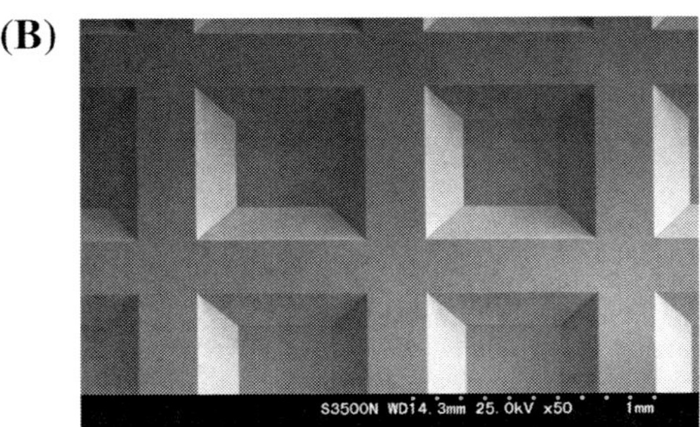

Fig. 13.2. (A) Photographic image of the microchamber array chip. The 1248 microchambers are integrated on the 1 × 3 inch. of a silicon chip. (B) Scanning electron micrograph (SEM) of a silicon microchamber array.

(KEYENCE, Japan) and scanning electron microscope (HITACHI, Japan). Each chamber was parallelepiped with dimensions of 650 × 650 × 200 μm, a pitch of each microchamber of ~900 μm, and accommodates 50 nL (Fig. 13.2). The total number of chambers on each chip is of 1248. To achieve precise introduction of sample mixture into the microchamber, only the inner wall surfaces of the microchamber were prepared as hydrophilic by leaving an oxidized layer on them by photolithographic techniques, as described recently by Felbel et al. [30].

13.2.1.2 **Sample Loading with a Nanoliter Dispenser**

The microchamber array chip was soaked in 1% (w/v) bovine serum albumin (BSA) solution overnight, then rinsed with deionized water, and dried to prevent nonspecific adsorption by coating the chamber wall. The chip was placed onto the dispensing stage of a nanoliter dispenser from Cartesian Technologies. The precise dispensing of nanoliter volumes of solutions exactly at previously determined locations had become very simple by using their technology. The volume of dispensed solution in a single microchamber was 40 nL. Mineral oil as a cover lid was coated onto the template DNA-modified chip, and 40 nL of PCR mixture, which included target-specific primers and probe, was dispensed into all of the microchambers

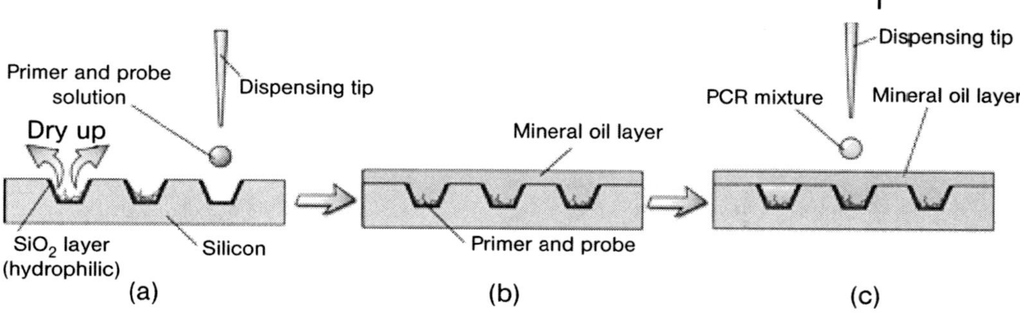

Fig. 13.3. Illustration of preparation steps for on-chip DNA amplification: (a) Different types of primers are dispensed into the microchamber and then dried. (b) Mineral oil is coated on the chip after the primer solution has dried. (c) A PCR mixture (40 nL) with no primers is dispensed in each chamber through the oil layer. The solution sinks to the bottom and then spreads to the microchamber walls.

through the oil layer. After preparing this setup, the chip was placed onto a conventional thermal cycling system to achieve PCR reaction. Thermal cycling was initiated at 94 °C and held for 10 min, followed by 40 cycles of 94 °C for 10 s and 60 °C for 60 s. After the end of PCR amplification, the amplified DNA was observed using a charge-coupled device (CCD) camera (Hamamatsu Photonics, Japan) mounted on a fluorescence microscope (Leica, Heidelberg, Germany). The inner walls of the microchamber were rendered hydrophilic with the formation of an oxidized layer on their surfaces. After coating the microchamber array with mineral oil, the remaining hydrophobic surface of the microchamber prevented the spread of the aqueous solution to the outside of the microchamber. After dispension of the aqueous sample solution, it first formed a droplet, which in time was replaced with the oil in the microchamber, and settled inside the microchamber with the convection of the oil (Fig. 13.3). The thickness of the oil layer had a significant effect on the protection of nanoliter-scale solutions from evaporation. The oil layer was adjusted by controlling the volume of the oil drop. As the thickness of the oil layer increased, the dispensing of the sample solution became more erroneous (data not shown). The optimum thickness of the oil layer for introduction of the sample mixture was chosen to be ~200 μm.

13.2.2 Multiplexed Detection of Different Target DNA on a Single Chip

Target DNA sequence was amplified specifically in a nanoliter-volume microchamber, and the microchambers, in which fluorescence signal was released, were counted in consequence to TaqMan PCR. Cross-contamination between chambers was tested by using alternate dispensings of wells containing template and those without template. A high concentration of the template DNA was introduced as alternate dispensings into the microchambers. Fluorescence signals were obtained only from the chambers, into which template DNA was introduced. No fluorescence was gained from the remaining chambers, into which no template DNA was dispensed. Thus, it was concluded that the selective distribution of the template DNA into the microchambers was achieved in our system.

There were several inhibition factors, such as nonspecific adhesion of biomaterials, variety in the distribution efficiency of the sample dispensing, and errors caused by PCR itself. Surface treatment of the microchamber was also an important factor affecting the efficiency of the PCR reaction. Shoffner et al. have reported several kinds of surface treatment methods that are useful for avoiding the adsorption of biomaterials on the silicon surface [31]. Erill et al. have reported a systematic analysis of material-related inhibition and adsorption phenomena in glass-silicon PCR chips [32]. Their results suggested that the previously reported inhibition of PCR by silicon-related materials was caused mainly by the adsorption of Taq polymerase at the walls of the chip due to increased surface-to-volume ratios; thus, direct chemical action of silicon-related materials on the PCR mixture was negligible. In contrast to Taq polymerase, DNA was not adsorbed in significant amounts. The net effect of polymerase adsorption could be prevented by the addition of a titrated amount of a competing protein, BSA, and the ensuing reactions could be kinetically optimized to yield efficient PCR amplifications. In our system, we combined these advantageous points of previous reports. The surface of the microchamber walls was first modified by an oxidized layer [30, 31] and then coated with BSA [32]. Only a very low fluorescence signal could be observed, when no BSA coating was employed. Thus, it was found necessary to coat the oxidized walls of the microchambers with BSA, in good agreement with the findings of Erill et al. [32]. To quantify DNA concentration, a certain number of microchambers were used as one region for only one concentration.

Figure 13.4 shows the fluorescence image of the chip after DNA amplification of three different target DNA sequences from three different DNA templates. Visual comparison of the positive fluorescence intensity signals with the negative ones greatly simplified the procedure of distinguishing which chamber contained the target DNA. If the target DNA sequence was present in the dispensed sample, a high fluorescence signal was easily obtained as a result of TaqMan PCR. Additionally, the background fluorescence intensity of the β-actin PCR system was much lower than that of the other two probes by using both our chip and the SmartCycler real-time PCR system (Fig. 13.4B). SmartCycler real-time PCR system results were in good agreement with the results of our chip. The difference in the background fluorescence intensity was caused by the bp distance between FAM and TAMRA dyes of the TaqMan probes. In the TaqMan probe for β-actin gene, FAM was only 6 bp away from TAMRA, but FAM and TAMRA were 26 and 31 bp apart in the probes for SRY and RhD genes, respectively. Such a short distance of 6 bp between the dyes caused the rapid quenching of the signal, and thus, the β-actin system could release much lower fluorescence signals after amplification in comparison with the other systems. Although the PCR systems in this experiment had such different background fluorescence intensities, accurate detection of TaqMan amplification for all systems was achieved by using our chip. Since Rh(-) human female genomic DNA did not contain SRY and RhD genes, almost none

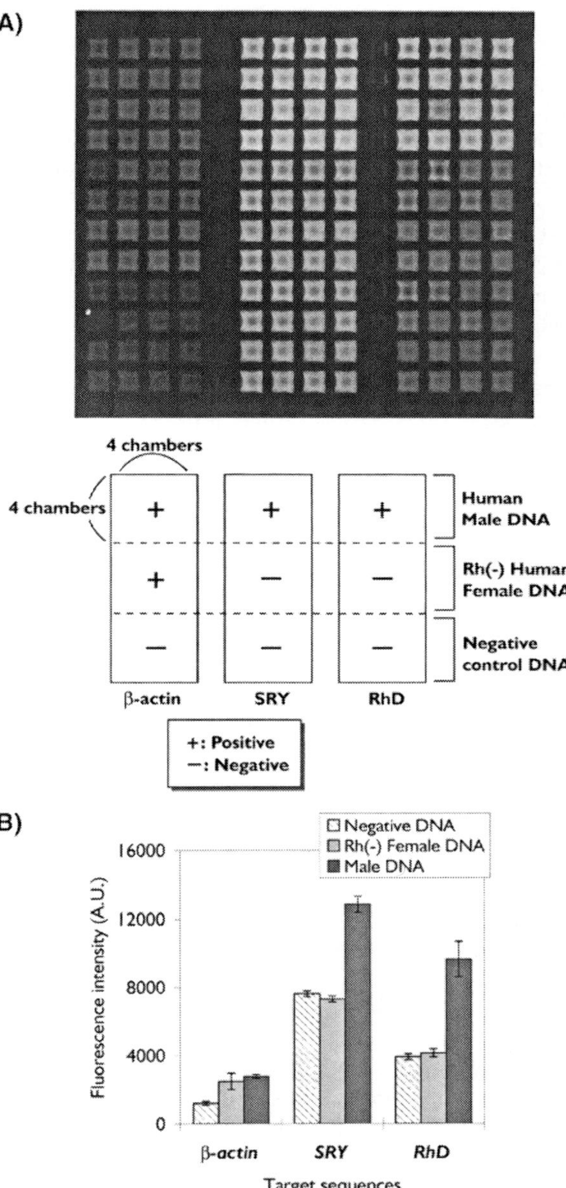

Fig. 13.4. Multiple PCR analysis of β-actin, SRY, and RhD genes using three different kinds of template DNA. (A) Photograph of the microchamber array after multiple TaqMan PCR. (B) Average fluorescence intensity values obtained from 16 microchambers.

of the microchambers showed a fluorescence signal. Both human male and female genomic DNA contained the β-actin gene; thus, a fluorescence signal could be observed successfully in all microchambers of the related area on the chip. The fluorescence intensity of the microchambers was also scanned and evaluated using a DNA microarray scanner and its analysis system. The fluorescence intensity values were obtained from 16 chambers. A remarkable difference between the fluorescence intensities of the positive and negative controls was observed clearly for all three DNA templates. This result indicated that our system can detect different kinds of target DNA sequences from different DNA sources simultaneously. Since TaqMan PCR required the same thermal cycling protocol for the amplification of many kinds of target sequences, it was found to be the most suitable detection technique for microchamber array PCR systems in this report. For example, the detection of genetic diseases such as Down's syndrome [33], 22q11.2 deletion syndrome [34], and β-thalassaemia [35] using TaqMan PCR has already been reported. These clinically important diseases can also be detected simultaneously using our microchamber array-based PCR chip.

13.2.3
On-chip Quantification of Amplified DNA

The initial RhD gene concentration was also quantified by using the microchamber array (Fig. 13.5A). Amplification of the RhD gene was performed by dispensing different concentrations of target DNA into the microchambers. As target DNA was increased from 0 to 12 copies per chamber, the number of the microchambers with positive fluorescence signal also increased. PCR amplification in almost the whole block of the chip was achieved by using eight copies of the target DNA. When 0.4 copies of the target DNA were used, an average of 2 out of 60 chambers ($n = 3$) showed a signal above the threshold level. The average fluorescence intensity value of 1000 AU was determined as the threshold. The high fluorescence released in these two chambers could also be visually detected. Since 0.4 copies of the target DNA were enough to give a readable signal, this concentration was determined as our limit-of-detection. The chambers with positive fluorescence signals, which mean successful PCR amplification, showed easily distinguishable fluorescence intensity (Fig. 13.5A).

Figure 13.5(B) plots the number of chambers with positive fluorescence signal versus input template DNA copy number for the amplification of RhD (B) sequence. As target DNA increased from 0 to 12 copies/chamber, the number of microchambers with a positive fluorescence signal also increased. It was possible to fit the data from 0 to 8 copies/chamber into a straight line with the regression coefficients of 0.9879 for RhD sequence. The system reached a saturation plateau after DNA concentration of 8 copies/chamber, indicating that a trace amount of target DNA was satisfactory for the detection process. PCR amplification in almost the whole block of the chip was achieved by using eight copies of the target DNA. Such a behavior indicated the high detection capacity of our system, so that even a trace amount of DNA copies would be satisfactory for a precise quantification.

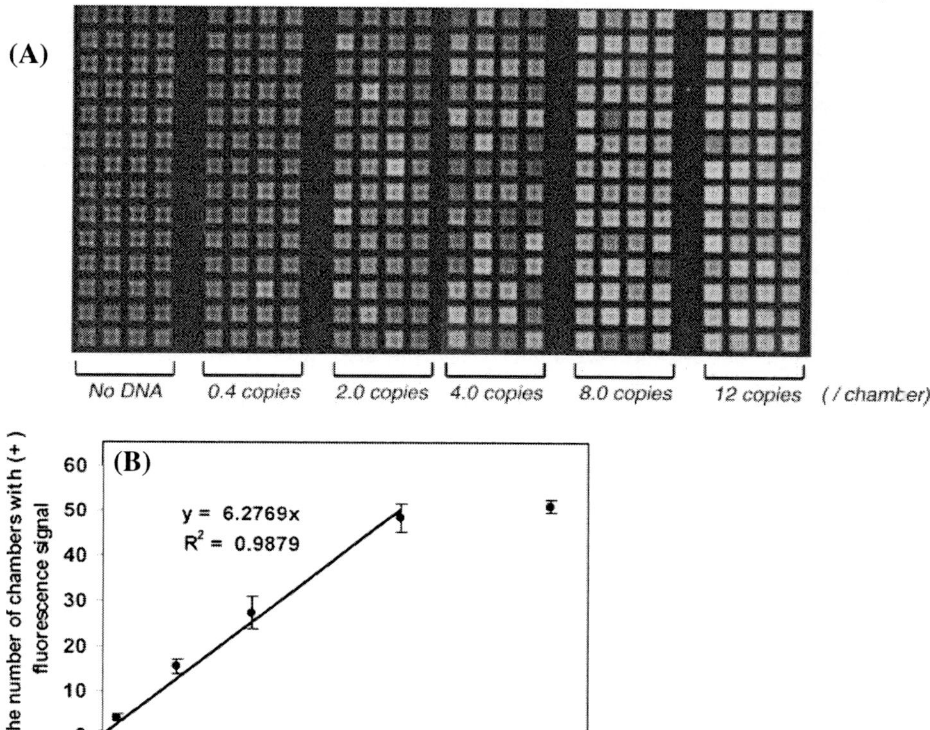

Fig. 13.5. (A) Photograph of fluorescence image for on-chip quantitative analysis of RhD gene. (B) Plot for the average number of chambers within a specific fluorescence signal range versus the number of target DNA copies related to RhD gene after TaqMan PCR.

Copy numbers above eight that would be amplified without any problems but, however, could not be quantified. Even 0.4 copies of the target DNA was enough to give a readable signal, which was determined as our limit of detection. The average fluorescence intensity value of 1000 AU was determined as the threshold. When 0.4 copies were dispensed, only two microchambers reached within the 1000–1200 AU level. One would have expected to get 4 out of 10 microchambers to be positive and 6 to be negative, if we had an average of 0.4 copies/ microchamber, and every copy was intact. The polymerase concentration, annealing temperature, $MgCl_2$ concentration, and the specific primers and TaqMan probes should have been kept under the optimum conditions in 40-nL volumes in the microchamber – a tedious task. Instrumental limitations also added to the inefficiency of PCR, when such a small volume containing such a trace amount of analyte was dispensed on a microchamber array. Thus, the combined negative effects of biochemical (resulting from the TaqMan PCR itself) and instrumental limitations caused the appearance of only two microchambers out of 60 with positive signals after TaqMan PCR on our system. Our method requires only counting the microchambers that show a positive fluorescence signal as a consequence of PCR amplification. No special equipment for the detection of real-time fluorescence intensity is required for determination of DNA copy numbers. Only a simple fluorescence microscope, or a transilluminator used for gel electrophoresis, need be employed for accurate observation of the fluorescence released microchambers with positive signals.

The microchamber array chip presented could be used to amplify multiple DNA targets in combination with a nanoliter dispenser. Theoretically, the chip could be used to amplify and detect ~1200 target DNA simultaneously. The size and total number of the microchambers are determined by the dispensing system. The minimum solution volume that could be dispensed with reliability was optimized as 40 nL for our experiments. If the dispensing instrument could be improved to provide the dispension of a lesser volume of solution, the microchamber size could become smaller, and the chip would become more integrated with a higher number of microchambers. Such further integration of our microarray PCR chips with a miniaturized thermal cycler unit is in progress in our laboratory. The microarray PCR chip reported here has significant potential to be implemented for a wide range of applications. Overall, this system is a promising candidate for mass microfabrication due to its low-cost and high-throughput detection ability.

13.3
On-chip Cell-free Protein Synthesis using A Picoliter Chamber Array

The progress in analyzing the human genome has shifted the focus of research from genes to proteins [36–40]. Although the number of human genes is reported to be 28 000–38 000 [41], the functions of most of them are remain unknown. A rapid and easy method for synthesizing gene products has yet to be developed. Thus, an *in vitro* protein synthesis system has been designed and constructed on the microarray to make a protein library chip. The chip has proteins arranged in an array, and can detect target molecules. Gene cloning and expression is widely used in the preparation of proteins. However, some kinds of proteins often cannot be expressed well in host cells. Our cell-free protein synthesis system could be suitable for expressing such proteins. This protein synthesis system has other advantages as well, such as labeling proteins with isotopes for detection by NMR spectroscopy [42], easy purification of the synthesized protein, and short protein synthesis time.

A highly integrated protein chip is a powerful tool for accelerating post-genomic research. Our aim is to develop protein chips directly from a DNA library using the *in vitro* protein synthesis system. Recently, a cell-free protein synthesis system from *Escherichia coli*, rabbit reticulocytes, and wheat germ has been commercialized [43]. In this research, a rapid translation system from *E. coli* was used for protein expression. Previously, we reported the development of a large-scale integrated pico-liter microchamber arrays for PCR [27], the introduction of a novel nano-liter dispensing system suitable for DNA amplification on microchamber array chip [28], the development of a simultaneous multianalyte immunoassay method for detecting human immunoglobulins based on a protein chip and imaging detection [44], and the development of a new approach for manufacturing encoded microstructures used as versatile building blocks for miniaturized multiplex bioassays [45]. Others have reported the construction of protein chips [46, 47] or cell-free protein synthesis in small chambers [48]. Kukar et al. have detected eight samples simultaneously on one chip [46], and Kojima et al. have constructed an electrochemical immunochip including an assembly of 36 electrodes [47].

There have been reports of high-throughput screening of a mutated anti-human serum albumin single-chain antibody (anti-HSA-scFv) using an *in vitro* protein synthesis system [49, 50]. In these reports, two amino acids were mutated randomly, and over 600 mutations were screened on 96-well plates. Our newly developed chip could also be a powerful tool in similar applications.

High-throughput screening is required for the rapid elucidation of protein functions. Microscale reactions have the advantages of short reaction time and the use of a small amount of samples and reagents. Especially in a high-throughput screening system, numerous samples must be analyzed simultaneously and, if possible, economically. Thus, we have made a highly integrated protein microchamber array chip by using micro-fabrication techniques and polydimethylsiloxane (PDMS) [6]. PDMS micro-molding techniques have been used to fabricate microfluidic systems [51, 52]. Unlike traditional micro-fabrication materials, such as silicon and glass, PDMS can be bonded and manufactured easily and efficiently [53]. In addition, PDMS has some properties that are advantageous for biochemical applications such as high transparency in the 230–700 nm wavelength range, and high permeability to gases.

13.3.1
Cell-free Protein Synthesis Chip Fabrication

Photolithography has been used to fabricate thin microarray sheets using PDMS [53]. A master pattern was formed on a silicon wafer using SU-8 photoresist. The PDMS prepolymer (Sylgard-184: Dow Corning, USA) mixture was poured onto the master and covered with a transparency film (overhead projector sheet). A multilayer stack of aluminum plates, the master pattern, PDMS, a transparency film, a glass wafer, and rubber sheets were clamped tightly and the PDMS prepolymer was baked and crosslinked at 80 °C for 2 h. The resulting thin PDMS sheet, which has over 200 000 microchambers (micro-holes), was put on a slide glass in acetone and treated with oxygen plasma to bind the sheet to the glass. A reactive ion etching (RIE) system was used for the oxygen plasma treatment. The PDMS sheet has

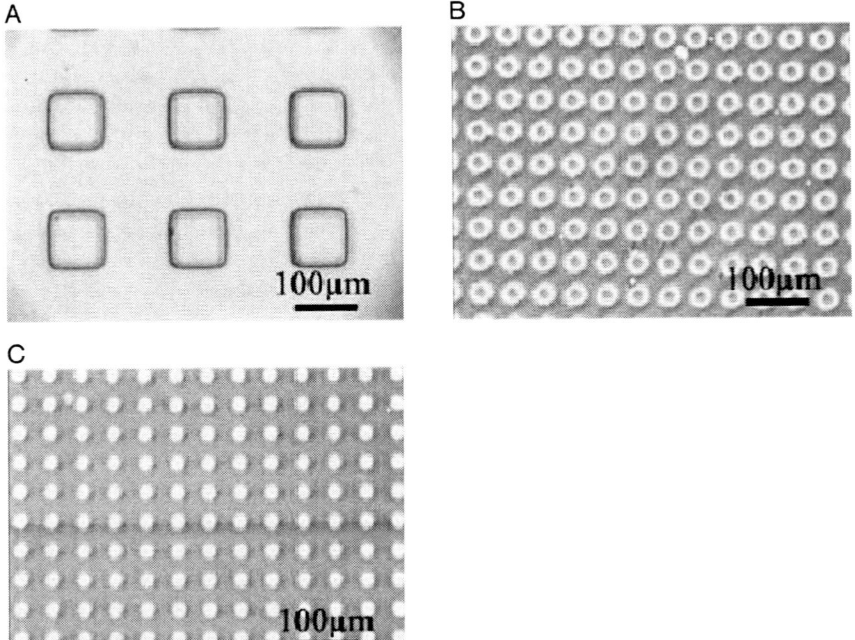

Fig. 13.6. Optical images of PDMS/glass complex chambers. (A) Rectangular chambers about 100 × 100 × 15 μm; the volume is about 150 pL. (B) Cylindrical chambers 20 μm in diameter and 15 μm deep; the volume is about 5 pL. (C) Cylindrical chambers 10 μm in diameter and 15 μm deep; the volume is about 1 pL.

micro-holes, and a hydrophobic surface. Thus, only the bottoms of the microchambers were hydrophilic and the solution easily remained in the chambers. Three different types of chips were designed and fabricated (Fig. 13.6).

The PDMS microchamber was used for *in vitro* protein synthesis. The ribosome source was based on a lysate from *E. coli* (RTS-500 kit: Roche, USA). The wild-type GFP gene contained in the kit was used as a reporter gene, and expressed on the chips. Cell-free protein synthesis reagents were prepared according to the supplier's directions. The reaction solution was composed of a mixture of *E. coli* lysate solution (0.25 mL), of the reconstituted reaction mixture (0.75 mL), of the enzyme mixture (50 μL), and the GFP vector at a final concentration of 10 μg mL^{-1}. First, we dripped the *in vitro* protein synthesis solution on the chip, and removed the surplus. Next, the microchamber chip was covered with a gap cover glass, and sealed to prevent evaporation. A 20 μm gap between the chip surface and the cover glass prevented capillary action among the chambers. The chip was then held at 30 °C, and GFP expression was detected by an optical fluorescence microscope with an FITC filter (excitation: 450–490 nm, emission: 515–565 nm).

13.3.2
Cell-free Protein Synthesis using a Microchamber Array

RIE treatment (oxygen plasma treatment) was used to bind PDMS to a slide glass. Three different types of PDMS chips were designed and fabricated (Fig. 13.6). Since the PDMS and glass construction gives the chamber structure a hydrophobic surface and a hydrophilic bottom substrate, an aqueous solution poured onto the chip enters through the holes of the array, and remains only in these microchambers. This phenomenon prevents the cross-contamination between the microchambers. This chip is also suitable for optical observations because of its transparency over a wide wavelength range. Protein synthesis was carried out on the microchamber chips with the GFP gene used as a reporter gene. Expression of GFP was detected by fluorescence using an optical microscope. Figure 13.7 shows the results of cell-free protein synthesis on the chip. Fluorescence intensity was detected within 1 h of incubation, and remained constant. In a batch system, protein synthesis is said to be inhibited by a lack of substrate or accumulated waste within 2 h [54]. In our system, cell-free protein synthesis stopped within 2 h. This result agrees with the report of Spirin et al. [54]. However, the formation of the GFP fluorescent group is known to take 1–2 h [54]; thus, it may be considered that protein synthesis stopped before the GFP fluorescence became constant. However, a similar shift in GFP fluorescence was shown in chambers with 10 (Fig. 13.8) or 20 μm i.d.

The lowest concentration of DNA template necessary for the detection of the GFP signal was determined to be only 10 molecules of DNA per chamber. A microchamber array chip with cylindrical chambers 10 μm wide and 15 μm deep was used in this experiment. The volume of this chamber is about 1 pL. Thus, the concentration of a solution containing 10 molecules of DNA is about 4×10^{-5}

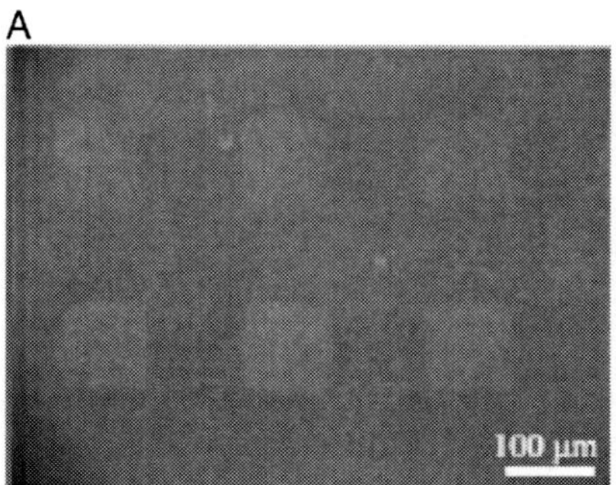

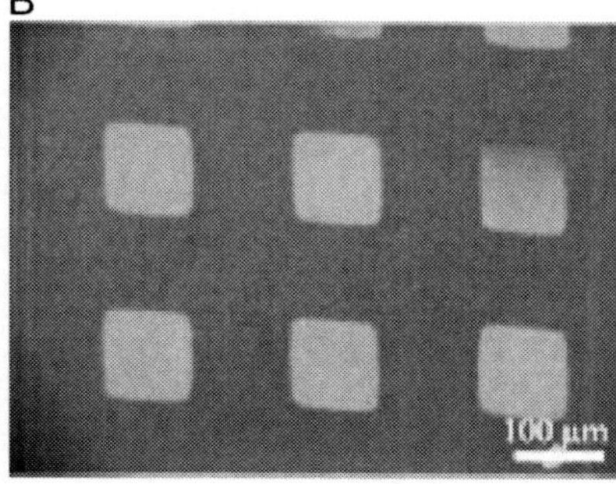

Fig. 13.7. Fluorescent images of microchips showing the expressed GFP protein. Chambers on the chip were filled with the cell-free protein synthesis reagents. Fluorescent image of the chambers after incubation for (A) 0 h and (B) 1 h; GFP expression was detected under an optical microscope with a FITC filter.

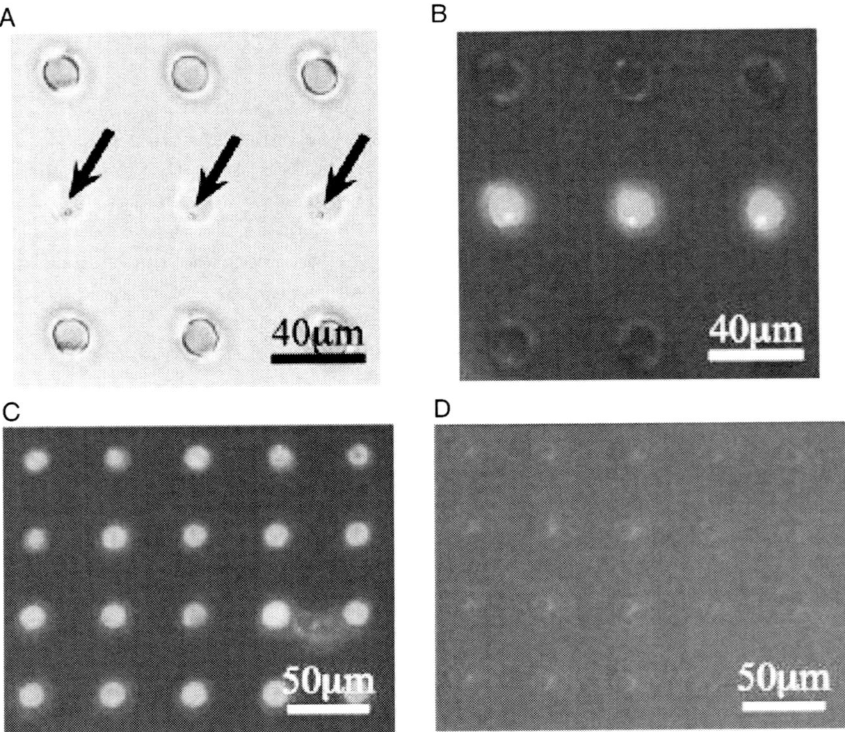

Fig. 13.8. Optical images of GFPuv expression from DNA-immobilized beads. (A) Optical image under white light; arrows indicate the presence of beads. (B) Fluorescence image of the chamber, $\lambda_{ex.} = 400$–440 nm, $\lambda_{em} = 475$ nm. (C) Fluorescence image of a positive control containing cell-free protein synthesis reagents with 10 fg per pL pGGFPH vector. (D) Fluorescence image of a negative control containing only cell-free protein synthesis reagents (no template DNA). Cylindrical chambers shown in the figure are 10 μm in diameter.

mg mL^{-1}. This concentration is about 1/100 to 1/500 compared with the DNA concentration utilized in conventional cell-free protein synthesis protocols. The use of a small volume chamber increases the possibility of contact between DNA and reagents, making it possible to express a protein using a trace amount of DNA.

The distribution of over 10 000 samples using a DNA spotter would take a very long time; therefore, self-layout of samples containing the DNA library was used in this study. DNA-immobilized beads were used as DNA carriers. The amount of DNA immobilized on one bead was about 200 molecules on Dynabeads M-270 Carboxylic Acid, and 10 000 molecules on Dynabeads M-280 Streptavidin (Dynal, USA). A microchamber array chip with a 10 μm i.d. chamber was used for bead arrangement. About 60% of the chambers contained one bead; however, some chambers had multiple beads because the diameter of a bead, which is 2.8 μm, is much smaller than that of the chamber. The design and fabrication of a new chip with smaller chambers to allow the entry of only one bead into each chamber are currently underway in our laboratory.

With the aid of lipid, it was possible to disperse the beads into the chambers. Rhodamine-modified lipid was used instead of phosphatidylcholine, and the chip covered with lipid solution was observed both in air and in water. Interestingly, the lipid moved into the chamber when the chip was soaked in water. A chip with beads containing the GFP gene was used for *in vitro* protein synthesis. As shown in Fig. 13.8, fluorescence was observed after 1 h of incubation only in chambers that contained DNA beads. As a positive control, pGGFPH vector solution was added to the *in vitro* protein synthesis reagents at 10 fg pL^{-1}; no template DNA was added to the *in vitro* protein synthesis reagents as a negative control. The results of these control experiments are shown in Fig. 13.8. A comparison between DNA immobilized beads and DNA in solution suggests that the amount of protein per DNA molecule in solution is greater than that on DNA-immobilized beads. Nevertheless, easy and fast manipulation of DNA immobilized beads prompted us to use this method. In the experiments shown in Fig. 13.8, DNA immobilized beads with biotin–streptavidin conjugate were used. Similar results were observed when primer-immobilized beads were used. The concentration of GFP solution was about 10 μg mL^{-1}, estimated from the intensity of the fluorescence. The amount of GFP protein per chamber was about 10 fg. The GFP solution did not diffuse from the chambers, indicating that the solution in each chamber is physically separated from that in other chambers.

Tabuchi et al. have made a microchamber chip with a chamber volume per chip of 10 μL [48]. In our case, the chamber number is much larger and the chamber volume is much smaller, which is advantageous for high-throughput applications. However, the analysis of over 10^4 chambers takes a long time. Therefore, a scanner-type analyzer for our system is being developed to enable automatic screening in the near future.

A new method for making a highly integrated protein chip from a DNA library using *in vitro* protein synthesis on a microchamber array has been demonstrated. The chambers are of three types, based on their volume capacity: 1, 5, and 150 pL, and the total number of chambers per chip is 10 000 (150 pL), and 250 000 (both 1 and 5 pL). The array has a hydrophobic surface of PDMS and a hydrophilic glass bottom. These structural properties provide the advantage of preventing cross-contamination among the chambers. *In vitro* protein synthesis using these chambers was achieved. The fluorescence of GFP expressed on the micro chamber was rapidly detected. GFP expression was also achieved using immobilized DNA molecules on polymer beads, which allows easy handling of the DNA molecules. Brenner et al. have described a method for cloning nucleic acid molecules onto the surfaces of 5 μm microbeads rather than in biological hosts [55]. A unique tag sequence was attached to each cDNA molecule, and the tagged library was amplified. A unique tag was also attached to each bead, and the tagged library was conjugated with the tagged beads.

This method allows the immobilization of one kind of DNA on a bead. Because such clones are segregated on microbeads, they can be manipulated simultaneously and then assayed separately. If this method can be applied to the chip described in this report, it will be possible to analyze easily a whole DNA library on a chip in a short time. In future, this system will be used for the exhaustive expression of proteins included in target cells, the functional analysis of proteins expressed from unknown genes, and the screening of artificially mutated proteins.

In this study, we have reported a highly integrated protein microarray chip using *in vitro* protein synthesis from DNA conjugated microbeads. The protein microarray system made it possible to perform high-throughput screening and analysis for multiplexed gene expression on single beads in each pico-litter chamber. In the future, this system will be applied for the expression analysis of proteins in multiplexed single-cells and the functional analysis of proteins expressed from unknown genes or cells.

13.4
High-throughput Single-cell Analysis System using Pico-liter Microarray

Cell-based assays are one of the newest tools being used to broaden and strengthen drug discovery for the identification of new therapeutic agents. They are fast becoming the assay format of choice, especially in target validation, lead identification and optimization. Cell-based assay screening is by nature more complex and automation is more difficult than biochemical screening. These assays hold the promise of increasing productivity in the discovery process and the ability to screen out compound failure earlier in the development process. The last few years have seen an increasing number of cell-based assays being used, driving the market to an anticipated $500 million mark by 2005 [56, 57]. Cell-based assays play a very important role in the post-genomic era focusing on high-throughput functional genomics and drug discovery. High-throughput screening assays play a pivotal role in the search of novel drugs and potential therapeutics. Over the last decade, various scientific advances include the growing number of potential therapeutic targets emerging from the field of functional genomics and the rapid development of large compound libraries derived from parallel and combinatorial chemical synthetic techniques, driven by the need for improved drug discovery screening technology [58, 59]. High-throughput cell-based assays reduce the total cost in screening the specific therapeutic target for a specific disease remarkably reducing the time. High-throughput drug screening methods employed so far involve use of 96- and 384-well microtiter plates and require at least a minimum of 100-μL-assay mixture, suggesting further miniaturized assay formats to reduce the total cost of drug screening. Combining miniaturized technology with developments in automation, sensitive signal-detection, plate formats, automated compound-delivery and data management results in highly efficient, and cost-effective, integrated miniaturized ultrahigh-throughput screening (uHTS) systems. The need to screen numerous compounds rapidly, in increasingly automated dispensing systems, and with very small reaction volumes prompted us to carry out this study. It was of interest to develop a high-throughput cell-based assay using miniaturized microarray chip formats.

For example, each B-cell clone expresses antigen-receptors, antibodies, with a unique antigen-specificity: an antigen-specific monoclonal antibody derived from a single B-cell clone finds applications in antibody medicine and clinical diagnosis. Though each B-cell has 10^7 to 10^8 varieties of monoclonal antibody on these surfaces only small percentage of B-cells respond and produce a specific monoclonal antibody. Only one or two cells in a total of 10 000 B-cells become active and produce antigen-specific antibodies after stimulation with a hepatitis B virus (HBV) surface antigen (HbsAg) [60]. It would be impossible to make a HBsAg specific antibody taking this ratio of positive B-cells into account using recently available technologies. A flow cytometer allows us to monitor individual cells that flow through sheeth fluid, but the signals of the cells become background noise, which consists of 0.1 to sometimes 1% of total cells, thus it is quite difficult to monitor a signal of a minor population of cells whose signals are buried in the noise by a flow cytometer. Further, we cannot compare the states of each cell before and after stimulation by using flow cytometer. In contrast, a fluorescence microscope allows us to observe the states of cells both before and after stimulation. However, it is difficult to observe signals of large number of cells under a microscope. Accordingly, it is difficult to monitor Ca^{2+} mobilization of a minor population of cells.

Therefore, it is necessary to construct a microarray platform that can confine a large number of single-cells and detect antigen-specific single B-cells before and after stimulation with an antigen from a bulk cell suspension. For high-throughput single-cell separation and analysis, Thorsen et al. have reported high-density microfluidic chips that contain plumbing networks with thousands of micromechanical valves and hundreds of individually addressable chambers and showed the separation of single *E. coli* cells in each chamber [61]. To achieve single-cell separation, they diluted cells to create a median distribution of 0.2 cells per compartment, so that reliable capturing of cells in each chamber is difficult. In another recent report, Anderson et al. have tested biomaterial microarrays for their effects on human embryonic stem cell growth and differentiation using populations of human embryonic stem cells [8]. However, single-cell based assay seemed to be impossible using this microarray format. Also we have reported recently a microchamber array and microfluidic chip for measuring high-throughput analysis of cellular fluorescence [62–64]. Here, we discuss an improved microchamber array to monitor Ca^{2+} mobilization of over 25 000 cells simultaneously at a single-cell level [13]. We have also developed a novel high-throughput screening and analysis system for antigen-specific single B-cells using microarray, which was carried out by detecting antigen-specific single B-cells against an antigen of interest and their retrieval by a micromanipulator for antibody DNA analysis [13]. The single-cell microarray system developed in this study does not need to use myeloma, as in the case of conventional hybridoma technique, and can screen the antigen-specific single B-cells directly from cell suspension and analyze antigen-specific antibody DNA at a single-cell level. This system is simple and easy in its operation, and quick enough for making monoclonal antibodies when compared with conventional techniques. Moreover it can perform high-throughput single-cell analysis using chip devices.

13.4.1 Single-cell Microarray Chip Fabrication

The microarray chip is made from polystyrene with over 200 000 microchambers (10 μm width, 12 μm depth, and 30 μm pitch) by using the Lithographie Galvanoformung Abformung (LIGA) process and was performed by the Starlight Co. Ltd., Japan (Fig. 13.9A–D). Using X-ray lithography from synchrotron radiation, a poly(methyl methacrylate) (PMMA) as a resist was exposed and patterned with Au metal mask. After development of the PMMA substrate, the resulting PMMA mold was used for nickel mold construction by electroforming. Finally, a polystyrene microarray chip was fabricated from the nickel mold by injection molding (Fig. 13.9A). Each microarray chip is consisted of 225 (15 × 15) clusters, and each cluster consisted of 900 (30 × 30) microchambers. Each microchamber is cylindrical and can accommodate only a single-cell.

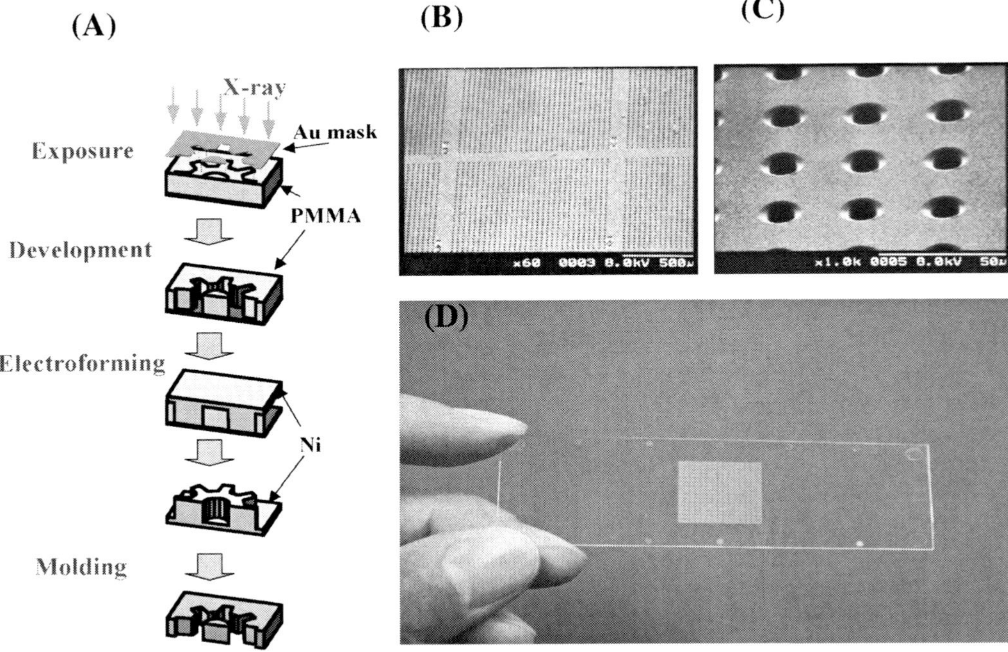

Fig. 13.9. Construction of single-cell microarray chip. (A) LIGA process for the fabrication of single-cell microarray chip. (B and C) SEM images and (D) a real picture of the microarray chip device. The microarray chip is made from polystyrene with over 200 000 microchambers (10 μm wide, 12 μm deep, 30 μm pitch).

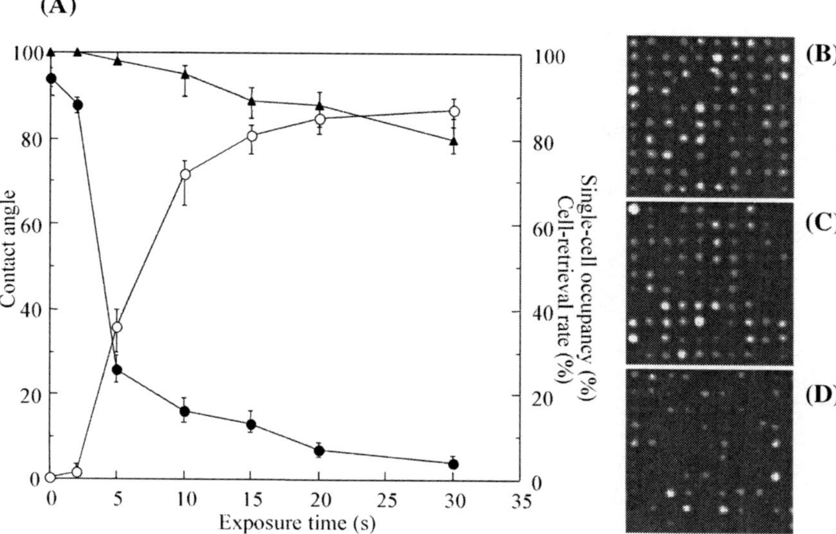

Fig. 13.10. Microarray chip surface treatment studies for single-cell confinement and retrieval. (A) Optimization of RIE exposure time of chip surface for single-cell confinement and retrieval from microchambers. Contact angle (●); single-cell occupancy (○); and cell-retrieval rate (▲ closed triangle). Scanned microarray image of single-cells in 100 microchambers after (B) 20, (C) 10 and (D) 5 s RIE exposure. Each treatment showed single-cell occupancy of 83, 73 and 38%, respectively.

The microarray surface was rendered hydrophilic by reactive ion etching (RIE) treatment to make it convenient for cell studies. RIE exposure time controlled the cell adhesion on the chip surface. An increase in RIE exposure time increases the hydrophilicity of the microarray chip surface, which is inversely proportional to the decrease in the contact angle of chip surface (Fig. 13.10A). This condition is suitable for increased levels of single-cell occupancy in the microchambers because RIE treatment keeps the confined single-cell in the microchamber during the washing process, which is shown in Fig. 13.10(B–D). The polystyrene microarray chip maintained hydrophilicity character without alteration for at least one week. However, with increasing hydrophilicity, it becomes difficult to retrieve positive single-cells from microchambers, e.g., over 30-s RIE restricted the retrieval rate to <80% (Fig. 13.10A). This suggests optimization in the exposure time that suits single B-cell occupancy as well as for cell retrieval – determined to be 20 s exposure, which resulted in over 80% of single-cell occupancy and cell-retrieval rate (Fig. 13.10B). To achieve over 80% single-cell occupancy, the cell density should always be 1×10^6 cells mL^{-1} or more, otherwise a decrease in the percentage of single-cell occupancy was observed. We succeeded in single-cell separation and retrieval on the plastic microarray chip by controlling the chip surface treatment. In contrast, we could not retrieve B-lymphocytes that adhered strongly to the glass and silicon chip substrates, because these materials have highly hydrophilic surfaces. Further, the transparent nature of a plastic microarray chip, along with its flat bottom-surfaced cylindrical microchambers, made by the LIGA process, is convenient for visualization of single B-cells as well as retrieval of cells from microchambers by a micromanipulator under a light microscope. Thus, the single-cell microarray chip developed is study was suitable for separation and retrieval of single cells to analyze the antigen-specific single B-cells.

13.4.2
Pico-liter Microarray for Single-cell Studies

A single-cell microarray chip that was constructed by using LIGA process has over 200 000 flat-bottomed cylindrical microchambers (10 μm diameter, 12 μm depth) in 1.4 cm^2 area on a plastic wafer (2×8 cm^2) (Fig. 13.9A–D). Characterization of such a microarray chip for single-cell studies was performed using mouse splenic B-lymphocytes. The microchamber design was made such that each chamber allowed entry of single B-cell (approximately 8 μm diameter) and hence there is no possibility of two cells in a single microchamber. To achieve single B-cell confinement in each microchamber as well as cell-retreival from microchambers, hydrophilicity of the microarray chip surface was controlled by adopting different timings of RIE exposure (Fig. 13.10). As the result of chip surfice treatment using RIE, lymphocytes derived from mouse spleen or human blood were spread on the microarray, and over 80% of the microchambers achieved single-cell status. In addition, this novel microarray system demonstrated easy retrieval of positive single B-cells from microchambers by a micromanipulator under microscope.

In this study, we report the data from approximately 30 000 microchambers for detecting antigen-specific single B-cells, because the assay we have adopted, i.e., measurement of increase in the intracellular calcium after antigenic stimulus using Fluo-4, lasts for few minutes. During this short period a microarray scanner could scan the analyzable area of 30 000–40 000 microchambers. This number of microchambers is sufficient to screen and identify below as little as 0.1% of antigen-specific single-cells in a total B-cell population, which is the limitation of flow cytometer, as described above. The ultimate potential of this single-cell microarray can be realized by improving the assay system and the scanning speed of microarray scanner for detecting the total number of microchambers on the microarray chip.

Recently, however, researchers have reported single-cell based cultivation on microchips with the help of optical tweezers [65] and specially designed micropipette [66]. By using these systems, it would be impossible to perform high-throughput single-cell screening and analysis. Other chip systems, microfluidic devices, have been reported to perform single-cell separation and analysis from bulk cell suspension in microchannels under the influence of integrated valves and pumps [67–70], which make these systems complex in handling compared with microarray formats. Despite their complexity these devices could not guarantee capturing a target cell, moreover they were not compatible for high-throughput analysis. The single-cell microarray developed in the present study could easily separate single-cells from a bulk cell suspension without using specialized tools.

13.4.3
Single-cell Microarray System for Analysis of Antigen-specific Single B-cells

To evaluate the utility of a novel single-cell microarray system for detecting intracellular calcium, the $(Ca^{2+})_i$ level of individual cells, we have employed anti-mouse IgM antibody as a stimulant for mouse splenic B-cells, which delivers the signals through B-cell antigen-receptors and induces a transient increase in the $(Ca^{2+})_i$ levels. Mouse lymphocytes loaded with Ca^{2+}-indicator, Fluo-4, were applied on the chip and their fluorescence was detected with a microarray scanner before and after stimulation with anti-mouse IgM antibody (Fig. 13.11). Fluo-4 is a standard fluorescent calcium indicator [71], which enters into the cells and generates fluorescence after binding to $(Ca^{2+})_i$. It is known that the concentration of $(Ca^{2+})_i$ increases after B-cells respond to antigenic stimulation [72]. In this experiment, major single B-cells on the microarray showed increases in the $(Ca^{2+})_i$ after stimulation (Fig. 13.11A). The fluorescence intensity reached to its maximum level after 1 min stimulation, and maintained a high magnitude for 2 min and then decreased gradually (data not shown). Thus, we used a 2 min stable duration for analysis. Scatter plot analysis of indvidual cell's fluorescence-intensity (Fig. 13.11B) enabled us to discriminate B-cells that were activated with anti-mouse IgM antibody. The data from over 30 000 microchambers in the microarray chip revealed 68% (18 130 cells) of total splenic lymphocytes (26 650 cells) showed more than twice the fluorescence after stimulation with anti-mouse IgM (Fig. 13.11B). While cells with over a 5× increase in fluorescence (3883 cells) existed in 14% of the total splenic lymphocytes [Fig. 13.11B(I)]. Whereas most lymphocytes incubated with control antibody (anti-human IgM) showed a less than 5× increase in fluorescence intensities [Fig. 13.11B(II)].

Each B-cell expresses membrane-bound antibodies with unique antigen-specificity as antigen receptors. When B-cells were stimulated with an antigen instead of anti-IgM antibody, only a minor population of total B-cells is stimulated and their $(Ca^{2+})_i$ increases. Previous studies reported diverse frequencies in the number of antigen-specific B-cells, ranging from 1 in 10^2 for rabies virus [73] to 7 in 10^5 for myelin basic protein [74]. However, it is quite difficult to screen and identify a minor cell population of antigen-specific single B-cells ($<0.1\%$) with a flow cytometer because fluorescent signals of minor populations of cells are buried in the noise of non-specific cells. Whereas the single-cell microarray platform developed in this study could successfully screen and detect a low frequency of antigen-specific single B-cells using a single chip in one run. In addition, it also analyzes the same single-cells before and after stimulations with antigen,which a flow cytometer could not.

If antibody cDNA is recovered from an antigen-specific single B-cell that is detected with single-cell microarray, antigen-specific monoclonal antibodies can be

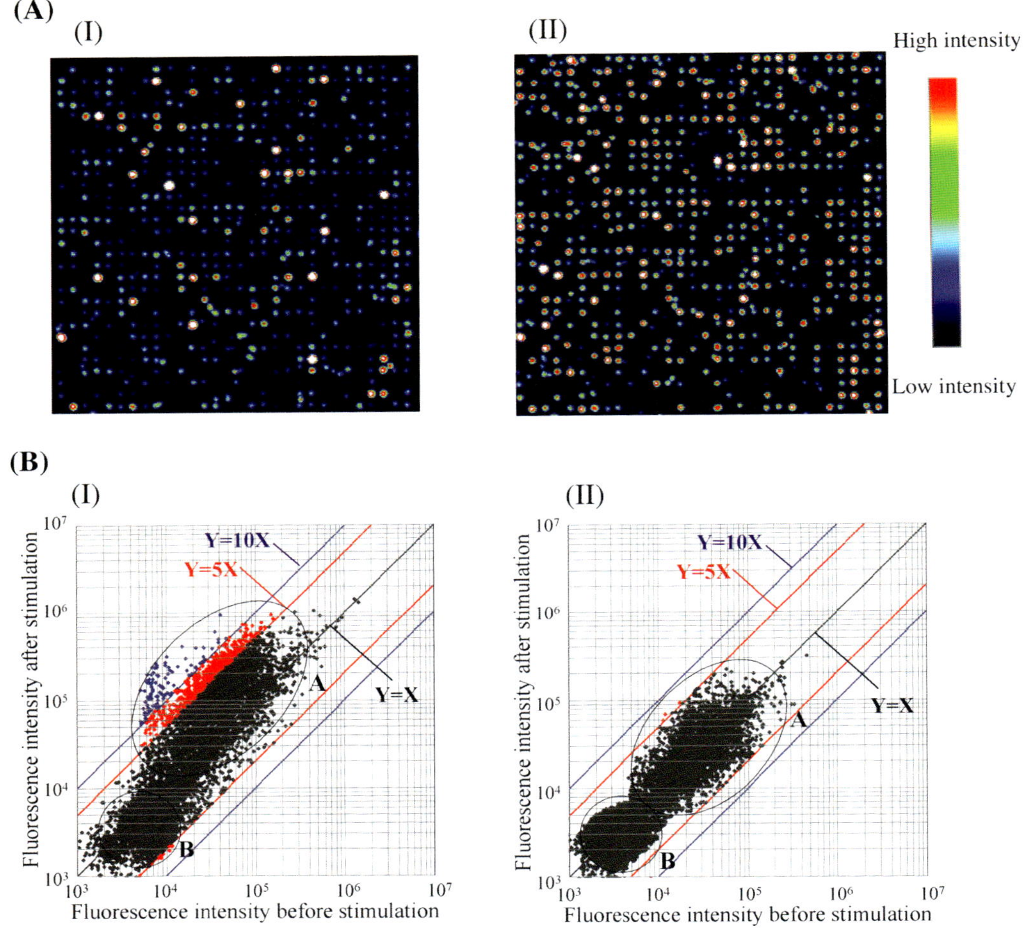

Fig. 13.11. Detection of activated single B-cells using Fluo-4 upon stimulation with anti-mouse IgM antibody. (A) Scanned images of single-cells microarray in a single cluster area of 900 (30 × 30) microchambers on a microarray (I) before and (II) after stimulation. Color scale represents the intensity of fluorescence emission. (B) Scattered plot analysis of single B-cell response in 32 400 microchambers area after stimulation with (I) anti-mouse IgM antibody and (II) negative control antibody. Circle A, microchambers with lymphocytes; circle B, empty microchambers; X, fluorescence intensity before stimulation; Y, fluorescence intensity after stimulation; blue line ($Y = 10X$), 10× higher fluorescence intensity; red line ($Y = 5X$), 5× higher fluorescence intensity.

developed. To this end, we have successfully performed retrieval of a single B-cell from the microarray by a micromanipulator system under a microscope (Fig. 13.12) and subsequent single-cell RT-PCR amplification to determine its antibody cDNA. Monoclonal antibodies are usually produced using hybridoma techniques

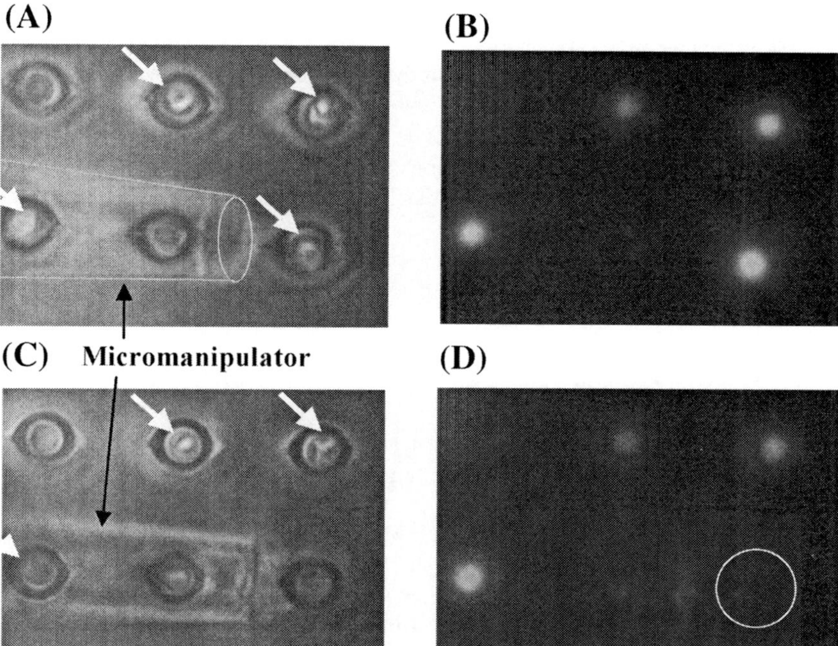

Fig. 13.12. Retrieval and antibody cDNA analysis of single B-cells derived from human blood. (A and B) Light and fluorescence microscopic images of the single B-cells on a microarray chip before cell-retrieval using a micromanipulator system. B-cells were treated with CellTracker Orange ($\lambda_{ex.} = 541$ nm, $\lambda_{em.} = 565$ nm) for observation of fluorescence from living cells. The arrows indicate a single-cell in each microchamber, observed under the 40× lens of an inverted microscope. (C and D) Light and fluorescence microscopic images after target single-cell retrieval using a micromanipulator. The circle in (D) indicates the disappearance of fluorescence of target a single-cell that was retrieved from the microchambers.

developed by Milstein and Koeller [75]. Using conventional hybridoma methods, one hybridoma is routinely produced from 10^5 splenocytes [57] and, therefore, not all B-lymphocytes are screened for antigen-specificity. However, the single-cell microarray system we have developed could screen directly all the B-cell population to detect antigen-specific single B-cells and analyze the antigen-specific antibody DNA at a single-cell level. Our chip system is simple and easy in its operation, and can perform high-throughput single-cells analysis. Moreover, it provides a superior screening system for antigen-specific monoclonal antibodies that provide the source for antibody medicines.

The single-cell microarray system we have developed made it possible to perform high-throughput screening and analysis of intracellular Ca^{2+}-response at a single-cell level and analyze multiplexed single-cell status before and after antigenic stimulation on a single platform, which flow cytometory could not. Further, this system facilitated easy retrieval of antigen-specific single B-cell from the microarray, demonstrating its novelty and superiority over conventional methods in making monoclonal antibody. This explorative study throws more light on a novel high-throughput single-cell analysis system that has tremendous potential to analyze antibody DNA of an antigen-specific single B-cell for developing highly specific monoclonal antibody as antibody medicines. It might also be applicable for detection of antigen-specific T cells, which could lead to immune therapy or gene therapy in the future.

13.5 Conclusions

Pico- and nano-liter chamber array chips have been developed that show great potential to detect and analyze DNA, protein and cell at single-molecule or single-cell level.

The microchamber (40 nL) array PCR chip presented in this chapter was used to amplify and detect quantitatively multiple DNA targets in combination with a nanoliter dispenser. This demonstrated that a microarray PCR chip can be used to amplify and detect ~1200 target DNAs simultaneously; in addition, target gene analysis at single copy level was achieved. Thus, the microarray PCR chip system has a significant potential to be implemented for a wide range of purposes such as detection of pathogenic microorganisms and clinical diagnosis.

The highly integrated protein microarray chip presented here successfully performed *in vitro* protein synthesis using a pico-liter chamber array. Furthermore, it made it possible to perform screening and analysis for multiplexed gene expression on single beads in each chamber. In future, this system will be applied for high-throughput screening and functional analysis of proteins expressed from unknown genes or cells.

The single-cell microarray system, which can accommodate single-cells in pico-liter chambers, made it possible to perform high-throughput screening and analysis of antigen-specific single B-cells and analyze multiplexed single-cell status before and after antigenic stimulation on a single platform. Further, this system facilitated easy retrieval of positive antigen-specific single B-cell from the microarray, demonstrating its novelty over conventional methods in making monoclonal antibody. The single-cell analysis system has tremendous potential to analyze antibody DNA of an antigen-specific single B-cell for developing highly specific monoclonal antibody as antibody medicines. It might also be applicable for cell therapy as personal diagnosis and development of personalized medicine in future.

Thus, the pico- and nano-liter chamber array systems discussed here could become a potential tool for high-throughput analysis of DNA, protein and cell at single-molecule or single-cell level, which could lead to clinical diagnosis and drug discovery.

Acknowledgments

Research into *in vitro* protein synthesis on a pico-liter chamber array was supported by the grants of New Energy and Industrial Technology Development Organization (NEDO) and the Japan Society for the Promotion of Science (JSPS). The single-cell microarray chip research was carried out under the Toyama Medical Biocluster project, which was sponsored by the Ministry of Education, Culture, Sports, and Science, Japan.

References

1 Oosterbroek, E., van den Berg, A., Lab-on-a-chip: Miniaturized systems for biochemical analysis and synthesis, Elsevier, Amsterdam, **2003**.

2 Figeys, D., Pinto, D., Lab-on-a-chip: A revolution in biological and medical science, *Anal. Chem.*, **2000**, 72, 330–335.

3 Heller, M.J., DNA microarray technology: Devices, systems, and applications. *Annu. Rev. Biomed. Eng.*, **2002**, 4, 129–153.

4 Eisen, M.B., Brown, P.O., DNA arrays for analysis of gene expression, *Methods Enzymol.*, **1999**, 303, 179–205.

5 Panicker, R.C., Huang, X., Yao, S.Q., Recent advances in peptide-based microarray technologies, *Comb. Chem. High. Throughput. Screen.*, **2004**, 7, 547–556.

6 Figeys, D., Adapting arrays and lab-on-a-chip technology for proteomics, *Proteomics*, **2002**, 2, 373–382.

7 Cottingham, K., The single-cell scene, *Anal. Chem.*, **2004**, 76, 235A–238A.

8 Anderson, D.G., Levenberg, S., Langer, R., Nanoliter-scale synthesis of arrayed biomaterials and application to human embryonic stem cells, *Nat. Biotechnol.*, **2004**, 22, 863–866.

9 Ziauddin, J., Sabatini, D.M., Microarrays of cells expressing defined cDNAs, *Nature*, **2001**, 411, 107–110.

10 Kinpara, T., Mizuno, R., Murakami, Y., Kobayashi, M., Yamamura, S., Hasan, Q., Morita, Y., Nakano, H., Yamane, T., Tamiya, E., A picoliter chamber array for cell-free protein synthesis, *J. Biochem. (Tokyo)*, **2004**, 136, 149–154.

11 Matsubara, Y., Kerman, K., Kobayashi, M., Yamamura, S., Morita, Y., Takamura, Y., Tamiya, E., On-chip nanoliter-volume multiplex TaqMan polymerase chain reaction from a single copy based on counting fluorescence released microchambers, *Anal. Chem.*, **2004**, 76, 6434–6439.

12 Matsubara, Y., Kerman, K., Kobayashi, M., Yamamura, S., Morita, Y., Tamiya, E., Microchamber array based DNA quantification and specific sequence detection from a single copy via PCR in nanoliter volumes, *Biosens. Bioelectron.*, **2005**, 20, 1482–1490.

13 Yamamura, S., Kishi, H., Tokimitsu, Y., Kondo, S., Honda, R., Ramachandra, R.S., Omori, M., Tamiya, E., Muraguchi, A., Single-cell microarray for analysing cellular response, *Anal. Chem.*, **2005**, 77, 8050–8056.

14 Saiki, R.K., Scharf, S., Faloona, F., Mullis, K.B., Horn, G.T., Erlich, H.A., Arnheim, N., Enzymatic amplification of beta-globin genomic sequences and restriction site analysis for diagnosis of sickle cell anemia, *Science*, **1985**, 230, 1350–1354.

15 Mullis, K.B., Faloona, F.A., Specific synthesis of DNA *in vitro* via a polymerase-catalyzed chain reaction, *Methods Enzymol.*, **1987**, 155, 335–350.

16 Irino, T., Kitoh, T., Koami, K., Kashima, T., Mukai, K., Takeuchi, E., Hongo, T., Nakahata, T., Schuster, S.M., Osaka, M., Establishment of real-time polymerase chain reaction method for quantitative analysis of asparagine synthetase expression, *J. Mol. Diagn.*, **2004**, 6, 217–224.
17 Hirt, C., Schuler, F., Dolken, L., Schmidt, C.A., Dolken, G., Low prevalence of circulating t(11,14)(q13,q32) positive cells in the peripheral blood of healthy individuals as detected by real-time quantitative PCR, *Blood*, **2004**, 104, 904–905.
18 Nurmi, J., Wikman, T., Karp, M., Lovgren, T., High-performance real-time quantitative RT-PCR using lanthanide probes and a dual-temperature hybridization assay, *Anal. Chem.*, **2002**, 74, 3525–3532.
19 Yang, J., Liu, Y., Rauch, C.B., Stevens, R.L., Liu, R.H., Lenigk, R., Grodzinski, P., High sensitivity PCR assay in plastic micro reactors, *Lab Chip*, **2002**, 2, 179–187.
20 Lee, T.M.H., Carles, M.C., Hsing, I.M., Microfabricated PCR-electrochemical device for simultaneous DNA amplification and detection, *Lab Chip*, **2003**, 3, 100–105.
21 Schneegass, I., Brautigam, R., Kohler, M., Miniaturized flow-through PCR with different template types in a silicon chip thermocycler, *Lab Chip*, **2001**, 1, 42–49.
22 Fukuba, T., Yamamoto, T., Naganuma, T., Fujii, T., Micro-fabricated flow-through device for DNA amplification – towards in situ gene analysis, *Chem. Eng. J.*, **2004**, 101, 151–156.
23 Lee, D.S., Park, S.H., Yang, H., Chung, K.H., Yoon, T.H., Kim, S.J., Kim, K., Kim, Y.T., Bulk-micromachined submicroliter-volume PCR chip with very rapid thermal response and low power consumption, *Lab Chip*, **2004**, 4, 401–407.
24 Lagally, E.T., Emrich, C.A., Mathies, R., Fully integrated PCR-capillary electrophoresis microsystem for DNA analysis, *Lab Chip*, **2001**, 1, 102–107.
25 Lagally, E.T., Medintz, I., Mathies, R.A., Single-molecule DNA amplification and analysis in an integrated microfluidic device, *Anal. Chem.*, **2001**, 73, 565–570.
26 Nagai, H., Murakami, Y., Yokoyama, K., Tamiya, E., High-throughput PCR in silicon based microchamber array, *Biosens. Bioelectron.*, **2001**, 16, 1015–1019.
27 Nagai, H., Murakami, Y., Morita, Y., Yokoyama, K., Tamiya, E., Development of a microchamber array for picoliter PCR, *Anal. Chem.*, **2001**, 73, 1043–1047.
28 Matsubara, Y., Kobayashi, M., Morita, Y., Tamiya, E., Application of a microchamber array for DNA amplification using a novel dispensing method, *Arch. Histol. Cytol.*, **2002**, 65, 481–488.
29 Leamon, J.H., Lee, W.L., Tartaro, K.R., Lanza, J.R., Sarkis, G.J., deWinter, A.D., Berka, J., Lohman, K.L., A massively parallel PicoTiter-PlateTM based platform for discrete picoliter-scale polymerase chain reactions, *Electrophoresis*, **2003**, 24, 3769–3777.
30 Felbel, J., Bieber, I., Pipper, J., Kohler, J.M., Investigations on the compatibility of chemically oxidized silicon (SiO*x*)-surfaces for applications towards chip-based polymerase chain reaction, *Chem. Eng. J.*, **2004**, 101, 333–338.
31 Shoffner, M.A., Cheng, J., Hvichia, G.E., Kricka, L.J., Wilding, P., Chip PCR. I. Surface passivation of microfabricated silicon-glass chips for PCR, *Nucleic Acids Res.*, **1996**, 24, 375–379.
32 Erill, I., Campoy, S., Erill, N., Barbe, J., Aguilo, J., Biochemical analysis and optimization of inhibition and adsorption phenomena in glass–silicon PCR-chips, *Sens. Actuators B*, **2003**, 96, 685–692.
33 Zimmermann, B., Holzgreve, W., Wenzel, F., Hahn, S., Novel real-time quantitative PCR test for Trisomy 21, *Clin. Chem.*, **2002**, 48, 362–363.
34 Kariyazono, H., Ohno, T., Ihara, K., Igarashi, H., Joh-o, K., Ishikawa, S.,

HARA, T., Rapid detection of the 22q11.2 deletion with quantitative real-time PCR, *Mol. Cell. Probes*, **2001**, 15, 71–73.

35 CHIU, R.W., LAU, T.K., LEUNG, T.N., CHOW, K.C., CHUI, D.H., LO, Y.M., Prenatal exclusion of β thalassaemia major by examination of maternal plasma, *Lancet*, **2002**, 360, 998–1000.

36 RUDERT, F., Genomics and proteomics tools for the clinic, *Curr. Opin. Mol. Ther.*, **2000**, 2, 633–642.

37 SRINIVAS, P.R., SRIVASTAVA, S., HANASH, S., WRIGHT, G.L., JR., Proteomics in early detection of cancer, *Clin. Chem.*, **2001**, 47, 1901–1911.

38 FIGEYS, D., PINTO, D., Proteomics on a chip: Promising developments, *Electrophoresis*, **2001**, 22, 208–216.

39 REID, G., GAN, B.S., SHE, Y.M., ENS, W., WEINBERGER, S., HOWARD, J.C., Rapid identification of probiotic lactobacillus biosurfactant proteins by ProteinChip tandem mass spectrometry tryptic peptide sequencing, *Appl. Environ. Microbiol.*, **2002**, 68, 977–980.

40 TABUCHI, M., BABA, Y., A novel injection method for high-speed proteome analysis by capillary electrophoresis, *Electrophoresis*, **2002**, 23, 1138–1145.

41 VENTER, J.C. et al., The sequence of the human genome, *Science*, **2001**, 291, 1304–1351.

42 PAVLOV, M.Y., FREISTROFFER, D.V., EHRENBERG, M., Synthesis of region-labelled proteins for NMR studies by *in vitro* translation of column-coupled mRNAs, *Biochimie*, **1997**, 79, 415–422.

43 http://www.roche-applied-science.com/sis/proteinexpression/

44 ZHI, Z.L., MURAKAMI, Y., MORITA, Y., HASAN, Q., TAMIYA, E., Multianalyte immunoassay with self-assembled addressable microparticle array on a chip, *Anal. Biochem.*, **2003**, 318, 236–243.

45 ZHI, Z.L., MORITA, Y., HASAN, Q., TAMIYA, E., Micromachining microcarrier-based biomolecular encoding for miniaturized and multiplexed immunoassay, *Anal. Chem.*, **2003**, 75, 4125–4131.

46 KUKAR, T., ECKENRODE, S., GU, Y., LIAN, W., MEGGINSON, M., SHE, J.X., WU, D., Protein microarrays to detect protein–protein interactions using red and green fluorescent proteins, *Anal. Biochem.*, **2002**, 306, 50–54.

47 KOJIMA, K., HIRATSUKA, A., SUZUKI, H., YANO, K., IKEBUKURO, K., KARUBE, I., Electrochemical protein chip with arrayed immunosensors with antibodies immobilized in a plasma-polymerized film, *Anal. Chem.*, **2003**, 75, 1116–1122.

48 TABUCHI, M., HINO, M., SHINOHARA, Y., BABA, Y., Cellfree protein synthesis on a microchip, *Proteomics*, **2002**, 2, 430–435.

49 RUNGPRAGAYPHAN, S., KAWARASAKI, Y., IMAEDA, T., KOHDA, K., NAKANO, H., YAMANE, T., High-throughput, cloning-independent protein library construction by combining single-molecule DNA amplification with *in vitro* expression, *J. Mol. Biol.*, **2002**, 318, 395–405.

50 NAKANO, H., OKUMURA, R., GOTO, C., YAMANE, T., *In vitro* combinatorial mutagenesis of the 65th and 222nd positions of the green fluorescent protein of *Aequarea Victoria*, *Biotechnol. Bioprocess Eng.*, **2002**, 7, 311–315.

51 MCDONALD, J.C., WHITESIDES, G.M., Poly(dimethylsiloxane) as a material for fabricating microfluidic devices, *Acc. Chem. Res.*, **2002**, 35, 491–499.

52 NG, J.M., GITLIN, I., STROOCK, A.D., WHITESIDES, G.M., Components for integrated poly(dimethylsiloxane) microfluidic systems, *Electrophoresis*, **2002**, 23, 3461–3473.

53 JO, B.H., VAN LERBERGHE, L.M., MOTSEGOOD, K.M., BEEBE, D.J., Three-dimensional micro-channel fabrication in polydimethylsiloxane (PDMS) elastomer, *J. MEMS*, **2000**, 9, 76–81.

54 SPIRIN, A.S., BARANOV, V.I., RYABOVA, L.A., OVODOV, S.Y., ALAKHOV, Y.B., A continuous cell-free translation system capable of producing polypeptides in high yield, *Science*, **1988**, 242, 1162–1164.

55 Brenner, S., Williams, S.R., Vermaas, E.H., Storck, T., Moon, K., McCollum, C., Mao, J.I., Luo, S., Kirchner, J.J., Eletr, S., DuBridge, R.B., Burcham, T., Albrecht, G., *In vitro* cloning of complex mixtures of DNA on microbeads: Physical separation of differentially expressed cDNAs, *Proc. Natl. Acad. Sci. U.S.A.*, **2000**, 97, 1665–1670.

56 Giese, K., Kaufmann, J., Pronk, G.J., Klippel, A., Unravelling novel intracellular pathways in cell-based assays, *Drug Discovery Today* **2002**, 7, 179–186.

57 Latraite, S., Bigot-Lasserre, D., Bars, R., Carmichael, N., Optimization of cell-based assays for medium throughput screening of oxidative stress, *Toxicol. In Vitro* **2003**, 17, 207–220.

58 Wolcke, J., Ullmann, D., Miniaturised HTS technologies-μHTS, *Drug Discovery Today*, **2001**, 6, 637–646.

59 Hertzberg, R.P., Pope, A.J., High-throughput screening: New technology for the 21st century, *Curr. Opin. Biotechnol.*, **2000**, 4, 445–451.

60 Shokrgozar, M.A., Shokri, F., Enumeration of hepatitis B surface antigen-specific B lymphocytes in responder and non-responder normal individuals vaccinated with recombinant hepatitis B surface antigen, *Immunology*, **2001**, 101, 75–79.

61 Thorsen, T., Maerkl, S.J., Quake, S.R., Microfluidic large-scale integration, *Science*, **2002**, 298, 580–584.

62 Akagi, Y., Ramachandra Rao, S., Morita, Y., Tamiya, E., Optimization of fluorescent cell-based assays for high-throughput analysis using microchamber array chip formats, *Sci. Technol. Adv. Mater.*, **2004**, 5, 343–349.

63 Yamamura, S., Ramachandra Rao, S., Omori, M., Tokimitsu, Y., Kondo, S., Kishi, H., Muraguchi, A., Takamura, Y., Tamiya, E., High-throughput screening and analysis for antigen specific single-cell using microarray, *Proceedings of Micro Total Analysis System (μTAS) 2004*, The Royal Society of Chemistry, Cambridge, **2004**, Vol. 1, pp. 78–80.

64 Ramachandra Rao, S., Yamamura, S., Takamura, Y., Tamiya, E., Multiplexed microfluidic devices for single-cell manipulation and analysis, *Proceedings of Micro Total Analysis System (μTAS) 2004*, The Royal Society of Chemistry, Cambridge, **2004**, Vol. 1, pp. 61–63.

65 Umehara, S., Wakamoto, Y., Inoue, I., Yasuda, K., On-chip single-cell microcultivation assay for monitoring environmental effects on isolated cells, *Biochem. Biophys. Res. Commun.*, **2003**, 305, 534–540.

66 Suzuki, I., Sugio, Y., Moriguchi, H., Jimbo, Y., Yasuda, K., Modification of a neuronal network direction using stepwise photo-thermal etching of an agarose architecture, *J. Nanobiotechnol.*, **2004**, 2, 7–14.

67 Wheeler, A.R., Throndset, W.R., Whelan, R.J., Leach, A.M., Zare, R.N., Liaor, Y.H., Farrell, K., Manger, I.D., Daridon, A., Microfluidic device for single-cell analysis, *Anal. Chem.*, **2003**, 75, 3581–3586.

68 Huang, W.H., Cheng, W., Zhang, Z., Pang, D.W., Cheng, J.K., Cui, D.F., Transport, location, and quantal release monitoring of single cells on a microfluidic device, *Anal. Chem.*, **2004**, 76, 483–488.

69 Peng, X.Y., Li, P.C.H., A three-dimensional flow control concept for single-cell experiments on a microchip. 1. Cell selection, cell retention, cell culture, cell balancing, and cell scanning, *Anal. Chem.*, **2004**, 76, 5273–5281.

70 Li, X., Li, P.C., Microfluidic selection and retention of a single cardiac myocyte, on-chip dye loading, cell contraction by chemical stimulation, and quantitative fluorescenct analysis of intracellular calcium, *Anal. Chem.*, **2005**, 77, 4315–4322.

71 Haugland, R.P., In *Handbook of Fluorescent Probes and Research Products*, 9th edn., Molecular Probes, Inc., USA, **2002**.

72 Nishida, M., Sugimoto, K., Hara, Y., Mori, E., Morii, T., Kurosaki, T.,

Mori, Y., Amplification of receptor signaling by Ca^{2+} entry-mediated translocation and activation of PLCgamma2 in B lymphocytes, *EMBO J.*, **2003**, 22, 4677–4688.

73 Ueki, Y., Goldfarb, I.S., Harindranath, N., Gore, M., Koprowski, H., Notkins, A.L., Casali, P., Clonal analysis of a human antibody response. Quantitation of precursors of antibody-producing cells and generation and characterization of monoclonal IgM, IgG, and IgA to rabies virus, *J. Exp. Med.*, **1990**, 171, 19–34.

74 Jingwu, Z., Henderikx, P., Ying, C., Medaer, R., Raus, J.C.M., A method to establish antibody secreting B cell lines and simultaneously perform frequency analysis, *J. Immunol. Methods*, **1989**, 123, 153–154.

75 Kohler, G., Milstein, C., Continuous cultures of fused cells secreting antibody of predefined specificity, *Nature*, **1975**, 256, 495–497.

Index

Nanotechnologies for the Life Sciences Vol. 8
Nanomaterials for Biosensors. Edited by Challa S. S. R. Kumar

ISBN: 978-3-527-31388-4

d

e

n

o

p

Related Titles

Willner, I., Katz, E. (eds.)

Bionanomaterials

Synthesis and Applications for Sensors, Electronics and Medicine

2007
Hardcover
ISBN-13: 978-3-527-31454-7
ISBN-10: 3-527-31454-7

Daunert, S., Deo, S. K. (eds.)

Photoproteins in Bioanalysis

2006
Hardcover
ISBN-13: 978-3-527-31016-9
ISBN-10: 3-527-31016-9

Hüser, J. (ed.)

High-Throughput Screening in Drug Discovery

2006
Hardcover
ISBN-13: 978-3-527-31283-8
ISBN-10: 3-527-31283-8

Kelsall, R., Hamley, I. W., Geoghegan, M. (eds.)

Nanoscale Science and Technology

2005
Hardcover
ISBN-13: 978-0-470-85086-8
ISBN-10: 0-470-85086-8

Willner, I., Katz, E. (eds.)

Bioelectronics

From Theory to Applications

2005
Hardcover
ISBN-13: 978-3-527-30690-0
ISBN-10: 3-527-30690-0

Kumar, C. S. S. R., Hormes, J., Leuschner, C. (eds.)

Nanofabrication Towards Biomedical Applications

Techniques, Tools, Applications, and Impact

2005
Hardcover
ISBN-13: 978-3-527-31115-7
ISBN-10: 3-527-31115-7

Öberg, P. Å., Togawa, T., Spelman, F. A. (eds.)

Sensors in Medicine and Health Care

Sensors Applications Vol. 3

2004
Hardcover
ISBN-13: 978-3-527-29556-9
ISBN-10: 3-527-29556-9

Goodsell, D. S.

Bionanotechnology

Lessons from Nature

2004
Hardcover
ISBN-13: 978-0-471-41719-X
ISBN-10: 0-471-41719-X